TRAUMATIC BRAIN INJURY

Edited by **Farid Sadaka** and **Tanya Quinn**

Traumatic Brain Injury
http://dx.doi.org/10.5772/57090
Edited by Farid Sadaka and Tanya Quinn

Contributors

David Tannehill, Rekha Lakshmanan, Farid Sadaka, Ashok Palagiri, Tanya Quinn, Deborah Zelinsky, Matthew Korobey, Benedict C. Albensi, Chris Cadonic, Renee Turner, Tobias Wensing, Jean Levasseur-Moreau, Alexander Sack, Shirley Fecteau, Birgitta Johansson, Lars Ronnback, Jehuda Soleman, Javier Fandino, Carl Muroi, Philipp Taussky, Edilene Curvelo Hora, Liane Viana Santana, Lyvia de Jesus Santos, Gizelle de Oliveira Souza, Analys Vasconcelos Pimentel, Natalia Tenorio Cavalcante Bezerra, Sylvia Rodrigues de Freitas Doria, Afonso Abreu Mendes Junior, Tiago Pinheiro Vaz de Carvalho, Renata Julie Porto Leite Lopes, Jessica Almeida Rodrigues, Ricardo Fakhouri, Frank Baas, Valeria Ramaglia, Narayan Verma, Arunima Verma Jayakar, Steven Wheeler, Michelle Albicini, Audrey McKinlay, Wafaa Al Yazeedi, Venkatachalam Loganathan, Somaya Al Molawi, Fatma Al Kuwari, Jason Huang, Matthew Dashnaw, Anthony Petraglia, T.M. Ayodele Adesanya, Rachael Sullivan, Stanislaw P. A. Stawicki, David C. Evans, Tanea Reed, Jesus Moreta, José Luis Martínez-de los Mozos, Giovanna Mioni, Franca Stablum, Shawn M. McClintock, Vafa Rahimi-Movaghar, Shayan Zadegan, Mogammad Shaheed Soeker, Mercedes Tamashiro, Maria Elisa Rivas, Pablo Etchandy, Diego Cozzo, Martin Mattei, Leonardo Corrotea, Veronica Jarmoluk, Fernando Salierno, Eliana Buffetti, Hans Von Holst, Svein Kleiven

Published by InTech

Janeza Trdine 9, 51000 Rijeka, Croatia

Publishing Process Manager Iva Lipovic

Technical Editor InTech DTP team

Cover InTech Design team

First published February, 2014

Printed in Croatia

Additional hard copies can be obtained from orders@intechopen.com

Traumatic Brain Injury, Edited by Farid Sadaka and Tanya Quinn

p. cm.

ISBN 978-953-51-1222-8

For EU product safety concerns:
IN TECH d.o.o., Prolaz Marije Krucifikse Kozulić 3, 51000 Rijeka, Croatia,
info@intechopen.com or visit our website at intechopen.com.

Meet the editors

Farid Sadaka is certified by the American Board of Internal Medicine in Internal Medicine and Critical Care Medicine and by the United Council of Neurologic Subspecialties in Neurocritical care. Dr. Sadaka is the medical director of the Neurologic and Trauma ICU at Mercy Hospital St. Louis, and is also on teaching faculty at St. Louis University. He is an active member of the Society of Critical Care Medicine and Neurocritical Care Society. Dr. Sadaka has been invited to speak at several academic centers nationally and locally, on topics pertaining to neuroscience, traumatic brain injury, sepsis and the microcirculation and therapeutic hypothermia. Dr. Sadaka is actively involved in clinical research. He has had several peer-reviewed publications and book chapters, especially pertaining to neuroscience, traumatic brain injury, sepsis and the microcirculation. Dr. Sadaka is also an editor of a recently released book "Therapeutic Hypothermia in Brain Injury" by InTech.

Tanya Quinn obtained her medical degree at the University of Missouri - Columbia and then went on to complete her internship in General Surgery and residency in Neurosurgery at the Medical University of South Carolina in Charleston, South Carolina. After her training, she joined the faculty at St. Louis University practicing general neurosurgery, with special interests in neurotrauma and cerebrovascular surgery. During this time, she also performed additional training in the area of cerebrovascular surgery. She joined the staff at Mercy Hospital in St. Louis Missouri in January 2013. She continues to have a busy general neurosurgical practice, with continued interest in cerebrovascular surgery and neurotrauma.

Contents

Preface

Traumatic brain injury is a major source of death and severe disability worldwide. Traumatic brain injury needs a multidisciplinary approach for improved outcomes. This book focuses on the management of patients with traumatic brain injury in detail, in a stepwise approach, starting with management in the intensive care unit, through discharge from the hospital, rehabilitation, physical therapy, and ending with recovery and assimilation in family and society.

This book discusses the pathophysiology of traumatic brain injury, and reviews the therapeutic modalities for this devastating injury at the basic science as well as clinical levels. It also tackles the importance of extensive rehabilitation of a patient recovering from traumatic brain injury, from the acute phase in the hospital all the way to returning the individual to home, family, and society.

This book also describes the multiple cognitive impairments experienced by a patient after traumatic brain injury, such as memory deficits, post-traumatic stress, information processing, mental fatigue, communication problems, and sleep/wake disorders. In addition, each chapter provides grounds for future research in its specific topic pertaining to traumatic brain injury.

Farid Sadaka and Tanya Quinn
Mercy Hospital St. Louis/St. Louis University,
Critical Care Medicine/Neurocritical Care,
St. Louis, USA

Section 1

Definitions, Pathophysiology, and Therapeutics

Chapter 1

Mild Traumatic Brain Injury: A Review of Terminology, Symptomatology, Clinical Considerations and Future Directions

Michelle Albicini and Audrey McKinlay

Additional information is available at the end of the chapter

http://dx.doi.org/10.5772/57208

1. Introduction

Traumatic Brain Injury (TBI) is a leading cause of morbidity and mortality, with estimates of prevalence varying between 100-1000 per 100,000 [1, 2]. Among these figures, 70-90% will be classified as mild TBI (mTBI) [2]. While only 10% of those with a history of mTBI will have any ongoing problems [2], the sheer volume of incidents means that these events represent a major health concern. Accurate identification and diagnosis of those with mTBI is the first step in providing care and treatment [3-6]. However, the evaluation, management and diagnosis of mTBI represent an ongoing challenge for clinicians [3, 4].

Many of those who experience mTBI do not seek medical attention [7] or do not consult their clinician for many days after the injury event [8]. Delays often prevent accurate identification of those with mTBI due to the commonly rapid resolution of symptoms [8], subtle neurological signs and symptoms [9], and typical absence of evidence on neuroimaging [8]. Diagnosis of mTBI involves the detection of injury characteristics and symptoms established during a clinical interview with the patient, and in the case of children, with their guardian/carer [8]. Therefore, the primary source of information obtained is generally subjective in nature, leading to further diagnostic issues [4]. In addition, in some cases the information regarding the TBI may be based on a historical event for which the patient must rely on their own memory to recall, which can lead to the reporting of incorrect information [4]. However, for most of those assessed for mTBI, the assessment is conducted in the emergency department (ED) [3, 10], which does not specialise in diagnosis or treatment of mTBI. This presents a further problem in the case of children who are frequently admitted to the ED for orthopaedic injuries but not assessed for mTBI, and any potential mTBI in such children therefore goes unrecognised [11]. Finally, less severe forms of injury, including those for which a loss of consciousness (LOC) was not sustained, may be overlooked or may receive little attention [5], despite potentially being a clinically important injury warranting assessment and treatment.

Currently, there is no "gold-standard" process of evaluation and diagnosis of mTBI, with structured clinical interviews as a commonly accepted standard procedure [5, 9]. It is mistakenly thought that anyone can detect and diagnose mTBI [12], however determining the clinical importance and potential effects this injury will have on each individual provides a distinct challenge. This chapter will highlight the issues associated with the evaluation, management and diagnosis of mTBI, and the factors that need to be considered during a structured clinical interview. These will include definitions and terminology, symptomatology, evaluating these symptoms during an interview, the importance of a concussive blow, brain imaging techniques, medical intervention, and complex concussion. Moreover, special consideration will be given to the evaluation of children with mTBI, given their heightened vulnerability to the negative effects of such an injury [13].

2. Definitions and terminology

A major clinical concern is the lack of a uniform definition for mTBI [8, 14]. Studies apply different definitions and inconsistent criteria to identify mTBI, resulting in varying findings and conclusions [15-17]. Furthermore, the terminology used to describe mTBI is inconsistent and may include concussion, minor head injury, minor brain injury, minor head trauma and minor TBI [6, 10]. This presents an issue for clinical diagnosis and evaluation as non-shared terminology across clinicians can lead to misinterpretation and misunderstanding among patients [4, 6]. Accurate and consistent use of terminology is essential to reduce confusion among clinicians and ensure the identification of patients with mTBI in an acute setting, so they can receive appropriate care [15] and are able to better understand their condition [4].

There are three main indicators most commonly used to identify mTBI – length of LOC, duration of post-traumatic amnesia (PTA), and a patient's Glasgow Coma Scale (GCS) score [18]. The GCS involves the summation of scores from three measures including eye opening, and best motor and verbal responses, with 15 indicating the best possible response (and lowest injury severity) [12]. Based on this information, the American Congress of Rehabilitation (ACRM) [18] advocated four specific criteria when defining mTBI [8]. Since then, additional definitions have been suggested, mostly consisting of some variation of the ACRM's initial criteria. Table 1 outlines definitions of mTBI put forward by the ACRM [18], the World Health Organisation (WHO) Collaborative Task Force [19], and the Centres for Disease Control and Injury Prevention (CDC) [11].

When comparing the definitions displayed in Table 1, similarities and inconsistencies are evident, particularly with terminology, across the standardised definitions of mTBI. The WHO's definition was derived from the ACRM and CDC definitions [8], with the addition of the presence of confusion/disorientation, and the ruling out of symptoms manifested from other problems. However, the purpose behind these additions was not discussed or explained, and although these definitions are similar, slight differences among the terminology may result in different clinical decisions. When considering the CDC definition, different terminology is

Defining Body	Definition
American Congress of Rehabilitation Medicine (ACRM)	Traumatically induced physiological disruption of brain function with at least one of: a) LOC (≤ 30mins) b) PTA - loss of memory for events before/after injury (≤ 24 hours) c) alteration in mental state at time of accident; transient or non-transient focal neurological deficits d) initial GCS score of 13 to 15 after 30 mins
World Health Organisation (WHO) Task Force	Brain injury due to mechanical energy to head from external physical forces and includes: a) 1 or more of confusion/disorientation, LOC ≤ 30 mins, PTA ≤ 24 hours, and/or other transient neurological abnormalities such as focal signs, seizure and intracranial lesions not requiring surgery; b) initial GCS score of 13-15 30 mins after injury c) manifestations not due to drugs, alcohol, medications, or other injuries, treatments for injuries or problems
Centres for Disease Control and Prevention (CDC)	An injury to the head due to blunt trauma or acceleration/deceleration forces resulting in one or more periods of observed or self-reported: a) Transient confusion, disorientation, or impaired consciousness; b) Dysfunction of memory around the time of injury; c) LOC ≤ 30 minutes.

Table 1. Comparison of definitions for mild traumatic brain injury

used again, with no use of the term PTA, no time frame of PTA symptoms and no emphasis placed on a GCS score. Therefore, even among the most used and standardised definitions of mTBI, there are important discrepancies which will impede the process of identifying patients with mTBI.

2.1. The Glasgow coma scale

The GCS has been deemed the best initial score of injury severity [12]; however, there are noted problems with this. The scoring system often results in ceiling effects, with 15 being the upper limit of mTBI and also the highest possible score obtained on the scale [15]. People often therefore mistake a score of 15 to represent normal neurological functioning, which is sometimes not the case. The GCS is arguably not sensitive to the defining criteria of mTBI [15] in that a patient without mTBI who receives a score of 15 will likely be very neurologically different to a patient who has sustained a mTBI and also obtains this score. Another problem resides with the definition criteria involving GCS, in that an initial score of 13-15 is required 30 minutes after injury. The issue here refers to the fact that most patients with mTBI will present to the ED and will not be evaluated within this time frame [20]. As such, a practical issue arises in that the GCS cannot capture symptoms retrospectively or reflect neurological status immediately following the mechanical blow [14], therefore less emphasis should be placed on the GCS for identifying mTBI, especially where patients are assessed more than 30 minutes following their injury.

2.2. Clinical considerations

To obtain an understanding of recovery and response to treatment for mTBI, there needs to be a consistent definition and definitively shared terminology used across clinicians [10]. However, as discussed above this is yet to be the case. Over-inclusive definitions may lead to false positives, where mTBI is mistakenly diagnosed in unaffected patients, whereas restrictive definitions may result in patients with mTBI going unrecognised [15]. While most clinicians and research studies refer to the ACRM or WHO definitions for identifying mTBI, it is clear that some will place more or less emphasis on GCS scores or duration of LOC. When considering structured clinical interviews, it is important that clinicians adopt a more uniformed approach for defining mTBI and for the terms used to describe it, to avoid confusion among patients regarding the importance of their injury and also misdiagnosis.

3. Short-term symptoms of mild TBI

When assessing mTBI, it is often difficult for clinicians to evaluate symptomatology and to relate these to the injury as they are characterised as broad and non-specific, and often experienced in other disorders [8, 9]. However, the short-term symptoms of mTBI are typically grouped according to four categories: 1) physical, such as headaches and fatigue, 2) cognitive, such as poor concentration, 3) emotional, such as poor emotional control, and 4) sleep problems [3, 8, 11]. In addition, there may be transient and focal neurological signs present in those who have experienced mTBI, including seizures, visual problems, balance and/or gait problems, acute aphasia, anosmia/hyposmia, cranial nerve defects and intracranial lesions [8, 9]. The most commonly referred to list of symptoms experienced by those following mTBI are those reported by the CDC [11], as displayed in Table 2.

	Physical	Cognitive	Emotional	Sleep
Adults	• Headache • Nausea • Vomiting • Balance Problems • Dizziness • Visual Problems • Fatigue • Light Sensitivity • Noise Sensitivity • Numbness/Tingling • Dazed/Stunned	• Feeling "foggy" • Feeling "slowed down" • Difficulty Concentrating • Forgetful • Confusion • Slower Information Processing	• Irritability • Sadness • More emotional • Nervousness	• Drowsiness • Sleeping less • Sleeping more • Trouble falling asleep
Additional symptoms in children	• Seizures • Change in nursing/eating habits	• Personality Change • Poor attention • Lack interest in favourite toys/items	• Crying • Cannot be consoled • Restlessness • Upset easily • Temper tantrums • Lethargic mood	• Change in sleep patterns

Table 2. Short-term symptoms of mild traumatic brain injury in adults and children proposed by the CDC

While the CDC provides a comprehensive list of symptoms commonly reported by those following mTBI, similarly to the definition of mTBI there are differences among criteria and terminology. Some are more broad when discussing the symptoms of mTBI, using terms such as differences in 'higher-cognitive functioning' [21] and 'cognitive deficits' [9]. Further, although the CDC has put forward a list of symptoms, not all clinicians will consult this list during assessments. While it is near impossible to create an exhaustive list of symptoms one is expected to experience after mTBI, it is important to consider the validity of certain symptoms when using them to ascertain a diagnosis.

3.1. Validity of specific symptoms for diagnosis

Initial diagnosis of mTBI is usually based on LOC, PTA, or neurological signs [8, 22], with often greatest emphasis placed on LOC [22]. LOC is believed to be a symptom essential for diagnosis [8], however with this come practical concerns. Many patients are usually unable to report a period of LOC and the associated time frame, which means they must rely on others' reports of the incidents or make assumptions [8]. Patients will often mistake periods where they are unable to recall information as periods of LOC [8], or alternatively, where LOC is only experienced for brief seconds, it will go unreported [22]. There are additional difficulties when determining confusion/disorientation or PTA, as a strong emotional reaction evoked by the traumatic event may also produce these symptoms as secondary to the insult, which cannot be attributed to the injury itself [8, 23]. Therefore, mTBI can often be misdiagnosed when basing decisions on symptoms alone [24].

3.2. Special considerations for children

As is demonstrated in Table 2, children and adults will present differently following a TBI. While children may experience the same general symptoms as adults, they may also experience additional symptoms or express their symptoms in a different way [11]. Headaches, dizziness and fatigue are common symptoms of mTBI [11], and are often seen in both children and adults; however these specific symptoms tend to resolve more rapidly in child patients [20]. Furthermore, even when children are suffering the same symptoms as adults, they may not be able to explain or express them [11]. For instance, a young child may feel nauseous but not experience vomiting, and without the verbal expression of this feeling this symptom would go unrecognised. It is therefore recommended that when assessing symptoms of patients following mTBI, the threshold should be lower for children when deciding whether to proceed with follow-up evaluations and management [11].

3.3. Clinical considerations

Considering the unreliable nature of symptom reporting from patients, using this as the basis for diagnosis is not ideal [23]. Due to the patient's psychological impairment following their injury, their reports of LOC and PTA would not provide accurate evidence for mTBI [23]. It is recommended that in the case of LOC, collateral reports from others close to the incident may provide more valid information and aid diagnosis and evaluation [8]. In addition, the WHO collaborating task force [25] state that for the identification of mTBI, vomiting, seizures and anterograde amnesia may be the important symptoms to consider following the injury. However, in addition to symptomatology, at a minimum, information pertaining to age, mechanism of injury, GCS score, skull fracture, and evidence of trauma above the clavicles should be assessed and recorded [25].

4. Evaluating symptoms during an interview

As discussed above, there are issues associated with the evaluation of symptoms during the diagnosis of mTBI. However, in addition to certain symptoms being unreliable for the identification of mTBI, there are also problems regarding how to elicit this list of symptoms from patients during a clinical interview [24]. The process of evaluation of mTBI requires patients to volunteer any symptoms they are experiencing and to answer questions in a subjective manner [9], but there are limits to these self-report methods such as problems with recall, stigma or secondary gains [24]. Furthermore, it is not uncommon that patients find it difficult to communicate their symptoms and problems, particularly in the early stages of their injury [11].

It has now been found in multiple studies that there is a difference in the number of symptoms reported by patients to their clinicians depending on whether their symptoms are assessed through self-report checklists or in an open-ended questioning format [24, 26-28]. For example [24], in a study with mTBI patients, it was found that participants were more likely to endorse more symptoms when completing a symptom checklist compared to during a clinical interview [24]. Similarly, in another study, patients reported significantly more symptoms per symptom category on a symptom checklist than when responding spontaneously [27]. Table 3 outlines studies investigating how the type of symptom evaluation can affect symptom reporting, and the results validate the finding that patients are more likely to report symptoms when completing a questionnaire than when answering questions during an interview [24, 26-28].

The findings in Table 3 raise the question of whether patients exaggerate their symptoms when using a checklist, or whether having a specified list reminds them of symptoms they would have otherwise forgotten to mention. It has been suggested that individuals may be less likely to report sensitive information face-to-face due to embarrassment, or simply believe that certain symptoms are trivial and not necessary to mention or are unrelated to their injury [24]. It is also possible that specific symptom questions may either result in a patient feeling pressured to provide a response, even if inaccurate [28], or alternatively, may cue an individual to identify and have the ability to label an otherwise ambiguous experience or problem [26]. While it appears that questionnaires may promote over-endorsement of symptoms, the issue may be that clinical interviews result in under-reporting of symptoms that appear unimportant or embarrassing, but are actually clinically important for diagnosis and evaluation.

4.1. Clinical considerations

The question is whether it is more important to have the patient over-report symptoms, even if not completely accurate, or to under-report their symptoms, thereby disregarding potentially important clinical problems. Over-reporting can lead to issues associated with malingering

Authors	Aims	Participants	Results
Gerber and Schraa (1995)	Compare mTBI patients with orthopaedic patients and healthy controls on measures of injury severity, symptoms and disability at time of injury and 6 months following	22 consecutive mTBI patients, 22 matched orthopaedic injury patients, 22 matched uninjured controls	The method for evaluating symptoms (volunteered versus elicited) had a significant effect on symptom frequency for mTBI patients, compared to the control participants
Nolin et al. (2006)	Identify role of spontaneous response versus suggested response methods for evaluating mTBI symptoms	108 mTBI patients contacted through follow-up phone interview	Participants reported significantly more symptoms and any given symptom when list was given to them Both the number and type of symptoms reported were different across the two methods
Stapleton and Mills (2008)	Explore the usefulness of open-ended questionnaires for evaluating balance symptoms	54 consecutive patients at specialist balance clinic	Open-ended questionnaire did not provide adequate information for diagnosis Patients more likely to respond affirmative if asked about symptoms directly
Iverson et al. (2010)	Examine the different outcomes between spontaneous, interview-based symptom reporting and standardised symptom questionnaires	61 patients in concussion clinic with mTBI	Patients endorsed 3.3 symptoms during open-ended interview compared to 9.1 on symptom questionnaire

Table 3. Overview of studies investigating how method of symptom evaluation affects symptom reporting

and financial gain, and also the use of unnecessary treatment, rehabilitation and management resources. However, under-reporting may result in patients not receiving the care they need, and it may therefore be more effective to have an over-inclusive list to avoid missing important symptoms. It has been suggested that clinicians have patients freely narrate their own stories [8] through answering open-ended questions and follow-up probes [4]. While encouraging a patient-centred response to specific symptom questions may result in an inaccurate diagnosis or decision [28], using both open-ended questioning and validated symptom checklists during a structured clinical interview may enhance the accuracy and value of clinical evaluations [27]. Although there remains to be inconsistencies in which approach clinicians use regarding the evaluation of symptoms following mTBI, it appears clear that neither will be a sufficient tool on their own in determining a diagnosis.

5. Importance of a concussive blow

Previously mentioned, the symptoms of mTBI are non-specific and are often also experienced by others with different injuries or disorders [8, 9]. Therefore, it is important to note that for mTBI to have occurred there must have existed a mechanically induced disruption to brain physiology due to external forces [8, 18, 19]. At the least, the signs and symptoms that may warrant evaluation of mTBI require evidence of direct or indirect force and/or blow to the head [3], potentially severe enough to alter brain functioning [3]. The main area of controversy for this topic involves the effects of whiplash in the evaluation of mTBI [29]. Consequentially, the importance of a 'minimum biomechanical threshold' for mTBI to have occurred has been put forward [15] when determining whether the injury event was sufficient to result in the presenting signs and symptoms in a patient with potential mTBI.

Mild TBI is described as a biomechanically induced neurological injury, whereby acceleration and deceleration forces to the head results in a cascade of neurochemical and metabolic events [30]. It is argued that to determine the presence of mTBI, there must be some uniform acceleration or deceleration applied to a fluid body, thereby generating a pressure gradient, and in extreme enough cases, causing trauma to the brain [31]. Pure translation of a rigid body, such as in the case of shaking, will not cause any strain to the head [31]. Further, it has been estimated that when the head is struck by a force, the impact is 50-100 times greater than the result of shaking (and therefore no blow to the head) alone [31, 32]. Extensive reviews have also concluded that no impact produced by translational movement alone will be greater than that caused by an acceleration/deceleration force to the head [29, 32, 33].

It has been mandated that for mTBI to be present, a credible force must have been applied to the head, and that trauma to the brain is more likely if the force is rotary in nature [29]. This is due to the fact that a pressure gradient is generated transiently through the brain when there is a concussive blow resulting in rotational movement [33]. In effect, this is the reason behind the necessity of the presence of a concussive blow when diagnosing mTBI, to avoid diagnosis based on symptoms which may otherwise be attributed to causes such as acquired brain injury, drug/alcohol induced intoxication, tumour, stroke, or other problems [18, 29].

5.1. Special considerations for children

While research has been conducted concerning adults when determining the biomechanical nature of mTBI, the biomechanics of paediatric mTBI are less well-defined [31]. The hypothesis that smaller brains may require more angular accelerational force to produce the effects of mTBI has been studied. With the use of primates, the results have revealed that larger brains are actually more vulnerable to lower levels of angular velocity and acceleration [31], meaning they require a larger concussive blow than smaller brains. It has been suggested that in infants and very young children, the skull is not as rigid and therefore upon impact the structure will undergo elastic and potentially plastic deformation, resulting in various skull fractures [31], however the meaning and effects of this are not well studied or understood. Further studies are required to clarify the differences in the biomechanics of paediatric mTBI compared to that of adults, especially considering the enhanced vulnerability of children to the long-term effects of mTBI [13, 31].

5.2. Clinical considerations

Therefore, during a structured clinical interview it is important to determine whether the mechanism of injury is likely to be associated with significant acceleration/deceleration force, and not pure translational movement [32]. Clinicians should confirm that the mTBI was induced by traumatic biomechanical forces, which are secondary to forces to the head [11]. Further, it must be ascertained how the injury occurred, the type of force sustained, and the location on the head or body where the force was received [11]. In particular, the clinician should search for evidence of injury or assault to regions above the clavicles to be certain the symptoms being reported were a result of a forcible blow to the head, potentially resulting in neurological disruption.

6. Brain imaging techniques for mild TBI

The use of neuroimaging techniques is sometimes adjunctive to other testing for diagnosis [8], however compared to their effectiveness in evaluating moderate and severe TBI they appear to provide the poorest sensitivity in detecting neurological abnormalities for milder forms of TBI [15]. This is because the results of imaging procedures for mTBI are often inconclusive and difficult to interpret [34], and while computerised tomography (CT) remains the technique of choice [15, 35], scans may still appear normal [18]. Typically, length of LOC and/or PTA, and results of the physical and neuropsychological examination, will guide the decision to order head scanning for mTBI [25]. However the challenge is how to detect which apparently intact patient has an intracranial lesion requiring surgical intervention [36]. Studies have therefore looked into how to predict which CT scans will result in positive findings such as intracranial lesions in mTBI patients [35].

When deciding whether to evaluate a patient for neurological abnormalities, clinicians will most often use a CT or magnetic resonance imaging (MRI) scan [12]. While MRI scanning will reveal more abnormalities, both early and late after mTBI, it is easier to use a CT scan for an unstable patient [12] due to the physical restraints and uncomfortable nature associated with an MRI. However, although MRI is useful for providing structural appearance of the brain, it cannot provide us with information that functional MRI (fMRI) may provide [21]. Moreover, symptoms of mTBI may more likely reflect functional as opposed to structural damage to the brain [16], and so fMRI could be used to map clinical symptoms to specific damaged brain regions; however, clinical access to CT scans is much easier to obtain than that of fMRI.

Studies have sought to determine the utility of brain imaging techniques following mTBI. For instance, the use of MRI and CT scanning for mTBI patients has been explored and compared [37]. Consecutive patients under 50 years presenting at an ED for mTBI were evaluated, and the results demonstrated a relatively high sensitivity of both CT and MRI for posttraumatic lesions [37]. However, when the negative predictive value of CT scanning and the necessity of hospital admission for patients without positive CT findings were examined in 2152 mTBI patients, the results indicated that no single variable, or combination of variables, could predict which patients would have a positive or negative CT scan [38]. Therefore, the utility of brain imaging techniques in the evaluation of mTBI remains to be questionable.

6.1. Prevalence of medical imaging use in mild TBI patients

The use of imaging for mTBI requires some consideration given that no standardised procedure exists, and the use of brain imaging techniques is extremely variable [25]. Nonetheless, it has been reported that in mTBI patients, the prevalence of CT scan abnormalities is 5% among those with a GCS score of 15, 20% for those with a GCS of 14, and over 30% for those with a GCS of 13 [25]. Furthermore, 6-8% of those with mTBI will have specific injuries displayed through CT, including subarachnoid haemorrhage, hematoma, cerebral contusion, intraparenchymal haemorrhage and evidence of axonal injury [16]. In a study exploring the prevalence of abnormal CT scans following mTBI [39], 15.8% of patients exhibited injuries on their 'day of injury' CT scan, but nearly 25% did not undergo scanning at all (without justification). The use of CT scanning should be regulated across clinicians, which may only be done so with the introduction of more standardised decisional criteria.

6.2. Special considerations for children

While the imaging techniques used to evaluate children will be the same as for adults, the procedures are likely to be more difficult with very young children [40]. These children may require sedation [40] to obtain accurate clinical information due to the need to be still while undergoing scanning. This can lead to further problems such as exacerbation of symptoms, prolonged lowered consciousness and problems with breathing [40]. A study [41] investigated CT scan results in children following mTBI, revealing that up to 38% exhibited abnormalities. Furthermore, even among those with a GCS score of 15, there was evidence of abnormalities on the CT scans [39]. This may suggest GCS score should not be a quantitative indicator of whether a child should undergo brain imaging. Furthermore, it has been reported in some children for whom an intracranial injury was present, some had no signs or symptoms before the imaging [40]. Therefore, the threshold for deciding whether to conduct brain imaging procedures should potentially be lower for children if they are likely to experience intracranial lesions without associated symptoms.

6.3. Clinical considerations

There is a commonly held perception that there is an undefined, but clinically important, false-negative rate when scanning patients with mTBI [36]. Also, it is hard to determine whether the expensive cost of using CT scans regularly can be outweighed by the diagnostic and intervention benefits that come with it. Moreover, remote regions and smaller practices may not even have regular access to such facilities. In an acute setting, one of the primary concerns is identifying intracranial lesions that may require surgical intervention [9]. Many have therefore put forward suggestive criteria in deciding which patients should undergo brain imaging, as displayed in Table 4.

As is clear in Table 4, emphasis has been placed on GCS scores [25]. However, as mentioned above the GCS may not be valid for determining this decision in children. Among the proposed criteria and decisional rules suggested by authors, difficulty still remains in making this decision as no clinical guidelines have been universally applied. This may result in a propor-

Authors	Suggestions
Borg et al. (2004)	• Patients with mTBI presenting with GCS of 15 and one or more risk factors should undergo CT scan; • mTBI with GCS 13 or 14 should undergo CT scan and be admitted for hospitalised observation
Jagoda et al. (2002) and Haydel et al. (2002)	Presence of one or more of: • Headache • Vomiting • Age greater than 60 • Drug/alcohol intoxication • Short-term memory deficits • Physical evidence of trauma above the clavicles • Seizure
Schutzman et al. (2001)	• Younger age may require lower threshold for imaging; • Greater the severity and number historical symptoms the stronger the consideration; • Discrete cut offs can't be provided for signs along continuum, however the greater the force involved, the greater the physical findings and the greater the risk of intracranial injuries; • All children with head trauma should be evaluated for extracranial injuries

Table 4. Suggestions of criteria for deciding to evaluate a mTBI patient using CT scan

tion of mTBI patients with potential intracranial injuries still going untested in the ED and acute settings [15, 40]. It is therefore important that a specified and universal set of guidelines or rules be applied for clinicians wanting to evaluate a patient with mTBI, so that there are no waste or resources and no patients needing intervention who go un-tested.

7. Medical and surgical intervention following mild TBI

During a structured clinical interview for evaluating a patient with potential mTBI, a clinician may be concerned with detecting intracranial injuries not detectable through the assessment of symptoms alone [25]. It is imperative clinicians be aware that certain intracranial complications may be present following mTBI, and how to detect which patients are likely to be affected [42]. There are many neurological disorders associated with mTBI, including seizures or cranial nerve defects, potentially requiring surgery [42]. The most common types of medical intervention required are craniotomy, elevation of depressed skull fracture, intracranial pressure monitoring and intubation [25]. However, other issues the clinician would need to address involve detecting operative lesions and cerebrospinal leaks [42], all of which are not easily detected and may therefore be missed in the ED.

7.1. Prevalence of cases requiring intervention

In patients with mTBI, it is estimated that surgical intervention is required in 1% [20, 25], with skull fractures seen in around 5% of treated mTBI cases [25]. A study [43] examining the prevalence of those with CT abnormalities and in need of medical intervention following mTBI revealed that 44 (1%) patients required neurological intervention, 254 (8%) displayed clinically important brain injury, and 94 (4%) had unimportant lesions. Furthermore, of those receiving intervention, 27.9% underwent craniotomy, 9.3% received elevation of skull fracture and 5.2% required intubation (4.1% died secondary to injury) [43]. Therefore, although only a small percentage, some individuals with mTBI will require some sort of surgical intervention following their injury. Considering the large number of individuals who will experience a mTBI in their lifetime [1, 2], this small percentage will still equate to a significant amount of morbidity and mortality, and therefore clinicians should be aware of which patients may be in need of this intervention.

7.2. Predicting the need for intervention

A small proportion of seemingly intact mTBI patients may deteriorate and require neurosurgical intervention [44] and so early diagnosis by CT followed by early surgery is important [43]. However, controversy exists whereby normal CT scans do not always equate to the patient not needing later intervention [43]. Indeed, a study [43] revealed that five patients with subdural hematoma did not undergo CT scanning [43] which means many medical problems requiring intervention may often go unnoticed at immediate assessment. A clinical concern is that clinicians will apply different rules and criteria to aid their decision in ordering scans [43, 44], which results in varying outcomes for patients [25, 45].

Researchers have consequentially aimed to provide criteria for classifying patients at risk of intracranial complications to aid early identification and intervention [25, 45]. Factors that have been associated with this likelihood is the mechanism of head injury, such as being hit by a vehicle, increasing age, and the finding of focal neurological deficits [45]. Others also discuss the presence of clinical 'high risk factors' for detecting the need for intervention, and clinical 'medium risk factors' for indicating the likelihood of a clinically important brain scan [25].

7.3. Special considerations for children

An additional issue regarding paediatric mTBI is that an epidural hematoma can develop in the absence of LOC or skull fracture [45], making the need for medical intervention even harder to detect. While clinical factors can be used to predict cranial abnormalities in adults, the evidence is lacking for children [20, 46]. Furthermore, although skull fracture has been reported as a significant risk for intracranial complications, up to 50% of paediatric intracranial complications can occur in the absence of skull fracture [46].

Therefore, the clinical factors used to predict individuals likely to need medical intervention will not be the same for children as they are for adults. It has been suggested that a neurosurgeon be consulted for any child displaying intracranial injury on a CT scan, and any skull fractures, so that a lower threshold for making the decision is applied than would be for adults [40]. It is particularly important that appropriate and specific guidelines be developed and applied for children due to the negative long-term effects of delayed surgery for acute hematoma or haemorrhage [40, 46].

http://dx.doi.org/10.5772/57208

7.4. Clinical considerations

Due to the variable clinical findings and criteria used when determining who will be likely to require medical intervention following mTBI, a number of studies have proposed guidelines for clinicians to consider [25, 43, 45]. Patients are often classified as 'high risk' or 'medium risk' for requiring later medical intervention, and depending on this risk these patients are either sent home or admitted for observation [45]. Those with high-risk factors are likely to need intervention and may require brain imaging, whereas those with medium risk-factors may have clinically important lesions apparent on CT but not at risk for medical intervention [43].Table 5 displays criteria proposed by two studies in predicting patients who will need intervention following mTBI [43, 45].

Authors	Proposed Criteria
Servadei et al. (2001)	a) adult patients with GCS 13-14, and 15 where there is skull fracture, or; b) GCS 13-14 and 15 with abnormal mental status, or; c) GCS 15 with skull fracture, or; d) GCS 13-14 and history of LOC, or; e) mandatory with GCS 13, recommended for 14-15, or; f) one or more of: - neurological deficits - basilar skull structure - scalp injury - > 60years
Stiell et al. (2001)	High Risk: a) GCS <15 2 hours after injury b) suspected open or depressed skull fracture c) sign of basal skull fracture d) vomiting (>twice) e) >65 years. Medium Risk: a) amnesia before impact >30min b) dangerous mechanism of injury

Table 5. Proposed guidelines for predicting patients that will require medical or surgical intervention following mTBI.

As evident in Table 5, differences are apparent in the literature regarding how to appropriately evaluate and assess patients presenting with potential mTBI. The inconsistencies in the literature prevent clinicians from following a standardised procedure when examining patients with mTBI during a structured clinical interview. Therefore, the process of decision-making is commonly quite subjective in nature. For accurate diagnosis and consistent decisions, a universally applied procedure and list of criteria needs to be developed for all clinicians.

8. Complex concussion

Complex concussion, or complicated mTBI, is where an individual has a GCS score of 13-15 but shows 1) specific problems such as concussive convulsions, 2) LOC of more than 1 minute, 3) persistent symptoms or prolonged impairment, and 4) history of multiple mTBIs [15, 39, 47, 48]. More specifically, mTBIs with associated symptoms lasting more than 10 days will be classified as complex [15, 47]. Patients with complex concussion are more likely to suffer from persistent cognitive and psychological symptoms, and their recovery pattern will be more similar to those with more moderate head injuries [39].

Since complex concussion is associated with an extended recovery time, and therefore more morbidity, it is important for clinicians to be able to identify those who are likely to suffer from the disorder, possibly through the use of 'concussion modifiers' [47]. Such concussion modifiers which may allude clinicians to the type of patients who will develop complex concussion include intrinsic factors (such as age of the patient, phenotype of symptoms, prolonged duration of LOC) and extrinsic factors (such as type of injury sustained, location of injury on head or neck) [47].

8.1. Risk factors and predictors for complex concussion

Many studies have attempted to delineate the specific risk factors and predictors for developing a complex concussion [22, 49-51]. In one study [50], 172 mTBI patients were assessed in the ED, 37 of which developed complex concussion symptomatology. Reported risk factors in this study were skull fracture, serum protein S-100B, dizziness, headache on admission, childhood psychiatric illness, LOC and PTA [50]. However, it was also found that LOC, PTA, extracranial injury, prior TBI, employment status, insurance, psychotropic drugs, current heavy drinking, smoking and prior use of illicit drugs were not independent risk factors for the development of prolonged mTBI symptoms [50]. Therefore, although LOC and PTA appear to at least partially predict who may suffer from complex concussion, they alone cannot be used as definitive markers for such patients. It was suggested that the association between the serum protein and prolonged symptoms of mTBI may indicate an organic aetiology for the disorder [50], however further research is needed. Table 6 displays commonly reported risk factors and predictors suggested to have a role in determining who will develop complex concussion following mTBI.

Although studies have sought to determine which factors are associated with complex concussion in mTBI patients, there is still uncertainty over how and why some patients do not recover after the 10 day mark [49]. However, it is likely that based on the above predictors, those at risk of experiencing prolonged symptoms can be identified as early as the ED room during acute care [50]. This is important to consider as a clinician when assessing mTBI patients during a structured clinical interview.

8.2. Special considerations for children

It has been noted that recovery time following mTBI may be longer for children and adolescents [11] compared to adults. However, persistent symptoms that need to be addressed and

Authors	Risk Factors/Predictors
Erlanger et al. (2003)	• Cognitive impairment at initial evaluation relative to baseline measurements
Williams et al. (2010)	• Abnormal brain imaging findings • Prior mTBIs • Greater LOC/PTA • Longer duration of initial symptoms • Younger age • Female • Previous psychiatric history
Bazarian et al. (1999)	• Female • Digit span forward scores • Verbal learning scores • Retrograde or anterograde amnesia -- all associated with the absence of complex concussion
Centres for Disease Control and Injury Prevention	• History of mTBI • History of headaches • Developmental disorders (e.g. Attention-Deficit/Hyperactivity Disorder) • Psychiatric problems (e.g. history of anxiety, depression, sleep disorders)

Table 6. Reported risk factors/predictors of developing complex concussion.

monitored when evaluating children at follow-up are persistent headaches, poor attention, change in nursing/eating habits, being easily upset/having tantrums, lethargic mood, lack of interest in favourite toys, and excessive crying [11]. Considering this, it is important that children only return to school and leisure-play when their symptoms fully resolve and they have completely recovered. Unfortunately, many children do in fact return to school without rehabilitation and almost 1/3 of schools are unaware that the child has suffered a TBI [13]. Furthermore, teachers are rarely aware of the potential long-term effects associated with mTBI and that it can lead to problems which will impede on a child's learning and development [13]. Therefore, it is important that the clinician take special care when assessing and evaluating a child for complex concussion to ensure they do not return to school or play too early, where they may not get the rest and care they need. Further, it may be beneficial for a clinician to notify the school of the child's injury to enable adequate recovery time.

8.3. Clinical considerations

Patients "at risk" of developing complex concussion need to be identified early, even in the ED [50]. Therefore, a functional evaluation by the clinician during a clinical interview should be conducted and include activities of daily living, mobility skills, linguistic-pragmatic abilities, sexuality issues, vocational/academic status and psychosocial issues [42]. This would ensure estimates of the patient's ability to return to school, work or recreational activity are obtained to avoid further injury and enhance recovery. It has been found that the presence of complaints of headaches, balance problems, dizziness, fatigue, depression, anxiety, irritability, and memory and attention difficulties may moderate a patient's outcomes following mTBI [52]. It is clear that a clinician must consider all factors when evaluating a patient with mTBI in the ED, to ensure they are not at heightened risk of experiencing complex concussion and therefore providing immediate prevention and intervention.

9. Conclusions

It is often falsely assumed that mTBI is not associated with any significant health burden, however considering the large number of individuals affected per year, accurate identification of the problem is paramount. Along with being difficult to define, mTBI is also difficult to evaluate, detect and diagnose. The non-specific symptomatology associated with the injury, and issues with self-report of these symptoms during clinical evaluation, suggest that mTBI cannot be diagnosed through the consideration of symptomatology alone. Moreover, it has been established that a certain threshold for concussive blow to the head, causing trauma and potential neurological disruption, is required for an accurate diagnosis. The use of brain imaging techniques such as CT and MRI scans may aid the diagnosis of mTBI and help identify those who require surgical and medical intervention; however, these remain inconclusive and more work needs to be done regarding their clinical utility for very mild injuries. And finally, the potential for mTBI patients to suffer from a complex concussion, whereby prolonged symptoms can impede on everyday life and functioning, is an important factor to consider when evaluating individuals in an acute setting. It is clear that controversies surround identifying the most effective and successful procedures for assessing mTBI in a structured clinical interview, and more work needs to be done so that patients with mTBI can be accurately detected and treated accordingly.

Author details

Michelle Albicini* and Audrey McKinlay

*Address all correspondence to: michelle.albicini@monash.edu

Monash University, Faculty of Medicine, Nursing and Health Sciences, School of Psychology and Psychiatry, Victoria, Australia

References

[1] Hawley C A, Ward A B, Long J, Owe, DW, Magnay AR. Prevalence of traumatic brain injury amongst children admitted to hospital in one health district: a population-based study. Injury 2003;34 256-260.

[2] McKinlay A, Grace RC, Horwood LJ, Fergusson DM, Ridder EM, MacFarlane MR. Prevalence of traumatic brain injury among children, adolescents and young adults: prospective evidence from a birth cohort. Brain Injury 2008;22(2) 175-181.

[3] Gioia GA, Collins M, Isquith PK. Improving identification and diagnosis of mild traumatic brain injury with evidence: Psychometric support for the Acute Concussion Evaluation. Journal of Head Trauma Rehabilitation 2008;23(4) 230-242.

[4] Vanderploeg RD, Groer S, Belanger HG. Initial developmental process of a VA semi-structured clinical interview for TBI identification 2012;49(4) 545-556.

[5] Schwab KA, Ivins B, Cramer G, Johnson W, Sluss-Tiller M, Kiley K, et al. Screening for traumatic brain injury in troops returning from deployment in Afghanistan and Iraq: Initial investigation of the usefulness of a short screening tool for traumatic brain injury. Journal of Head Trauma Rehabilitation 2007;22(6) 377-389.

[6] Powell JM, Ferraro JV, Dikmen SS, Temkin NR, Bell K. Accuracy of mild traumatic brain injury diagnosis. Archives of Physical Medicine and Rehabilitation 2008;89 1550-1557.

[7] Willer B, Leddy, JJ. Management of concussion and post-concussion syndrome. Psychiatric Manifestations of Neurologic Disease 2006; 8 415-426.

[8] Ruff RM, Iversion GL, Barth JT, Bush SS, Broshek DK, NAN Policy and Planning Committee. Recommendations for diagnosing mild traumatic brain injury: A national academy of neuropsychology education paper. Archives of Clinical Neuropsychology 2009; 24 3-10.

[9] Marion DW. Current diagnostic and therapeutic challenges. In: Tsao, JW. (ed.)Traumatic Brain Injury: A Clinician's Guide to Diagnosis, Management, and Rehabilitation. Springer Science; 2012. P313-322.

[10] Esselman PC, Uomoto JM. Classification of the spectrum of mild traumatic brain injury. Brain Injury 1995; 9(4) 417-424.

[11] Centers for Disease Control and Injury Prevention. Heads up: Facts for physicians about mild traumatic brain injury. 2006.

[12] Greenwood, R. Head injury for neurologists. Journal of Neurology, Neurosurgery and Psychiatry 2002;73(Suppl I) i8-i16.

[13] Hawley CA, Ward AB, Magnay AR, Mychalkiw W. Return to school after brain injury. Archives of Disease in Childhood 2004;89 136-142.

[14] Ruff RM, Jurica P. In search of a unified definition for mild traumatic brain injury. Brain Injury 1999;13(12) 943-952.

[15] McCrea MA. Mild traumatic brain injury and postconcussion syndrome: The new evidence base for diagnosis and treatment. New York; Oxford University Press; 2008.

[16] Haydel M. Management of mild traumatic brain injury in the emergency department. Emergency Medicine Practice 2012;14(9) 1-24.

[17] National Centre for Injury Prevention and Control. Report to Congress on mild traumatic brain injury in the United States: Steps to prevent a serious public health problem. Atlanta, GA: Centres for Disease Control and Prevention; 2003.

[18] American Congress of Rehabilitation Medicine. Definition of mild traumatic brain injury. Journal of Head Trauma Rehabilitation 1993;8(3) 86-87.

[19] Carroll LJ, Cassidy JD, Holm L, Kraus J, Coronado V. Methodological issues and research recommendations for mild traumatic brain injury: The WHO Collaborating Task Force on Mild Traumatic Brain Injury. Journal of Rehabilitation Medicine 2004; (suppl 43) 113–125.

[20] Holm L, Cassidy JD, Carrol LJ, Borg J. Summary of the WHO collaborating centre for neurotrauma task force on mild traumatic brain injury. Journal of Rehabilitation Medicine 2005;37 137-141.

[21] Kosaka B. Neuropsychological assessment in mild traumatic brain injury: A clinical overview. BC Medical Journal 2006;48(9) 447-452.

[22] Erlanger D, Kaushik T, Cantu R, Barth JT, Broshek DK, Freeman JR, et al. Symptom-based assessment of the severity of a concussion. Journal of Neurosurgery 2003;98 477-484.

[23] De Monte VE, Geffen GM, May CR, McFarland K. Double-cross validation and improved sensitivity of the rapid screen of mild traumatic brain injury. Journal of Clinical and Experimental Neuropsychology 2004;26(5) 628-644.

[24] Iverson GL, Brooks BL, Ashton VL, Lange RT. Interview versus questionnaire symptom reporting in people with the postconcussion syndrome. Journal of Head Trauma Rehabilitation 2010;25(1) 23-30.

[25] Borg J, Holm L, Cassidy JD, Peloso PM, Carroll LJ, von Holst H, Ericson K. Diagnostic procedures in mild traumatic brain injury: Results of the WHO collaborating centre task force on mild traumatic brain injury. Journal of Rehabilitation Medicine 2004;Suppl 43 s61-s75.

[26] Gerber DJ, Schraa JC. Mild traumatic brain injury: searching for the syndrome. Journal of Head Trauma Rehabilitation 1995; 10(4) 28-40.

[27] Nolin P, Villemure R, Heroux L. Determining long-term symptoms following mild traumatic brain injury: Method of interview affects self-report. Brain Injury 2006;20(11) 1147-1154.

[28] Stapleton E, Mills R. Role of open-ended questionnaires in patients with balance symptoms. The Journal of Laryngology and Otology 2008;122 139-144.

[29] Rees PM. Contemporary issues in mild traumatic brain injury. Archives of Physical Medicine and Rehabilitation 2003;84 1885-1894.

[30] Barkhoudarian G, Hovda DA, Giza CC. The molecular pathophysiology of concussive brain injury. Clinical Sports Medicine 2011;30 33-48.

[31] Ommaya AK, Goldsmith W, Thibault L. Biomechanics and neuropathology of adult and pediatric head injury. British Journal of Neurosurgery 2002;16(3) 220-242.

[32] Alexander MP. Mild traumatic brain injury: Pathophsyiology, natural history, and clinical management. Neurology 1995;45 1253-1260.

[33] Meaney DF, Smith DH. Biomechanics of concussion. Clinical Sports Medicine 2001;30 19-31.

[34] Hoge CW, McGurk D, Thomas JL, Cox AL, Engel CC, Castro CA. Mild traumatic brain injury in U.S. soldiers returning from Iraq. The New England Journal of Medicine 2008;358(5) 453-462.

[35] Haydel MJ, Preston CA, Mills TJ, Luber S, Blaudeau E, DeBlieux PMC. Indications for computed tomography in patients with minor head injury. The New England Journal of Medicine 2000;343 100-105.

[36] Jagoda AS, Cantrill SV, Wears RL, Valadka A, Gallagher EJ, Gottesfeld SH, Pietrzak MP, Bolden J, Bruns JJ Jr, Zimmerman R. Clinical policy: neuroimaging and decision making in adult mild traumatic brain injury in the acute setting. Annals of Emergency Medicine 2002;40 231-249.

[37] Livingston DH, Lavery RF, Passannante MR, Skurnick JH, Baker S, Fabian TC et al. Emergency department discharge of patients with a negative cranial computed tomography scan after minimal head injury. Annals of Surgery 2000;232(1) 126-132.

[38] Hofman PAM, Stapert SZ, van Kroonenburgh MJPG, Jolles J, de Kruijk J, Wilmink JT. MR imaging, single-photon emission CT, and neurocognitive performance after mild traumatic brain injury. American Journal of Neuroradiology 2001;22 441-449.

[39] Iverson GL, Lovell MR, Smith S, Franzen MD. Prevalence of abnormal CT scans following mild head injury. Brain Injury 2000;14(12) 1057-1061

[40] Schutzman SA, Barnes P, Duhaime AC, Greenes D, Homer C, Jaffe D et al. Evaluation and management of children younger than two years old with apparently minor head trauma: Proposed guidelines. Pediatrics 2001;107(5) 983-993.

[41] Davis RL, Mullen N, Makela M, Taylor JT, Cohen W, Rivara FP. Cranial computed tomography scans in children after minimal head injury with loss of consciousness. Pediatrics 1994;24(4) 640-645.

[42] Zasler ND. Mild traumatic brain injury: Medical assessment and intervention. Journal of Head Trauma Rehabilitation 1993;8(3) 13-29.

[43] Stiell IG, Wells GA, Vandemheen K, Clement C, Lesiuk H, Laupacis A et al. The Canadian CT head rule for patients with minor head injury. Lancet 2001;357 1391-196.

[44] Stiell IG, Wells GA, Vandemheen K, Laupacis A, Brison R, Eisenhauer MA et al. Variation in ED use of computed tomography for patients with minor head injury. Annals of Emergency Medicine 1997;30(1) 14-22.

[45] Servadei F, Teasdale G, Merry G. Defining acute mild head injury in adults: A proposal based on prognostic factors, diagnosis, and management. Journal of Neurotrauma 2001;18(7) 657-664.

[46] Simon B, Letourneau P, Vitorino E, McCall J. Pediatric minor head trauma: Indications for computed tomographic scanning revisited. The Journal of Trauma 2001;51(2) 231-238.

[47] Makdissi M. Is the simple versus complex classification of concussion a valid and useful differentiation?. British Journal of Sports Medicine 2009;43(Suppl I) i23-i27.

[48] Chen JK, Johnston KM, Collie A, McCroy P, Ptito A. A validation assessment of the postconcussion symptom scale in the assessment of complex concussion using cognitive testing and functional MRI. Journal of Neurology, Neurosurgery and Psychiatry 2007;78 1231-1238.

[49] Williams WH, Potter S, Ryland H. Mild traumatic brain injury and postconcussion syndrome: A neuropsychological perspective. Journal of Neurology, Neurosurgery and Psychiatry 2010;81 1116-1122.

[50] Savola O, Hillbom M. Early predictors of post-concussion symptoms in patients with mild head injury. European Journal of Neurology 2003;10 175-181.

[51] Bazarian JJ, Wong T, Harris M, Leahey N, Mookerjee S, Dombovy M. Epidemiology and predictors of post-concussive syndrome after minor head injury in an emergency population. Brain Injury 1999;13(3) 173-189.

[52] Belanger HG, Curtiss G, Demery JA, Lebowitz BK, Vanderploeg RD. Factors moderating neuropsychological outcomes following mild traumatic brain injury: A meta-analysis. Journal of the International Neuropsychological Society 2005;11 215-227.

Chapter 2

Indications for Brain Computed Tomography Scan After Mild Traumatic Brain Injury

Shayan Abdollah Zadegan and
Vafa Rahimi-Movaghar

Additional information is available at the end of the chapter

http://dx.doi.org/10.5772/57308

1. Introduction

Trauma to the head may cause damage to the brain parenchyma. This intracranial damage is more probable in severe than mild forms of head trauma. Despite this knowledge, physicians could not disregard the risk of brain injury in mild forms; thus, in many cases of head trauma, computed tomography (CT) scans are performed to detect possible intracranial injuries. This excessive CT scanning could cause unnecessary radiation exposure for patients, many cases of which are children and impose a heavy burden on the society. Different studies tried to decrease unnecessary CT scans by establishing clinical rules to predict possible injuries. These rules will help clinicians to safely recognize at risk patients with clinical signs and symptoms and perform CT scans when it is really necessary. The need for these rules is more prominent in mild forms of traumatic brain injury, where patients usually come with normal level of consciousness. This chapter will introduce different clinical studies related to indications for brain CT scan after Mild traumatic brain injury.

2. Mild Traumatic Brain Injury (MTBI)

When it comes to mild traumatic brain injury (MTBI), general expectation is a self-limited situation with no intracranial structural damages. Well, it is not true. According to the American Congress of Rehabilitation Medicine (ACRM) and World Health Organization (WHO), MTBI is described as Glasgow Coma Scale (GCS) score of 13 to 15 due to brain damage resulting from blunt trauma or acceleration/deceleration forces. One or more of the following symptoms and signs could be accompanied by MTBI: "confusion/disorientation, loss of consciousness for 30 min or less, posttraumatic amnesia for less than 24 hr, and/or other transient neurological abnormalities such as focal signs, seizure, and intracranial lesion" [1]. Some other signs and symptoms like headache, dizziness, irritability, lethargy, vomiting, blurred vision, fatigue, and poor concentration have also been reported [1-5]. However, the definition has been changing for the last decade and there is no fully agreed definition of MTBI [6]. Also, there is a body of evidence suggesting that MTBI could be a misnomer, because in traumatic brain injury (TBI) cases with GCS score of 13-15, some structural changes and long-term sequels are documented. However the rate of neurosurgical intervention is very low and the effect of these structural changes on long-term prognosis is not fully understood [7-11]. So, the term "complicated MTBI" was suggested by some authors to fill this defect, but this area is still challenging [8;12]. Possible pathological features including intracranial hemorrhage, diffuse axonal injury and brain contusion might be present in MTBI and could be detected by different neuroimaging modalities. The first-line modality is computed tomography (CT) scan; skull radiographs are not recommended as a useful screen test in TBI due to low sensitivity and misleading false negatives [13]. Emergency head CT studies showed that up to 15% of head trauma patients with GCS score of 15 could have acute lesions; however, the rate of neurosurgical intervention in these patients is very low (less than 1%) [13-22]. In consequence, there is a chance to reduce unnecessary imaging studies in MTBI with certain clinical rules. Using such rules could reduce imaging charges, radiation exposure and time in emergency departments (ED) [2;6;8;12;13;15;23]. In this chapter, we will focus on the most frequently used clinical rules for reducing CT scans after MTBI.

3. Leading clinical rules

The most validated rule in minor head injury is the Canadian CT head rule (CCHR), presented by Stiell and colleagues in 2001 [24]. Their prospective cohort study was conducted in ED of 10 Canadian large hospitals and 3,121 patients were enrolled. The enrollment criteria were based on MTBI description. They also excluded patients below 16 years old and those with minimal injuries (no loss of consciousness, amnesia, or disorientation), primary events like primary seizure or syncope, obvious penetrating skull injury or depressed fracture, acute focal neurological deficit, unstable vital signs, bleeding disorder or used oral anticoagulants and pregnancy as well as patients returned for reassessment of the same head injury. The main outcome measures for this rule were the need for neurological intervention and clinically important brain injury (detected by CT scan). From 3,121 patients, 2,078 underwent CT scanning according to the clinical judgment of the physician and the remaining 1,043 patients underwent the 14 day telephone proxy outcome measure. High risk patients in this guideline were patients with GCS score of less than 15 (2 hours after injury), suspected open or depressed skull fracture, signs of basal skull fracture (hemotympanum, 'raccoon' eyes, cerebrospinal fluid otorrhea/rhinorrhea, Battle's sign), two or more episodes of vomiting and elderly patients (≥65 years). The medium risk group in this criteria include patients with more than 30 minutes of amnesia before impact and dangerous mechanism of injury (like pedestrian struck by motor vehicle, occupant ejected from motor vehicle, fall from height >3 feet or five stairs) [24] (see

http://dx.doi.org/10.5772/57308

table 1). It was suggested that patients with any of the high-risk factors would probably need neurosurgical intervention and the CT is necessary in this group. Those MTBI patients with medium risk factors may have a lesion that could be seen on CT scan but probably do not require neurosurgical intervention (even with either of the two medium risk factors) and managing these patients could be performed by CT scan or direct monitoring according to local policy. The sensitivity of the test was 100% for high-risk factors and 98 4% for medium-risk factors. The test was 49 6% specific [24].

High-risk criteria*	– GCS score <15 – Suspected open or depressed skull fracture – Signs of basal skull fracture (hemotympanum, 'raccoon' eyes, cerebrospinal fluid otorrhea/rhinorrhea, Battle's sign) – Vomiting ≥ 2 episodes – Elder patients (≥65 years)
Medium-risk criteria**	– Amnesia before impact (>30 min) – Dangerous mechanism of injury • Pedestrian struck by motor vehicle • Occupant ejected from motor vehicle • Fall (>3 feet or five stairs)

* Neurosurgical intervention is likely to be needed in high-risk group and a head CT scan is mandatory.

** Medium-risk group probably do not require neurosurgical intervention and could be managed by CT scanning or direct monitoring.

Table 1. The Canadian CT head rule (CCHR) for brain CT scan. (Originally described by Stiell et al. Reproduced with permission [24])

After the Stiell's study, many researchers started to validate the CCHR rule. The results were very diverse. Few studies came up with great questions in effectiveness of CCHR [25;26]. Even increased numbers of CT scans were reported in one British study. The study was a retrospective case note review. This study design may raise some questions about recording all important clinical variables. However, the authors of the article were confident about data collection due to standardized forms that were used in their hospital [25]. Another retrospective study in Australia showed that CCHR would decrease the number of CT scanning without missing the patients requiring neurosurgical intervention but in this study, it missed two patients with clinically significant CT abnormality. Similar to the previous, this study was retrospective and relies on the accuracy of medical records [26].

Despite these studies, a larger number of studies found it beneficial or cost-effective to replace current protocols with CCHR [27-35]. A prospective cohort study with 4,551 patients from 10 hospital EDs of the United States reported that CCHR could accurately identify all patients requiring neurosurgical intervention [30]. Another multicenter prospective validation study was conducted in 4 university hospitals of the Netherlands. Among 3,181 patients, CCHR identified all cases requiring neurosurgical intervention (sensitivity of 100%); however, it did not show high sensitivity for neurocranial lesions of CT scan (approximately 85%). The specificity of the test was 37.2%-39.7%. This study did not use the exact predictors that Stiell defined previously. For instance, they evaluated GCS of patients at 1 hour after presentation instead of 2 hours. They also used post-traumatic amnesia of more than 30 minutes as a risk factor instead of retrograde amnesia and vomiting was defined as any period of emesis instead of more than 1 episode of emesis [17]. Another cost effective analysis research showed the cost effectiveness of CCHR with US $120 million savings annually. However, this finding will be valid only if the CCHR is highly sensitive for neurosurgical intervention prediction and at sensitivities lower than 97%, it will be more cost-effective to perform CT scan for all patients [36].

There are also few trials that aimed to implement CCHR at emergency departments [37;38]. In the opinion of Stiell and colleagues [38], the failure of such trials was the result of physicians' non-compliance. Some examples of such non-compliance of the physicians were forgetting the details of the rules, not believing in the nature of the research, being forced to order CT scans due to patient/family expectations, having inconsistencies among different services in ordering CT scans and insufficient time in a busy ED and finally the idea that the application of the rule would either take too much time or would not be safe for the patients [38]. Other epidemiological studies have confirmed this opinion [39;40].

Two other famous criteria are the NEXUS-II (National Emergency X-Ray Utilization Study) rule and the NOC (New Orleans Criteria). According to NOC, presented by Haydel and colleagues in 2000 [15], minor head injury is defined as loss of consciousness in patients with normal cranial nerves, normal strength and sensation in arms and legs and a GCS score of 15 at arrival. Red flags for head injury are the following seven factors: headache, vomiting, age, drug or alcohol intoxication, short-term memory deficit, physical evidence of trauma above the clavicles, and seizure. These seven factors showed a sensitivity of 100% for detecting patients with positive CT scans, negative predictive value of 100% and specificity of 25%. Coagulopathy is absent in NOC because in the first phase of the study conducted by Haydel and colleagues, only 1 patient out of 520 had coagulopathy; so, the researchers could not evaluate this item [15]. Also, imaging for all patients with evidence of trauma above the clavicles would be excessive, because small contusions or lacerations and minor facial injuries would be included and this will increase the rate of CT imaging. Prospective evaluation of NOC showed lower reduction rate of CT scanning and lower specificity (3% to 31%) compared to CCHR (48% to 77% for high-risk criteria and 37% to 48% for high and medium-risk criteria), but the sensitivity of NOC was similar to CCHR (99% to 100%) [16;17;26;32;33;41-43]. Although a 2013 research by Bouida and colleagues in Tunisia has reported a slightly lower sensitivity for NOC (86% sensitivity for clinically significant head CT findings and 82% sensitivity for the need for neurosurgical intervention) [34]. This difference in sensitivity could be the result of NOC application in GCS lower than 15. As mentioned above, NOC was defined for MTBI patients with GCS scores of not lower than 15. This conclusion will be strengthened by the knowledge that the difference was diminished in a subgroup of patients with GCS score of 15 [34].

In 2002, a prospective multicenter study called NEXUS-II (National Emergency X-Radiography Utilization Study II) was designed to prepare a decision rule for CT imaging of patients with acute blunt head trauma [23;44]. NEXUS II decision instrument enrolled 13,728 patients and presented eight factors significantly associated with intracranial injuries: 1. evidence of significant skull fracture, 2. hematoma of scalp, 3. neurological deficit, 4. altered level of alertness, 5. abnormal behavior, 6. coagulopathy, 7. persistent vomiting and 8. patients aged 65 years or more [23]. NEXUS II criteria provided guidance for patients without any episode of loss of consciousness. This great advantage was absent in the two previous criteria. A recent prospective cohort study reported a sensitivity of 95.1% and specificity of 41.4% for neurosurgical intervention for NEXUS II criteria. This study showed that NEXUS II criteria were less sensitive but more specific than NOC and CCHR (sensitivity of 100% for NOC and CCHR and specificity of 38.3% for CCHR and 20.4% for NOC were documented). The potential reduction rate in CT scanning was higher in NEXUS II (39.6%) compared to CCHR (27%) and NOC (20.2%) [45]. Another retrospective analysis reported the sensitivity and specificity of NEXUS II criteria at respectively 97% (versus 99% sensitivity for CCHR and NOC) and 47% (versus specificity of 47% for CCHR and 33% for NOC) for detecting any lesions. The sensitivity of NEXUS II criteria for detecting hematoma was higher than the two other criteria (100% vs. 99%); however, researchers conducting the study could not demonstrate that this higher sensitivity was statistically significant [43]. In conclusion, NEXUS II criteria showed the highest reduction rate for CT scanning compared to the other two rules, but it failed to recognize some patients requiring neurosurgical intervention. Table 2 shows all characteristics included in the three above-mentioned leading rules.

4. Other clinical rules

Other studies conducted on indications of brain CT scan after MTBI are briefly cited. In 1993, Reinus et al. [46] developed criteria by reviewing medical records and CT scans of 373 patients. The four variable criteria were positive neurologic examination, intoxication, amnesia and a history of focal neurologic deficit. The criteria had a sensitivity of 90.1% and a negative predictive value of 98.1% for detecting abnormality in CT scan. The criteria missed 4 patients. None of them needed neurosurgical intervention [46]. In 1994, Duus and colleagues [47] performed a prospective cohort study on 2,204 patients in Denmark. They admitted patients with confusion or aggression in ED, impaired consciousness, focal neurological sign, suspected skull fracture in a clinical exam, alcohol intoxication or other medical conditions that interfere with assessment, history of convulsions, amnesia (more than 15 minutes), history of unconsciousness (more than 15 minutes and witnessed by a competent observer), 3-year-old or younger children with symptoms and no responsible adult at home. The study proposed the use of the guideline instead of skull radio graphs [47].

In 1995, in a retrospective descriptive study of 1,448 American patients with GCS score of 13 or more, Borczuk [48] found the following factors as high-risk: soft tissue injury, focal neurologic deficit, signs of basilar skull fracture and being older than 60 years. The sensitivity was 91.6% and specificity was 46.2% for detecting a CT abnormality. None of the patients missed by these criteria, required medical or neurosurgical management [48]. In 1997, Miller et al. [49] enrolled 2143 patients with the GCS score of 15 and a history of loss of consciousness (LOC) to develop simple clinical criteria to safely reduce the number of CT scans. Miller's 4 high-risk criteria included headache, nausea, vomiting and signs of depressed skull fracture. The criteria showed 61% reduction in the number of CT scans and also identified all patients who required neurosurgical intervention. Holmes et al. examined the criteria in patients with a lower GCS score (GCS=14). Of the total 264 patients, 35 had abnormal CT scans and Miller criteria failed to detect 17 of them including 2 intoxicated (with ethanol) patients that needed neurosurgical intervention [49;50]. Another study was performed by Arienta et al. (1977) [51] on 10,000 patients in Italy. In this retrospective study, risk factors for intracranial lesions were loss of consciousness, amnesia after trauma, vomiting (repeated episodes), neurologic deficits, and signs of basal skull fracture, seizure, penetrating or perforating wounds, lack of cooperation, previous intracranial operations, coagulopathy or anticoagulant therapy, epileptic or alcoholic patients [51]. In 2003, Falimirski and colleagues [52] prospectively included 331 MTBI patients with GCS score of 14 and 15 and a history of LOC. The GCS score of 13 was excluded. The study examined patients for the presence of 10 typical constitutional signs and symptoms (CSS) for head injury including headache, somnolence, confusion, nausea/vomiting, seizure, perseveration, neurologic deficit, blurred/double vision, vertigo and hemotympanum. The results showed that loss of consciousness alone (i.e. LOC without CSS) could not predict intracranial injury. Of 195 patients without CSS, only 11 (5.6%) had intracranial lesion and of 136 remaining patients (with CSS) 29 (21.3%) had positive CT. However, it is unusual that none of the CT positive patients in both groups (with and without CSS) required neurosurgical intervention [52]. In 2007, Smits et al. [53] attempted to make a decision instrument for MTBI patients regardless of the presence or absence of a history of loss of consciousness. In a prospective study called CHIP (CT in Head Injury Patients) in four academic hospitals in the Netherlands, 3181 patients with MTBI and the GCS score of 13 and more (GCS score of 15 with at least 1 risk factor) were observed. Of 243 (7.6%) patients with intracranial CT findings, 17 (0.5%) patients underwent neurosurgical intervention. The major criteria were pedestrian or cyclist (vs. vehicle), ejection from vehicle, vomiting, post-traumatic amnesia for more than 4 hours, clinical signs of skull fracture, GCS score of less than15, GCS deterioration of more than 2 points, use of anticoagulants, seizure, age of more than 60 years. The minor criteria were fall from any elevation, persistent anterograde amnesia, post-traumatic amnesia for 2–4 hours, skull contusion, neurological deficit, LOC, GCS score drop of one point, age of 40–60 years. The sensitivity was 100% for neurosurgical interventions with the specificity of 23-30%. For intracranial lesions, the sensitivity and specificity were 94-96% and 25-32% in the original study. The reduction rate was 23-30% [53]. The CHIP criteria were validated as a cost-effective rule. However, the annual savings were less than the CCHR rule ($71 million savings in CHIP vs. 120$ savings in CCHR) [36]. In 2009, Saadat and colleagues [54] aimed to develop the criteria for CT scanning of MTBI patients in developing countries. Three hundred and eighteen Iranian patients with the GCS score of 13 and more were included. The major criteria were GCS score of less than 14, raccoon sign, failure to remember the impact, vomiting and age of 65 years or more. The minor criteria were wound at the scalp and GCS score of less than 15. Patients with one major criterion or 2 minor criteria had an abnormal CT scan with the sensitivity of 100% and specificity of 46% [54]. Another recent study on 642 patients in Iran showed headache, vomiting, LOC or amnesia, and alcohol intoxication as main indicators for CT scanning of MTBI patients. In the previous Iranian study, headache was not a criterion [11].

http://dx.doi.org/10.5772/57308

	CCHR (medium- and high-risk)	NOC	NEXUS II
GCS score/Level of consciousness	<15 at 2 hours after injury	-	Altered level of consciousness, Abnormal behavior
Evidence of significant trauma to the calvarium	Suspected open or depressed skull fracture, Any sign of basal skull fracture – Hemotympanum – Raccoon' eyes, – Cerebrospinal fluid otorrhea/ rhinorrhea – Battle's sign	Physical evidence of trauma above the clavicles	Any evidence of significant skull fracture, Scalp hematoma
Vomiting	≥ 2 episodes	Any	Persistent
Age	≥ 65 years	> 60 years	≥ 65 years
Amnesia*	> 30 minutes, retrograde	Short-term, anterograde	-
Mechanism of injury*	Dangerous: – Pedestrian struck by motor vehicle – Occupant ejected from motor vehicle – Fall from height >3 feet or five stairs	-	-
Focal neurological deficit	-	-	Any
Coagulopathy	-	-	Any
Headache	-	Any	-
Intoxication	-	Drug, Alcohol	-
Seizure	-	Any	-
Test sensitivity (neurosurgical intervention)	– 100% [16;17;24;30;33;34;45]	– 100% [16;17;33;45] 82% [34]	– 95.1% [45]
Test sensitivity (clinically important intracranial injury)	– 100% [16;33] – 98.4% [24] – 97 to 99% [43] – 95% [34] – 83.4% to 87.2% [17] – 79.2% [45]	– 100% [16;33] – 99% [43] – 97.7% to 99.4% [17] – 91.9% [45] – 86% [34]	– 98.3% [23] – 97% [43] – 88.7% [45]
Test specificity	– 65% [34]	– 33% [43]	– 47% [43]

	CCHR (medium- and high-risk)	NOC	NEXUS II
(clinically important intracranial injury)	- 50.6% [16] - 49.6% [24] - 47 to 51% [43] - 41.3% [45] - 37.2% to 39.7% [17] - 35% [33]	- 28% [34] - 22.4% [45] - 12.7% [16] - 9.9% [33] - 3.0% to 5.6% [17]	- 46.5% [45] - 13.7% [23]

CCHR, Canadian CT head rule; NOC, New Orleans Criteria; NEXUS II, National Emergency X-Ray Utilization Study II; GCS, Glasgow Coma Scale

*Medium-risk factors for CCHR

Table 2. Comparison of three leading clinical decision rules for CT imaging in mild traumatic brain injury in adults

5. Pediatric clinical rules

Clinical rules for pediatric patients are less developed. It is reasonable that any of the above mentioned criteria could reduce CT imaging in pediatrics as well as adults, but further validation of pediatric head CT decision aids is needed to guarantee patients' safety [18]. In 2003, Haydel and colleagues [55] tried to determine the efficacy of the clinical rule previously developed for adults (the NOC) in 5 to 17 year-old patients. They also used the same definition of minor head injury that was used for adults. Among 175 children that were enrolled, fourteen had intracranial injury and all of them had one or more of the six following findings: headache, emesis, intoxication, seizure, short-term memory deficit and physical evidence of trauma above the clavicles. Although the sensitivity of these factors was 100%, the small sample size caused a confidence interval of 74% to 100% [55]. Another study by Palchak and colleagues in 2003 [56], enrolled 2,043 children presented with traumatic brain injury (with any degree of severity) to a Level I trauma center. This observational cohort study introduced abnormal mental status, clinical signs of skull fracture, vomiting, scalp hematoma in children ≤ 2 years of age and headache as main factors associated with traumatic brain injuries on CT scan. The rule identified all 105 patients who needed acute neurosurgical intervention and 97/98 of the patients with signs of brain injury on CT scan [56]. In 2006, NEXUS-II investigators enrolled 1,666 children including very young children (<3 years) with blunt head trauma [57]. This pediatric cohort study suggested the following seven factors as the inclusion criteria for pediatrics head CT: 1. evidence of significant skull fracture, 2. altered level of alertness, 3. neurological deficit, 4. persistent vomiting, 5. presence of scalp hematoma, 6. abnormal behavior, 7. coagulopathy. All clinically important intracranial injuries were successfully identified by the rule but two patients (in 1,666 patients) were missed. However none of them needed neurosurgical interventions [57]. Children's head injury algorithm for the prediction of important clinical events (CHALICE) is another 2006 study that was performed by Dunning et al. [58] (see table 3). A large number of children (22,772) from 10 hospitals in UK with any severity of head injury were included. A sensitivity of 98% and specificity of 87% for prediction

History	
	– Witnessed loss of consciousness > 5-minute duration
	– History of amnesia (anterograde/retrograde) > 5-minute duration
	– Abnormal drowsiness
	– Vomiting after head injury (≥ 3 episodes)
	– Suspicion of non-accidental injury
	– Seizure after head injury in patients with no history of epilepsy
Examination	
	– Glasgow Coma Score (GCS) <14, (GCS<15 if <1 year old)
	– Suspicion of penetrating or depressed skull injury or tense fontanel
	– Signs of a basal skull fracture (blood or cerebrospinal fluid from ear or nose, panda eyes, battles sign, hemotympanum, facial crepitus, serious facial injury)
	– Positive focal neurological sign
	– Presence of bruise, swelling or laceration >5 cm if <1 year old
Mechanism	
	– High-speed road traffic accident (with speed >40 m/h) either as pedestrian, cyclist or occupant
	– Fall of >3 m in height
	– High-speed injury from a projectile or an object

Table 3. Children's head injury algorithm for the prediction of important clinical events (CHALICE). (Originally described by Dunning et al. Reproduced with permission [58])

of clinically significant head injury were documented in the study. The sensitivity was high but the rule was more comprehensive than the previous ones [58]. A retrospective survey from one hospital in Finland compared three pediatric clinical rules. NEXUS II, CHALICE and the rule presented by Palchak were compared. Each rule had 100% sensitivity in detecting severely complicated head trauma. Severely complicated head trauma was defined as patients who required neurosurgical intervention, patients who succumbed, epidural hematoma, subdural hematoma, subarachnoid hematoma and intracerebral hematoma. NEXUS II showed the best specificity (21% versus 12% for Palchak and 5% for CHALICE), although it was not high enough to detect patients who do not need neurosurgical interventions [59].

The retrospective study of Schachar and colleagues [18] on 2,101 pediatric patients (0-21 years old) has compared NOC, CCHR and NEXUS II. Sensitivities of these tests were 96.7%, 65.2% and 78.3%, respectively. Specificities were 11.2%, 64.2% and 34.2%, respectively. The study emphasized on the prospective ability of these tests on reduction of CT scans and concluded that further validation is needed [18].

Canadian Assessment of Tomography for Childhood Head injury (CATCH) was developed in 2010 by Osmond et al. [60]. This prospective cohort study was performed in 10 Canadian academic centers and 3,866 patients (0-16 years old) with GCS score of 13 or more were included. High-risk factors (with 100% sensitivity for predicting the need for neurologic intervention) include failure to reach GCS score of 15 within two hours, suspicion of open skull fracture, worsening headache and irritability. Three additional medium-risk factors (with 98.1% sensitivity for predicting brain injury on CT scan) are large, boggy hematoma of the scalp, signs of basal skull fracture and dangerous mechanism of injury (defined as motor vehicle crash, fall from elevation ≥ 3 ft [91 cm] or 5 stairs, fall from bicycle with no helmet) [60]. Future validation studies are needed to prove the feasibility of CATCH.

6. Conclusion

In this chapter, we reviewed different clinical rules developed for patients with mild TBI. The differences among the presented guidelines are based on diversity in main outcome measures. Main outcome measure might be the following items: 1. positive brain CT scan for TBI, 2. positive clinical findings suggesting the need for neurosurgical/medical intervention, 3. a combination of the two above-mentioned findings. Brain CT scan could demonstrate small lesions related to mild TBI but it may be of a questionable practical importance. For example, the presence of a linear fracture in right frontal bone with 2 ml. of epidural hematoma below the fracture a few hours after trauma, confirms mild injury and might be valuable for legal issues, but does not predict surgical intervention requirement or even may not need anti-convulsion therapy. Therefore, if the main outcome measure was any CT findings related to the injury, we consider this case as a positive case, but if the main outcome measure was the need for neurosurgical/medical intervention, we consider the case as negative. Thus, it is important to note the fact that in many patients with traumatic brain injury on CT scan, acute neurosurgical intervention is not required and detection of intracranial injury is important when considering the need for neurosurgical intervention.

Among the above-mentioned rules, CCHR is the most validated rule with appropriate sensitivity, but the authors believe that using this rule without considering patients' history (i.e. coagulopathy, drug and alcohol intoxication) would result in missing some patients. Also progressive severe headache, focal neurological deficit and seizure should be noticed. With reference to various sensitivities and specificities documented for different clinical rules, we need more studies with this topic to reach a safe practical guideline that could be applied worldwide.

Acknowledgements

The authors would like to acknowledge Mrs. Bita Pourmand (Sina Hospital, Research Development Center) for her careful editing.

http://dx.doi.org/10.5772/57308

Author details

Shayan Abdollah Zadegan[1] and Vafa Rahimi-Movaghar[1,2,3*]

*Address all correspondence to: v_rahimi@sina.tums.ac.ir; v_rahimi@yahoo.com

1 Sina Trauma and Surgery Research Center, Tehran University of Medical Sciences, Tehran, Iran

2 Department of Neurosurgery, Shariati Hospital, Tehran University of Medical Sciences, Tehran, Iran

3 Research Centre for Neural Repair, University of Tehran, Tehran, Iran

References

[1] Carroll LJ, Cassidy JD, Holm L, Kraus J, Coronado VG. Methodological issues and research recommendations for mild traumatic brain injury: the WHO Collaborating Centre Task Force on Mild Traumatic Brain Injury. J Rehabil Med 2004 Feb;(43 Suppl):113-25.

[2] McCrea M, Iverson GL, McAllister TW, Hammeke TA, Powell MR, Barr WB, et al. An integrated review of recovery after mild traumatic brain injury (MTBI): implications for clinical management. Clin Neuropsychol 2009 Nov;23(8):1368-90.

[3] Moore EL, Terryberry-Spohr L, Hope DA. Mild traumatic brain injury and anxiety sequelae: a review of the literature. Brain Inj 2006 Feb;20(2):117-32.

[4] Ruff RM, Iverson GL, Barth JT, Bush SS, Broshek DK. Recommendations for diagnosing a mild traumatic brain injury: a National Academy of Neuropsychology education paper. Arch Clin Neuropsychol 2009 Feb;24(1):3-10.

[5] Kirkwood MW, Yeates KO, Taylor HG, Randolph C, McCrea M, Anderson VA. Management of pediatric mild traumatic brain injury: a neuropsychological review from injury through recovery. Clin Neuropsychol 2008 Sep;22(5):769-800.

[6] Tavender EJ, Bosch M, Green S, O'Connor D, Pitt V, Phillips K, et al. Quality and consistency of guidelines for the management of mild traumatic brain injury in the emergency department. Acad Emerg Med 2011 Aug;18(8):880-9.

[7] Tellier A, la Malva LC, Cwinn A, Grahovac S, Morrish W, Brennan-Barnes M. Mild head injury: a misnomer. Brain Inj 1999 Jul;13(7):463-75.

[8] Kushner D. Mild traumatic brain injury: toward understanding manifestations and treatment. Arch Intern Med 1998 Aug 10;158(15):1617-24.

[9] Washington CW, Grubb RL, Jr. Are routine repeat imaging and intensive care unit admission necessary in mild traumatic brain injury? J Neurosurg 2012 Mar;116(3): 549-57.

[10] Zare MA, Ahmadi K, Zadegan SA, Farsi D, Rahimi-Movaghar V. Effects of brain contusion on mild traumatic brain-injured patients. Int J Neurosci 2013 Jan;123(1):65-9.

[11] Sharif-Alhoseini M, Khodadadi H, Chardoli M, Rahimi-Movaghar V. Indications for brain computed tomography scan after minor head injury. J Emerg Trauma Shock 2011 Oct;4(4):472-6.

[12] Ruff RM. Mild traumatic brain injury and neural recovery: rethinking the debate. NeuroRehabilitation 2011;28(3):167-80.

[13] Jagoda AS, Bazarian JJ, Bruns JJ, Jr., Cantrill SV, Gean AD, Howard PK, et al. Clinical policy: neuroimaging and decisionmaking in adult mild traumatic brain injury in the acute setting. J Emerg Nurs 2009 Apr;35(2):e5-40.

[14] Mittenberg W, Canyock EM, Condit D, Patton C. Treatment of post-concussion syndrome following mild head injury. J Clin Exp Neuropsychol 2001 Dec;23(6):829-36.

[15] Haydel MJ, Preston CA, Mills TJ, Luber S, Blaudeau E, DeBlieux PM. Indications for computed tomography in patients with minor head injury. N Engl J Med 2000 Jul 13;343(2):100-5.

[16] Stiell IG, Clement CM, Rowe BH, Schull MJ, Brison R, Cass D, et al. Comparison of the Canadian CT Head Rule and the New Orleans Criteria in patients with minor head injury. JAMA 2005 Sep 28;294(12):1511-8.

[17] Smits M, Dippel DW, de Haan GG, Dekker HM, Vos PE, Kool DR, et al. External validation of the Canadian CT Head Rule and the New Orleans Criteria for CT scanning in patients with minor head injury. JAMA 2005 Sep 28;294(12):1519-25.

[18] Schachar JL, Zampolin RL, Miller TS, Farinhas JM, Freeman K, Taragin BH. External validation of the New Orleans Criteria (NOC), the Canadian CT Head Rule (CCHR) and the National Emergency X-Radiography Utilization Study II (NEXUS II) for CT scanning in pediatric patients with minor head injury in a non-trauma center. Pediatr Radiol 2011 Aug;41(8):971-9.

[19] Mack LR, Chan SB, Silva JC, Hogan TM. The use of head computed tomography in elderly patients sustaining minor head trauma. J Emerg Med 2003 Feb;24(2):157-62.

[20] Ibanez J, Arikan F, Pedraza S, Sanchez E, Poca MA, Rodriguez D, et al. Reliability of clinical guidelines in the detection of patients at risk following mild head injury: results of a prospective study. J Neurosurg 2004 May;100(5):825-34.

[21] Fabbri A, Servadei F, Marchesini G, Dente M, Iervese T, Spada M, et al. Clinical performance of NICE recommendations versus NCWFNS proposal in patients with mild head injury. J Neurotrauma 2005 Dec;22(12):1419-27.

[22] Iverson GL, Lovell MR, Smith S, Franzen MD. Prevalence of abnormal CT-scans following mild head injury. Brain Inj 2000 Dec;14(12):1057-61.

[23] Mower WR, Hoffman JR, Herbert M, Wolfson AB, Pollack CV, Jr., Zucker MI. Developing a decision instrument to guide computed tomographic imaging of blunt head injury patients. J Trauma 2005 Oct;59(4):954-9.

[24] Stiell IG, Wells GA, Vandemheen K, Clement C, Lesiuk H, Laupacis A, et al. The Canadian CT Head Rule for patients with minor head injury. Lancet 2001 May 5;357(9266):1391-6.

[25] Boyle A, Santarius L, Maimaris C. Evaluation of the impact of the Canadian CT head rule on British practice. Emerg Med J 2004 Jul;21(4):426-8.

[26] Rosengren D, Rothwell S, Brown AF, Chu K. The application of North American CT scan criteria to an Australian population with minor head injury. Emerg Med Australas 2004 Jun;16(3):195-200.

[27] Sultan HY, Boyle A, Pereira M, Antoun N, Maimaris C. Application of the Canadian CT head rules in managing minor head injuries in a UK emergency department: implications for the implementation of the NICE guidelines. Emerg Med J 2004 Jul; 21(4):420-5.

[28] van Essen TA, Heeringa JJ, Muizelaar JP. (In)appropriate neurosurgical consultation. Clin Neurol Neurosurg 2010 Nov;112(9):775-80.

[29] Syed AT, Lone NA, Wani MA, Bhat AS. Clinical management of patients with minor head injuries. Int J Health Sci (Qassim) 2007 Jan;1(1):131-40.

[30] Clement CM, Stiell IG, Schull MJ, Rowe BH, Brison R, Lee JS, et al. Clinical features of head injury patients presenting with a Glasgow Coma Scale score of 15 and who require neurosurgical intervention. Ann Emerg Med 2006 Sep;48(3):245-51.

[31] Fong C, Chong W, Villaneuva E, Segal AY. Implementation of a guideline for computed tomography head imaging in head injury: a prospective study. Emerg Med Australas 2008 Oct;20(5):410-9.

[32] Perry JJ, Stiell IG. Impact of clinical decision rules on clinical care of traumatic injuries to the foot and ankle, knee, cervical spine, and head. Injury 2006 Dec;37(12): 1157-65.

[33] Papa L, Stiell IG, Clement CM, Pawlowicz A, Wolfram A, Braga C, et al. Performance of the Canadian CT Head Rule and the New Orleans Criteria for predicting any traumatic intracranial injury on computed tomography in a United States Level I trauma center. Acad Emerg Med 2012 Jan;19(1):2-10.

[34] Bouida W, Marghli S, Souissi S, Ksibi H, Methammem M, Haguiga H, et al. Prediction Value of the Canadian CT Head Rule and the New Orleans Criteria for Positive

Head CT Scan and Acute Neurosurgical Procedures in Minor Head Trauma: A Multicenter External Validation Study. Ann Emerg Med 2013 May;61(5):521-7.

[35] Holmes MW, Goodacre S, Stevenson MD, Pandor A, Pickering A. The cost-effectiveness of diagnostic management strategies for adults with minor head injury. Injury 2012 Sep;43(9):1423-31.

[36] Smits M, Dippel DW, Nederkoorn PJ, Dekker HM, Vos PE, Kool DR, et al. Minor head injury: CT-based strategies for management--a cost-effectiveness analysis. Radiology 2010 Feb;254(2):532-40.

[37] Stiell IG, Clement CM, Grimshaw JM, Brison RJ, Rowe BH, Lee JS, et al. A prospective cluster-randomized trial to implement the Canadian CT Head Rule in emergency departments. CMAJ 2010 Oct 5;182(14):1527-32.

[38] Stiell IG, Bennett C. Implementation of clinical decision rules in the emergency department. Acad Emerg Med 2007 Nov;14(11):955-9.

[39] Curran JA, Brehaut J, Patey AM, Osmond M, Stiell I, Grimshaw JM. Understanding the Canadian adult CT head rule trial: use of the theoretical domains framework for process evaluation. Implement Sci 2013 Feb 21;8:25. doi: 10.1186/1748-5908-8-25.:25-8.

[40] Eagles D, Stiell IG, Clement CM, Brehaut J, Taljaard M, Kelly AM, et al. International survey of emergency physicians' awareness and use of the Canadian Cervical-Spine Rule and the Canadian Computed Tomography Head Rule. Acad Emerg Med 2008 Dec;15(12):1256-61.

[41] Bruns JJ, Jr., Jagoda AS. Mild traumatic brain injury. Mt Sinai J Med 2009 Apr;76(2): 129-37.

[42] Harnan SE, Pickering A, Pandor A, Goodacre SW. Clinical decision rules for adults with minor head injury: a systematic review. J Trauma 2011 Jul;71(1):245-51.

[43] Stein SC, Fabbri A, Servadei F, Glick HA. A critical comparison of clinical decision instruments for computed tomographic scanning in mild closed traumatic brain injury in adolescents and adults. Ann Emerg Med 2009 Feb;53(2):180-8.

[44] Mower WR, Hoffman JR, Herbert M, Wolfson AB, Pollack CV, Jr., Zucker MI. Developing a clinical decision instrument to rule out intracranial injuries in patients with minor head trauma: methodology of the NEXUS II investigation. Ann Emerg Med 2002 Nov;40(5):505-14.

[45] Ro YS, Shin SD, Holmes JF, Song KJ, Park JO, Cho JS, et al. Comparison of clinical performance of cranial computed tomography rules in patients with minor head injury: a multicenter prospective study. Acad Emerg Med 2011 Jun;18(6):597-604.

[46] Reinus WR, Wippold FJ, Erickson KK. Practical selection criteria for noncontrast cranial computed tomography in patients with head trauma. Ann Emerg Med 1993 Jul; 22(7):1148-55.

[47] Duus BR, Lind B, Christensen H, Nielsen OA. The role of neuroimaging in the initial management of patients with minor head injury. Ann Emerg Med 1994 Jun;23(6): 1279-83.

[48] Borczuk P. Predictors of intracranial injury in patients with mild head trauma. Ann Emerg Med 1995 Jun;25(6):731-6.

[49] Miller EC, Holmes JF, Derlet RW. Utilizing clinical factors to reduce head CT scan ordering for minor head trauma patients. J Emerg Med 1997 Jul;15(4):453-7.

[50] Holmes JF, Baier ME, Derlet RW. Failure of the Miller criteria to predict significant intracranial injury in patients with a Glasgow Coma Scale score of 14 after minor head trauma. Acad Emerg Med 1997 Aug;4(8):788-92.

[51] Arienta C, Caroli M, Balbi S. Management of head-injured patients in the emergency department: a practical protocol. Surg Neurol 1997 Sep;48(3):213-9.

[52] Falimirski ME, Gonzalez R, Rodriguez A, Wilberger J. The need for head computed tomography in patients sustaining loss of consciousness after mild head injury. J Trauma 2003 Jul;55(1):1-6.

[53] Smits M, Dippel DW, Steyerberg EW, de Haan GG, Dekker HM, Vos PE, et al. Predicting intracranial traumatic findings on computed tomography in patients with minor head injury: the CHIP prediction rule. Ann Intern Med 2007 Mar;%20;146(6): 397-405.

[54] Saadat S, Ghodsi SM, Naieni KH, Firouznia K, Hosseini M, Kadkhodaie HR, et al. Prediction of intracranial computed tomography findings in patients with minor head injury by using logistic regression. J Neurosurg 2009 Oct;111(4):688-94.

[55] Haydel MJ, Shembekar AD. Prediction of intracranial injury in children aged five years and older with loss of consciousness after minor head injury due to nontrivial mechanisms. Ann Emerg Med 2003 Oct;42(4):507-14.

[56] Palchak MJ, Holmes JF, Vance CW, Gelber RE, Schauer BA, Harrison MJ, et al. A decision rule for identifying children at low risk for brain injuries after blunt head trauma. Ann Emerg Med 2003 Oct;42(4):492-506.

[57] Oman JA, Cooper RJ, Holmes JF, Viccellio P, Nyce A, Ross SE, et al. Performance of a decision rule to predict need for computed tomography among children with blunt head trauma. Pediatrics 2006 Feb;117(2):e238-e246.

[58] Dunning J, Daly JP, Lomas JP, Lecky F, Batchelor J, kway-Jones K. Derivation of the children's head injury algorithm for the prediction of important clinical events decision rule for head injury in children. Arch Dis Child 2006 Nov;91(11):885-91.

[59] Klemetti S, Uhari M, Pokka T, Rantala H. Evaluation of decision rules for identifying serious consequences of traumatic head injuries in pediatric patients. Pediatr Emerg Care 2009 Dec;25(12):811-5.

[60] Osmond MH, Klassen TP, Wells GA, Correll R, Jarvis A, Joubert G, et al. CATCH: a clinical decision rule for the use of computed tomography in children with minor head injury. CMAJ 2010 Mar 9;182(4):341-8.

Chapter 3

Current Therapeutic Modalities, Enzyme Kinetics, and Redox Proteomics in Traumatic Brain Injury

Zachariah P. Sellers, ReBecca A. Williams,
Jonathan W. Overbay, Jooyoung Cho,
Moses Henderson and Tanea T. Reed

Additional information is available at the end of the chapter

http://dx.doi.org/10.5772/57306

1. Introduction

Traumatic brain injury (TBI) is an event that occurs suddenly and has long lasting effects to the brain that are dependent on the severity of insult. Symptoms can range from mild to severe. It is a well-established notion that imbalances in the production of reactive oxygen species (ROS), reactive nitrogen species (RNS), and native antioxidant mechanisms have been shown to increase oxidative stress. Multiple research efforts have evaluated the use of pre-injury therapies on moderate TBI; however since traumatic brain injuries are sudden, treatment should begin post-injury. There is no known cure for TBI, although flavonoids, neurosteroids, statins, gamma-glutamylcysteine ethyl ester, and novel histone deacteylase inhibitors show therapeutic promise. Hence, therapeutic strategies that improve outcomes following injury and the time at which treatment is most beneficial is paramount. As traumatic brain injury has been shown to alter energy metabolism and the use of redox proteomics and enzyme kinetics has been used to investigate the oxidative modification of proteins that may lead to reduced cognition observed in TBI patients. This chapter will evaluate the current therapeutic modalities for traumatic brain injury in various animal models as well as affected patients. It will also discuss the role of redox proteomics and enzyme kinetics in traumatic brain injury, most importantly how these oxidatively modified proteins play a role in learning and memory, both of which are effected in those affected by traumatic brain injury.

2. Current models

The field of traumatic brain injury research employs multiple animal models to investigate consequences of this devastating injury. Glutathione (GSH) is the major antioxidant in the brain, therefore treatment of neurodegenerative diseases including traumatic brain injury (TBI) with GSH is of great interest. Alzheimer's disease and other classically diagnosed neurodegenerative diseases have similar patterns of decreased GSH as found in blast induced TBI by fluorescence assay [1]. Proposed studies utilizing rat models incorporate prevention of secondary insults post-TBI with a glutathione mimetic. Omega-3 fatty acids from fish oils as a dietary supplement are used in multiple mild TBI rat models as a preventative measure [2]. This study was found to show cognitive improvement and weight loss recovery post-injury when a 6% fish oil supplement containing omega-3 fatty acids was given as a regular dietary supplement prior to the injury [2]. Axonal damage can cause detrimental effects on the brain. Axonal damage due to TBI results in loss of elasticity in axons making them brittle, swollen, and dramatically stretched allowing calcium to enter the axons leading to extensive damage by initiating a protease cascade [3]. In order to determine the efficiency and safety of future human TBI treatment for restoration of axonal damage, recently rat models have been used as a preliminary study for stem cell treatment for TBI [4]. While understanding axonal damage is important when researching TBI, metabolic rat model studies are exploring cerebral glucose uptake in TBI metabolic pathways [5]. Results show that a decrease in cerebral glucose uptake occurs in TBI rat models with axonal damage and glial activation [5]. Therefore, glucose is unable to be transported out of the cell for use as fuel to the body, much like insulin resistance in Type-2 diabetes. Additionally, rat models are utilized to determine histone deacetylase inhibitors (HDACi) in post injury treatment as a neuroprotectant in acute TBI [6]. HDAC inhibitors activate neurotrophic tyrosine kinase receptor type 1 (TrKA) and nerve growth factor (NGF) expression present significant results in preventing cell apoptosis [6]. Consequently, HDAC inhibitors are able to prevent cell death post-TBI. Findings support a decrease of glial fibrillary acidic protein (GFAP), a biomarker for neurotoxicity in damaged tissues [6, 7]. As findings in rat studies are gaining positive results, progressive animal models with physiology closer to humans are being incorporated into TBI research.

TBI is a growing issue among military personnel, athletes, and victims of all ages and backgrounds; therefore, it is of great importance to include a variety of animal models with physiological properties close to humans. Use of a neonatal piglet TBI model has increased due to their similarities in human myelin sheath, tissue morphology, cerebral structure, and development [8]. Neonatal piglet TBI research allows for future research in behavioral and cognitive assessment and is presumed to be utilized as a tool for preventing secondary injury, such as stroke and resuscitation strategies [9]. Canine epilepsy due to traumatic brain injury has brought insight into therapeutic strategies for posttraumatic epilepsy (PTE) as secondary insult of human TBI [10]. Researchers hope to find treatments in the canine model that will greatly benefit clinical research of TBI as epilepsy and seizures are a severe secondary insult that can lead to detrimental cognitive issues, specifically due to the seizures increasing weeks after the initial injury [10, 11]. Hypothermia treatment has been considered for post-TBI therapy, although clinical trials have not been completed at this time [12]. Previous research demonstrated that rat and swine models have shown positive results with hypothermia experiments as preventative measures against neurodegeneration and currently non-human primates are the prime model for this study [12-14]. A product known as the "chiller-pad" has been used in trials to determine its effectiveness on neuroprotection. Results are encouraging, although the need for a craniotomy to allow hypothermia to be addressed at an exact locality is a perceived downside [12, 13]. It is unclear how hypothermia protects neurodegeneration; however it is necessary to localize cooling, as to not reduce overall body temperature [12]. Rhesus monkey models ranging from ages 3.9 to 14 years old have been used in a long-term study to determine the effects of microglia as defense against injured neurons in the brain and spinal cord after TBI [15]. Most studies appear to monitor the crucial timeframe of TBI, which typically falls immediately after the injury's occurrence [15, 16]. This study allows scientists to view the effects of progression or degression of each monkey for one year [15, 17]. The results encourage further study into therapy with genetic manipulation of microglial phenotype to determine the functionality of impairment on the recovery process as well as anti-inflammatory treatment [15, 17].

Overall, the previous and current studies for traumatic brain injury have been beneficially positive for future studies and treatment for TBI. Researchers are aware that there is still an informational gap in understanding the mechanisms behind TBI and therapeutic strategies, but with future studies and the scientific, as well as public communities' increasing awareness, there will be a bridge to close that gap of uncertainty [18].

3. Pre and post injury strategies

Pre-injury traumatic brain injury (TBI) strategies occur prior to the injury as preventative measures, while post-injury TBI strategies are therapeutic measures taken to prevent secondary insult from occurring after the injury, as there is no way to reverse the initial injury, but further damage can be prevented. With the increasing awareness of TBI, research for preventative and post injury measures is on the rise. There are numerous studies with a variety of computational and animal models that depict a range of TBI severities. There are definite correlations between past and present studies as well as connections between different animal models, with a hopeful finality for positive treatment of humans with TBI.

Researchers are developing statistical models to compare different TBI patient populations in order to determine which new therapies would be most beneficial. The Traumatic Brain Injury National Data and Statistical Center has developed a longitudinal prospective cohort study utilizing Extended Glasgow Outcome scale (GOS-E) scores to determine the furtherance of secondary insult over a period time after the TBI has occurred [19]. The parameters for this model include initial (GOS-E) scores, Functional measure of independence (FIMTM), race, gender, and reusable learning objects (RLOS) [19, 20]. In addition to statistical data, many of the testing strategies require the use of psychological testing to determine pre and post coping models of TBI in humans. The testing mechanism for these models include Coping Scale for Adults (Short Version), Quality of Life Inventory, Sydney Psychosocial Reintegration Scale, and Hospital Anxiety and Depression Scale during 36 months post-injury [21]. In 2008, the Department of Veterans Affairs Polytrauma System of Care partnered with the Department of Education to perform ongoing research with longitudinal patterns using the Traumatic Brain Injury Model Systems of Care (TBIMS) project in order to keep track of rehabilitating veterans with TBI [22].

Where some studies investigate computational approaches to evaluating traumatic brain injury progression in patients, others are utilizing various animal models and combinatory models to determine the most effective way to treat TBI in humans. Patients with a combination of TBI and acute respiratory distress syndrome (ARDS) have symptoms of intracranial pressure, cerebral perfusion pressure, partial pressure of carbon dioxide in the blood ($PaCO_2$), and fractional inspired oxygen [23]. High-frequency oscillation (HFO) and tracheal gas insufflation (TGI) were found to ameliorate gas-exchange and alleviate the pressures caused by these symptoms [23]. Studies suggest monitoring beyond intracranial pressure symptoms by using the brain tissue oxygenated pressure ($PbtO_2$) targeted therapy, transcranial doppler, and cerebral microdialysis to individualize treatment plans of TBI patients [24]. A link between Alzheimer's disease (AD) development and post-TBI in humans is being hypothesized due to ongoing research. Cleavage of amyloid precursor protein (APP), which produces amyloid-β (Aβ) peptide, is a common hallmark in AD. At autopsy, TBI patients have reported Aβ peptides and neurofibrillary tangles of hyperphosphorylated tau proteins, thus leading to the hypothesis that AD can possibly be a secondary effect of TBI [25]. A past study showed that an increase of neuronal C5a receptors and C5b-9 terminal complex, a promoter of the cell cycle and cell lysis, in diffuse axonal injury (DAI) of TBI, could lead to possible secondary neuronal cell death [26, 27]. New studies show that the C5a receptors and C5b-9 terminal complex should be used as targeted treatment areas, but are not necessarily a source of secondary axotomy in DAI [28].

Alternate studies have utilized medication to treat and prevent secondary insults due to the TBI. Progesterone, a cholesterol derived hormone, has been used in rat and human models as a neuroprotectant [29]. Progesterone (PROG) is converted into allopregnanolone (ALLO), which is beneficial to an injured brain. Although it is unknown how the PROG metabolic process assists in the recovery of TBI patients; progesterone is one of the first applicable medicinal treatments for TBI in humans [29]. The ProTECT III study has reached human clinical trials for progesterone treatment in traumatic brain injuries [30]. The ProTECT program has shown great benefits of progesterone as a neuroprotective agent due to its ability to enter the blood brain barrier quickly to decrease damage, downregulate inflammation, and limit inclusive cellular necrosis, apoptosis, and cerebral edema [31, 32]. This program has been carried out in several medical institutions including Emory University, University of Kentucky, and The Ohio State University. A useful tool for monitoring TBI in human patients is the time-resolved optical method [33]. This method is used to obtain data regarding reflected photons and fluorescence signals with indocyanine green (ICG) to determine brain perfusion in TBI patients [33]. During this optical method study researchers were able to differentiate between healthy patients with TBI [33]. As the science behind human studies of TBI becomes progressively clearer, increasing animal model studies are developing promising results. With the collaboration of scientists, clinical studies representatives, statisticians, and induced TBI studies, therapies used to fight TBI will continue to move in a positive direction.

4. Neurosteroids

The nervous system is capable of synthesizing its own reservoir of steroids, termed neurosteroids due to their origins, both independently of serum steroid levels as well by utilizing serum steroids. These were first observed by Baulieu in 1981 [34, 35]. Using cholesterol as a starting material, endogenous steroid production in the brain and other organs amounts to a highly involved pathway map of enzymatic action employing a number of cytochrome P450 enzymes among others. Overall, this biosynthesis pathway is called neurosteroidogenesis (Figure 1). The majority of these steroidogenic enzymes are expressed at different locations around the nervous system and at multiple sites within neurons, indicating that a high amount of coordination is involved in neurosteroid production [36]. This amount of coordination highlights the importance of regulating such highly potent compounds. The pervasive functions of neurosteroids involving binding and modulating neurotransmitter receptor, in addition to influencing the genome via nuclear steroid receptors, implicate neurosteroids in a number of neurological processes and thus possible candidates for post-injury treatment of traumatic brain injury (TBI) [29, 36].

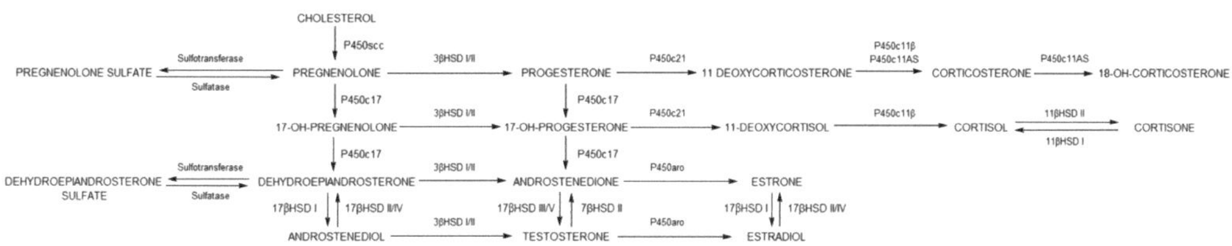

Figure 1. Pathway of neurosteroidogenesis and neurosteroidogenic enzymes. Cholesterol is initially converted into pregnenolone. Neurosteroids are then synthesized from pregnenolone through different pathways involving various enzymes. P450scc, mitochondrial cholesterol side chain cleavage enzyme, mediates c20-hydroxylation, 22-hydroxylation, and scission of the c20–22 bond; P450c17, mitochondrial 17-hydroxylase, mediates 17α-hydroxylation and scission between c17-20 bond; P450c 21, mitochondrial 21-hydroxylase, mediates 21-hydroxylation; P450aro, mitochondrial aromatase, mediates two 19-hydroxylations and one 2-hydroxylation; P450c11β, mitochondrial 11β-hydroxylase, mediates 11-hydroxylation; P450c11AS, mitochondrial aldosterone synthase, mediates c11,18-hydroxylation and 18-oxidation; P450c11B3, the third mitochondrial 11β-hydroxylase, mediates 11β and18-hydroxylation; 3β-HSD, 3β-hydroxysteroid dehydrogenases, mediates both 3β -hydroxysteroid dehydrogenase and D5-D4-isomerase activities; 17β-HSD, 17-ketosteroid reductase (KSR), mediates c17β reduction or c17 oxidation; HST, sulfotransferase, mediates 3β-hydroxylation with sulfate groups; STS, sulfatase, hydrolyzes sulfate groups at 3β.

A number of neurotransmitter receptors are modulated by neurosteroids, including the γ-aminobutyric acid ($GABA_A$), N-methyl-D-aspartate (NMDA), α-amino-3-hydroxy-5-methyl-4-isoxazolepropionic acid (AMPA), kainate, glycine, 5-hydroxytryptamine ($5\text{-}HT_3$ or serotonin), sigma type 1 (σ_1), nicotinic acetylcholine, and oxytocin receptors, and the modulation can be positive or negative depending on the steroid identity [36]. Although some steroids produced in other endocrine organs are still found in the brain, the physiological role and significance of neurosteroids are different in the brain. Neurosteroids affect the development, structure, and function of the central nervous system. This diverse population of neurotransmitter receptors illustrates how influential neurosteroids are on the brain. The complexity of these interactions is illustrated by considering how similar steroid structures can have different effects on the same receptor and how these minor structure alterations are tightly regulated. In particular, the neurosteroids progesterone, dehydroepiandrosterone (DHEA), pregnenolone, and allopregnenolone (ALLO), illustrated in Figure 2, are potential post-TBI therapeutic candidates due to their roles in various aspects of neurogenesis and the repair and survival of neurons [37].

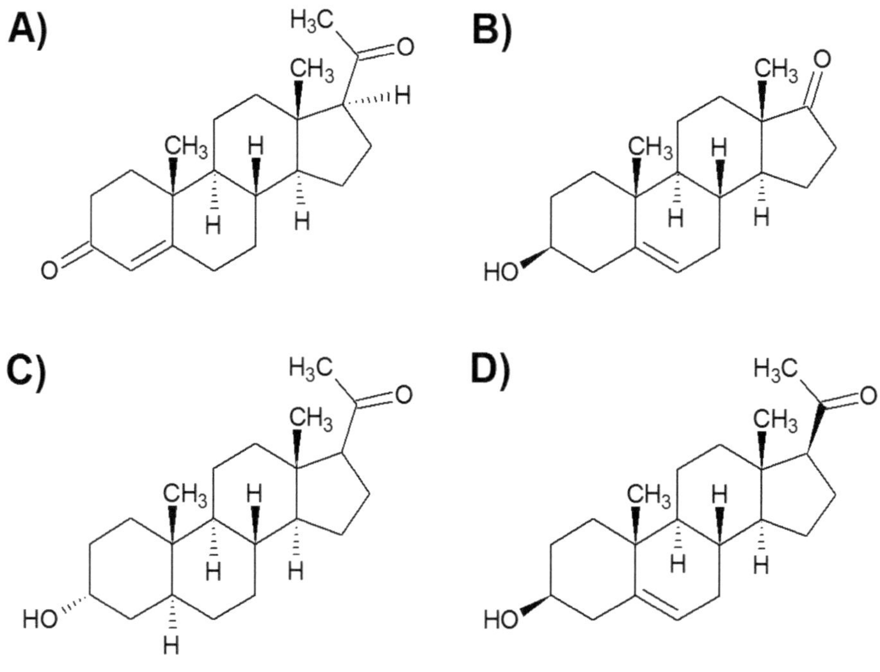

Figure 2. Progesterone (A), dehydroepiandrosterone (B), pregnenolone (C), and allopregnenolone (D) are potentially beneficial neurosteroids for the treatment of TBI.

DHEA promotes axonal growth and functional activity of specific neuronal networks, while ALLO can induce axonal regression in the developing hippocampus. Progesterone can regulate myelin formation, improve the myelination of injured nerves *in vivo*, facilitate myelination *in vitro*, and induce oligodendrocyte maturation *in* vitro. Moreover, neurosteroids can modulate neuronal excitability and activity by interacting with different types of receptors. There are mainly two different action mechanisms that neurosteroids undergo to carry out their biological functions in the brain: genomic actions and nongenomic actions. Genomic actions are mediated by nuclear steroid receptors, such as progesterone, glucocorticoid, and mineralocorticoid receptors, while nongenomic actions are mediated by neurotransmitter receptors, such as $GABA_A$, N-methyl-D-aspartate (NMDA), and sigma receptors. Classical nuclear steroid receptors regulate gene transcription so it is a long-term process. PROG binds both to progesterone and neurotransmitter receptors. Corticosterone binds to glucocorticoid receptors with a low affinity, while both corticosterone and aldosterone bind to mineralocorticoid receptors with a high affinity. However, the acute effects of neurosteroids are not related to nuclear steroid receptors, but the majority of neurosteroids interact with neuronal membrane receptors and ion channels [38].

As positive modulator(s) of the $GABA_A$ and kainate receptors and negative modulator(s) of the glycine, 5-HT_3, nicotinic, acetylcholine, and oxytocin receptors, progesterone and its derivatives could play potential roles in neuroregeneration involving reducing inflammation, helping remyelinate neurons and reducing inflammation [36, 37]. In part by negatively modulating the $GABA_A$ receptor and positively modulating the NMDA and σ_1 receptors, DHEA and its sulfate derivative (DHEAS) promote dendritic growth and branching and form synapses as well as help with memory [36]. DHEA protects against oxidative damage in hippocampal neurons and glutamate toxicity by inhibiting nitric oxide synthase and G-protein-coupled receptors that activate cell-survival pathways [39-41]. Pregnenolone sulfate is a negative modulator of $GABA_A$, kainate, AMPA, glycine, and σ_1 receptors and a positive modulator of NMDA receptors that has no effect on 5-HT_3 receptors, helping this neurosteroid play an important role in benefiting memory [36]. ALLO, a potent positive allosteric modulator of $GABA_A$, binds to discrete site on the $GABA_A$ receptor, thereby increasing the mean open time and decreasing the mean closed time of the channel, resulting in the increase in the chloride current through the channel and consequently in a reduction of neuronal excitability. NMDA receptors exhibit at least two distinct sites for neurosteroid modulation, mediates the effects of either positive or negative modulators. A significant number of studies have shown the neuroprotective effects of neurosteroids against many pathologic conditions (Figure 3). PROG and ALLO reduce cell damage and improve outcomes of focal ischemia in a stroke model [42]. Allopregnenolone, while contrasting pregnenolone's benefits on memory, does help relieve anxiety [36]. Many studies focus on the neuroprotective role of estrogen specifically in the hippocampus and as it improves cognitive function against disease, aging, and stroke, mitigating cell death and stimulating neuronal proliferation in the hippocampus and other structures [43]. Testosterone also shows a neuroprotective effect, but it is partly due to its conversion to estrogen in the CNS [44]. Testosterone protects primary human fetal neurons against serum-deprivation-induced cell death and oxidative stress through androgen receptors [45, 46].The structures for these described neurosteroids can be seen in Figure 3.

Neurosteroids can affect brain function by not only interacting with neurons, but also acting on glia cells, such as microglia, Schwann cells, oligodendrocytes, and astroglia. By acting on microglia and Schwann cells myelination is modulated, while acting on oligodendroglia and astroglia the response of nerve tissue to pathological insults is altered. Specifically, the effect of neurosteroids on astroglia is crucial because astroglia play critical roles both in the central nerve system (CNS) and neural signaling by regulating extracellular ion concentrations and local cerebral blood flow, and by modulating synaptic transmission and plasticity [35]. Due to their wide variety of influences on important neurotransmitter receptors, neurosteroids have a large potential benefit for neuronal protection and repair following TBI. The various roles of

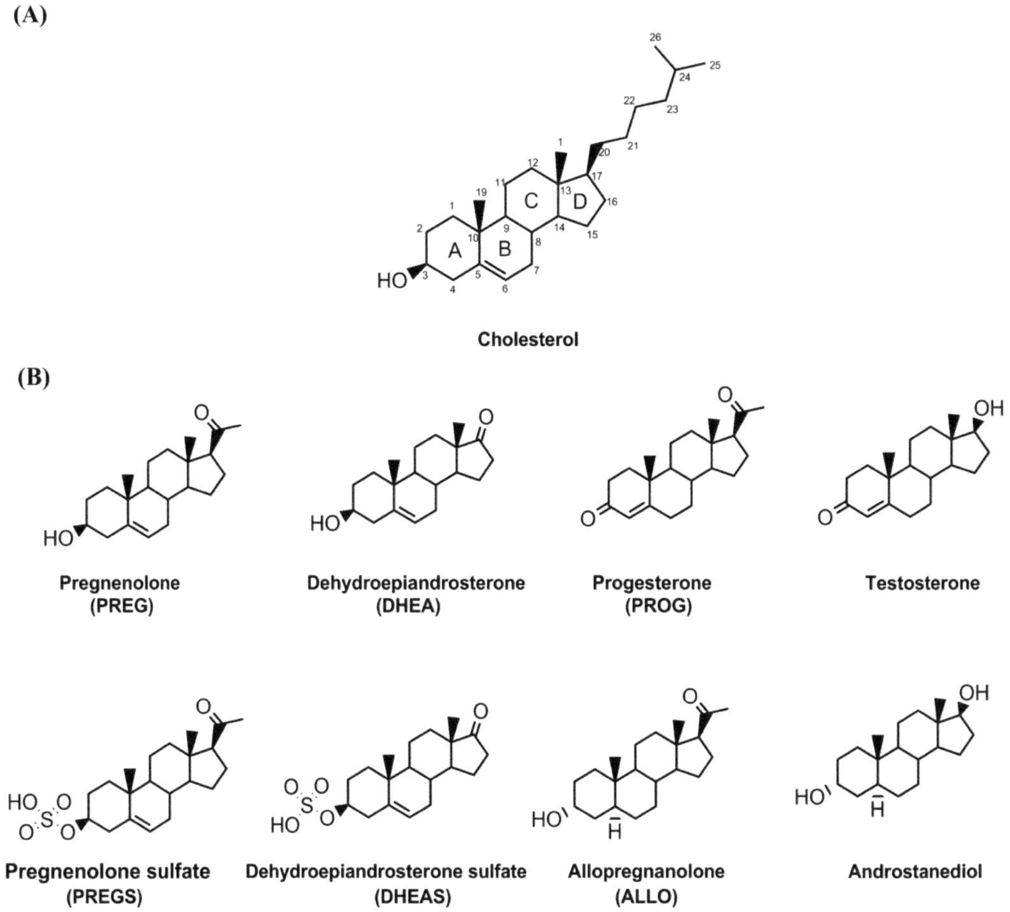

Figure 3. Chemical structure of cholesterol and neurosteroids. Cholesterol (A) is converted into various neurosteroids (B). Each neurosteroid has significant roles in the brain and nerve system, but these neurosteroids, specifically, have shown neuroprotective properties.

neurosteroids such as progesterone, dehydroepiandrosterone, pregnenolone, and allopregnenolone give them special consideration as post-TBI therapeutics, and progesterone research in particular is offering promising results [29, 37, 47]. Studies in rodents and humans have shown beneficial effects of PROG on both mortality and functional outcomes following TBI by its neuroprotective effect [48]. In addition to PROG, its metabolite, ALLO also has shown beneficial effects equal to or in excess of those of PROG in TBI by modulating the $GABA_A$ receptor, not classic nuclear receptors [29]. Both PROG and ALLO have beneficial effects in reducing blood-brain barrier dysfunction and intracranial pressure following TBI, and also in reducing brain swelling and edema that are associated with TBI [49]. DHEAS and its analog, fluasterone (DHEF), can also improve recovery of function in male rats after TBI [50, 51]. These findings altogether bolster the neuroprotective effects of neurosteroids and their analogs thereby, leading to plausible future therapeutic strategies in the clinical treatment of TBI.

5. Flavonoids

Secondary events following a traumatic brain injury (TBI) account for the majority of neurodegeneration, beginning within the first hour post-TBI and lasting for days [52]. The elevation of ROS levels within neurons has been recognized as an early event following TBI occurring within minutes [53].The overproduction of ROS due to TBI cannot be eliminated by the limited amount of antioxidants, leading to oxidative stress, one of the most common secondary injury mechanisms. These remaining ROS begin to interact with proteins, lipids, carbohydrates, nucleic acids, or signaling molecules. The oxidative modification of neuronal molecules changes their structures and function as well as signaling pathways, subsequently leading to irreversible neuronal inflammation, dysfunction, and death. Due to the initial rush of rapid superoxide ($O_2^{\cdot-}$) production during the first hour and the shift to a pro-oxidant environment that occurs as early as 3 hours post-TBI, free radicals play an integral part in the initial events following a TBI and continue to react with cellular components for several days, with their prevalence peaking between one and two days after the initial injury [52-54]. As revealed by Ansari, et al. there is a narrow window of opportunity for post-TBI therapeutic intervention due to this shift towards an oxidative environment [54]. Thus, supplementing the brain with substances having potent antioxidant capacities, such as flavonoid compounds, as early as possible could be a beneficial treatment for mitigating the otherwise inevitable cascade of secondary damage.

The flavonoid class of compounds describes molecules that share one of the three similar backbones illustrated in Figures 4 and 5. Flavonoids are polyphenolic compounds with a diphenylpropane skeleton which consists of two aromatic rings, (A ring and B ring) bound together by three carbon atoms (C_1, C_2, and C_3) forming an oxygenated heterocycle (C ring) as depicted in Figure 5A. These compounds are divided into six major classes depending on their saturation and the number and arrangement of either the carbonyl group or hydroxyl groups in the C ring (Figure 5B). Flavanones contain a carbonyl group at C_3, flavones have a carbonyl group at C_3 and a double bond between C_1 and C_2, isoflavones have a carbonyl group at C_3, a double bond between C_1 and C_2, but the B ring is connected to C_2 instead of C_1, flavanols contain a hydroxyl group at C_2, while anthocyanidins have a hydroxyl group at C_2, one double bond between oxygen and C_1 and another between C_2 and C_3, lastly flavonols have a carbonyl group at C_3 and a hydroxyl group at C_2. Furthermore, there are a variety of flavonoids present within each group due to the differences in the number and arrangement of hydroxyl group in the A and B rings in the structure (Figure 5B). The relative antioxidant ability of individual flavonoids to scavenge ROS is correlated with these different chemical structures [55]. The unsaturation of the C ring enhances the scavenging ability of flavonoids by increasing their stability via electron delocalization. The effects of the number and the arrangement of these hydroxyl groups also affect antioxidant activity. For example, the absence of a hydroxyl group at C_2 of the C ring decreases the activity, while the presence of two hydroxyl groups in the *ortho*-diphenolic arrangement in the B ring increases the activity against superoxide anion. In addition, the presence of a catechol group also enhances the ability against peroxynitrite scavengers, because catechol can afford a two-electron reduction of peroxynitrite to nitrite coupled to their oxidation to the corresponding *o*-quinones [56].

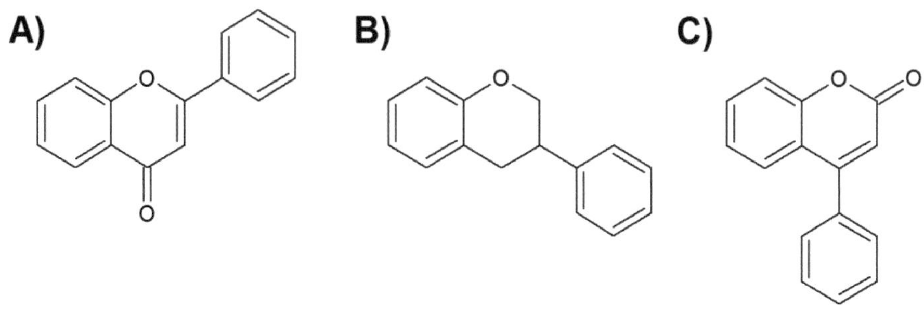

Figure 4. The flavone (A), isoflavan (B), and neoflavonoid (C) structures serve as the chemical backbones for flavonoid compounds.

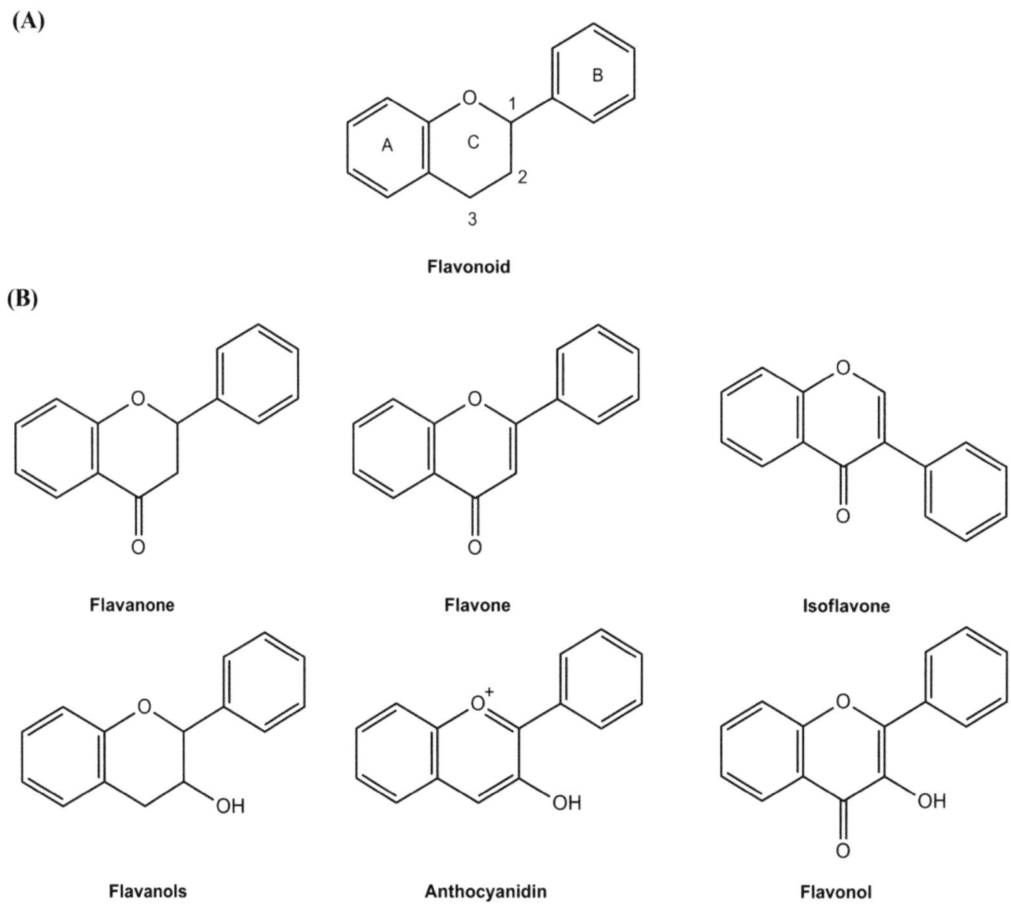

Figure 5. Chemical structure of flavonoids. Basic chemical structure (A) and six classes (B) including: flavanones, flavones, isoflavones, flavanols, anthocyanidins, and flavonols.

Many natural flavonoids can be found in plant material such as fruits, vegetables, and plant-derived beverages, such as green or black tea. Once introduced into the body, flavonoids are converted into their metabolites which can have substantially different bioactivity properties. These structures have been well established as antioxidants that reduce oxidative stress not only by directly scavenging reactive oxygen species (ROS), but also by increasing the activity of antioxidant enzymes or decreasing the activity of redox enzymes respectively [57]. Moreover, flavonoids can modulate cellular homeostasis helping to reduce inflammation and cell toxicity [57].These compounds have neuroprotective properties in addition to their abilities to scavenge free radicals [34, 58, 59]. Due in part to the these properties and the abilities of certain flavonoids to cross the blood-brain barrier (BBB), flavonoids appear to be excellent treatment opportunities for TBI models [59]. Indeed, extensive therapeutic use of flavonoids and similar polyphenolic compounds post-TBI have proven this to be true, with these compounds serving to remedy many TBI-induced maladies such as edema, neuroinflammation, cognitive and neuronal functions, cell survival, oxidative stress, and cellular energy homeostasis [60-63]. Catechin, epicatechin, and epigallocatechin gallate (EGCG), chemicals all found in green tea (Figure 6A), reduce the production of nitric oxide and hydrogen peroxide, lipid peroxidation, and DNA oxidation induced by ischemia/reperfusion in rats [64]. Several flavones including baicalein, baicalin, and wogonin (Figure 6B) components of the *Scutellaria* plant, reduce lipid peroxidation and nitric oxide production in the cortex and hippocampus of rats as well as inhibit pro-inflammatory cytokines [65]. The down-regulation of the NF_kB pathway confirmed with medicinal extracts rich in flavonoids, such as *Crataegus,* orally administrated and protected the brain against delayed cell death by increasing the levels of antioxidants in the brain, Ginkgo biloba extract (EGb 761) [66]. It has been well documented that metabolites of flavonoids also have neuroprotective properties. The death of both cortical and striatal neurons induced by oxidative stress has been shown to be inhibited by the methylated metabolite of epicatechin with a protection capacity similar to epicatechin [67]. Specifically, caffeic acid has shown neuroprotective properties and is a component of a patented mixture of flavonoid compounds, known as Pycnogenol® (PYC), which has produced very promising results in rat models of TBI [58, 68].

Based on studies showing that flavonoids have potent neuroprotective and anti-inflammatory properties, there are several recent studies in which flavonoids were used as post injury therapeutic strategies for TBI. These include epicatechin, wogonin, baicalein, and pycnogenol. Water containing EGCG, a major component of green tea possessing strong antioxidant properties, was given to rats after TBI and demonstrated the ability to inhibit free radical induced degradation that have the potential to differentiate into neurons and glia around the injured area [69]. The effects of wogonin and baicalein, flavonoids possessing potent anti-inflammatory properties, were investigated on mice or rats subjected to controlled cortical impact injury as post-injury treatments [61]. Their neuroprotective effects were suggested because they improve functional and histological outcomes, and reduced induction of pro-inflammatory cytokines. Prior to its use in TBI models, the bioflavonoid, PYC has exhibited various antioxidant properties as well as the abilities to prevent neurotoxicity and apoptosis [54, 70-74]. Scheff, et al. recently reported a number of benefits from post-TBI intraperitoneal

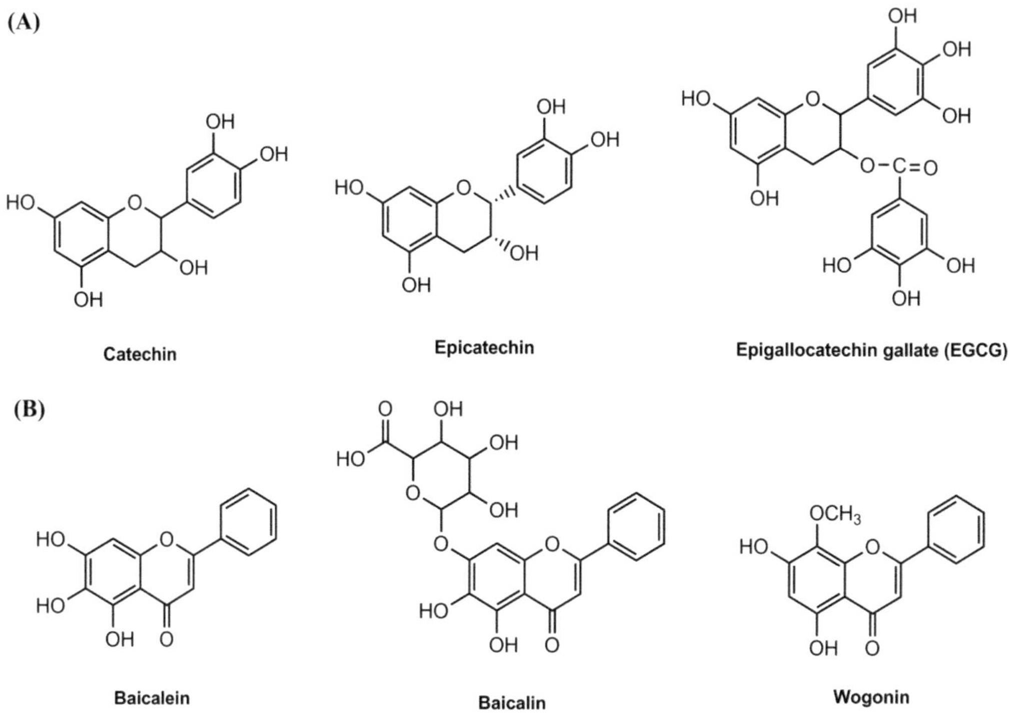

Figure 6. Flavonoids possess neuroprotective properties and therapeutic effects in TBI. Flavanols, such as catechin, epicatechin, and epigallocatechin gallate (EGCG), found in green tea (A). Flavones, such as baicalein, baicalin, and wogonin, are found in celery (B).

administration of PYC in rat models, including an increased presence of endogenous antioxidants in the hippocampus and cortex at 48 hours post-TBI, increased GSH:GSSG ratio in both regions at 48 hours post-TBI, and increased synaptic proteins at 96 hours post-TBI [58]. The neuroinflammatory cytokines interleukin-6 and tumor necrosis factor α were also significantly decreased at 48 hours post-TBI using this treatment. Because of the substantial amount of free radical-induced oxidative damage that occurs post-TBI, these increases in antioxidants and decreases in inflammatory markers not only highlight the benefits of the particular mixture of flavonoids in PYC, but also serves to further evidence the abilities of flavonoids in general to help re-establish the balance of antioxidants and oxidants to offset the development and progression of oxidative stress post-TBI.

Flavonoids such as those found in PYC offer a unique approach to TBI treatment, which aims to remedy the initial post-TBI cascade of oxidative damage and early decreases in antioxidant capacity. PYC has helped to prove the potential of flavonoids in mitigating the secondary events of TBI by supplementing the brain with antioxidants that cross the BBB. Because the initial injury of TBI is virtually untreatable, therapeutic intervention within the first three hours post-TBI with compounds such as flavonoids offer promising insight into treatments that will result in future clinical significance.

http://dx.doi.org/10.5772/57306

6. Statins

Statins are a class of drugs that inhibit 3-hydroxy-3-methyl-glutaryl-CoA (HMG-CoA) reductase, the rate-limiting enzyme in the endogenous production of cholesterol, as illustrated in Figure 7. As such, statins are prescribed to lower serum cholesterol levels in patients for the purpose of preventing heart disease. However, a number of benefits of statin use have also been reported for the treatment of traumatic brain injury (TBI) which are not related to lower cholesterol levels, including increasing synaptogenesis and angiogenesis, increasing neuronal survival, improving cognitive abilities, and decreasing lipid peroxidation, among others [75-77]. This versatility of statins gives them unique advantages over other potential treatments for TBI that are specialized for only certain injury-related symptoms [76].

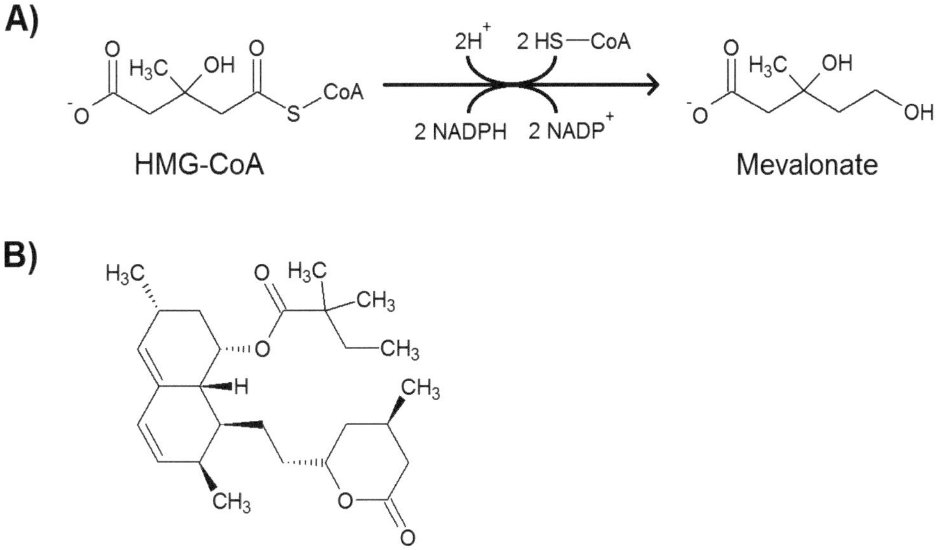

Figure 7. HMG-CoA reductase catalyzes the rate-limiting step of cholesterol formation (A). The structure for simvastatin, a statin drug that inhibits HMG-CoA reductase to decrease endogenous cholesterol synthesis (B).

Most of the neurodegeneration that results from a TBI is generated over the days following the initial insult [52]. The events causing this neurodegeneration, collectively referred to as the secondary injury, offer several opportunities for therapeutic intervention. There are currently no approved therapeutic treatments for TBI, so care is limited to non-chemotherapeutic methods of supporting normal cranial pressure and cerebral perfusion [6]. Because the primary injury of TBI creates a contusion around which cerebral blood flow is significantly limited, there is a high potential for ischemia and thus infarction as a secondary injury [78, 79]. In addition, many microscopic inflammatory processes are mediated largely by oxidative stress from excess free radical formation, which occurs in significant amounts within the first hour post-injury [52, 53]. Thus, because of the many benefits of statin use on endothelial function and the role(s) of statins in preventing oxidative stress, this avenue of research offers significant promise for TBI treatment [80, 81].

Since Lu, et al. first began investigating the therapeutic potential of statins in 2004, a growing body of evidence illustrating the neuroprotective effects of statin use in both pre- and post-TBI treatment has been established [75, 77, 82-85]. However, the exact mechanism(s) by which statins execute this neuroprotection is not completely understood. Statins are known to provide many benefits to endothelial function independent of cholesterol levels, many of which involve nitric oxide (NO), a second messenger molecule deeply rooted in signaling pathways related to endothelial functions. These NO–related functions of statins include both upregulating and triggering activation of the enzyme endothelial NO synthase (eNOS), recovering eNOS function when oxidized low density lipoprotein is present, and recovering eNOS function during hypoxia [86-88]. Studies aimed at elucidating the exact mechanisms underlying the relationship between statins and eNOS have determined that inhibiting the synthesis of mevalonate has downstream inhibitory effects on the isoprenylation of the Rho protein, a member of a multifunctional family of GTPase proteins [89, 90].

As the direct relationships between statins, Rho, and eNOS begin to unravel, the importance of statins in potential post-TBI treatment will likely increase. Statins differ in their lipophilic properties, HMG-CoA reductase binding abilities, and their abilities to cross the blood-brain barrier (BBB), which will make the search for an ideal post-TBI candidate very involved. Simvastatin, depicted in Figure 7B, has proven useful in TBI treatments due in part to its BBB penetration, leading to significant benefits on BBB permeability, cerebral edema, and learning and memory [83, 91].

7. Gamma-glutamylcysteine Ethyl Ester

Glutathione (γ-glutamylcysteineglycine), illustrated in its reduced (GSH) and oxidized (GSSG) forms in Figure 8, plays a key role in cellular defense and repair, where it serves to maintain intracellular redox status by conjugating toxic species for export, regulating the activities of redox-sensitive enzymes, and neutralizing oxidative species via antioxidant enzyme mechanisms [92]. To highlight the significance of these responsibilities, glutathione exists at millimolar concentrations in the cell and its reduced state is carefully maintained to keep the GSH:GSSG ratio high [93]. The combination of high polyunsaturated fatty acid content, high oxygen respiration and glucose metabolism, and relatively low antioxidant capacity in the brain establishes an environment that is highly susceptible to oxidative stress. Thus, because the brain is not adequately prepared to defend against the almost immediate surge of reactive oxygen species (ROS) formation that is brought about by a traumatic brain injury (TBI), utilizing the native antioxidant mechanisms of the brain by increasing glutathione availability is a treatment route whose ability to curb initial inflammatory processes in the brain appears very promising [94, 95].

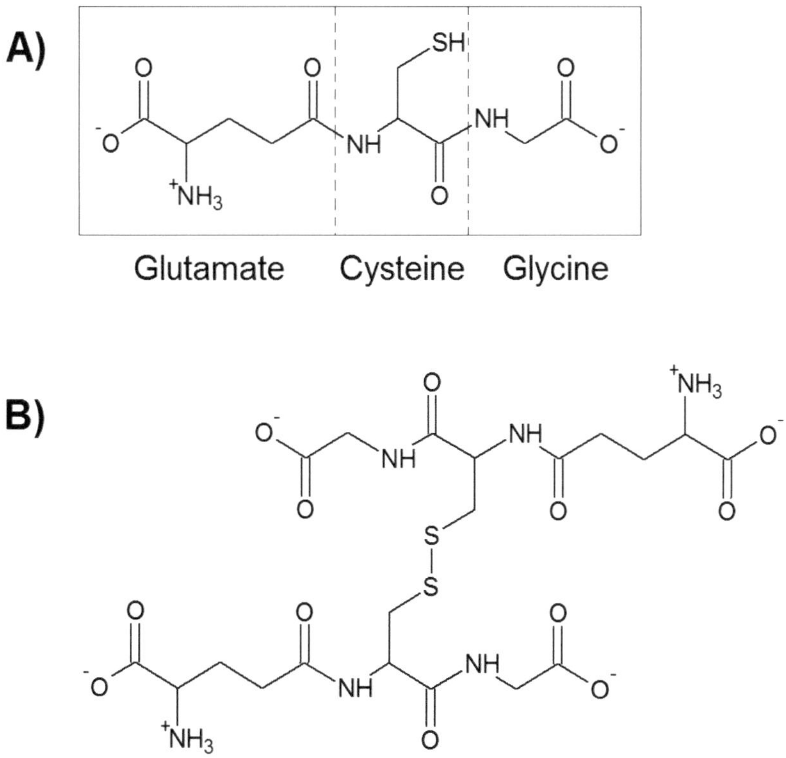

Figure 8. Glutathione can exist in its reduced form (A) and oxidized form (B) at physiological pH.

Synthesis of glutathione, illustrated in Figure 9, occurs in a two-step mechanism where glutamate and cysteine are first covalently bound via γ-linkage to produce γ-glutamylcysteine (γ-GC) by the enzyme γ-glutamylcysteine synthetase, and then glycine is covalently bound to γ-GC via α-linkage to cysteine by the enzyme glutathione synthetase. This mechanism offers a prime opportunity to increase glutathione synthesis since regulation only occurs at the first step. Glutathione is a feedback inhibitor of γ-glutamylcysteine synthetase, prohibiting excess production of γ-GC. This step is also rate-limited by the availability of cysteine, whose intracellular concentration is lower than the concentrations of glutamate or glycine. However, glutathione synthetase is not inhibited by glutathione, so glutathione concentrations can be increased when excess γ-GC is supplemented [93]. In fact, intracerebroventricular administration of γ-GC increases brain glutathione content more so than intracerebroventricular administration of GSH ethyl ester or intraperitoneal administration of cysteine in rats [96]. The γ-GC analog, γ-glutamylcysteine ethyl ester (GCEE), shown in Figure 9B, has also proven to be an excellent candidate for supplementing intracellular γ-GC availability by increasing glutathione concentrations, and has been shown to decrease markers of oxidative stress both *in vitro* and *in vivo* [95, 97-99].

Various models of TBI have offered valuable insight into the mechanisms by which the brain incurs substantial damage following traumatic injury. Following the initial TBI injury,

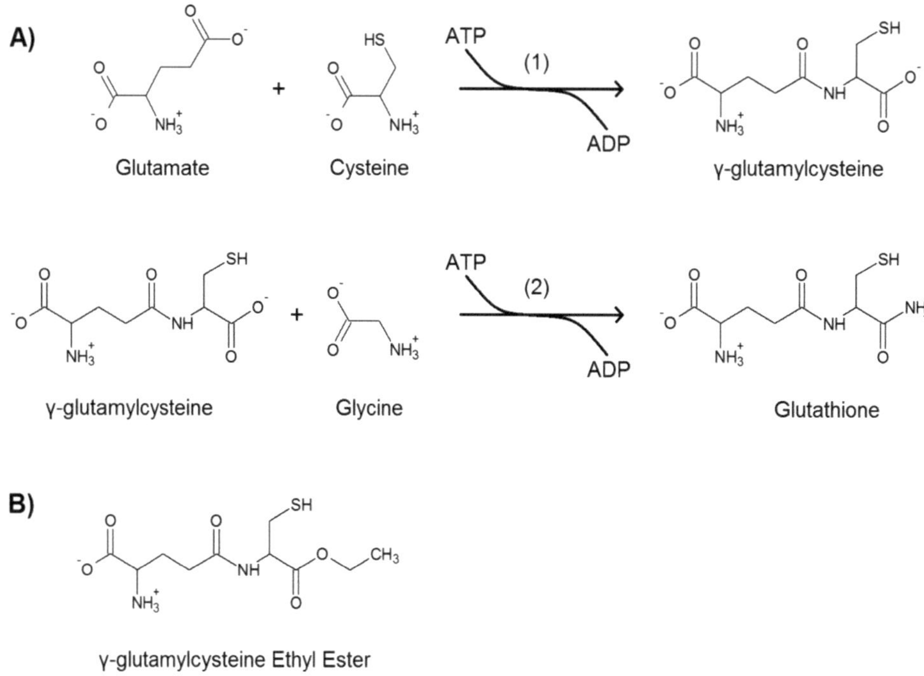

Figure 9. The enzymes γ-glutamylcysteine synthetase (1) and glutathione synthetase (2) synthesize glutathione in a two-step mechanism (A). The γ-glutamylcysteine analog γ-glutamylcysteine ethyl ester can increase glutathione synthesis (B).

secondary events can occur for days post-injury and these events account for the majority of TBI-induced neurodegeneration [52]. Due to the role of the superoxide radical ($O_2^{\cdot-}$) in creating an oxidative stress environment, the significant production of $O_2^{\cdot-}$ in cat brain during the first hour following a fluid percussion model of TBI implicates free radical-induced oxidative stress in early mechanisms of TBI secondary injury [53]. The large capacity of GSH for detoxifying reactive oxygen species (ROS) and reactive nitrogen species (RNS), which are produced directly from native reactions of $O_2^{\cdot-}$ with various intracellular components, emphasizes the importance of maintaining adequate GSH availability during such rapid increases in $O_2^{\cdot-}$ production. Indeed, GSH availability is significantly decreased at 2 hours and even 24 hours post-injury using a rat model of blast induced mild TBI [1]. By replenishing GSH reserves via bypassing native feedback inhibition of GSH synthesis, GSH becomes available when it is needed most following TBI, and GCEE treatment post-TBI has so far proven to be very favorable by decreasing protein nitration, decreasing autophagy, and decreasing blood-brain barrier permeability [95, 100, 101].

In contrast to the various beneficial mechanisms of ROS and RNS scavenging and detox by glutathione, it is also worth noting that GSH can react directly with $O_2^{\cdot-}$ to produce the glutathione radical (GS•), which can abstract hydrogen atoms polyunsaturated fatty acids, carbohydrates, and peptides [102-105]. Also, mitochondrial manganese superoxide dismutase

(Mn-SOD) is known to be a better scavenger of $O_2^{\cdot -}$ than GSH and should thus limit the amount of GS• formed, but the activities of Mn-SOD and copper/zinc-SOD (Cu/Zn-SOD) are at their minimums at 24 hours post injury in the ipsilateral hippocampus of rat brain following a cortical contusion model of mild TBI and stay decreased for several days post-injury [54, 105]. Indeed, the presence of the lipid peroxidation product 4-hydroxynonenal (HNE) is significantly elevated as protein-bound HNE in a TBI rat model, and post-TBI treatment of GCEE does not decrease its presence [95].

Post-TBI treatment and treatment of oxygen-glucose deprivation, a possible effect of ischemia, with GCEE have proven to be very beneficial, offering promising avenues of TBI research to determine how these effects translate into ameliorating clinical symptoms [95, 101]. While glutathione does have the potential to cause lipid peroxidation and models have yet to show a beneficial effect in this regard, increased glutathione availability via post-TBI treatment with GCEE has been shown to significantly decrease the presence of nitrated and carbonylated proteins, which are known to be formed by peroxynitrite ($ONOO^-$), a product of $O_2^{\cdot -}$ reacting with nitric oxide (NO) in the presence of carbon dioxide [95, 104]. *In vitro* models using synaptosomes indicate that up regulated glutathione synthesis via GCEE administration can detoxify $ONOO^-$ directly [97]. However, due to the reaction of GSH and $O_2^{\cdot -}$ in forming GS•, there is also potential for a trade-off of ROS formation that allows unreacted NO to carry out its beneficial endothelial functions: elevated glutathione could potentially out-compete NO for reacting with $O_2^{\cdot -}$, increasing lipid peroxidation but allowing NO to promote blood flow and thus encourage neuron survival [81, 106]. Because glutathione has the potential to cause lipid peroxidation but is also known to export products of lipid peroxidation, such as HNE, out of the cell via multidrug resistant proteins, the lack of a significant increase in protein-bound HNE levels in GCEE-treated rats post-TBI could be a result of this trade-off. While further studies are required to determine the exact mechanisms and therapeutic significance of GCEE treatment post-TBI, the many functions of glutathione and initial positive results indicate a high therapeutic potential for this compound [92, 95, 101].

8. Histone deacetylase inhibitors

Treatment of traumatic brain injury (TBI) using histone deacetylase inhibitors (HDACi) is a relatively new therapeutic strategy that has so far provided some very encouraging results for invoking neuroprotective mechanisms via epigenetic changes. By inhibiting histone deacetylase, HDACi suppress the removal of acetyl groups from histone lysine residues, freeing DNA segments for transcription. Like statins, the exact mechanisms by which these compounds provide benefits in TBI models are largely unknown. This could be due in part to the large number of proteins affected by such epigenetic changes as well as the intricate relationships between affected proteins and cognition, although the development of novel HDACis is beginning to shed light on proteins of particular importance such as nerve growth factor (NGF) and heat-shock protein 70 (HSP70) [6].

Epigenetic changes are chemical modifications to the genome changes that result in altered gene regulation. These epigenetic changes occur throughout one's life due in part to environmental interactions [107, 108]. Following such a drastic change to one's internal environment as that which results from a TBI, it should be expected that one's epigenetic profile would change accordingly. Indeed, TBI results in substantial epigenetic changes, resulting in increased transcription of several proteins involved in pro-survival pathways and anti-apoptotic pathways [6]. Epigenetic changes in the hippocampus histone-H3 have been found to occur for the first days after a pediatric rat model of TBI, during the same time frame that the secondary injury of TBI is known to occur [52, 109]. Changes in gene regulation are key responses to TBI, and like statins, HDACis are novel therapeutics that can treat several consequences of a TBI at once [6, 110].

Gene arrays of TBI models indicate that there are substantial changes to the gene expression following injury [1, 111-113]. Post-TBI treatment with the HDACi LB-205 was able to increase histone-H3 acetylation, a key histone that has been found to have reduced acetylation following TBI, at safer therapeutic doses than a previously employed therapeutic. In addition, the up regulation of pro-survival proteins nerve growth factor (NGF), p-TrkA, p-AKT, NF-kB, and Bcl-2, as well as down regulation of pro-apoptotic proteins p-JNK, BAX, and p75(NTR) were results of post-TBI LB-205 treatment [6]. Of particular importance to neuron survival is the up regulation of NGF. This neuropeptide is essential to the survival of specific neurons and could help with regeneration of neurons post-TBI as higher CSF levels of NGF are associated with better clinical outcomes following TBI [114-119]. While further research into the potential therapeutic uses of HDACis for treatment of TBI is necessary to fully understand their clinical value, current understanding of post-TBI gene regulation and initial testing of HDACis such as LB-103 have uncovered a multifunctional treatment opportunity that may prove to be of significant importance in the future [1, 6].

9. Enzyme kinetics and TBI

Enzyme kinetics focuses on chemical reactions that are catalyzed by enzymes. Enzymes are proteins that alter other molecules by forming an enzyme-substrate complex, transforming the initial substrate bound to the active site of the enzyme into a product via an enzymatic mechanism in which the enzyme is not consumed within the catalyzed reaction. Enzymes do not alter the equilibrium position of a reaction, but continue to increase the reaction rate as the substrate concentration of the respective enzyme increases until all active sites of the enzyme are saturated. From this point, the reaction rate will depend upon the turnover rate of the enzyme. Numerous enzymes bind to one substrate and produce a single product. However, many enzymes are capable of binding to multiple substrates. Kinetic studies on single-substrate binding enzymes can typically show measurements of enzyme affinity to substrate binding, as well as turnover rate. Kinetic studies on multi-substrate binding enzymes usually can show substrate binding sequences, as well as product release sequences. There is typically one rate-determining step that indicates overall kinetics of the enzyme, which could include a conformational change or chemical reaction. Enzyme kinetics can be used to measure the reaction rate of a diverse range of processes for investigation of varying conditions in an experimental design. Enzyme kinetics can uncover metabolic roles, control mechanisms, catalytic mechanisms, and how a drug/agonist may inhibit an enzyme.

10. Alteration of energy metabolism

Alteration of energy metabolism in specific enzymatic reactions involving varying conditions of oxidative stress placed on the enzymes following moderate traumatic brain injury (TBI) has been investigated as possible biomarkers of moderate TBI, which can be observed by enzyme kinetics. Moderate TBI is a common source of oxidative stress in biological systems, caused by secondary damage due to the presence of reactive oxygen and nitrogen species (ROS and RNS, respectively). This leads to a decrease in overall antioxidant capacity [120, 121]. This oxidative stress can disrupt protein structure, leading to a loss of function. The body can attempt to prevent or overcome this event in numerous ways including the use of antioxidants and epigenetic factors. Antioxidant enzymes that are found to be more active following TBI include catalase and glutathione peroxidase. Catalase catalyzes the decomposition of hydrogen peroxide to water and oxygen (Figure 10), while glutathione peroxidase reduces hydroperoxidases to their respective alcohols (including the conversion of hydrogen peroxides to water), using reduced glutathione as a cofactor, which will be oxidized to glutathione disulfide (Figure 11). Catalase was found to have a threefold increase, while glutathione peroxidase had a twofold increase in a time course study reflecting temporal increase in nerve growth factor (NGF) following a cortical contusion model [122]. Enzyme activity levels of energy-producing enzymes such as enolase and creatine phosphokinase are of interest as possible indicators of moderate TBI. Enolase and creatine phosphokinase both play significant roles in energy production. Enolase, an integral part of the glycolytic pathway, is highly expressed in neuronal cytoplasm. It converts 2-phosphoglycerate to phosphoenolpyruvic acid (Figure 12). Neuron-specific enolase, found only in the brain, can be used for the possible detection of neuronal cell death with high sensitivity and specificity. Increased cerebrospinal fluid (CSF) and serum levels of enolase are associated with TBI and could be used to gauge the degree of injury. This finding could play a key role in future research, as upregulated proteins after TBI injury are typically identified as possible diagnostic biomarkers [123]. S100β proteins, a component of astroglia and Schwann cells, are known to increase post-injury in CSF and serum. Creatine phosphokinase can act as an energy reservoir for rapid buffering of ATP concentration in cells, as well as an intracellular mode of energy transport via a creatine phosphokinase shuttle circuit [124]. Creatine phosphokinase acts by converting creatine to creatine phosphokinase (Figure 13). Creatine kinase isoenzyme BB (CK-BB), neuron specific enolase, and S-100β proteins have been widely studied as biochemical serum markers of TBI. It has been suggested that CK-BB sensitivity and specificity is inadequate for use as a TBI indicator [125]. It is also suggested that serum levels of neuron specific enolase do not correspond to the extent of TBI damage, most likely due to a long half-life of 20 hours. S-100β serum levels have correlated to injury severity in several studies [125].

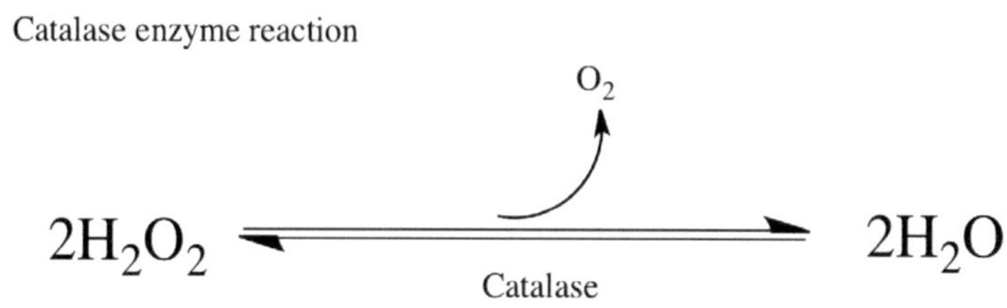

Figure 10. Catalase enzymatic reaction.

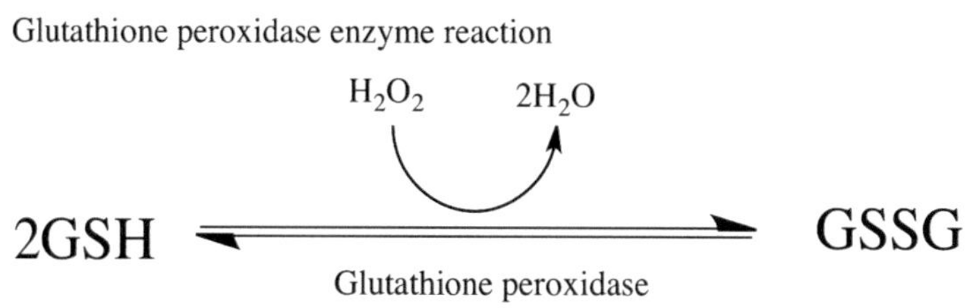

Figure 11. Glutathione peroxidase enzymatic reaction.

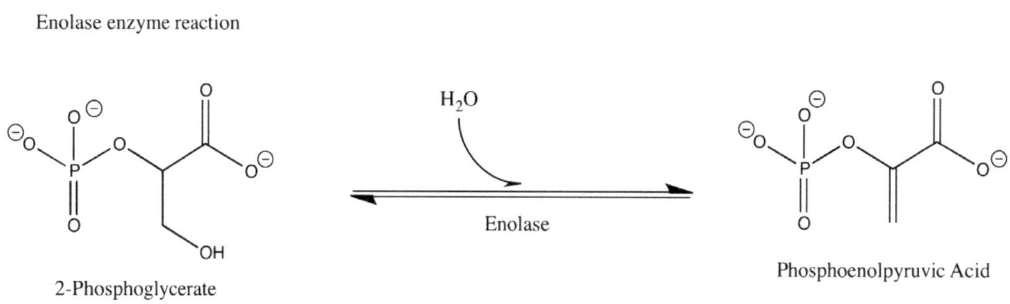

Figure 12. Enzymatic reaction for enolase.

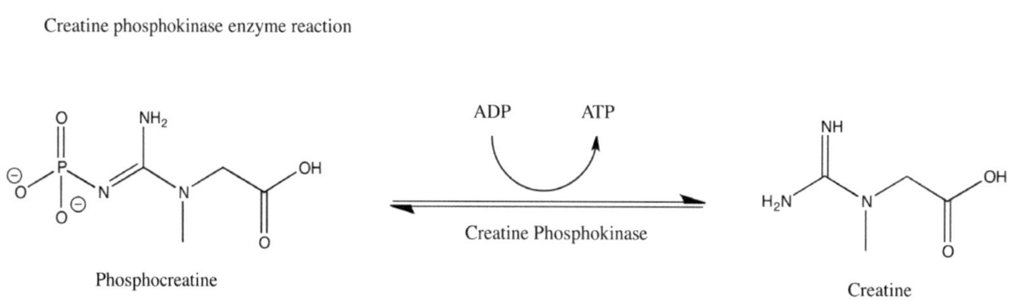

Figure 13. Enzymatic reaction for creatine kinase.

Numerous other proteins such as enolase are known to be upregulated post-injury including C-reactive proteins, transferrin, and breakdown products of collapsin response mediator proteins (CRMP-2), synaptotagmin, and α2-spectrin. C-reactive proteins bind phosphotidylcholine on the surface of dead cells, which is a sign of inflammatory response. Transferrin, an iron-binding glycoprotein, transfers iron from transferrin receptors on the surface of cells to cytoplasmic regions. Expressed in the nervous system, CRMP-2 is essential for axon formation from neurites and growth cone guidance. As a result of trauma, the protease calpain targets CRMPs for cleavage. Both calpain and caspase-3 proteases cleave α2-spectrin, indicating necrotic and apoptotic cell death. Synaptotagmin act as a calcium sensor throughout the brain, thereby monitoring calcium homeostasis. In brain-specific systems biology analysis, over 320 proteins have been found to be upregulated 24 hours post-injury in a penetrating ballistic-like brain injury model (PBBI). The majority of these proteins are involved in protein metabolism. These proteins are essential in neurite outgrowth and cell differentiation. Calcineurin, also known as protein phosphatase 3, is a calcium-stimulated phosphatase. It was significantly increased in post-injury hippocampal and cortical homogenates of a moderate, central fluid percussion TBI injury model for 2-3 weeks post-injury. Substrate affinity was not changed, but maximal dephosphorylation rate did increase [126]. The increased rate has many physiological consequences, resulting in a greater possibility of neuronal cell death. Enolase, phosphoglycerate mutase, and ATP synthase are nitrated in TBI and all are associated with energy metabolism in the brain [127]. Phosphoglycerate mutase, a key glycolytic enzyme, catalyzes a phosphoryl group transfer allowing 3-phosphoglycerate to be converted to 2-phosphoglycerate (Figure 14). ATP synthase provides energy to cells through the production of ATP from ADP and inorganic phosphate. Protein dysfunction of these key enzymes can lead to an accumulation of glycolytic intermediates in several metabolic pathways. This ultimately signifies a reduction in pyruvate formation and overall ATP production in both cytoplasmic and mitochondrial regions of the cell. Since ATP is such an important component of cells it would be beneficial for the cells to upregulate ATP producing enzymes, particularly at points where neural communication is established. The acute phase after TBI shows a regionally heterogeneous metabolic state. The glucose transporter protein family and hexokinase activities were studied, showing consistently reduced hexokinase activity throughout the whole brain following contusional TBI. Overall, glucose transporter proteins appeared to only be impaired in the immediate area around the contusion [128].

Many other candidate biomarker proteins are in the process of receiving clinical validation. An abundance of glial fibrillary acidic proteins (GFAPs), acting as intermediate filament proteins, can indicate damaged glia. GFAP, found only in the central nervous system, forms a network support in astroglial cytoskeletons. GFAP serum levels have been found to increase post-injury and rapidly decrease after the first six hours, making GFAP an ideal biomarker due to the specificity of the protein [125]. A recent study found that cerebrospinal fluid levels of neuron specific enolase, brain specific creatine kinase, glial fibrillary acidic protein, and S100β were all significantly increased six hours post-TBI in a swine model [129]. These enzyme levels then decreased 24 hours post-injury in all markers except S100β at 72 hours post-injury. Increased tau and spectrin protein breakdown products indicate axonal damage. Other enzymes consistently showing upregulation in post-TBI injuries include ubiquitin carboxy-

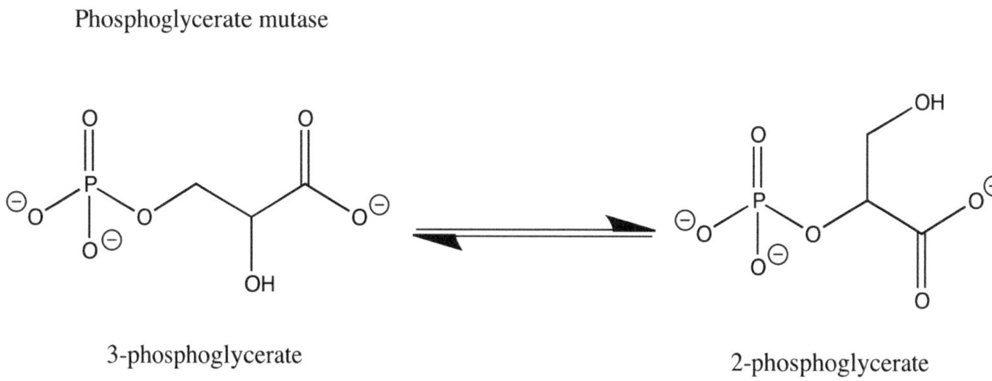

Figure 14. Enzymatic reaction for phosphoglycerate mutase.

terminal hydrolase L1 (UCHL1), tyrosine hydroxylase, and syntaxin-6. UCHL1 is a deubiquitinating protease enzyme, hydrolyzing the C-terminal bond of ubiquitin or unfolded polypeptides. It showed significant elevation in TBI patients [130]. Calpain and caspase-3 proteases are intimately involved in necrotic and apoptotic cell death; their respective spectrin breakdown products appear in significantly elevated levels in mortality and severe cases involving TBI [123]. Additional mRNA expression testing has been done on caspase-1 and caspase-3, both members of the cysteine-aspartic acid protease family. The mRNA expression of caspase-3 was increased fivefold in the injured zone of the cortex and twofold in the injured zone of the hippocampus on fluid percussion-induced TBI rat models at 24 hours post-injury; caspase-1 mRNA expression is also increased to a lesser extent [131]. This suggests elevated caspase protein activity during neuronal apoptosis.

In mitochondria, a crucial location for metabolic enzymes, moderate TBI has negative effects on function including mitochondrial glucose utilization and oxidative phosphorylation. Dysfunction of the cytochrome oxidase complex in mitochondrial metabolism has been observed including cytochrome c oxidase, which is essential for oxidative phosphorylation [121]. Cytochrome c oxidase mRNA expression was investigated in rat impact accelerator models of diffuse TBI (DTBI), also known as "Marmarou" weight-dropping trauma models, followed by real-time quantitative PCR assessment of tissue sections from ipsilateral and contralateral cortex zones. Reductions of cytochrome c oxidase I, II, and III were detected in injured cortex zones. The lowest expression was seen three days post-injury. Contralateral cortex zones were detected to have slightly increased mRNA expression at times ranging from six hours to seven days [121]. *In vivo* structural damage to the mitochondria stimulates free radical generation, causing mitochondrial population depletion depending on injury severity. The remaining mitochondrial population provides metabolic support that the body needs for survival and repair [132]. This appears to be one of many complex multifunctioning mechanisms the brain is capable of in terms of neuroregeneration. It is important that nitrated and oxidized enzymes that can be used as biomarkers for TBI, as well as upregulated or increasingly active enzymes during post-injury, continue to be explored for expansion of neurological insight and therapeutic strategies for harmful brain injuries.

11. Redox proteomics and TBI

The field of proteomics encompasses the study of the proteome that provides specific details regarding its structure and function. Proteins are vital aspects of living organisms, as they are the primary factors of the metabolic pathways of cells. Proteomics assesses the expression level of proteins, posttranslational modifications, and interactions of proteins within a tissue, cell, subcellular compartment, or biofluid [123]. The objective of proteomics is to observe and thoroughly understand the physiological processes that are occurring in normal and diseased cells.

12. Proteins oxidatively modified by TBI

Remarkable efforts have been invested into the discovery of biomarkers of specific neurological disorders. Although there are no clinically validated biomarkers to diagnose TBI, several candidate biomarkers of TBIs are being investigated and multiple studies are being conducted to confirm these protein biomarkers as clinically accurate. Protein oxidation is the modification of a protein induced either directly by reactive oxygen species or indirectly by reaction with secondary by-products of oxidative stress [133]. Because proteins have many different and unique functions, oxidative modifications of proteins can lead to diverse functional consequences. Modifications of proteins can lead to a loss of function, inhibition of enzymatic activity, or potential risk of diseases. Kobeissy et al. identified several proteins decreased in abundance in traumatic brain injury including, but not are limited to, collapsin response mediator protein-2 (CRMP-2), glyceraldehyde-3-phosphate dehydrogenase, microtubule-associated proteins MAP2A/2B, and hexokinase [134]. Upregulated proteins included C-reactive proteins, transferrin, and breakdown products of CRMP-2, synaptotagmin, and α2-spectrin [134]. Furthermore, ubiquitin carboxy-terminal hydrolase L1 protein (UCHL-1), spectrin breakdown products (SBDPs), and neuron-specific enolase (NSE) are presented as the most studied candidate protein biomarkers for TBI as neuronal and axonal protein-specific, while glial protein-specific markers include glial fibrillary acidic protein (GFAP) and S100β [123].

UCHL-1, a cysteine protease, hydrolyzes the C-terminal bond of ubiquitin or unfolded polypeptides [135]. UCHL-1 is predominantly expressed in neurons, although it is also expressed in small amounts in neuroendocrine cells [123]. Compared to control, the serum UCHL-1 levels of TBI patients were significantly elevated after the acute phase of TBI and beyond (> 1 week) [136]. Mondello et al., 2012 demonstrated that the parameters of UCHL-1 could be used as a biomarker for severely injured TBI patients.

α2-spectrin, primarily found in neurons, is concentrated in axons and presynaptic terminals [137]. Upon activation in TBI, calpain cleaves the protein to SBDPs of molecular weights 145 kDa (SBDP145), 150 kDa (SBDP150), while casapse-3 cleaves α2-spectrin to a 120-kDa (SBDP120) product. Calpain and caspase-3 are major executioners of necrotic and apoptotic cell death, respectively, during ischemia or TBI [138-140]. SBDPs indicate calpain and caspase-3 proteolysis of α2-spectrin, providing crucial information on the underlying cell death mechanisms [123]. Elevated levels of SBDPs in CSF from adults with severe TBI were reported and their significant relationships with severity of injury and outcome [139]. Increased CSF SBDP levels were found to be significantly associated with mortality in patients with severe TBI. The temporal profile of SBDPs in non-survivors was also found to be different compared to those of survivors [140]. Taken together, these findings suggest that SBDPs may provide crucial information not only on severity of brain injury, but also on underlying pathophysiological mechanisms associated with necrotic and apoptotic cell death [123].

Neuron-specific enolase (NSE), a glycolytic pathway enzyme, is highly expressed in neuronal cytoplasm. NSE has been shown to have the sensitivity and specificity to detect neuronal cell death [141]. Increased CSF and serum levels of NSE have been reported after TBI. NSE concentrations were also associated with severity of injury, CT scan findings, and outcome [141-143]. Of the numerous candidate biomarkers for TBI, GFAP holds the most promise. One of the strengths of GFAP as an ideal biomarker for TBI is that this protein is not found outside the central nervous system [144]. GFAP, an intermediate filament protein that forms networks that support the astroglial cells, is found only in astroglial cytoskeleton. Damage to the astroglial cells (astrogliosis) shows subsequent upregulation of GFAP [123]. Even if the body is subjected to multiple forms of trauma, GFAP doesn't rise without brain injury [145, 146]. The specificity of GFAP as a biomarker makes it a formidable indicator of glial injury.

S100β is mainly found in astroglia and Schwann cells and is one of the most well established biomarkers of brain damage [147-149]. The concentration of S100β is known to increase in CSF and serum after injury making this protein a potential biomarker for TBI [150]. However, S100β is not specific to the brain, as it is observed in non-nervous cells such as adipocytes, epidermal, chondrocytes, melanoma, and Langerhans cells [151]. Furthermore, general trauma without brain injury can increase S100β. Although a possible candidate as a biomarker for TBI, it seems S100β is not independently accurate to determine brain damage and prognosis, but rather in comparison to other biomarkers.

3-Nitrotyrosine (3-NT) is one of the most frequently observed byproducts from reactive nitrogen species (RNS) reacting with proteins. As a biomarker of nitrosative stress, elevated levels of 3-NT signify the presence of oxidative stress and decreased levels of antioxidant enzymes [152]. The formation of RNS from oxidative stress is assumed to play a major role in neuronal death and 3-NT is a marker for this biochemical event. Therefore, 3-NT can be utilized as an *in vivo* marker of oxidative nitric oxide damage following TBI [153]. 3-NT is formed *in vivo* by the reaction of tyrosine with nitrating oxidants, superoxide and nitric acid [154]. Research has shown that elevated 3-NT levels are directly related to traumatic brain injury (TBI), protein nitration, and oxidative stress. A listing of proteins identified as being nitrated can be seen in Table 1. The lack of consensus in the definition of mild TBI further complicates the matter and the challenge lies in accurate diagnosis in managing post-injury [123]. The role of 3-NT formation as an intermediate will predict the involvement of protein nitration and oxidative stress in the brain.

Protein	Function
Synapsin 1	Regulation of axonogenesis and synaptogenesis; Encode neuronal phosphoproteins which associate with the cytoplasmic surface of synaptic vesicles
Gamma-enolase	Neurotrophic and neuroprotective properties on a broad spectrum of central nervous system (CNS) neurons
Guanosine diphosphate dissociation	Regulate the GDP-GTP exchange reaction of members of the Rab family, small GTP-binding proteins of the Ras superfamily, that are involved in vesicular trafficking of molecules between cellular organelles
Phosphoglycerate mutase (PGM)	Catalyzes the conversion of 3-phosphoglycerate (3PG) to 2-phosphoglycerate (2PG) through a 2,3-bisphosphoglycerate intermediate in glycolysis
Heat shock protein 70 (Hsp70)	Molecular chaperone and folding catalyst that assist in a variety of protein folding processes in the cell by transient association of their substrate binding domain with short hydrophobic peptide segments within their substrate proteins
ATP synthase	Provides energy for the cell to use through the synthesis of adenosine triphosphate (ATP)
Alpha-spectrin	Primary spectrin in blood that supports cell structure and critical for platelet activation

Table 1. Nitrated proteins in TBI

Proteomics and the analysis of potential biomarkers for TBI have provided insight into the mechanism and biochemistry of TBI, which have enabled opportunities to elucidate protein behavior. The field of proteomics has assisted in yielding better insight to the progression of injury, assessment of accurate diagnostic criteria for TBI, as well as the development of possible therapies for TBI. The investigation and discovery of many candidate biomarkers for TBI will continue to increase targeted proteomic experiments in the future.

13. Learning and memory

Traumatic brain injury (TBI) is a costly medical crisis for which no clinically proven pharmacological therapies currently exist. TBI, also known as the "silent epidemic", is profoundly related to oxidative stress, which has been indicated as a mechanism of secondary neuronal injury in TBI that can ultimately result in numerous related neurological maladies [155]. Oxidative stress has been associated in the pathogenesis of numerous neurological disorders such as Alzheimer's disease, Parkinson's disease, Huntington's disease, and stroke. TBIs could impair coordination of respiratory muscles, pharyngeal function, and tongue/lip movement involved with speech, which is one of the most complex functions humans can perform. Incoordination of these precise processes can be detrimental and cause slurred or disjointed speech. Problems can arise with expressive and receptive language, literacy abilities, and diminished social communication skills. Furthermore, TBI in humans can cause an extensive range of cognitive impairments including, but are not limited to, a variety of deficits in learning, memory, retrograde amnesia, anterograde amnesia, impairments in new learning or acquisition, and deficits in working memory, reference memory, and spatial memory.

Impairments in memory are a core element of the cognitive deficits associated with TBI and are likely related to disruption of cholinergic functioning in the hippocampus. Under oxidative stress conditions, antioxidant levels fluctuate significantly. Certain environmental or external factors may be the cause of an imbalance and as a result, reactive nitrogen species (RNS) are created. RNS have the ability to react with biomolecules including carbohydrates, proteins, lipids, DNA and RNA which leads to oxidative damage and ultimately cellular dysfunction [156]. The brain is vulnerable to oxidative stress due to the high content of peroxidizable unsaturated fatty acids, increased consumption of oxygen, elevated levels of free radicals, and moderately low levels of antioxidant defense systems [157]. Damage to the brain from a traumatic event can have life-long, lingering effects resulting in the deformation of normal physiological processes. As a result of these biochemical events results, TBIs serve as a source of significant and persistent cognitive dysfunction.

14. Conclusion

Traumatic brain injuries are a sudden, severe trauma to the brain. Presently, there is no cure for these occurrences. These harmful injuries are the causative factors in a multitude of events including protein dysfunction, altered energy metabolism, loss of enzymatic activity, and increases in ROS and RNS. The field of proteomics has demonstrated the importance of how oxidatively modified proteins play a pivotal role in both learning and memory, two key features effected in traumatic brain injury. Current research efforts are introducing new animal models and pre and post therapeutic strategies which show promise in delaying and possibly preventing further neuronal damage for primary and secondary injuries associated with TBI, making a clinical treatment for TBI a strong prospect in the near future.

Acknowledgements

The enzymatic work was supported in part by NIH grants to T.T.R. [R15 NS 072870-01A].

Author details

Zachariah P. Sellers, ReBecca A. Williams, Jonathan W. Overbay, Jooyoung Cho, Moses Henderson and Tanea T. Reed*

*Address all correspondence to: tanea.reed@eku.edu

Eastern Kentucky University, Department of Chemistry, Richmond, KY, USA

References

[1] Kochanek PM, Dixon CE, Shellington DK, Shin SS, Bayir H, Jackson EK, et al. Screening of biochemical and molecular mechanisms of secondary injury and repair in the brain after experimental blast-induced traumatic brain injury in rats. J Neurotrauma. 2013;30(11):920-37.

[2] Wang T, Van KC, Gavitt BJ, Grayson JK, Lu YC, Lyeth BG, et al. Effect of fish oil supplementation in a rat model of multiple mild traumatic brain injuries. Restorative neurology and neuroscience. 2013.

[3] Smith DH, Meaney DF. Axonal damage in traumatic brain injury. Neuroscientist. 2000;6(6):483-95.

[4] Gold EM, Su D, Lopez-Velazquez L, Haus DL, Perez H, Lacuesta GA, et al. Functional assessment of long-term deficits in rodent models of traumatic brain injury. Regenerative medicine. 2013;8(4):483-516.

[5] Selwyn R, Hokenbury N, Jaiswal S, Mathur S, Armstrong RC, Byrnes K. Mild Traumatic Brain Injury Results in Depressed Cerebral Glucose Uptake: An 18-FDG PET Study. J Neurotrauma. 2013.

[6] Lu J, Frerich JM, Turtzo LC, Li S, Chiang J, Yang C, et al. Histone deacetylase inhibitors are neuroprotective and preserve NGF-mediated cell survival following traumatic brain injury. Proceedings of the National Academy of Sciences of the United States of America. 2013;110(26):10747-52.

[7] O'Callaghan JP. Assessment of neurotoxicity: use of glial fibrillary acidic protein as a biomarker. Biomedical and environmental sciences : BES. 1991;4(1-2):197-206.

[8] Gieling ET, Schuurman T, Nordquist RE, van der Staay FJ. The pig as a model animal for studying cognition and neurobehavioral disorders. Current topics in behavioral neurosciences. 2011;7(Current Topics in Behavioral Neurosciences):359-83.

[9] Sullivan S, Friess SH, Ralston J, Smith C, Propert KJ, Rapp PE, et al. Improved behavior, motor, and cognition assessments in neonatal piglets. J Neurotrauma. 2013.

[10] Steinmetz S, Tipold A, Loscher W. Epilepsy after head injury in dogs: a natural model of posttraumatic epilepsy. Epilepsia. 2013;54(4):580-8.

[11] Benardo LS. Prevention of epilepsy after head trauma: do we need new drugs or a new approach? Epilepsia. 2003;44 Suppl 10:27-33.

[12] King C, Robinson T, Dixon CE, Rao GR, Larnard D, Nemoto CE. Brain temperature profiles during epidural cooling with the ChillerPad in a monkey model of traumatic brain injury. J Neurotrauma. 2010;27(10):1895-903.

[13] Wagner KR, Beiler S, Beiler C, Kirkman J, Casey K, Robinson T, et al. Delayed profound local brain hypothermia markedly reduces interleukin-1beta gene expression

and vasogenic edema development in a porcine model of intracerebral hemorrhage. Acta neurochirurgica Supplement. 2006;96:177-82.

[14] Clark RS, Kochanek PM, Marion DW, Schiding JK, White M, Palmer AM, et al. Mild posttraumatic hypothermia reduces mortality after severe controlled cortical impact in rats. Journal of cerebral blood flow and metabolism : official journal of the International Society of Cerebral Blood Flow and Metabolism. 1996;16(2):253-61.

[15] Nagamoto-Combs K, Morecraft RJ, Darling WG, Combs CK. Long-term gliosis and molecular changes in the cervical spinal cord of the rhesus monkey after traumatic brain injury. J Neurotrauma. 2010;27(3):565-85.

[16] Lotze M, Grodd W, Rodden FA, Gut E, Schonle PW, Kardatzki B, et al. Neuroimaging patterns associated with motor control in traumatic brain injury. Neurorehabilitation and neural repair. 2006;20(1):14-23.

[17] Nagamoto-Combs K, McNeal DW, Morecraft RJ, Combs CK. Prolonged microgliosis in the rhesus monkey central nervous system after traumatic brain injury. J Neurotrauma. 2007;24(11):1719-42.

[18] Finnie JW. Comparative approach to understanding traumatic injury in the immature, postnatal brain of domestic animals. Australian veterinary journal. 2012;90(8): 301-7.

[19] Pretz CR, Dams-O'Connor K. A Longitudinal Description of the GOS-E for Individuals in the Traumatic Brain Injury Model Systems National Database: A National Institute on Disability and Rehabilitation Research Traumatic Brain Injury Model Systems Study. Archives of physical medicine and rehabilitation. 2013.

[20] Kozlowski AJ, Pretz CR, Dams-O'Connor K, Kreider S, Whiteneck G. An introduction to applying individual growth curve models to evaluate change in rehabilitation: a national institute on disability and rehabilitation research traumatic brain injury model systems report. Archives of physical medicine and rehabilitation. 2013;94(3):589-96.

[21] Gregorio GW, Gould KR, Spitz G, van Heugten CM, Ponsford JL. Changes in Self-Reported Pre- to Postinjury Coping Styles in the First 3 Years After Traumatic Brain Injury and the Effects on Psychosocial and Emotional Functioning and Quality of Life. The Journal of head trauma rehabilitation. 2013.

[22] Beck L. United States Department of Veterans Affairs. 2013 [updated April 1, 2013]; Available from: http://www.polytrauma.va.gov/research-and-advancements/.

[23] Vrettou CS, Zakynthinos SG, Malachias S, Mentzelopoulos SD. High frequency oscillation and tracheal gas insufflation in patients with severe acute respiratory distress syndrome and traumatic brain injury: an interventional physiological study. Critical care. 2013;17(4):R136.

[24] Bouzat P, Sala N, Payen JF, Oddo M. Beyond intracranial pressure: optimization of cerebral blood flow, oxygen, and substrate delivery after traumatic brain injury. Annals of intensive care. 2013;3(1):23.

[25] Tsitsopoulos PP, Marklund N. Amyloid-beta Peptides and Tau Protein as Biomarkers in Cerebrospinal and Interstitial Fluid Following Traumatic Brain Injury: A Review of Experimental and Clinical Studies. Front Neurol. 2013;4:79.

[26] Stahel PF, Kossmann T, Morganti-Kossmann MC, Hans VH, Barnum SR. Experimental diffuse axonal injury induces enhanced neuronal C5a receptor mRNA expression in rats. Brain research Molecular brain research. 1997;50(1-2):205-12.

[27] Rus H, Cudrici C, Niculescu F. C5b-9 complement complex in autoimmune demyelination and multiple sclerosis: dual role in neuroinflammation and neuroprotection. Annals of medicine. 2005;37(2):97-104.

[28] Rostami E, Davidsson J, Gyorgy A, Agoston DV, Risling M, Bellander BM. The complement terminal pathway is activated in focal penetrating but not in mild diffuse Traumatic Brain Injury. J Neurotrauma. 2013.

[29] Cooke PS, Nanjappa MK, Yang Z, Wang KK. Therapeutic effects of progesterone and its metabolites in traumatic brain injury may involve non-classical signaling mechanisms. Frontiers in neuroscience. 2013;7:108.

[30] Wright D. Progesterone for the Treatment of Traumatic Brain Injury (ProTECT III). 2013.

[31] Stein DG, Wright DW, Kellermann AL. Does progesterone have neuroprotective properties? Ann Emerg Med. 2008;51(2):164-72.

[32] Vandromme M, Melton SM, Kerby JD. Progesterone in traumatic brain injury: time to move on to phase III trials. Critical care. 2008;12(3):153.

[33] Weigl W, Milej D, Gerega A, Toczylowska B, Kacprzak M, Sawosz P, et al. Assessment of cerebral perfusion in post-traumatic brain injury patients with the use of ICG-bolus tracking method. Neuroimage. 2013.

[34] Spencer JP. Flavonoids: modulators of brain function? The British journal of nutrition. 2008;99 E Suppl 1:ES60-77.

[35] Baulieu, Robel P, Schumacher M. Neurosteroids : a new regulatory function in the nervous system1999. xiii, 378 p.

[36] Mellon SH, Griffin LD. Neurosteroids: biochemistry and clinical significance. Trends in endocrinology and metabolism: TEM. 2002;13(1):35-43.

[37] Morrow AL. Recent developments in the significance and therapeutic relevance of neuroactive steroids--Introduction to the special issue. Pharmacology & therapeutics. 2007;116(1):1-6.

[38] Reddy DS. Neurosteroids: endogenous role in the human brain and therapeutic potentials. Prog Brain Res. 2010;186:113-37.

[39] Bastianetto S, Ramassamy C, Poirier J, Quirion R. Dehydroepiandrosterone (DHEA) protects hippocampal cells from oxidative stress-induced damage. Brain research Molecular brain research. 1999;66(1-2):35-41.

[40] Kurata K, Takebayashi M, Morinobu S, Yamawaki S. beta-estradiol, dehydroepiandrosterone, and dehydroepiandrosterone sulfate protect against N-methyl-D-aspartate-induced neurotoxicity in rat hippocampal neurons by different mechanisms. The Journal of pharmacology and experimental therapeutics. 2004;311(1):237-45.

[41] Charalampopoulos I, Alexaki VI, Lazaridis I, Dermitzaki E, Avlonitis N, Tsatsanis C, et al. G protein-associated, specific membrane binding sites mediate the neuroprotective effect of dehydroepiandrosterone. FASEB journal : official publication of the Federation of American Societies for Experimental Biology. 2006;20(3):577-9.

[42] Jiang N, Chopp M, Stein D, Feit H. Progesterone is neuroprotective after transient middle cerebral artery occlusion in male rats. Brain Res. 1996;735(1):101-7.

[43] Veiga S, Melcangi RC, Doncarlos LL, Garcia-Segura LM, Azcoitia I. Sex hormones and brain aging. Experimental gerontology. 2004;39(11-12):1623-31.

[44] Garcia-Segura LM, Veiga S, Sierra A, Melcangi RC, Azcoitia I. Aromatase: a neuroprotective enzyme. Progress in neurobiology. 2003;71(1):31-41.

[45] Ahlbom E, Prins GS, Ceccatelli S. Testosterone protects cerebellar granule cells from oxidative stress-induced cell death through a receptor mediated mechanism. Brain Res. 2001;892(2):255-62.

[46] Hammond J, Le Q, Goodyer C, Gelfand M, Trifiro M, LeBlanc A. Testosterone-mediated neuroprotection through the androgen receptor in human primary neurons. Journal of neurochemistry. 2001;77(5):1319-26.

[47] Roof RL, Duvdevani R, Braswell L, Stein DG. Progesterone facilitates cognitive recovery and reduces secondary neuronal loss caused by cortical contusion injury in male rats. Exp Neurol. 1994;129(1):64-9.

[48] Margulies S, Hicks R. Combination therapies for traumatic brain injury: prospective considerations. J Neurotrauma. 2009;26(6):925-39.

[49] Ishrat T, Sayeed I, Atif F, Hua F, Stein DG. Progesterone and allopregnanolone attenuate blood-brain barrier dysfunction following permanent focal ischemia by regulating the expression of matrix metalloproteinases. Exp Neurol. 2010;226(1):183-90.

[50] Hoffman SW, Virmani S, Simkins RM, Stein DG. The delayed administration of dehydroepiandrosterone sulfate improves recovery of function after traumatic brain injury in rats. J Neurotrauma. 2003;20(9):859-70.

[51] Malik AS, Narayan RK, Wendling WW, Cole RW, Pashko LL, Schwartz AG, et al. A novel dehydroepiandrosterone analog improves functional recovery in a rat traumatic brain injury model. J Neurotrauma. 2003;20(5):463-76.

[52] Hall ED, Vaishnav RA, Mustafa AG. Antioxidant therapies for traumatic brain injury. Neurotherapeutics : the journal of the American Society for Experimental NeuroTherapeutics. 2010;7(1):51-61.

[53] Kontos HA, Povlishock JT. Oxygen radicals in brain injury. Central nervous system trauma : journal of the American Paralysis Association. 1986;3(4):257-63.

[54] Ansari MA, Roberts KN, Scheff SW. Oxidative stress and modification of synaptic proteins in hippocampus after traumatic brain injury. Free radical biology & medicine. 2008;45(4):443-52.

[55] Rice-Evans CA, Miller NJ, Paganga G. Structure-antioxidant activity relationships of flavonoids and phenolic acids. Free radical biology & medicine. 1996;20(7):933-56.

[56] Santos MR, Mira L. Protection by flavonoids against the peroxynitrite-mediated oxidation of dihydrorhodamine. Free radical research. 2004;38(9):1011-8.

[57] Potapovich AI, Kostyuk VA. Comparative study of antioxidant properties and cytoprotective activity of flavonoids. Biochemistry Biokhimiia. 2003;68(5):514-9.

[58] Scheff SW, Ansari MA, Roberts KN. Neuroprotective effect of Pycnogenol(R) following traumatic brain injury. Exp Neurol. 2013;239:183-91.

[59] Vauzour D, Vafeiadou K, Rodriguez-Mateos A, Rendeiro C, Spencer JP. The neuroprotective potential of flavonoids: a multiplicity of effects. Genes & nutrition. 2008;3(3-4):115-26.

[60] Chen SF, Hsu CW, Huang WH, Wang JY. Post-injury baicalein improves histological and functional outcomes and reduces inflammatory cytokines after experimental traumatic brain injury. British journal of pharmacology. 2008;155(8):1279-96.

[61] Chen CC, Hung TH, Wang YH, Lin CW, Wang PY, Lee CY, et al. Wogonin improves histological and functional outcomes, and reduces activation of TLR4/NF-kappaB signaling after experimental traumatic brain injury. PloS one. 2012;7(1):e30294.

[62] Itoh T, Imano M, Nishida S, Tsubaki M, Hashimoto S, Ito A, et al. (-)-Epigallocatechin-3-gallate protects against neuronal cell death and improves cerebral function after traumatic brain injury in rats. Neuromolecular medicine. 2011;13(4):300-9.

[63] Schultke E, Kamencic H, Zhao M, Tian GF, Baker AJ, Griebel RW, et al. Neuroprotection following fluid percussion brain trauma: a pilot study using quercetin. J Neurotrauma. 2005;22(12):1475-84.

[64] Abd El Mohsen MM, Kuhnle G, Rechner AR, Schroeter H, Rose S, Jenner P, et al. Uptake and metabolism of epicatechin and its access to the brain after oral ingestion. Free radical biology & medicine. 2002;33(12):1693-702.

[65] Parajuli P, Joshee N, Chinni SR, Rimando AM, Mittal S, Sethi S, et al. Delayed growth of glioma by Scutellaria flavonoids involve inhibition of Akt, GSK-3 and NF-kappaB signaling. Journal of neuro-oncology. 2011;101(1):15-24.

[66] Longpre F, Garneau P, Christen Y, Ramassamy C. Protection by EGb 761 against beta-amyloid-induced neurotoxicity: involvement of NF-kappaB, SIRT1, and MAPKs pathways and inhibition of amyloid fibril formation. Free radical biology & medicine. 2006;41(12):1781-94.

[67] Spencer JP, Schroeter H, Crossthwaithe AJ, Kuhnle G, Williams RJ, Rice-Evans C. Contrasting influences of glucuronidation and O-methylation of epicatechin on hydrogen peroxide-induced cell death in neurons and fibroblasts. Free radical biology & medicine. 2001;31(9):1139-46.

[68] Jatana M, Singh I, Singh AK, Jenkins D. Combination of systemic hypothermia and N-acetylcysteine attenuates hypoxic-ischemic brain injury in neonatal rats. Pediatric research. 2006;59(5):684-9.

[69] Itoh T, Imano M, Nishida S, Tsubaki M, Mizuguchi N, Hashimoto S, et al. (-)-Epigallocatechin-3-gallate increases the number of neural stem cells around the damaged area after rat traumatic brain injury. J Neural Transm. 2012;119(8):877-90.

[70] Kobayashi MS, Han D, Packer L. Antioxidants and herbal extracts protect HT-4 neuronal cells against glutamate-induced cytotoxicity. Free radical research. 2000;32(2): 115-24.

[71] Kobuchi H, Virgili F, Packer L. Assay of inducible form of nitric oxide synthase activity: effect of flavonoids and plant extracts. Methods in enzymology. 1999;301:504-13.

[72] Nardini M, Scaccini C, Packer L, Virgili F. In vitro inhibition of the activity of phosphorylase kinase, protein kinase C and protein kinase A by caffeic acid and a procyanidin-rich pine bark (Pinus marittima) extract. Biochimica et biophysica acta. 2000;1474(2):219-25.

[73] Packer L, Rimbach G, Virgili F. Antioxidant activity and biologic properties of a procyanidin-rich extract from pine (Pinus maritima) bark, pycnogenol. Free radical biology & medicine. 1999;27(5-6):704-24.

[74] Peng QL, Buz'Zard AR, Lau BH. Pycnogenol protects neurons from amyloid-beta peptide-induced apoptosis. Brain research Molecular brain research. 2002;104(1): 55-65.

[75] Lu D, Mahmood A, Goussev A, Schallert T, Qu C, Zhang ZG, et al. Atorvastatin reduction of intravascular thrombosis, increase in cerebral microvascular patency and integrity, and enhancement of spatial learning in rats subjected to traumatic brain injury. Journal of neurosurgery. 2004;101(5):813-21.

[76] Lu D, Qu C, Goussev A, Jiang H, Lu C, Schallert T, et al. Statins increase neurogenesis in the dentate gyrus, reduce delayed neuronal death in the hippocampal CA3 re-

gion, and improve spatial learning in rat after traumatic brain injury. J Neurotrauma. 2007;24(7):1132-46.

[77] Turkoglu OF, Eroglu H, Okutan O, Gurcan O, Bodur E, Sargon MF, et al. Atorvastatin efficiency after traumatic brain injury in rats. Surgical neurology. 2009;72(2): 146-52; discussion 52.

[78] Graham DI, Ford I, Adams JH, Doyle D, Teasdale GM, Lawrence AE, et al. Ischaemic brain damage is still common in fatal non-missile head injury. Journal of neurology, neurosurgery, and psychiatry. 1989;52(3):346-50.

[79] von Oettingen G, Bergholt B, Gyldensted C, Astrup J. Blood flow and ischemia within traumatic cerebral contusions. Neurosurgery. 2002;50(4):781-80.

[80] Ahn MJ, Sherwood ER, Prough DS, Lin CY, DeWitt DS. The effects of traumatic brain injury on cerebral blood flow and brain tissue nitric oxide levels and cytokine expression. J Neurotrauma. 2004;21(10):1431-42.

[81] Liao JK, Laufs U. Pleiotropic effects of statins. Annual review of pharmacology and toxicology. 2005;45:89-118.

[82] Lu D, Goussev A, Chen J, Pannu P, Li Y, Mahmood A, et al. Atorvastatin reduces neurological deficit and increases synaptogenesis, angiogenesis, and neuronal survival in rats subjected to traumatic brain injury. J Neurotrauma. 2004;21(1):21-32.

[83] Chauhan NB, Gatto R. Restoration of cognitive deficits after statin feeding in TBI. Restorative neurology and neuroscience. 2011;29(1):23-34.

[84] Schneider EB, Efron DT, MacKenzie EJ, Rivara FP, Nathens AB, Jurkovich GJ. Premorbid statin use is associated with improved survival and functional outcomes in older head-injured individuals. The Journal of trauma. 2011;71(4):815-9.

[85] Jungner M, Lundblad C, Bentzer P. Rosuvastatin in experimental brain trauma: Improved capillary patency but no effect on edema or cerebral blood flow. Microvascular research. 2013;88:48-55.

[86] Laufs U, La Fata V, Plutzky J, Liao JK. Upregulation of endothelial nitric oxide synthase by HMG CoA reductase inhibitors. Circulation. 1998;97(12):1129-35.

[87] Kureishi Y, Luo Z, Shiojima I, Bialik A, Fulton D, Lefer DJ, et al. The HMG-CoA reductase inhibitor simvastatin activates the protein kinase Akt and promotes angiogenesis in normocholesterolemic animals. Nature medicine. 2000;6(9):1004-10.

[88] Laufs U, Fata VL, Liao JK. Inhibition of 3-hydroxy-3-methylglutaryl (HMG)-CoA reductase blocks hypoxia-mediated down-regulation of endothelial nitric oxide synthase. The Journal of biological chemistry. 1997;272(50):31725-9.

[89] Singh R, Wang B, Shirvaikar A, Khan S, Kamat S, Schelling JR, et al. The IL-1 receptor and Rho directly associate to drive cell activation in inflammation. The Journal of clinical investigation. 1999;103(11):1561-70.

[90] Laufs U, Liao JK. Post-transcriptional regulation of endothelial nitric oxide synthase mRNA stability by Rho GTPase. The Journal of biological chemistry. 1998;273(37): 24266-71.

[91] Beziaud T, Ru Chen X, El Shafey N, Frechou M, Teng F, Palmier B, et al. Simvastatin in traumatic brain injury: effect on brain edema mechanisms. Critical care medicine. 2011;39(10):2300-7.

[92] Pompella A, Visvikis A, Paolicchi A, De Tata V, Casini AF. The changing faces of glutathione, a cellular protagonist. Biochemical pharmacology. 2003;66(8):1499-503.

[93] Dringen R. Metabolism and functions of glutathione in brain. Progress in neurobiology. 2000;62(6):649-71.

[94] Kontos HA, Wei EP. Superoxide production in experimental brain injury. Journal of neurosurgery. 1986;64(5):803-7.

[95] Reed TT, Owen J, Pierce WM, Sebastian A, Sullivan PG, Butterfield DA. Proteomic identification of nitrated brain proteins in traumatic brain-injured rats treated postinjury with gamma-glutamylcysteine ethyl ester: insights into the role of elevation of glutathione as a potential therapeutic strategy for traumatic brain injury. Journal of neuroscience research. 2009;87(2):408-17.

[96] Pileblad E, Magnusson T. Increase in rat brain glutathione following intracerebroventricular administration of gamma-glutamylcysteine. Biochemical pharmacology. 1992;44(5):895-903.

[97] Drake J, Kanski J, Varadarajan S, Tsoras M, Butterfield DA. Elevation of brain glutathione by gamma-glutamylcysteine ethyl ester protects against peroxynitrite-induced oxidative stress. Journal of neuroscience research. 2002;68(6):776-84.

[98] Drake J, Sultana R, Aksenova M, Calabrese V, Butterfield DA. Elevation of mitochondrial glutathione by gamma-glutamylcysteine ethyl ester protects mitochondria against peroxynitrite-induced oxidative stress. Journal of neuroscience research. 2003;74(6):917-27.

[99] Joshi G, Hardas S, Sultana R, St Clair DK, Vore M, Butterfield DA. Glutathione elevation by gamma-glutamyl cysteine ethyl ester as a potential therapeutic strategy for preventing oxidative stress in brain mediated by in vivo administration of adriamycin: Implication for chemobrain. Journal of neuroscience research. 2007;85(3):497-503.

[100] Lai Y, Hickey RW, Chen Y, Bayir H, Sullivan ML, Chu CT, et al. Autophagy is increased after traumatic brain injury in mice and is partially inhibited by the antioxidant gamma-glutamylcysteinyl ethyl ester. Journal of cerebral blood flow and metabolism : official journal of the International Society of Cerebral Blood Flow and Metabolism. 2008;28(3):540-50.

[101] Lok J, Leung W, Zhao S, Pallast S, van Leyen K, Guo S, et al. gamma-glutamylcysteine ethyl ester protects cerebral endothelial cells during injury and decreases blood-

brain barrier permeability after experimental brain trauma. Journal of neurochemistry. 2011;118(2):248-55.

[102] Nauser T, Schoneich C. Thiyl radicals abstract hydrogen atoms from the (alpha)C-H bonds in model peptides: absolute rate constants and effect of amino acid structure. Journal of the American Chemical Society. 2003;125(8):2042-3.

[103] Pogocki D, Schoneich C. Thiyl radicals abstract hydrogen atoms from carbohydrates: reactivity and selectivity. Free radical biology & medicine. 2001;31(1):98-107.

[104] Schoneich C, Dillinger U, von Bruchhausen F, Asmus KD. Oxidation of polyunsaturated fatty acids and lipids through thiyl and sulfonyl radicals: reaction kinetics, and influence of oxygen and structure of thiyl radicals. Archives of biochemistry and biophysics. 1992;292(2):456-67.

[105] Jones CM, Lawrence A, Wardman P, Burkitt MJ. Kinetics of superoxide scavenging by glutathione: an evaluation of its role in the removal of mitochondrial superoxide. Biochemical Society transactions. 2003;31(Pt 6):1337-9.

[106] Wada K, Chatzipanteli K, Busto R, Dietrich WD. Role of nitric oxide in traumatic brain injury in the rat. Journal of neurosurgery. 1998;89(5):807-18.

[107] Levenson JM, Sweatt JD. Epigenetic mechanisms in memory formation. Nature reviews Neuroscience. 2005;6(2):108-18.

[108] Jaenisch R, Bird A. Epigenetic regulation of gene expression: how the genome integrates intrinsic and environmental signals. Nature genetics. 2003;33 Suppl:245-54.

[109] Gao WM, Chadha MS, Kline AE, Clark RS, Kochanek PM, Dixon CE, et al. Immunohistochemical analysis of histone H3 acetylation and methylation--evidence for altered epigenetic signaling following traumatic brain injury in immature rats. Brain Res. 2006;1070(1):31-4.

[110] Kim HJ, Rowe M, Ren M, Hong JS, Chen PS, Chuang DM. Histone deacetylase inhibitors exhibit anti-inflammatory and neuroprotective effects in a rat permanent ischemic model of stroke: multiple mechanisms of action. The Journal of pharmacology and experimental therapeutics. 2007;321(3):892-901.

[111] Yoshiya K, Tanaka H, Kasai K, Irisawa T, Shiozaki T, Sugimoto H. Profile of gene expression in the subventricular zone after traumatic brain injury. J Neurotrauma. 2003;20(11):1147-62.

[112] Matzilevich DA, Rall JM, Moore AN, Grill RJ, Dash PK. High-density microarray analysis of hippocampal gene expression following experimental brain injury. Journal of neuroscience research. 2002;67(5):646-63.

[113] Long Y, Zou L, Liu H, Lu H, Yuan X, Robertson CS, et al. Altered expression of randomly selected genes in mouse hippocampus after traumatic brain injury. Journal of neuroscience research. 2003;71(5):710-20.

[114] Aloe L, Rocco ML, Bianchi P, Manni L. Nerve growth factor: from the early discoveries to the potential clinical use. Journal of translational medicine. 2012;10:239.

[115] Mahmood A, Lu D, Chopp M. Intravenous administration of marrow stromal cells (MSCs) increases the expression of growth factors in rat brain after traumatic brain injury. J Neurotrauma. 2004;21(1):33-9.

[116] Zhang Y, Chopp M, Meng Y, Zhang ZG, Doppler E, Mahmood A, et al. Improvement in functional recovery with administration of Cerebrolysin after experimental closed head injury. Journal of neurosurgery. 2013;118(6):1343-55.

[117] DeKosky ST, Goss JR, Miller PD, Styren SD, Kochanek PM, Marion D. Upregulation of nerve growth factor following cortical trauma. Exp Neurol. 1994;130(2):173-7.

[118] Holtzman DM, Sheldon RA, Jaffe W, Cheng Y, Ferriero DM. Nerve growth factor protects the neonatal brain against hypoxic-ischemic injury. Annals of neurology. 1996;39(1):114-22.

[119] Chiaretti A, Barone G, Riccardi R, Antonelli A, Pezzotti P, Genovese O, et al. NGF, DCX, and NSE upregulation correlates with severity and outcome of head trauma in children. Neurology. 2009;72(7):609-16.

[120] Butterfield DA, Sultana R. Identification of 3-Nitrotyrosine Modified Brain Proteins by Redox Proteomics. Methods of Enzymology: Elsevier; 2008. p. 295-308.

[121] Dai W, Cheng H-l, Huang R-q, Zhuang Z, Shi J-X. Quantitative detection of the expression of mitochondrial cytochrome c oxidase subunits mRNA in the cerebral cortex after experimental traumatic brain injury. Brain Research. 2009;1251(1):287-95.

[122] Goss JR, Taffe KM, Kochanek PM, DeKosky ST. The Antioxidant Enzymes Glutathione Peroxidase and Catalase Increase Following Traumatic Brain Injury in the Rat. Experimental Neurology. 1997;146(1):291-4.

[123] Guingab-Cagmat JD, Cagmat EB, Hayes RL, Anagli J. Integration of proteomics, bioinformatics, and systems biology in traumatic brain injury biomarker discovery. Front Neurol. 2013;4:61.

[124] Wallimann T, Tokarska-Schlattner M, Schlattner U. The creatine kinase system and pleiotropic effects of creatine. Amino Acids. 2011;40(5):1271-96.

[125] Ingebrigtsen T, Romner B. Biochemical serum markers for brain damage: A short review with emphasis on clinical utility in mild head injury. Restorative Neurology & Neuroscience. 2003;21(3/4):171-6.

[126] Kurz JE, Parsons JT, Rana A, Gibson CJ, Hamm RJ, Churn SB. A significant increase in both basal and maximal calcineurin activity following fluid percussion injury in the rat. J Neurotrauma. 2005;22(4):476-90.

[127] Casoni F, Basso M, Massignan T, Gianazza E, Cheroni C, Salmona M, et al. Protein nitration in a mouse model of familial amyotrophic lateral sclerosis: possible multi-

functional role in the pathogenesis. The Journal of biological chemistry. 2005;280(16): 16295-304.

[128] Hattori N, Huang S-C, Wu H-M, Liao W, Glenn TC, Vespa PM, et al. Acute Changes in Regional Cerebral 18F-FDG Kinetics in Patients with Traumatic Brain Injury. The Journal of Nuclear Medicine. 2004;45(5):775-83.

[129] Ahmed F, Gyorgy A, Kamnaksh A, Ling G, Tong L, Parks S, et al. Time-dependent changes of protein biomarker levels in the cerebrospinal fluid after blast traumatic brain injury. ELECTROPHORESIS. 2012;33(24):3705-11.

[130] Brophy GM, Mondello S, Papa L, Robicsek SA, Gabrielli A, Tepas J, 3rd, et al. Biokinetic analysis of ubiquitin C-terminal hydrolase-L1 (UCH-L1) in severe traumatic brain injury patient biofluids. J Neurotrauma. 2011;28(6):861-70.

[131] Yakovlev AG, Knoblach SM, Fan L, Fox GB, Goodnight R, Faden AI. Activation of CPP32-like caspases contributes to neuronal apoptosis and neurological dysfunction after traumatic brain injury. The Journal of neuroscience : the official journal of the Society for Neuroscience. 1997;17(19):7415-24.

[132] Lifshitz J, Sullivan PG, Hovda DA, Wieloch T, McIntosh TK. Mitochondrial damage and dysfunction in traumatic brain injury. Mitochondrion. 2004;4(5–6):705-13.

[133] Shacter E. Quantification and significance of protein oxidation in biological samples. Drug metabolism reviews. 2000;32(3-4):307-26.

[134] Kobeissy FH, Ottens AK, Zhang Z, Liu MC, Denslow ND, Dave JR, et al. Novel differential neuroproteomics analysis of traumatic brain injury in rats. Molecular & cellular proteomics : MCP. 2006;5(10):1887-98.

[135] Setsuie R, Wada K. The functions of UCH-L1 and its relation to neurodegenerative diseases. Neurochemistry international. 2007;51(2-4):105-11.

[136] Mondello S, Linnet A, Buki A, Robicsek S, Gabrielli A, Tepas J, et al. Clinical utility of serum levels of ubiquitin C-terminal hydrolase as a biomarker for severe traumatic brain injury. Neurosurgery. 2012;70(3):666-75.

[137] Riederer BM, Zagon IS, Goodman SR. Brain spectrin(240/235) and brain spectrin(240/235E): two distinct spectrin subtypes with different locations within mammalian neural cells. The Journal of cell biology. 1986;102(6):2088-97.

[138] Ringger NC, O'Steen BE, Brabham JG, Silver X, Pineda J, Wang KK, et al. A novel marker for traumatic brain injury: CSF alphaII-spectrin breakdown product levels. J Neurotrauma. 2004;21(10):1443-56.

[139] Pineda JA, Lewis SB, Valadka AB, Papa L, Hannay HJ, Heaton SC, et al. Clinical significance of alphaII-spectrin breakdown products in cerebrospinal fluid after severe traumatic brain injury. J Neurotrauma. 2007;24(2):354-66.

[140] Mondello S, Robicsek SA, Gabrielli A, Brophy GM, Papa L, Tepas J, et al. alphaII-spectrin breakdown products (SBDPs): diagnosis and outcome in severe traumatic brain injury patients. J Neurotrauma. 2010;27(7):1203-13.

[141] Selakovic V, Raicevic R, Radenovic L. The increase of neuron-specific enolase in cerebrospinal fluid and plasma as a marker of neuronal damage in patients with acute brain infarction. Journal of clinical neuroscience : official journal of the Neurosurgical Society of Australasia. 2005;12(5):542-7.

[142] Ross SA, Cunningham RT, Johnston CF, Rowlands BJ. Neuron-specific enolase as an aid to outcome prediction in head injury. British journal of neurosurgery. 1996;10(5): 471-6.

[143] Herrmann M, Jost S, Kutz S, Ebert AD, Kratz T, Wunderlich MT, et al. Temporal profile of release of neurobiochemical markers of brain damage after traumatic brain injury is associated with intracranial pathology as demonstrated in cranial computerized tomography. J Neurotrauma. 2000;17(2):113-22.

[144] Galea E, Dupouey P, Feinstein DL. Glial fibrillary acidic protein mRNA isotypes: expression in vitro and in vivo. Journal of neuroscience research. 1995;41(4):452-61.

[145] Pelinka LE, Kroepfl A, Schmidhammer R, Krenn M, Buchinger W, Redl H, et al. Glial fibrillary acidic protein in serum after traumatic brain injury and multiple trauma. The Journal of trauma. 2004;57(5):1006-12.

[146] Vos PE, Lamers KJ, Hendriks JC, van Haaren M, Beems T, Zimmerman C, et al. Glial and neuronal proteins in serum predict outcome after severe traumatic brain injury. Neurology. 2004;62(8):1303-10.

[147] Donato R, Battaglia F, Cocchia D. Effects of S-100 proteins on assembly of brain microtubule proteins: correlation between kinetic and ultrastructural data. Journal of neurochemistry. 1986;47(2):350-4.

[148] Donato R, Prestagiovanni B, Zelano G. Identity between cytoplasmic and membrane-bound S-100 proteins purified from bovine and rat brain. Journal of neurochemistry. 1986;46(5):1333-7.

[149] Donato R. S-100 proteins. Cell calcium. 1986;7(3):123-45.

[150] Townend W, Dibble C, Abid K, Vail A, Sherwood R, Lecky F. Rapid elimination of protein S-100B from serum after minor head trauma. J Neurotrauma. 2006;23(2): 149-55.

[151] Zimmer DB, Cornwall EH, Landar A, Song W. The S100 protein family: history, function, and expression. Brain research bulletin. 1995;37(4):417-29.

[152] Butterfield DA, Reed T, Sultana R. Roles of 3-nitrotyrosine- and 4-hydroxynonenal-modified brain proteins in the progression and pathogenesis of Alzheimer's disease. Free radical research. 2011;45(1):59-72.

[153] Darwish RS, Amiridze N, Aarabi B. Nitrotyrosine as an oxidative stress marker: evidence for involvement in neurologic outcome in human traumatic brain injury. The Journal of trauma. 2007;63(2):439-42.

[154] Beckman JS, Ischiropoulos H, Zhu L, van der Woerd M, Smith C, Chen J, et al. Kinetics of superoxide dismutase- and iron-catalyzed nitration of phenolics by peroxynitrite. Archives of biochemistry and biophysics. 1992;298(2):438-45.

[155] Tyurin VA, Tyurina YY, Borisenko GG, Sokolova TV, Ritov VB, Quinn PJ, et al. Oxidative stress following traumatic brain injury in rats: quantitation of biomarkers and detection of free radical intermediates. Journal of neurochemistry. 2000;75(5):2178-89.

[156] Sultana R, Perluigi M, Butterfield DA. Oxidatively modified proteins in Alzheimer's disease (AD), mild cognitive impairment and animal models of AD: role of Abeta in pathogenesis. Acta neuropathologica. 2009;118(1):131-50.

[157] Poon HF, Vaishnav RA, Getchell TV, Getchell ML, Butterfield DA. Quantitative proteomics analysis of differential protein expression and oxidative modification of specific proteins in the brains of old mice. Neurobiology of aging. 2006;27(7):1010-9.

Chapter 4

Uncovering the Path to Neurodegeneration from Playingfield to Battlefield

Frank Baas and Valeria Ramaglia

Additional information is available at the end of the chapter

http://dx.doi.org/10.5772/57186

1. Introduction

It is becoming clear that a history of traumatic brain injury predisposes individuals to neurodegeneration later in life. Even mild recurrent brain concussions, often neglected, can have serious consequences. For example, people engaged in contact-sports such as American football, ice hockey or boxing, and also military personnel, suffer from recurrent brain concussions with no loss of consciousness and no need for hospitalization. However, these people face the possibility of long-term neurocognitive problems with increased risk of developing dementia, including Alzheimer's disease, Parkinson's disease and amyotrophic lateral sclerosis [1-7]. The mechanisms responsible for the post-traumatic neurodegenerative sequelae are not understood. Understanding the mechanisms, which render the traumatized brain susceptible to neurodegeneration, will guide the development of timed and targeted therapies to prevent the severe neurological decline seen in people at risk.

Recent hypothesis, based on post-mortem observations, attribute the mechanisms underlying the risk relationship between brain injury and neurodegeneration to disturbed metabolism, protein aggregation and axonal injury caused by the impact. Our hypothesis is that the increased susceptibility to neurodegeneration for people with a history of traumatic brain injury is caused by early injury-induced changes in the reactivity of highly specialized immune cells in the brain, called microglia. According to our hypothesis, microglia in the traumatized brain are sensitized to respond more vigorously to a subsequent injurious event. This phenomenon is named microglia priming [8]. Primed microglia are in a "ready-to-go" state which, if triggered by a challenge – for example a systemic infection, surgery or vaccination occurring even later in life –, may outburst in a severe pro-inflammatory event which results in prolonged delirium and cognitive deficit. The pathways that regulate microglia priming are largely unknown. Our group has recently demonstrated that activated complement, a part of the innate immune system, is one of the triggers to switch the microglial phenotype from resting to primed, sensitizing the brain for accelerated pathology. It is apparent that knowledge of the regulatory pathways that control microglia priming will help to identify new therapeutic targets and guide in the development of novel strategies to treat patients at risk.

In this chapter we review the clinical evidence supporting the increased risk of many athletes and military personnel to develop a neurodegenerative disease. We report the neuropathological changes observed in the post-mortem brain tissue of donors who died of chronic traumatic encephalopathy, considered today as the disease which best represents the neuropathology of the concussive brain. We review the inflammatory changes which occur in the post-mortem brain. We present the current hypothesis of tau pathology and our novel hypothesis of microglia priming as putative mechanisms underlying the increased susceptibility of the traumatized brain to early on and accelerated neurodegeneration, and we discuss how these two hypotheses may co-exist. We focus on the candidate pathways, which may regulate the transition from the resting to the primed state and those, which may regulate the transition from the primed to the active state.

Our effort over the past 10 years has been directed to study the role of the complement cascade in the nervous system [8-14]. To date we have collected evidence which support a role for complement in microglia priming [8] as well as the protective role of inhibitors of complement activation in nerve damage [10,13]. Therefore, in this chapter we focus on the complement system as a potential target for therapy to fight neurodegeneration. We propose our conclusions and future directions towards which research should invest to identify targets, design drugs and test therapies to interfere with the changes occurring in the brain after recurrent trauma and prevent the neurodegenerative sequelae of events which take the life of many athletes and military forces.

2. Clinical evidence

Over the past decades, clinical studies have shown that traumatic brain injury (TBI) is a risk factor for the development of neurodegenerative diseases, including Alzheimer's disease (AD), Parkinson's disease (PD) and amyotrophic lateral sclerosis (ALS), manifesting even decades after the trauma [15-19]. For example, a retrospective case-control study showed a link between head trauma with loss of consciousness and AD, with the strongest association in cases without a history of familial AD and in males [15]. Another study on 196 subjects, who developed PD, showed that the frequency of head trauma was significantly higher in cases than controls. Furthermore, the association was higher for men than women [20]. Also in studies performed on twins, the sibling with the onset of PD at a younger age was more likely to have sustained a head injury [16]. Another case-control study of 109 ALS cases and 255 matched controls found that multiple head injuries increased the risk of developing ALS by 11 fold [17].

Additional evidence suggests that also mild forms of TBI, including repetitive concussive injury without loss of consciousness, can trigger chronic neurological problems. The first description of traumatic chronic brain damage dates back to 1928 and was termed punch drunk syndrome, due to its association with boxing. In 1937, Millspaugh renamed this syndrome *dementia pugilistica* but only forty-five years later Corsellis et al [21] described the associated neuropathological changes in a case series of fifteen retired boxers. Their findings included cerebral atrophy, neurofibrillary tangles and hypopigmentation of the substantia nigra. More recent studies have shown that such neuropathological changes are not restricted to boxers but also occur in athletes who practiced other contact-sports such as American football, hockey, soccer and professional wrestling. In addition, the same neuropathological changes also occur in victims of physical abuse, in soldiers engaged in military operations and in circus acrobats repeatedly shot out of a cannon [1,6,22-25]. For example, a study in a case series of postmortem brains from U.S. military veterans exposed to blast and/or concussive injury, found evidence of neuropathology similar to that observed in young amateur American football players and a professional wrestler with histories of concussive injuries [26]. In addition, neuropathologically confirmed *dementia pugilistica* was found in American football players with asymptomatic concussions but who played positions, such as lineman, with the greatest exposure to repeated head blows [27], suggesting that also subconcussive trauma leads to chronic neurodegeneration [28].Therefore the neuropathological changes associated with *dementia pugilistica* are the general outcome for individuals with a history of repeated head trauma.

Over the last decade, the chronic brain damage caused by multiple concussive injuries has been renamed chronic traumatic encephalopathy (CTE). CTE is defined as a neurodegenerative disorder, which occurs years or decades after concussive head trauma. In the case of athletes, it often manifests when the sporting career has already ended [22]. The clinical symptoms of CTE vary from mild behavioral changes, such as apathy, to severe cognitive deficits accompanied by movement disorders (e.g. parkinsonism) and neuropsychiatric problems such as disinhibition, aggressiveness and hypomania, often culminating in violent behavior [29] and suicidal acts [22-24]. The heterogeneity of the symptoms is likely reflective of the brain regions affected. Because of the heterogeneity of clinical symptoms of CTE, its extensive overlap with other causes of dementia and the lack of definite guidelines for a clinical diagnosis of CTE, it has been difficult to distinguish between CTE, AD, frontotemporal lobar degeneration (FTLD) or aging, especially in advanced disease. Therefore, the prevalence of CTE in the demented population has been difficult to establish. Early studies suggest that the prevalence of CTE in professional boxers is about 20% [30]. More recent MRI studies showed brain abnormalities, including atrophy, dilated perivascular spaces and diffuse axonal injury, in 76% of professional boxers [31]. Another recent study in a sample of 513 retired players of the American National Football League (NFL) Association, indicated possible cognitive impairment in 35.1% of retirees [32]. A large study completed on 2552 retired players of the American NFL Association found that retired players with three or more reported concussions (34.4%) had a threefold prevalence of significant neuropsychiatric problems compared with retirees without a history of concussion. They also observed an earlier onset of AD in the retirees than in the general American male population [4]. Overall, these findings suggest a link between recurrent concussions and increased risk of dementia. However, definite guidelines for the clinical diagnosis of CTE and future prospective longitudinal studies on large populations would probably provide more accurate incidence and prevalence of CTE over the coming years.

3. Neuropathological changes

Fourty years ago, Corsellis *et al.* [21] performed the first series of autopsies in professional boxers with *dementia pugilistica* and reported neuropathological changes which were confirmed in 2005 by Omalu *et al.* in the first autopsy from a NFL player [6]. In 2008, the Center for the Study of Traumatic Encephlopathy (CSTE) at Boston University School of Medicine established the CSTE brain bank at the Bedford VA Hospital to collect post-mortem brain and spinal cords of athletes, military veterans and civilians who experienced repetitive concussive injury. In 2009, McKee *et al.* at the CSTE performed a retrospective study on the archival literature and the three new cases available at the brain bank at that time. They verified the neuropathological findings by Corsellis and Omalu in 49 additional cases of confirmed CTE [22]. Today the CSTE collected over 100 brains with histories of repetitive mild traumatic brain injury. Of these, over 60 cases have been fully analyzed and diagnosed with neuropathologically confirmed CTE [33]. Overall, the neuropathological findings are consistent and together make CTE a disorder distinctive from other forms of neurodegeneration.

3.1. Gross pathological changes

Macroscopic examination of the CTE brain shows generalized atrophy of the frontal and temporal cortices, median temporal lobe, diencephalon and mammillary bodies; thinning of the corpus callosum and atrophy of the cerebral subcortical white matter; pallor of the substantia nigra and locus coeruleus. Frequent findings also include cavum septum pellucidum with fenestrations; dilation of the lateral and third ventricles [22]. These latter changes are probably caused by the mechanical shearing forces of the traumatic impact, which produce a fluid wave through the ventricular system.

3.2. Microscopic neuropathological changes

Microscopically, CTE is characterized by abundant neurofibrillary inclusions in the form of neurofibrillary tangles (NFTs), neuropil threads (NTs) and glial tangles (GTs) [22,34-36]. Tangles are intracellular thread-like aggregates of hyperphosphorylated tau protein[37]. Tau is an axonal protein whose normal function is to promote microtubule assembly and stability by binding to microtubules via its microtubule-binding domains conserved in all of its six isoforms. Hyperphosphorylation of tau at its several threonine or serine phosphorylation sites makes tau prone to aggregation in the form of tangles, thereby reducing microtubule binding. This causes disassembly of the microtubules, ultimately resulting in impaired axonal transport and compromised neuronal function [38].

Like in CTE, also in AD and other tauopathies, tau is found in a hyperphosphorylated form and tangle aggregates [38]. In addition, the specific tau isoforms found in CTE are indistinguishable from those found in AD [35]. However, tau deposition in CTE is topographically and quantitatively distinct from AD. In CTE NFTs deposition is often found in greater densities compared to severe AD. The distribution of tau NFTs in CTE is irregular and affects primarily the superficial cortical layers with foci at the depths of the sulci, in the diencephalon, basal ganglia and brainstem, and surrounding blood vessels [5,22]. By contrast, NFTs in AD are preferentially distributed in the hippocampus and in large projection neurons in layers III and V of the cortex [39].

Deposition of β-amyloid (Aβ), a major feature of AD pathology, is only found in 40-45% of CTE cases [22]. Furthermore, Aβ deposits in CTE are generally found in the form of diffuse plaques rather than neuritic plaques typically present in the AD brain. Aβ is generated from amyloid precursor protein (APP) by the action of β- and γ- secretases [40]. APP is highly expressed by neurons and is thought to promote axonal sprouting, neurite outgrowth and synaptogenesis, critical for axonal survival after damage [41-44]. However, APP rapidly accumulates in neurons and axons after injury, especially in truncated axonal bulbs [45]. This may stimulate overproduction of Aβ and accumulation in the extracellular space in the form of diffuse plaques [46].

The TAR-DNA-binding protein 43 (TDP-43) is another protein, which aggregates in the neurodegenerative brain. It was initially considered to be a specific neuropathological feature of frontotemporal lobar degeneration with ubiquitin-positive inclusions (FTLD-U) and ALS [47] but it is now apparent that TDP-43 accumulation occurs in several other neurodegenerative diseases including AD [48], dementia with Lewy bodies [49] and CTE [18]. In CTE, TDP-43 immunoreactivity is commonly seen in the frontal and medial temporal cortices, brainstem, diencephalon, substantia nigra pars compacta, amygdala, hippocampus, caudate, putamen, thalamus and hypothalamus [1,18].

Accumulation of α-synuclein, the main protein component of the Lewy body aggregates seen in PD, has not been reported in CTE [22].

Studies in man and animal models have also shown that the complement system is a key component of the neuropathology of severe TBI [50]. Although studies to verify complement activation in repetitive mild TBI or in CTE are lacking, the proved neuropathological role of complement in severe TBI [50] , AD [51,52], prion disease [53], ALS [14,54] and traumatic peripheral nerve injury [9-13], makes complement a likely player also in the neuropathology of concussive injuries and CTE. Complement is a key component of innate immunity. Its activation induces chemotaxis of inflammatory cells, facilitates solubilisation and clearance of opsonised immune complexes, mediates cell lysis, and influences adaptive immunity [55,56]. Activation occurs via three distinct routes: the classical, the lectin and the alternative pathways. The classical pathway is activated by the binding of C1q to antigen-antibody complexes or directly to "danger" epitopes. The lectin pathway is triggered by binding of mannose binding lectin to certain carbohydrates on the pathogen surface, whereas the alternative pathway starts by spontaneous low-rate hydrolysis of C3 which binds to activated factor B on a surface lacking complement inhibitors. Irrespective of the initial recognition pathway, all three routes converge in the cleavage of C3 and downstream in the assembly of the membrane attack complex (MAC) a pore through the cell membrane that can result in lysis of the target cell [55, 56]. Because of the tendency of complement to drift from its target site of activation to adjacent areas, healthy tissue is equipped with a battery of regulators which keep complement activation in check [55,56].

Deposits of the MAC component C9 have been found on injured neurons after TBI [50] and high levels of fB, C3 and sC5b-9 (soluble MAC) have been detected in the cerebrospinal fluid (CSF) of patients with severe TBI during the first 10 days after trauma [57]. Mice deficient in the C3 or C5 components subjected to traumatic brain cryoinjury showed reduced neutrophils and secondary tissue damage compared to their wildtype littermates [58]. TBI on $fB^{-/-}$ transgenic mice or mice treated with a monoclonal anti-fB antibody showed reduced post-traumatic neuronal cell death, a strong upregulation of the anti-apoptotic mediator Bcl-2 and downregulation of the pro-apoptotic Fas receptor compared to the $fB^{+/+}$ littermates, implicating the alternative complement pathway in the progression of the secondary neuronal death [59,60]. Neurological function after TBI was improved in transgenic mice with brain-targeted overexpression of complement receptor 1-related protein y (Crry), a potent inhibitor of the C3 convertase [61] or in mice treated with a recombinant Crry molecule (Crry-Ig) in which Crry is fused to the non-complement fixing mouse IgG1 Fc region [62]. Further, mice treated with C1 inhibitor, a serine protease inhibitor of the classical and lectin complement pathways, showed reduced neurobehavioural deficits and contusion volume after controlled cortical impact brain injury [63]. Involvement of the terminal complement pathway in experimental TBI has been proven because mice lacking the sole membrane bound regulator of the MAC, CD59a, display increased neuronal cell death and brain tissue destruction compared to CD59a $^{+/+}$ littermates [64]. Our group has initially shown that MAC is a key component of axonal injury, driving Wallerian degeneration after peripheral nerve trauma, and we have also recently proven that exogenous blockers of MAC assembly are neuroprotective in an experimental mouse model of severe closed head injury (unpublished observations).

4. The tau hypothesis

Neuropathological evidence suggest that CTE begins focally at the cortical sulci and perivascularly, spreading slowly over decades to involve the cortex, medial temporal lobe, diencephalon, basal ganglia, brainstem and spinal cord [33]. A recent post-mortem study on 85 donors of the CSTE brain bank classified CTE into four stages based on topographically predictable pattern of tau pathology. Stage I is characterized by perivascular NFTs, and is clinically associated with attention and concentration disturbances. Stage II is characterized by NFTs in superficial cortical layers, in the nucleus basalis of Meynert and locus coeruleus, and is clinically associated with depression, mood swings and short-term memory loss. Stage III is characterized by cerebral atrophy, enlargement of the ventricles and pallor of the substantia nigra. In addition, tau pathology is widespread in the cortex, diencephalon, brainstem and spinal cord. Stage III is clinically associated with memory impairment, executive dysfunction (e.g. problems with planning, organizing, multi-tasking, judgment), depression and irritability. Stage IV is characterized by further brain atrophy, ventricular dilation, pallor of the substantia nigra and locus coeruleus. Tau pathology is widespread also to white matter regions and is accompanied by neuronal loss, gliosis of the cerebral cortex and hippocampal sclerosis.

Clinically, stage IV presents with severe cognitive problems, memory loss, executive dysfunction, depression, irritability and increased violence [33].

According to the current hypothesis on the initiation and propagation of CTE pathology, the initial brain changes are caused by acceleration and deceleration forces provoked by the traumatic injury which stretch and disrupt neuronal and axonal membranes [65]. These acute cellular changes trigger a cascade of neurochemical dysfunctions, including a deregulated influx of calcium ions and efflux of potassium ions, which depolarize the axonal cell membrane. Depolarization in turn results in the release of excitatory neurotransmitters, including glutamate, which binds to its receptors to fire more action potentials, mediating the influx of more calcium ions. These events trigger an excitotoxic cascade, which depletes energy stores, impairs oxidative metabolism and results in the release of calpains, calcium-dependent proteases, which cause further tissue injury [66,67]. Mechanical shearing forces can also cause disruption of axons and dissolution of microtubules, impairing axonal transport and causing initial swelling followed by axotomy and in the end by Wallerian degeneration [66,67]. These changes are typically located in the cortical sulci, which is where tau pathology begins.

The initial trigger of tau phosphorylation, misfolding, truncation and aggregation into NFTs may be induced by caspases and calpains, released by the stretching of axons at the moment of the traumatic impact. Dissolution of microtubules, as a result of axonal injury, will also release tau, which is then phosphorylated and polymerizes into toxic filaments. The toxic gain of function of tau protein is supported by studies in which human tau overexpressed in lamprey central neurons becomes phosphorylated, accumulate in filamentous deposits and causes microtubule and synapse loss [68,69].

A recent hypothesis suggests that the predictable spreading of tau pathology from stage I to stage IV may be due to transfer of toxic tau species between neurons [70-73]. This can occur via a prion-like propagation of misprocessed tau (reviewed in [74]) and/or via the internalization of small misfolded tau species by bulk endocytosis at the somatodendritic compartment or at the axon terminals. Once exogenous tau is taken up by the neuron, it can be transported anterogradely and retrogradely, accumulating and enhancing tauopathy [75]. In addition to the trans-synaptical transfer of tau, paravenous drainage pathways may also regulate the extracellular levels of tau, Aβ and TDP-43. For example, a recent study using two-photon imaging of small fluorescent tracers showed that the CSF enters the brain parenchyma along paravascular spaces that surround penetrating arteries and that the brain interstitial fluid is cleared along these paravenous drainage pathways. In addition, fluorescent-tagged Aβ was transported along this route whereas impairment of the paravascular pathway suppressed clearance of soluble Aβ [76,77]. These findings suggest that impairment of the paravenous flow by the trauma may contribute to the accumulation of soluble proteins in neurodegenerative diseases. For example, APP may be inefficiently cleared by paravascular pathways and aberrantly cleaved to form Aβ. To date, the role of Aβ in the pathogenesis of CTE is unclear. On the other hand, there is evidence that accumulation of TDP-43 in CTE may be neurotoxic. Recent *in vivo* and *in vitro* studies suggested that overexpression of human TDP-43 and its relocation from the neuron nuclear compartment to the cytoplasm are associated with neurodegeneration and cell death [78-80].

5. Inflammatory changes

In brain donors and experimental animal models, TBI has been associated with increased reactivity of microglia [81,82]. Microglia are the resident macrophages of the brain. Similar to their peripheral equivalent, they are able of phagocytosis, antigen presentation and secretion of inflammatory mediators [83]. They continuously survey the local environment to detect and alert neighbouring cells of non-physiological changes but, despite their very dynamic phenotype, microglia are constantly suppressed in the normal brain [84]. In rodents, primates and humans, microglial hyper-reactivity persists for months to years after TBI, indicative of chronic neuroinflammation [85-89]. This phenotype is remarkably similar to the chronic microglial reactive state of the AD [90] or ageing brain [91].

6. The microglia priming hypothesis

The hypothesis of microglia priming proposes that in the ageing or neurodegenerating brain, microglial cells are chronically sensitized to respond more vigorously to a subsequent injurious event (reviewed in [92,93]). This is profoundly discordant with the behavior of peripheral macrophages, despite their common monocytic lineage origin with microglia. In fact, in the periphery, an initial challenge (e.g. lipopolysaccharide, LPS) conditions macrophages to produce a suppressed response upon restimulation with a second challenge [94].

The notion of microglia priming was initially proposed by Perry in 2007 [95] and later by van Gool [96] based on clinical observations, which reported a deleterious effect of systemic infections on the clinical outcome of elderly people or individuals with neurodegenerative diseases. For example, when elderly patients become delirious during infections, treatment of the infection may go well but the patients emerge with dementia. Similar observations have been made after postoperative delirium in elderly hip fracture patients free from preexisting dementia [97,98]. Other studies have shown that new episodes of cognitive decline in AD cases were concomitant to systemic inflammatory events [99,100]. Systemic infections have also been linked to worse outcome in multiple sclerosis patients [101]. Likewise, lung inflammation is associated with worse neurological outcome in patients with severe brain trauma [102]. According to the priming model, microglia are the cells responsible for the increased susceptibility of the brain to peripheral events.

In prion disease models, microglial priming is evident even in the preclinical stage and LPS challenge exacerbates neuronal death, induces acute cognitive impairment and accelerates disease progression [103-105]. In AD models, repeated LPS challenges exacerbate tau pathology [106], amyloid deposition [107] and inflammation [108]. In the superoxide dismutase (SOD) mouse model of familial ALS, chronic LPS administration exaggerates motor neuron degeneration, precipitates disease progression and elevates production of pro-inflammatory cytokines [109]. In animal models of mild TBI, sepsis worsen post-traumatic mortality and weight loss, motor deficit and cognitive impairments, and further exacerbates neuronal cell death concomitant with over-activation of microglial cells in the peri-lesional area [110]. These studies all suggest that microglia priming places subjects at risk for exacerbated neuropathology from an early stage of disease.

Despite the likelihood that microglia priming is an important event in neurodegeneration, its triggers are just starting to be uncovered.

6.1. Transition from resting to primed state

In the normal brain, microglia are dynamic but suppressed by the concerted action of regulatory proteins [84]. We propose that in the aged or traumatized or neurodegenerating brain, microglia become primed due to the abrogation of this suppression or due to the stimulation by pathological proteins.

Suppressing mechanisms include the interaction between neuronally derived proteins such as the glycoprotein CD200 or fractalkine, and their receptors on microglial cells. For example, deletion of CD200 in mice, results in exacerbated microglial activation in several models of inflammation [111,112] whereas enhancing CD200 signaling, by intracerebral injection of the CD200 fusion protein, rescues the overactivated microglial phenotype in aged mice challenged with LPS [113]. *In vitro*, treatment of fractalkine ligand, CX_3CL, to LPS-stimulated microglia attenuates production of inflammatory mediators including IL-1β, IL-6, TNF-α and inducible nitric oxide synthase (iNOS) [114-116]. *In vivo*, fractalkine receptor deficient mice challenged with LPS show amplified microglial expression of IL-1β and prolonged depressive-like behavior compared to wildtype or heterozygote mice [117] whereas inhibition of IL-1β associated inflammatory enzymes blocked the depressive behavior and restored microglia homeostasis [118]. More recent studies proposed that the brain specific microRNA-124 (miR-124) is a potential suppressor of microglial function [119]. Overexpression of miR-124 in bone marrow-derived macrophages resulted in lower levels of TNF-α and iNOS, and a higher level of the anti-inflammatory cytokine TGF-β1, reflecting a quiescent phenotype. On the other hand, knock down of miR-124 attenuated the downregulation of MHC class II and CD45 in microglia and macrophages, again pointing towards a role for miR-124 in maintaining the quiescent phenotype. Ponomarev *et. al.* [119] proposed that the inhibiting activity of miR-124 on microglial function is mediated via C/EBP-α and its downstream effector PU.1. Both transcription factors are involved in differentiation of myeloid and monocytic lineage cells and are downregulated in miR-124 transfected cells.

Microglia priming may also be induced by stimulation with pathological proteins, including protein aggregates such as Aβ and NFTs formed after TBI. For example, *in vitro* studies have shown that exposure to Aβ primes microglia to produce an exaggerated respiratory burst during phagocytosis [120]. In addition, our group has recently shown that deposition of activated complement proteins, known to opsonize damaged axons after trauma [9,11], can also regulate the microglial phenotype [8,121]. In mice, deletion of the major regulator of the C3 convertase, Crry, induces deposition of active C3 in tissue and induces microglia priming, likely by binding to complement receptor 3 (CD11b/CD14) on microglia. Crry deficient mice – with primed microglia – showed exaggerated microglia expression of pro-inflammatory molecules following LPS challenge. Furthermore, we identified evidence of microglia priming in the normal appearing white matter of brain donors with progressive multiple sclerosis and showed that, in an experimental model of multiple sclerosis, microglia priming is responsible for earlier disease onset and more severe clinical course compared to unprimed mice [8,121].

After recurrent mild TBI, deposition of active complement proteins, including C3, may occur in response to diffuse axonal injury, provoked by accelerating and decelerating forces. Disrupted axonal membranes expose epitopes, which are recognized as "danger" and are targeted by complement proteins. C3 activation products may then bind to CR3 on microglia and induce priming. Other sensors of endogenous tissue damage are toll-like receptors (TLRs) which, in concert with complement proteins [122], recognize as "danger" conserved molecular patterns, whether it is exogenous pathogens or endogenous tissue damage that are detected [123]. TLR2 and 4 are expressed by microglia [124,125] and have been shown to drive inflammation in response to the alarmin high mobility group box-1 protein (HMGB1) [126], a danger-associated molecular pattern (DAMP) molecule which is released by injured cells in a C3-dependent manner [127] within 30 minutes to 6 hours after severe trauma [128]. Although the TLR signaling is predicted to initiate an acute pro-inflammatory response [129], ligation of TLRs may be involved in priming of microglia or in mild (recurrent) trauma.

6.2. Transition from primed to active state

Exposure of primed microglia to an injurious stimulus – i.e. infections, surgery or vaccination – can lead to microglia overactivation, which trigger or accelerate dementia.

Peripheral infections are a widely recognized trigger of cognitive decline and delirium in elderly and AD patients. For example, AD patients who experienced systemic infections, showed concomitant precipitating cognitive decline. The periods of infection were associated with elevated levels of TNF-α. Notably, subjects with low TNF-α levels did not show cognitive decline over the 6-month investigated period [100]. Another study showed that cognitive impairment in AD patients was concomitant to a systemic inflammatory event and was preceded by raised levels of IL-1β [99]. Surgery can also activate the peripheral immune system and trigger an exaggerated inflammatory response in the brain [130]. Postoperative cognitive dysfunction (POCD) is the decline of cognitive performance observed after surgery, with possible long-term effects, such as changes in personality and social integration [131]. POCD has an estimated prevalence of 15-25% in patients over the age of 60, with age as main risk factor [130]. In experimental surgery, aged mice produce significantly higher levels of IL-1β compared to wildtype mice, supporting a role for surgery as a trigger for activation on a brain primed by aging [130]. Vaccination may also be a potential trigger for the transition from primed to active microglial state [132]. This would be especially relevant in view of research into developing vaccinations for AD [133-135]. Drug addictions, such as those reported for some cases of CTE who died from drug overdose, can also regulate the activation of microglia. For example in mice, microglial activation is an early step in methamphetamine-induced neurotoxicity[136]. The use of steroids could also regulate microglial function, but to our knowledge there are no studies which have tested this hypothesis.

An essential question is how the communication between the periphery and the brain occurs. It is well known that peripheral inflammation can activate CNS centers by a number of routes, including the circumventricular organs, vagal afferents, and the brain endothelium (see [137] for review). Blood-borne cytokines are thought to diffuse into the brain through fenestrated capillaries and stimulate parenchymal astrocytes to release secondary mediators within the brain such nitric oxide and prostaglandins [138]. Circulating cytokines could also induce the production of pro-inflammatory cytokine by macrophage-like cells in the circumventricular organs and choroid plexus, acting directly or indirectly on neurons which project to the brain parenchyma [139]. Another route of transmission of the peripheral immune message to the brain is via neural afferent pathways. This has been functionally demonstrated by vagal nerve resection experiments that abrogated LPS-induced inflammation and sickness behavior in rodents [140,141]. TLRs, expressed by many immune cells and also by endothelial cells [142], are the obvious responders to systemic infections especially to LPS via the TLR4 and its co-receptors MD2 and CD14 [143]. Therefore they are likely involved in mediating inflammation from the periphery to the brain. For a comprehensive review on TLR signaling see [144].

Overactivation of microglia can ultimately precipitate neuropathology and the clinical outcome of elderly or patients with ongoing neurodegeneration [99,100,145-147]. For example, microglial pro-inflammatory cytokines can stimulate γ-secretase activity and enhance APP levels and amyloidogenic APP processing, potentially exacerbating Aβ pathology [146,147]. Another example comes from studies on the triple transgenic mouse model of AD in which overexpression of IL-1β increases tau phosphorylation [145]. The hypothesis of microglia priming in recurrent mild brain trauma is shown in Figure 1.

7. Therapeutical targets

Currently there is no treatment or therapy to stop or reverse the neurodegenerative sequelae triggered by repetitive concussions or CTE. It is apparent that knowledge of the regulatory pathways that control tau or Aβ pathology, as well as the mechanisms that regulate microglia priming will help to identify new therapeutic targets and guide in the development of novel strategies to treat patients at risk. Therapies could be timed and targeted to intervene early after brain trauma to prevent accumulation of protein aggregates and priming of microglia, for example by administering therapies as prophylactic cover before a sporting match or a military operation. Alternatively, therapies could be targeted to reduce the impact of peripheral insults on primed cells, by monitoring traumatized patients for years post-trauma and administering therapies as prompt interventions for example during systemic infections.

Since the initiating events of CTE include accumulation of hyperphosphorylated tau and in some cases of Aβ, targeting the formation or processing of these protein aggregates would seem to be a promising strategy for treatment. However, AD research is still controversial on the pathological role of tau and APP/Aβ metabolism. Specifically, it is still unclear whether it triggers chronic neurodegeneration or it is simply the epiphenomenon of underlying pathogenic responses to neuronal damage. Over the past decade, our group has collected evidence which support a protective role for complement inhibitors in axonal damage [8-10,13]. We demonstrated that both genetic and pharmacological inhibition of MAC formation protects the peripheral nerve from early axon loss after traumatic injury [9,10,148], and stimulates post-

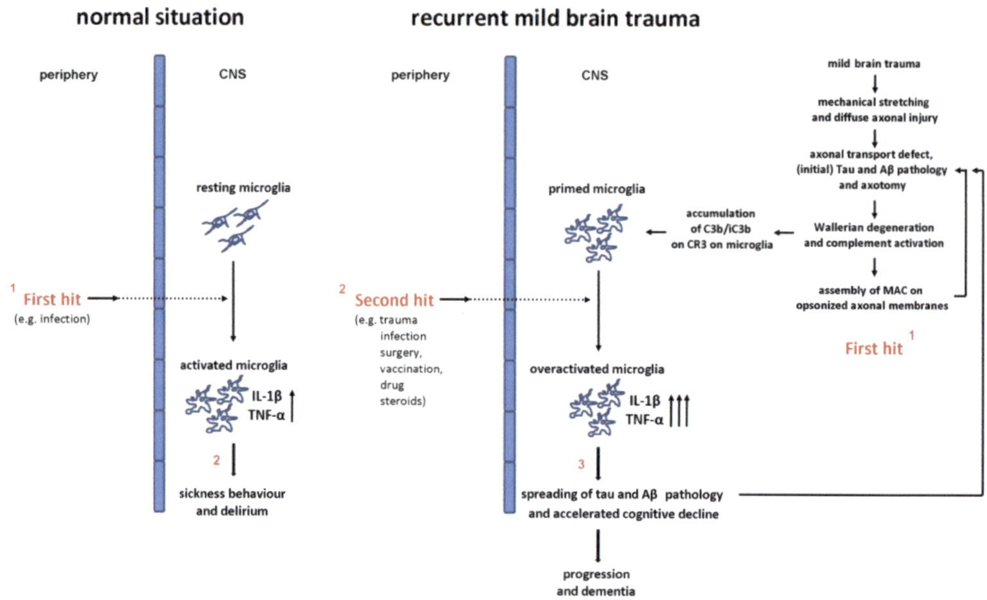

Figure 1. Hypothesis of microglia priming in recurrent mild brain trauma. In the normal situation, a peripheral event, such as an infection, is the first hit to the brain, activating resting microglia to produce inflammatory mediators affecting neuronal function and producing sickness behavior or delirium. In the context of recurrent mild brain injury, mild TBI provokes mechanical accelerating and decelerating forces, which cause axonal stretching and damage fragile axons, producing diffuse axonal injury. Diffuse axonal injury is characterized by disrupted axonal transport and microtubule disassembly, which culminate in the accumulation of tau and APP deposits, forming in some cases Aβ plaques around damaged axons. Impaired axonal transport also results in the formation of axonal swelling and axotomy, triggering Wallerian degeneration. Degenerating axons activate complement, which accumulates on targeted damaged membranes. MAC amplifies tissue injury whereas activated C3 products, C3b/iC3b, bind to their receptor CR3 on microglia. These injury-provoked cellular and neurochemical changes are the first hit which initiates diffuse axonal pathology and triggers priming of microglia for over-activation in response to a second hit. Under these circumstances, a peripheral injurious event, such a subsequent brain trauma, an infection, surgery, vaccination, use of recreational drugs or steroids, is the second hit to the brain that over-activates primed microglia to produce an elevated amount of inflammatory mediators. Pro-inflammatory molecules facilitate spreading of tau and Aβ pathology, driving neuropathology, accelerating cognitive decline and ultimately determining dementia.

traumatic axonal regeneration and functional recovery [149]. In addition, blockade of C3 activation in a mouse model of microglia priming reversed priming and suppressed experimental autoimmune encephalitis (EAE)-induced inflammation [8]. Studies from other groups have also shown that inhibition of C3 activation via the alternative pathway is protective in the experimental mouse model of closed head injury [60,62]. The complement system plays a vital role because it is the first line of defence against pathogens; it generates chemoattractants for inflammatory cells, facilitates solubilisation and clearance of opsonised immune complexes, mediates cell lysis, and influences adaptive immunity [150,151]. Therefore, the stage at which the complement system should be inhibited to block its detrimental effects but maintain its protective functions, is of considerable importance. The problem of systemic inhibition of complement might be overcome by targeting the agent to a specific site or choosing an agent, which allows certain physiological functions of complement while inhibiting others. For example, specific blockers of the CR3 receptors on microglia could be useful in modulating microglia priming whereas targeting the terminal complement pathway could be an attractive proposition because it maintains function of the initial steps of the complement cascade, including opsonisation, but it blocks the detrimental effect of the MAC on neuronal membranes. The therapeutic effects of complement inhibitors remain to be tested in experimental models of CTE but it is of high relevance since the recent advent of complement drugs into the clinic makes complement therapy a real option for patients.

Ultimately, prevention is the logical option. To this end, the World Medical Association (WMA) has recommended guidelines to make sport safer. For example, changing the boxing rules to reduce the number of fighting rounds and to introduce the mandatory use of thickly-padded head gear and gloves to diminish the impact from a punch may lower the risk of CTE in boxers (WMA, 2005). However, in sports where repeated blows to the head are unavoidable, appropriate assessment and management of the concussive injury may be critical for preventing long-term consequences. Therefore, a number of recommendations were developed following the 1st (Vienna 2001), 2nd (Prague 2004), 3rd (Zurich 2008) and 4th (Zurich 2012) international consensus conference on concussion in sport [152]. These include removal of an injured athlete from the field and strict monitoring over the first few hours following concussion, whereas returning to the field should follow a stepwise protocol from no activity to light aerobic exercise to full practice and playing.

8. Conclusions and future directions

Although public awareness and media attention on the long-term effects of repetitive brain trauma - especially in professional athletes - have increased in recent years, knowledge of the neurobiology and pathogenesis of this condition is still limited and treatment is not available. However, as the number of professional athletes increases and sports get more competitive, the expectation is that the number of sports-related CTE cases may increase in the future. This prediction together with the potential of CTE to impact a broad population from athletes to veterans to victims of abuse, make CTE an important public health issue. Therefore research should invest in this area. It is clear that accurate clinical diagnostic criteria for CTE alone or mixed disease and prospective longitudinal studies, terminating in autopsy, would be essential to identify patients, develop biomarkers, further our knowledge on the temporal evolution and mechanisms of the disease. It would be also important to determine whether severity of the trauma, number of exposures, age at first exposure, gender, age and race may play a role in the development of CTE.

Functional magnetic resonance imaging (fMRI) and diffusion tensor imaging (voxel-based morphometry) seem a promising tool to monitor brain damage [153]. For example, studies carried out at baseline and repeated immediately after concussion and later until symptom resolution, have shown that fMRI is sensitive enough to detect abnormal activation patterns in varsity hockey and football athletes who have suffered a concussion. Therefore, fMRI could provide an objective way to measure the severity of a concussion and subsequent recovery [153].

In the future, it would also be important to learn from advances made in other fields of neurodegeneration research, based on the consideration that CTE and other neurodegenerative disorders share many neuropathological and neurocognitive traits. For example, CTE patients could be screened for genes associated with AD, or FTLD or ALS. Epidemiological data have already implicated the apolipoprotein E epsilon 4 (APOE ε4) allele, important genetic risk factor for AD, in the development of AD after TBI [154,155] and carriers of the APOE ε4 allele were found to be at increased risk of Aβ accumulation following brain injury [156]. However, large epidemiological studies on the influence of the APOE ε4 allele on the risk of CTE are still missing. Other candidate genes which may turn out to be risk factors for CTE, and therefore predict poor outcome after trauma, may include TARDBP, encoding for TDP-43 involved in FTLD and ALS [157]; the GRN gene, encoding granulin and associated with FTLD [158]; the MAPT gene, encoding for tau and mutated in some cases of FTLD [158].

In addition, the future development of experimental animal models which best mimic the neuropathology seen in man after repeated mild brain injury or CTE, would be crucial to identify the mechanisms of trauma-induced neurodegeneration and test neuroprotective therapies. Currently available animal models of CTE involve the triple transgenic AD mouse because it develops hyperphosphorylated tau and Aβ plaques. However it lacks the widespread neuronal loss which is observed in the human disease [159,160], making experimental observation difficult to translate into the clinic. A recently published transgenic rat model of AD, line TgF344-AD, may represent a better model to study CTE in rodents [161]. This transgenic rat expresses mutant human APP and presenilin-1, each independently associated with early onset familial AD. This AD model expresses all hallmarks of human AD, including Aβ plaques, hyperphosphorylated tau, gliosis and loss of cortical and hippocampal neurons, including age-dependent cognitive deficit [161]. Therefore, these rats may be a good model to monitor Aβ and tau pathology as well as neuronal loss and cognitive impairment after repetitive mild TBI, and also test therapies to stop or reverse post-traumatic neurological deficit. Insights into the mechanisms and treatment options for repeated mild brain injury or CTE, will assist policy makers (e.g. sports league officials or military commanders) to draw accurate guidelines for the prevention and the treatment of brain trauma whether in the playingfield or in the battlefield.

Abbreviations

Aβ β-amyloid

AD Alzheimer's disease

ALS amyotrophic lateral sclerosis

APOEε4 Apolipoprotein E epsilon 4

APP amyloid precursor protein

C1q complement 1q

C3 complement 3

C5 complement 5

C9 complement 9

CR3 complement receptor 3

CNS central nervous system

Crry complement receptor 1-related protein y

CSF cerebrospinal fluids

CSTE center for the study of traumatic encephalopathy

CTE chronic traumatic encephalopathy

EAE experimental autoimmune encephalomyelitis

fB complement factor B

fMRI functional magnetic resonance imaging

FTLD frontotemporal lobar dementia

FTLD frontotemporal lobar dementia with ubiquitin-positive inclusions

GTs glial tangles

HMGB1 high mobility group box-1

IL-1β interleukin 1 beta

IL-6 interleukin 6

iNOS inducible nitric oxide synthase

LPS lipopolysaccharide

MAC membrane attack complex

miR-124 microRNA-124

NFTs neurofibrillary tangles

NFL national football league

NTs neuropil threads

PD Parkinson's disease

POCD postoperative cognitive dysfunction

sC5b-9 soluble complement 5b-9, MAC

SOD superoxide dismutase

TBI traumatic brain injury

TDP-43 transactivation responsive region deoxyribonucleic acid-binding protein 43

TGF-β1 tumor growth factor beta 1

TLRs toll-like receptors

TNF-α tumor necrosis factor alpha

WMA world medical association

Acknowledgements

Our work on the closed head injury model of TBI was supported by the Hersenstichting Nederlands Fellowship F2010(1)-05 to V.R.

Author details

Frank Baas and Valeria Ramaglia*

Department of Genome Analysis, Academic Medical Center, University of Amsterdam, The Netherlands

References

[1] Stern RA, Riley DO, Daneshvar DH, Nowinski CJ, Cantu RC, McKee AC. Long-term consequences of repetitive brain trauma: chronic traumatic encephalopathy. PM R 2011;3(10 Suppl 2):S460-S467.

[2] Hesdorffer DC, Rauch SL, Tamminga CA. Long-term psychiatric outcomes following traumatic brain injury: a review of the literature. J Head Trauma Rehabil 2009;24(6): 452-9.

[3] Areza-Fegyveres R, Rosemberg S, Castro RM, Porto CS, Bahia VS, Caramelli P, et al. Dementia pugilistica with clinical features of Alzheimer's disease. Arq Neuropsiquiatr 2007;65(3B):830-3.

[4] Guskiewicz KM, Marshall SW, Bailes J, McCrea M, Cantu RC, Randolph C, et al. Association between recurrent concussion and late-life cognitive impairment in retired professional football players. Neurosurgery 2005;57(4):719-26.

[5] Gavett BE, Stern RA, McKee AC. Chronic traumatic encephalopathy: a potential late effect of sport-related concussive and subconcussive head trauma. Clin Sports Med 2011;30(1):179-88, xi.

[6] Omalu BI, DeKosky ST, Minster RL, Kamboh MI, Hamilton RL, Wecht CH. Chronic traumatic encephalopathy in a National Football League player. Neurosurgery 2005;57(1):128-34.

[7] Baugh CM, Stamm JM, Riley DO, Gavett BE, Shenton ME, Lin A, et al. Chronic traumatic encephalopathy: neurodegeneration following repetitive concussive and subconcussive brain trauma. Brain Imaging Behav 2012;6(2):244-54.

[8] Ramaglia V, Hughes TR, Donev RM, Ruseva MM, Wu X, Huitinga I, et al. C3-dependent mechanism of microglial priming relevant to multiple sclerosis. Proc Natl Acad Sci U S A 2012;

[9] Ramaglia V, King RH, Nourallah M, Wolterman R, de Jonge R, Ramkema M, et al. The membrane attack complex of the complement system is essential for rapid Wallerian degeneration. J Neurosci 2007;27(29):7663-72.

[10] Ramaglia V, Wolterman R, de Kok M, Vigar MA, Wagenaar-Bos I, King RH, et al. Soluble complement receptor 1 protects the peripheral nerve from early axon loss after injury. Am J Pathol 2008;172(4):1043-52.

[11] Ramaglia V, Daha MR, Baas F. The complement system in the peripheral nerve: Friend or foe? Mol Immunol 2008;

[12] Ramaglia V, King RH, Morgan BP, Baas F. Deficiency of the complement regulator CD59a exacerbates Wallerian degeneration. Mol Immunol 2009;46(8-9):1892-6.

[13] Ramaglia V, Tannemaat MR, de KM, Wolterman R, Vigar MA, King RH, et al. Complement inhibition accelerates regeneration in a model of peripheral nerve injury. Mol Immunol 2009;47(2-3):302-9.

[14] Heurich B, El Idrissi NB, Donev RM, Petri S, Claus P, Neal J, et al. Complement upregulation and activation on motor neurons and neuromuscular junction in the SOD1 G93A mouse model of familial amyotrophic lateral sclerosis. J Neuroimmunol 2011;

[15] Mortimer JA, van Duijn CM, Chandra V, Fratiglioni L, Graves AB, Heyman A, et al. Head trauma as a risk factor for Alzheimer's disease: a collaborative re-analysis of case-control studies. EURODEM Risk Factors Research Group. Int J Epidemiol 1991;20 Suppl 2:S28-S35.

[16] Goldman SM, Tanner CM, Oakes D, Bhudhikanok GS, Gupta A, Langston JW. Head injury and Parkinson's disease risk in twins. Ann Neurol 2006;60(1):65-72.

[17] Chen H, Richard M, Sandler DP, Umbach DM, Kamel F. Head injury and amyotrophic lateral sclerosis. Am J Epidemiol 2007;166(7):810-6.

[18] McKee AC, Gavett BE, Stern RA, Nowinski CJ, Cantu RC, Kowall NW, et al. TDP-43 proteinopathy and motor neuron disease in chronic traumatic encephalopathy. J Neuropathol Exp Neurol 2010;69(9):918-29.

[19] Schmidt S, Kwee LC, Allen KD, Oddone EZ. Association of ALS with head injury, cigarette smoking and APOE genotypes. J Neurol Sci 2010;291(1-2):22-9.

[20] Bower JH, Maraganore DM, Peterson BJ, McDonnell SK, Ahlskog JE, Rocca WA. Head trauma preceding PD: a case-control study. Neurology 2003;60(10):1610-5.

[21] Corsellis JA, Bruton CJ, Freeman-Browne D. The aftermath of boxing. Psychol Med 1973;3(3):270-303.

[22] McKee AC, Cantu RC, Nowinski CJ, Hedley-Whyte ET, Gavett BE, Budson AE, et al. Chronic traumatic encephalopathy in athletes: progressive tauopathy after repetitive head injury. J Neuropathol Exp Neurol 2009;68(7):709-35.

[23] Omalu BI, Bailes J, Hammers JL, Fitzsimmons RP. Chronic traumatic encephalopathy, suicides and parasuicides in professional American athletes: the role of the forensic pathologist. Am J Forensic Med Pathol 2010;31(2):130-2.

[24] Omalu B, Hammers JL, Bailes J, Hamilton RL, Kamboh MI, Webster G, et al. Chronic traumatic encephalopathy in an Iraqi war veteran with posttraumatic stress disorder who committed suicide. Neurosurg Focus 2011;31(5):E3.

[25] Hof PR, Knabe R, Bovier P, Bouras C. Neuropathological observations in a case of autism presenting with self-injury behavior. Acta Neuropathol 1991;82(4):321-6.

[26] Goldstein LE, Fisher AM, Tagge CA, Zhang XL, Velisek L, Sullivan JA, et al. Chronic traumatic encephalopathy in blast-exposed military veterans and a blast neurotrauma mouse model. Sci Transl Med 2012;4(134):134ra60.

[27] Greenwald RM, Gwin JT, Chu JJ, Crisco JJ. Head impact severity measures for evaluating mild traumatic brain injury risk exposure. Neurosurgery 2008;62(4):789-98.

[28] Gavett BE, Stern RA, McKee AC. Chronic traumatic encephalopathy: a potential late effect of sport-related concussive and subconcussive head trauma. Clin Sports Med 2011;30(1):179-88, xi.

[29] Volavka J. Violent crime, epilepsy, and traumatic brain injury. PLoS Med 2011;8(12):e1001148.

[30] Jordan BD, Relkin NR, Ravdin LD, Jacobs AR, Bennett A, Gandy S. Apolipoprotein E epsilon4 associated with chronic traumatic brain injury in boxing. JAMA 1997;278(2):136-40.

[31] Orrison WW, Hanson EH, Alamo T, Watson D, Sharma M, Perkins TG, et al. Traumatic brain injury: a review and high-field MRI findings in 100 unarmed combatants using a literature-based checklist approach. J Neurotrauma 2009;26(5):689-701.

[32] Randolph C, Karantzoulis S, Guskiewicz K. Prevalence and Characterization of Mild Cognitive Impairment in Retired National Football League Players. J Int Neuropsychol Soc 2013;1-8.

[33] McKee AC, Stein TD, Nowinski CJ, Stern RA, Daneshvar DH, Alvarez VE, et al. The spectrum of disease in chronic traumatic encephalopathy. Brain 2013;136(Pt 1):43-64.

[34] Dale GE, Leigh PN, Luthert P, Anderton BH, Roberts GW. Neurofibrillary tangles in dementia pugilistica are ubiquitinated. J Neurol Neurosurg Psychiatry 1991;54(2): 116-8.

[35] Schmidt ML, Zhukareva V, Newell KL, Lee VM, Trojanowski JQ. Tau isoform profile and phosphorylation state in dementia pugilistica recapitulate Alzheimer's disease. Acta Neuropathol 2001;101(5):518-24.

[36] Geddes JF, Vowles GH, Robinson SF, Sutcliffe JC. Neurofibrillary tangles, but not Alzheimer-type pathology, in a young boxer. Neuropathol Appl Neurobiol 1996;22(1):12-6.

[37] Grundke-Iqbal I, Iqbal K, Tung YC, Quinlan M, Wisniewski HM, Binder LI. Abnormal phosphorylation of the microtubule-associated protein tau (tau) in Alzheimer cytoskeletal pathology. Proc Natl Acad Sci U S A 1986;83(13):4913-7.

[38] Noble W, Hanger DP, Miller CC, Lovestone S. The importance of tau phosphorylation for neurodegenerative diseases. Front Neurol 2013;4:83.

[39] Braak H, Braak E, Bohl J. Staging of Alzheimer-related cortical destruction. Eur Neurol 1993;33(6):403-8.

[40] Blennow K, de Leon MJ, Zetterberg H. Alzheimer's disease. Lancet 2006;368(9533): 387-403.

[41] Xie Y, Yao Z, Chai H, Wong WM, Wu W. Potential roles of Alzheimer precursor protein A4 and beta-amyloid in survival and function of aged spinal motor neurons after axonal injury. J Neurosci Res 2003;73(4):557-64.

[42] Small DH, Clarris HL, Williamson TG, Reed G, Key B, Mok SS, et al. Neurite-outgrowth regulating functions of the amyloid protein precursor of Alzheimer's disease. J Alzheimers Dis 1999;1(4-5):275-85.

[43] Alvarez J, Moreno RD, Llanos O, Inestrosa NC, Brandan E, Colby T, et al. Axonal sprouting induced in the sciatic nerve by the amyloid precursor protein (APP) and other antiproteases. Neurosci Lett 1992;144(1-2):130-4.

[44] Wang Z, Yang L, Zheng H. Role of APP and Abeta in synaptic physiology. Curr Alzheimer Res 2012;9(2):217-26.

[45] Johnson MW, Stoll L, Rubio A, Troncoso J, Pletnikova O, Fowler DR, et al. Axonal injury in young pediatric head trauma: a comparison study of beta-amyloid precursor protein (beta-APP) immunohistochemical staining in traumatic and nontraumatic deaths. J Forensic Sci 2011;56(5):1198-205.

[46] Chen SF, Richards HK, Smielewski P, Johnstrom P, Salvador R, Pickard JD, et al. Relationship between flow-metabolism uncoupling and evolving axonal injury after experimental traumatic brain injury. J Cereb Blood Flow Metab 2004;24(9):1025-36.

[47] Neumann M, Sampathu DM, Kwong LK, Truax AC, Micsenyi MC, Chou TT, et al. Ubiquitinated TDP-43 in frontotemporal lobar degeneration and amyotrophic lateral sclerosis. Science 2006;314(5796):130-3.

[48] Kadokura A, Yamazaki T, Lemere CA, Takatama M, Okamoto K. Regional distribution of TDP-43 inclusions in Alzheimer disease (AD) brains: their relation to AD common pathology. Neuropathology 2009;29(5):566-73.

[49] King A, Sweeney F, Bodi I, Troakes C, Maekawa S, Al-Sarraj S. Abnormal TDP-43 expression is identified in the neocortex in cases of dementia pugilistica, but is mainly confined to the limbic system when identified in high and moderate stages of Alzheimer's disease. Neuropathology 2010;30(4):408-19.

[50] Bellander BM, Singhrao SK, Ohlsson M, Mattsson P, Svensson M. Complement activation in the human brain after traumatic head injury. J Neurotrauma 2001;18(12): 1295-311.

[51] Kolev MV, Ruseva MM, Harris CL, Morgan BP, Donev RM. Implication of complement system and its regulators in Alzheimer's disease. Curr Neuropharmacol 2009;7(1):1-8.

[52] Lambert JC, Heath S, Even G, Campion D, Sleegers K, Hiltunen M, et al. Genome-wide association study identifies variants at CLU and CR1 associated with Alzheimer's disease. Nat Genet 2009;41(10):1094-9.

[53] Mitchell DA, Kirby L, Paulin SM, Villiers CL, Sim RB. Prion protein activates and fixes complement directly via the classical pathway: implications for the mechanism of scrapie agent propagation in lymphoid tissue. Mol Immunol 2007;44(11):2997-3004.

[54] Sta M, Sylva-Steenland RM, Casula M, de Jong JM, Troost D, Aronica E, et al. Innate and adaptive immunity in amyotrophic lateral sclerosis: Evidence of complement activation. Neurobiol Dis 2011;42(3):211-20.

[55] Walport MJ. Complement. First of two parts. N Engl J Med 2001;344(14):1058-66.

[56] Walport MJ. Complement. Second of two parts. N Engl J Med 2001;344(15):1140-4.

[57] Stahel PF, Morganti-Kossmann MC, Perez D, Redaelli C, Gloor B, Trentz O, et al. Intrathecal levels of complement-derived soluble membrane attack complex (sC5b-9) correlate with blood-brain barrier dysfunction in patients with traumatic brain injury. J Neurotrauma 2001;18(8):773-81.

[58] Sewell DL, Nacewicz B, Liu F, Macvilay S, Erdei A, Lambris JD, et al. Complement C3 and C5 play critical roles in traumatic brain cryoinjury: blocking effects on neu-

trophil extravasation by C5a receptor antagonist. J Neuroimmunol 2004;155(1-2): 55-63.

[59] Leinhase I, Holers VM, Thurman JM, Harhausen D, Schmidt OI, Pietzcker M, et al. Reduced neuronal cell death after experimental brain injury in mice lacking a functional alternative pathway of complement activation. BMC Neurosci 2006;7:55.

[60] Leinhase I, Rozanski M, Harhausen D, Thurman JM, Schmidt OI, Hossini AM, et al. Inhibition of the alternative complement activation pathway in traumatic brain injury by a monoclonal anti-factor B antibody: a randomized placebo-controlled study in mice. J Neuroinflammation 2007;4:13.

[61] Rancan M, Morganti-Kossmann MC, Barnum SR, Saft S, Schmidt OI, Ertel W, et al. Central nervous system-targeted complement inhibition mediates neuroprotection after closed head injury in transgenic mice. J Cereb Blood Flow Metab 2003;23(9): 1070-4.

[62] Leinhase I, Schmidt OI, Thurman JM, Hossini AM, Rozanski M, Taha ME, et al. Pharmacological complement inhibition at the C3 convertase level promotes neuronal survival, neuroprotective intracerebral gene expression, and neurological outcome after traumatic brain injury. Exp Neurol 2006;199(2):454-64.

[63] Longhi L, Perego C, Ortolano F, Zanier ER, Bianchi P, Stocchetti N, et al. C1-inhibitor attenuates neurobehavioral deficits and reduces contusion volume after controlled cortical impact brain injury in mice. Crit Care Med 2009;37(2):659-65.

[64] Stahel PF, Flierl MA, Morgan BP, Persigehl I, Stoll C, Conrad C, et al. Absence of the complement regulatory molecule CD59a leads to exacerbated neuropathology after traumatic brain injury in mice. J Neuroinflammation 2009;6:2.

[65] Spain A, Daumas S, Lifshitz J, Rhodes J, Andrews PJ, Horsburgh K, et al. Mild fluid percussion injury in mice produces evolving selective axonal pathology and cognitive deficits relevant to human brain injury. J Neurotrauma 2010;27(8):1429-38.

[66] Giza CC, Hovda DA. The Neurometabolic Cascade of Concussion. J Athl Train 2001;36(3):228-35.

[67] Barkhoudarian G, Hovda DA, Giza CC. The molecular pathophysiology of concussive brain injury. Clin Sports Med 2011;30(1):33-iii.

[68] Hall GF, Yao J, Lee G. Human tau becomes phosphorylated and forms filamentous deposits when overexpressed in lamprey central neurons in situ. Proc Natl Acad Sci U S A 1997;94(9):4733-8.

[69] Hall GF, Chu B, Lee G, Yao J. Human tau filaments induce microtubule and synapse loss in an in vivo model of neurofibrillary degenerative disease. J Cell Sci 2000;113 (Pt 8):1373-87.

[70] Clavaguera F, Bolmont T, Crowther RA, Abramowski D, Frank S, Probst A, et al. Transmission and spreading of tauopathy in transgenic mouse brain. Nat Cell Biol 2009;11(7):909-13.

[71] de CA, Polydoro M, Suarez-Calvet M, William C, Adamowicz DH, Kopeikina KJ, et al. Propagation of tau pathology in a model of early Alzheimer's disease. Neuron 2012;73(4):685-97.

[72] Liu L, Drouet V, Wu JW, Witter MP, Small SA, Clelland C, et al. Trans-synaptic spread of tau pathology in vivo. PLoS One 2012;7(2):e31302.

[73] Kim W, Lee S, Jung C, Ahmed A, Lee G, Hall GF. Interneuronal transfer of human tau between Lamprey central neurons in situ. J Alzheimers Dis 2010;19(2):647-64.

[74] Hall GF, Patuto BA. Is tau ready for admission to the prion club? Prion 2012;6(3): 223-33.

[75] Wu JW, Herman M, Liu L, Simoes S, Acker CM, Figueroa H, et al. Small misfolded Tau species are internalized via bulk endocytosis and anterogradely and retrogradely transported in neurons. J Biol Chem 2013;288(3):1856-70.

[76] Iliff JJ, Wang M, Liao Y, Plogg BA, Peng W, Gundersen GA, et al. A paravascular pathway facilitates CSF flow through the brain parenchyma and the clearance of interstitial solutes, including amyloid beta. Sci Transl Med 2012;4(147):147ra111.

[77] Gentleman SM, Nash MJ, Sweeting CJ, Graham DI, Roberts GW. Beta-amyloid precursor protein (beta APP) as a marker for axonal injury after head injury. Neurosci Lett 1993;160(2):139-44.

[78] Barmada SJ, Skibinski G, Korb E, Rao EJ, Wu JY, Finkbeiner S. Cytoplasmic mislocalization of TDP-43 is toxic to neurons and enhanced by a mutation associated with familial amyotrophic lateral sclerosis. J Neurosci 2010;30(2):639-49.

[79] Tatom JB, Wang DB, Dayton RD, Skalli O, Hutton ML, Dickson DW, et al. Mimicking aspects of frontotemporal lobar degeneration and Lou Gehrig's disease in rats via TDP-43 overexpression. Mol Ther 2009;17(4):607-13.

[80] Wils H, Kleinberger G, Janssens J, Pereson S, Joris G, Cuijt I, et al. TDP-43 transgenic mice develop spastic paralysis and neuronal inclusions characteristic of ALS and frontotemporal lobar degeneration. Proc Natl Acad Sci U S A 2010;107(8):3858-63.

[81] Davalos D, Grutzendler J, Yang G, Kim JV, Zuo Y, Jung S, et al. ATP mediates rapid microglial response to local brain injury in vivo. Nat Neurosci 2005;8(6):752-8.

[82] Koshinaga M, Katayama Y, Fukushima M, Oshima H, Suma T, Takahata T. Rapid and widespread microglial activation induced by traumatic brain injury in rat brain slices. J Neurotrauma 2000;17(3):185-92.

[83] Garden GA, Moller T. Microglia biology in health and disease. J Neuroimmune Pharmacol 2006;1(2):127-37.

[84] Nimmerjahn A, Kirchhoff F, Helmchen F. Resting microglial cells are highly dynamic surveillants of brain parenchyma in vivo. Science 2005;308(5726):1314-8.

[85] Csuka E, Hans VH, Ammann E, Trentz O, Kossmann T, Morganti-Kossmann MC. Cell activation and inflammatory response following traumatic axonal injury in the rat. Neuroreport 2000;11(11):2587-90.

[86] Gentleman SM, Leclercq PD, Moyes L, Graham DI, Smith C, Griffin WS, et al. Long-term intracerebral inflammatory response after traumatic brain injury. Forensic Sci Int 2004;146(2-3):97-104.

[87] Nagamoto-Combs K, McNeal DW, Morecraft RJ, Combs CK. Prolonged microgliosis in the rhesus monkey central nervous system after traumatic brain injury. J Neurotrauma 2007;24(11):1719-42.

[88] Ramlackhansingh AF, Brooks DJ, Greenwood RJ, Bose SK, Turkheimer FE, Kinnunen KM, et al. Inflammation after trauma: microglial activation and traumatic brain injury. Ann Neurol 2011;70(3):374-83.

[89] Shitaka Y, Tran HT, Bennett RE, Sanchez L, Levy MA, Dikranian K, et al. Repetitive closed-skull traumatic brain injury in mice causes persistent multifocal axonal injury and microglial reactivity. J Neuropathol Exp Neurol 2011;70(7):551-67.

[90] Kalaria RN. Microglia and Alzheimer's disease. Curr Opin Hematol 1999;6(1):15-24.

[91] Conde JR, Streit WJ. Microglia in the aging brain. J Neuropathol Exp Neurol 2006;65(3):199-203.

[92] Norden DM, Godbout JP. Review: microglia of the aged brain: primed to be activated and resistant to regulation. Neuropathol Appl Neurobiol 2013;39(1):19-34.

[93] Perry VH, Nicoll JA, Holmes C. Microglia in neurodegenerative disease. Nat Rev Neurol 2010;6(4):193-201.

[94] Biswas SK, Lopez-Collazo E. Endotoxin tolerance: new mechanisms, molecules and clinical significance. Trends Immunol 2009;30(10):475-87.

[95] Perry VH, Cunningham C, Holmes C. Systemic infections and inflammation affect chronic neurodegeneration. Nat Rev Immunol 2007;7(2):161-7.

[96] van Gool WA, van de Beek D, Eikelenboom P. Systemic infection and delirium: when cytokines and acetylcholine collide. Lancet 2010;375(9716):773-5.

[97] Kat MG, de Jonghe JF, Vreeswijk R, van der Ploeg T, van Gool WA, Eikelenboom P, et al. Mortality associated with delirium after hip-surgery: a 2-year follow-up study. Age Ageing 2011;40(3):312-8.

[98] Krogseth M, Wyller TB, Engedal K, Juliebo V. Delirium is an important predictor of incident dementia among elderly hip fracture patients. Dement Geriatr Cogn Disord 2011;31(1):63-70.

[99] Holmes C, El-Okl M, Williams AL, Cunningham C, Wilcockson D, Perry VH. Systemic infection, interleukin 1beta, and cognitive decline in Alzheimer's disease. J Neurol Neurosurg Psychiatry 2003;74(6):788-9.

[100] Holmes C, Cunningham C, Zotova E, Woolford J, Dean C, Kerr S, et al. Systemic inflammation and disease progression in Alzheimer disease. Neurology 2009;73(10): 768-74.

[101] Buljevac D, Flach HZ, Hop WC, Hijdra D, Laman JD, Savelkoul HF, et al. Prospective study on the relationship between infections and multiple sclerosis exacerbations. Brain 2002;125(Pt 5):952-60.

[102] Holland MC, Mackersie RC, Morabito D, Campbell AR, Kivett VA, Patel R, et al. The development of acute lung injury is associated with worse neurologic outcome in patients with severe traumatic brain injury. J Trauma 2003;55(1):106-11.

[103] Cunningham C, Wilcockson DC, Campion S, Lunnon K, Perry VH. Central and systemic endotoxin challenges exacerbate the local inflammatory response and increase neuronal death during chronic neurodegeneration. J Neurosci 2005;25(40):9275-84.

[104] Cunningham C, Campion S, Lunnon K, Murray CL, Woods JF, Deacon RM, et al. Systemic inflammation induces acute behavioral and cognitive changes and accelerates neurodegenerative disease. Biol Psychiatry 2009;65(4):304-12.

[105] Combrinck MI, Perry VH, Cunningham C. Peripheral infection evokes exaggerated sickness behaviour in pre-clinical murine prion disease. Neuroscience 2002;112(1): 7-11.

[106] Kitazawa M, Oddo S, Yamasaki TR, Green KN, LaFerla FM. Lipopolysaccharide-induced inflammation exacerbates tau pathology by a cyclin-dependent kinase 5-mediated pathway in a transgenic model of Alzheimer's disease. J Neurosci 2005;25(39): 8843-53.

[107] Brugg B, Dubreuil YL, Huber G, Wollman EE, Delhaye-Bouchaud N, Mariani J. Inflammatory processes induce beta-amyloid precursor protein changes in mouse brain. Proc Natl Acad Sci U S A 1995;92(7):3032-5.

[108] Sly LM, Krzesicki RF, Brashler JR, Buhl AE, McKinley DD, Carter DB, et al. Endogenous brain cytokine mRNA and inflammatory responses to lipopolysaccharide are elevated in the Tg2576 transgenic mouse model of Alzheimer's disease. Brain Res Bull 2001;56(6):581-8.

[109] Nguyen MD, D'Aigle T, Gowing G, Julien JP, Rivest S. Exacerbation of motor neuron disease by chronic stimulation of innate immunity in a mouse model of amyotrophic lateral sclerosis. J Neurosci 2004;24(6):1340-9.

[110] Venturi L, Miranda M, Selmi V, Vitali L, Tani A, Margheri M, et al. Systemic sepsis exacerbates mild post-traumatic brain injury in the rat. J Neurotrauma 2009;26(9): 1547-56.

[111] Hoek RM, Ruuls SR, Murphy CA, Wright GJ, Goddard R, Zurawski SM, et al. Down-regulation of the macrophage lineage through interaction with OX2 (CD200). Science 2000;290(5497):1768-71.

[112] Lyons A, Downer EJ, Crotty S, Nolan YM, Mills KH, Lynch MA. CD200 ligand receptor interaction modulates microglial activation in vivo and in vitro: a role for IL-4. J Neurosci 2007;27(31):8309-13.

[113] Cox FF, Carney D, Miller AM, Lynch MA. CD200 fusion protein decreases microglial activation in the hippocampus of aged rats. Brain Behav Immun 2012;26(5):789-96.

[114] Mizuno T, Kawanokuchi J, Numata K, Suzumura A. Production and neuroprotective functions of fractalkine in the central nervous system. Brain Res 2003;979(1-2):65-70.

[115] Zujovic V, Benavides J, Vige X, Carter C, Taupin V. Fractalkine modulates TNF-alpha secretion and neurotoxicity induced by microglial activation. Glia 2000;29(4):305-15.

[116] Lyons A, Lynch AM, Downer EJ, Hanley R, O'Sullivan JB, Smith A, et al. Fractalkine-induced activation of the phosphatidylinositol-3 kinase pathway attentuates microglial activation in vivo and in vitro. J Neurochem 2009;110(5):1547-56.

[117] Corona AW, Huang Y, O'Connor JC, Dantzer R, Kelley KW, Popovich PG, et al. Fractalkine receptor (CX3CR1) deficiency sensitizes mice to the behavioral changes induced by lipopolysaccharide. J Neuroinflammation 2010;7:93.

[118] Corona AW, Norden DM, Skendelas JP, Huang Y, O'Connor JC, Lawson M, et al. Indoleamine 2,3-dioxygenase inhibition attenuates lipopolysaccharide induced persistent microglial activation and depressive-like complications in fractalkine receptor (CX(3)CR1)-deficient mice. Brain Behav Immun 2013;31:134-42.

[119] Ponomarev ED, Veremeyko T, Barteneva N, Krichevsky AM, Weiner HL. MicroRNA-124 promotes microglia quiescence and suppresses EAE by deactivating macrophages via the C/EBP-alpha-PU.1 pathway. Nat Med 2011;17(1):64-70.

[120] Van Muiswinkel FL, Veerhuis R, Eikelenboom P. Amyloid beta protein primes cultured rat microglial cells for an enhanced phorbol 12-myristate 13-acetate-induced respiratory burst activity. J Neurochem 1996;66(6):2468-76.

[121] Lewis S. Neurodegenerative disorders: Microglia get ready, set... Nat Rev Neurosci 2012;13(3):154-5.

[122] Barratt-Due A, Pischke SE, Brekke OL, Thorgersen EB, Nielsen EW, Espevik T, et al. Bride and groom in systemic inflammation--the bells ring for complement and Toll in cooperation. Immunobiology 2012;217(11):1047-56.

[123] Bianchi ME. DAMPs, PAMPs and alarmins: all we need to know about danger. J Leukoc Biol 2007;81(1):1-5.

[124] Visintin A, Mazzoni A, Spitzer JH, Wyllie DH, Dower SK, Segal DM. Regulation of Toll-like receptors in human monocytes and dendritic cells. J Immunol 2001;166(1): 249-55.

[125] Babcock AA, Wirenfeldt M, Holm T, Nielsen HH, Dissing-Olesen L, Toft-Hansen H, et al. Toll-like receptor 2 signaling in response to brain injury: an innate bridge to neuroinflammation. J Neurosci 2006;26(49):12826-37.

[126] Yu M, Wang H, Ding A, Golenbock DT, Latz E, Czura CJ, et al. HMGB1 signals through toll-like receptor (TLR) 4 and TLR2. Shock 2006;26(2):174-9.

[127] Cai C, Gill R, Eum HA, Cao Z, Loughran PA, Darwiche S, et al. Complement factor 3 deficiency attenuates hemorrhagic shock-related hepatic injury and systemic inflammatory response syndrome. Am J Physiol Regul Integr Comp Physiol 2010;299(5):R1175-R1182.

[128] Cohen MJ, Brohi K, Calfee CS, Rahn P, Chesebro BB, Christiaans SC, et al. Early release of high mobility group box nuclear protein 1 after severe trauma in humans: role of injury severity and tissue hypoperfusion. Crit Care 2009;13(6):R174.

[129] Harry GJ. Microglia during development and aging. Pharmacol Ther 2013;139(3): 313-26.

[130] Rosczyk HA, Sparkman NL, Johnson RW. Neuroinflammation and cognitive function in aged mice following minor surgery. Exp Gerontol 2008;43(9):840-6.

[131] Dodds C, Allison J. Postoperative cognitive deficit in the elderly surgical patient. Br J Anaesth 1998;81(3):449-62.

[132] Vollmar P, Kullmann JS, Thilo B, Claussen MC, Rothhammer V, Jacobi H, et al. Active immunization with amyloid-beta 1-42 impairs memory performance through TLR2/4-dependent activation of the innate immune system. J Immunol 2010;185(10): 6338-47.

[133] Orgogozo JM, Gilman S, Dartigues JF, Laurent B, Puel M, Kirby LC, et al. Subacute meningoencephalitis in a subset of patients with AD after Abeta42 immunization. Neurology 2003;61(1):46-54.

[134] Robinson SR, Bishop GM, Munch G. Alzheimer vaccine: amyloid-beta on trial. Bioessays 2003;25(3):283-8.

[135] Munch G, Robinson SR. Potential neurotoxic inflammatory responses to Abeta vaccination in humans. J Neural Transm 2002;109(7-8):1081-7.

[136] Thomas DM, Walker PD, Benjamins JA, Geddes TJ, Kuhn DM. Methamphetamine neurotoxicity in dopamine nerve endings of the striatum is associated with microglial activation. J Pharmacol Exp Ther 2004;311(1):1-7.

[137] Dantzer R. Cytokine-induced sickness behaviour: a neuroimmune response to activation of innate immunity. Eur J Pharmacol 2004;500(1-3):399-411.

[138] Katsuura G, Arimura A, Koves K, Gottschall PE. Involvement of organum vasculosum of lamina terminalis and preoptic area in interleukin 1 beta-induced ACTH release. Am J Physiol 1990;258(1 Pt 1):E163-E171.

[139] Vitkovic L, Konsman JP, Bockaert J, Dantzer R, Homburger V, Jacque C. Cytokine signals propagate through the brain. Mol Psychiatry 2000;5(6):604-15.

[140] Bluthe RM, Walter V, Parnet P, Laye S, Lestage J, Verrier D, et al. Lipopolysaccharide induces sickness behaviour in rats by a vagal mediated mechanism. C R Acad Sci III 1994;317(6):499-503.

[141] Bluthe RM, Michaud B, Kelley KW, Dantzer R. Vagotomy attenuates behavioural effects of interleukin-1 injected peripherally but not centrally. Neuroreport 1996;7(9): 1485-8.

[142] Chang ZL. Important aspects of Toll-like receptors, ligands and their signaling pathways. Inflamm Res 2010;59(10):791-808.

[143] Akashi-Takamura S, Miyake K. TLR accessory molecules. Curr Opin Immunol 2008;20(4):420-5.

[144] Beutler B. Inferences, questions and possibilities in Toll-like receptor signalling. Nature 2004;430(6996):257-63.

[145] Ghosh S, Wu MD, Shaftel SS, Kyrkanides S, LaFerla FM, Olschowka JA, et al. Sustained interleukin-1beta overexpression exacerbates tau pathology despite reduced amyloid burden in an Alzheimer's mouse model. J Neurosci 2013;33(11):5053-64.

[146] Blasko I, Stampfer-Kountchev M, Robatscher P, Veerhuis R, Eikelenboom P, Grubeck-Loebenstein B. How chronic inflammation can affect the brain and support the development of Alzheimer's disease in old age: the role of microglia and astrocytes. Aging Cell 2004;3(4):169-76.

[147] Liao YF, Wang BJ, Cheng HT, Kuo LH, Wolfe MS. Tumor necrosis factor-alpha, interleukin-1beta, and interferon-gamma stimulate gamma-secretase-mediated cleavage of amyloid precursor protein through a JNK-dependent MAPK pathway. J Biol Chem 2004;279(47):49523-32.

[148] Ramaglia V, King RH, Morgan BP, Baas F. Deficiency of the complement regulator CD59a exacerbates Wallerian degeneration. Mol Immunol 2009;46(8-9):1892-6.

[149] Ramaglia V, Tannemaat MR, de Kok M, Wolterman R, Vigar MA, King RH, et al. Complement inhibition accelerates regeneration in a model of peripheral nerve injury. Mol Immunol 2009;47(2-3):302-9.

[150] Walport MJ. Complement. Second of two parts. N Engl J Med 2001;344(15):1140-4.

[151] Walport MJ. Complement. First of two parts. N Engl J Med 2001;344(14):1058-66.

[152] McCrory P, Meeuwisse WH, Aubry M, Cantu RC, Dvorak J, Echemendia RJ, et al. Consensus statement on concussion in sport--the 4th International Conference on Concussion in Sport held in Zurich, November 2012. PM R 2013;5(4):255-79.

[153] Gosselin N, Saluja RS, Chen JK, Bottari C, Johnston K, Ptito A. Brain functions after sports-related concussion: insights from event-related potentials and functional MRI. Phys Sportsmed 2010;38(3):27-37.

[154] Mayeux R, Ottman R, Maestre G, Ngai C, Tang MX, Ginsberg H, et al. Synergistic effects of traumatic head injury and apolipoprotein-epsilon 4 in patients with Alzheimer's disease. Neurology 1995;45(3 Pt 1):555-7.

[155] Mayeux R, Ottman R, Tang MX, Noboa-Bauza L, Marder K, Gurland B, et al. Genetic susceptibility and head injury as risk factors for Alzheimer's disease among community-dwelling elderly persons and their first-degree relatives. Ann Neurol 1993;33(5):494-501.

[156] Nicoll JA, Roberts GW, Graham DI. Apolipoprotein E epsilon 4 allele is associated with deposition of amyloid beta-protein following head injury. Nat Med 1995;1(2): 135-7.

[157] Janssens J, Van BC. Pathological mechanisms underlying TDP-43 driven neurodegeneration in FTLD-ALS spectrum disorders. Hum Mol Genet 2013;

[158] Van LT, van der Zee J, Gijselinck I, Engelborghs S, Vandenberghe R, Vandenbulcke M, et al. Distinct clinical characteristics of C9orf72 expansion carriers compared with GRN, MAPT, and nonmutation carriers in a Flanders-Belgian FTLD cohort. JAMA Neurol 2013;70(3):365-73.

[159] Tran HT, Sanchez L, Esparza TJ, Brody DL. Distinct temporal and anatomical distributions of amyloid-beta and tau abnormalities following controlled cortical impact in transgenic mice. PLoS One 2011;6(9):e25475.

[160] Tran HT, LaFerla FM, Holtzman DM, Brody DL. Controlled cortical impact traumatic brain injury in 3xTg-AD mice causes acute intra-axonal amyloid-beta accumulation and independently accelerates the development of tau abnormalities. J Neurosci 2011;31(26):9513-25.

[161] Cohen RM, Rezai-Zadeh K, Weitz TM, Rentsendorj A, Gate D, Spivak I, et al. A transgenic Alzheimer rat with plaques, tau pathology, behavioral impairment, oligomeric abeta, and frank neuronal loss. J Neurosci 2013;33(15):6245-56.

Chapter 5

Substance P: A Novel Target in the Treatment of Cerebral Oedema and Elevated Intracranial Pressure Following Traumatic Brain Injury

Renée J. Turner and Robert Vink

Additional information is available at the end of the chapter

http://dx.doi.org/10.5772/57329

1. Introduction

Traumatic brain injury (TBI) is the biggest killer of individuals under the age of 44 years [1], affecting over 5 million individuals worldwide each year. Many individuals are left with permanent neurological deficits caused by a number of complex biochemical cascades, referred to as secondary injury, that are initiated by the traumatic event and evolve over the hours to days thereafter. Secondary injury encompasses a wide variety of injury factors including excitotoxicity, loss of ion homeostasis, oxidative stress, inflammation, apoptosis, increased vascular permeability and cerebral oedema, amongst many others [2-3], and has been well documented to exacerbate injury and worsen outcome following trauma. Nevertheless, the delayed fashion in which the injury progresses provides a window of opportunity for therapeutic intervention to potentially halt secondary injury, reduce neuronal loss and improve outcome. As such, countless studies have now focused on characterising the secondary injury that occurs following trauma in order to develop therapies to reduce or ameliorate such pathways [4].

However, the findings of neuroprotective studies have to date produced unfavourable outcomes in clinical trials [2]. Although the reasons for such failures are multifactorial, it is apparent that targeting only a single injury factor is of limited benefit when numerous cascades contribute to the resultant injury. Alternatively, if a target is identified that modulates many aspects of secondary injury then this may produce favourable results [5]. Of the secondary injury pathways, disruption to the blood-brain barrier (BBB) and subsequent development of cerebral oedema are of particular concern due to the potential effect on intracranial pressure (ICP) dynamics [6]. In this review, we highlight recent data delineating a potentially crucial role for the neuropeptide substance P (SP) in the pathogenesis of cerebral oedema formation and the development of elevated ICP following TBI, and the multi-potential nature of SP antagonists as a therapeutic intervention.

2. Blood-brain barrier disruption

Under normal conditions the BBB provides a selectively permeable barrier between the vasculature and the brain that regulates the entry of blood-borne substances, thereby maintaining an optimal environment within the brain [7]. It is comprised of a complex network of cells that make up the cerebral capillaries and post-capillary venules, resting on the basal lamina [8]. The gate function of the BBB is provided by the tight and adherin junctions, composed of a complex network of transmembrane and cytosolic proteins [7]. The BBB functions to ensure a constant supply of nutrients, preserves ion homeostasis within the brain microenvironment and protect against noxious chemicals, variations in blood composition and breakdown of concentration gradients.

After trauma, alterations in BBB permeability have been well documented [9-11], with the temporal profile of BBB disruption largely dependent upon the type of injury. Specifically, early increases in BBB permeability are observed following models of diffuse TBI [12], whilst a biphasic opening of the BBB has been reported in more focal injuries [13]. Nevertheless, BBB dysfunction is permissive to the generation of cerebral oedema, specifically of the vasogenic type. In this type of cerebral oedema extravasation of plasma proteins occurs followed by a net movement of fluid from the vascular compartment into the brain parenchyma, leading to a disruption of both fluid and ionic homeostasis. Given the increase in the volume of the brain tissue under these circumstances, vasogenic oedema has the potential to markedly alter intracranial pressure dynamics [6] and negatively influence patient outcomes [13]. In addition, the loss of barrier integrity following acute injury to the brain allows peripheral immune cells to cross the barrier and further contribute to and exacerbate in the inflammatory processes within the brain [14].

3. Development of cerebral oedema

Of the secondary injury factors, cerebral oedema is of particular importance in terms of patient morbidity and mortality [15]. Indeed, cerebral oedema is a leading cause of death following TBI and a predictor of poor outcome in those individuals that survive. Specifically, it accounts for as much as half of all morbidity and mortality [15-16], largely because it increases intracranial pressure (ICP), resulting in reduced cerebral blood flow, initiation or exacerbation of an ischaemic state, deformation and herniation of brain tissue and a substantial increase in morbidity and mortality [17]. As such, brain oedema with increased ICP is widely recognised as a major clinical management target [18]. Despite this, there is currently no effective pharmacological treatment that reduces the considerable mortality and morbidity associated with cerebral oedema [15, 17]. Indeed, conventional treatments targeting cerebral oedema and elevated ICP, using for example hyperventilation, mannitol, diuretics, or barbituates [17], do not address the mechanisms associated with oedema formation but rather focus on the net result. Although these agents and interventions have, under ideal conditions, been shown to reduce ICP, their capacity to produce sustained decreases in ICP is inadequate. Furthermore, their effectiveness is limited as once cerebral oedema produces evidence of mass effect with midline shift on imaging, fatality rates become high, irrespective of treatment.

Decompressive craniectomy is a surgical procedure in which a large flap of bone overlying the swollen brain is removed, creating space to accommodate the increased volume associated with cerebral oedema [17]. This is currently one of the most powerful tools that clinicians have to combat elevated ICP, although it has been shown to improve survival is some studies, whilst associated with an increase in moderate to severe disability in others [19, 20], thereby emphasising the need for further clinical studies to clearly determine the effect of decompressive surgery following TBI. Furthermore, as decompressive craniectomy is a major operation, it is complicated in the gravely ill and the efficacy is markedly decreased in patients over 60 years of age; thus many patients are ineligible for the surgery [17]. Overall, such treatments and interventions have proven to be largely ineffective in combating cerebral oedema, mainly because they do not actually address the specific mechanisms that produce swelling of brain tissue. Recent studies in experimental TBI have identified that release of the neuropeptide substance P (SP) is a feature of acute injury to the brain and have revealed a crucial role for SP in the increases in vascular permeability and brain water content which are observed following TBI. As such, they may represent a novel pharmacological target for the treatment of oedema and increased ICP.

4. Substance P

SP is an 11 amino acid peptide and member of the tachykinin peptide family, which also includes neurokinin A (NKA), neurokinin B (NKB) and neuropeptide γ, amongst others [21]. Originally identified by von Euler and Gaddum in the 1930's for its potent smooth muscle and hypotensive properties [21], it is now known that SP is released from primary afferent nerves in both the peripheral and central nervous systems where it functions as a neurotransmitter [22]. SP is also release from non-neuronal cells such as endothelial and inflammatory cells [23]. Specifically, within the nervous system, SP is localised in capsaicin-sensitive neurons and is released in response to calcium-dependent depolarisation induced by various stimuli including electrical stimulation, pH changes and ligand-receptor binding [24]. Following release, SP can exert direct post-synaptic actions as a neurotransmitter, or modulate other non-neuronal targets [25], via binding to tachykinin NK receptors. The NK receptors are members of the rhodopsin family of 7-transmembrane domain G-protein coupled receptors. To date, 3 mammalian tachykinin receptors have been identified, namely the NK1, NK2 and NK3 receptors [22]. There is some cross-reactivity amongst the receptors, with each tachykinin able to bind all receptors types depending on neuropeptide concentration and receptor availability [22]. However, under normal conditions SP has the highest affinity for the NK1 receptor, NKA for the NK2 receptor and NKB for the NK3 receptor. Furthermore, the predominance of the NK1 receptor in the human adult brain [26] makes SP the main tachykinin of interest in the pathophysiology of CNS injury. Transduction of the SP signal through the NK1 receptor occurs via G protein signalling and the secondary messenger cAMP, ultimately leading to the regulation of ion channels, enzyme activity and alterations in gene expression [25].

5. Neurogenic inflammation

The release of neuropeptides, including SP and calcitonin gene-related peptide (CGRP), leads to the development of neurogenic inflammation, a neurally elicited, painful local inflammatory response that is characterised by vasodilation, increased vascular permeability, mast cell degranulation and protein extravasation [27]. Such changes in blood vessel size and permeability lead to the development of tissue swelling [27]. In addition, there are also other responses that are specific to individual tissues, including, for example, smooth muscle contraction/relaxation in the bladder and bronchoconstriction in the airways, amongst others. Although other neuropeptides such as CGRP are involved, SP is considered to be the most potent initiator of neurogenic inflammation. Nevertheless, CGRP can potentiate the effects of SP by increasing the expression of the NK1 tachykinin receptor and enhancing the bioavailability of SP by competing with SP for catabolism by endopeptidases [28]. Indeed, neurogenic inflammation leads to an increase in the PPT and NK1 receptor mRNA transcript, which encodes SP and its receptor.

5.1. Peripheral nervous system

It has been well documented that neurogenic inflammation occurs in peripheral tissues such as the oral, nasal, facial and ocular tissue, with release of SP known to initiate increased microvascular permeability and tissue swelling [29]. Indeed, SP, NKA or NKB injected into the paws of rats leads to a profound increase in paw swelling with a similar response observed following the administration of exogenous NK1, NK2 or NK3 agonists [30]. Furthermore, administration of NK1, NK2 or NK3 tachykinin receptor antagonists inhibits such oedema formation in a dose-dependent manner. These findings confirm the involvement of SP, NKA and NKB in the genesis of neurogenic inflammation and tissue swelling, and clearly implicates all three tachykinin ligand-receptor pairs in the observed neurogenic inflammatory responses. Nonetheless, the predominant role of SP in neurogenic inflammation has been confirmed in NK1 tachykinin receptor negative mice [31].

5.2. Central nervous system

The involvement of classical inflammation in the evolution of injury following TBI has been known for some time, however the concept of neurogenic inflammation in the brain has until recently remained unexplored. Originally described in peripheral tissues, it is now known that neurogenic inflammation occurs in the brain following injury [12, 32-38]. In stroke,

http://dx.doi.org/10.5772/57329

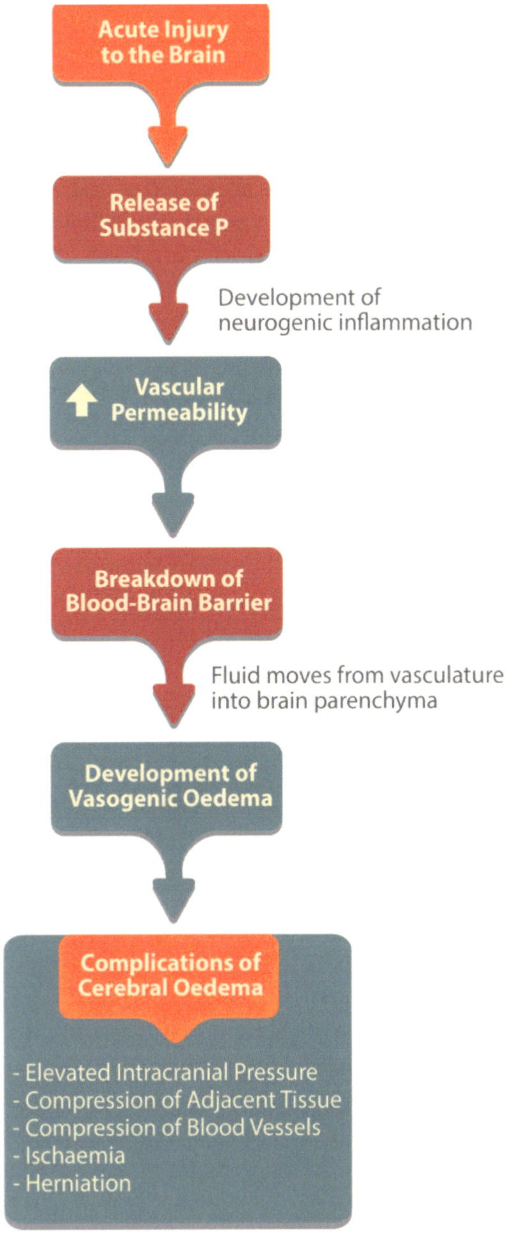

Figure 1. Release of substance P and the development of neurogenic inflammation following acute brain injury.

Stumm and colleagues (2002) proposed that activation of NK1 tachykinin receptors on vascular endothelium may contribute to cerebral oedema [39]. Indeed, chemical or electrical stimulation of the dura mater or treatment with capsaicin produces a neurogenic inflammatory response in the dura mater but was not observed in the pia mater or within the brain parenchyma itself [40]. Furthermore, administration of SP produced a marked increase in plasma extravasation within the dura mater of rats, an effect that was blocked by administration of an NK1 tachykinin receptor antagonist [41]. Such studies confirm the presence of neurogenic inflammation within the brain and the involvement of SP in changes in vascular permeability in the setting of injury. More recently, neurogenic inflammation in the brain has been widely characterised in a variety of acute injuries to the central nervous system (Figure 1) in animal models, including trauma [12, 32-33, 37, 42], stroke [34-36, 38] and spinal cord injury [43]. Furthermore, activation of the multimodal transient receptor potential vanilloid 1 (TRPV1) receptor initiates neurogenic inflammation [44] and is associated with increased BBB permeability, an effect abolished by the TRPV1 antagonist capsazepine [45]. Given that TRPV1 receptors are co-localised with both SP and CGRP suggests that it plays a role in BBB dysfunction following acute injury as a facilitator of neurogenic inflammation.

6. Substance P in traumatic brain injury

Our studies have shown that SP release is a ubiquitous feature of acute injury to the brain and is associated with marked increases in BBB permeability, cerebral oedema and functional deficits [12]. Specifically, an increase in SP was observed in brain following diffuse TBI that was particularly profound in the perivascular tissue. Such increases in SP immunoreactivity were observed at 5h and shown to persist to at least 24h following trauma in rats [12]. PCR studies later confirmed that SP levels mRNA remained elevated until 3 days post-trauma [46]. Serum levels of SP were also shown to be elevated following trauma, with significant increases observed at 30 mins [12], although levels declined quite rapidly after this time this most likely reflecting the rapid proteolysis of SP within the serum by non-specific proteases. Interestingly, when SP breakdown is inhibited through the administration of an angiotensin-converting enzyme inhibitor, an increase in SP immunoreactivity is observed with an exacerbation of injury and neurological dysfunction [47]. Taken together, these studies confirm that SP release is a feature of acute injury to the brain.

Increased SP levels following trauma have been associated with changes in cerebral vascular permeability and cerebral oedema. Specifically, in rodent TBI increased SP immunoreactivity within injured brain tissue was shown to co-localise with exogenously administered Evan's Blue dye, a marker of BBB breakdown [12]. Such alterations in vascular permeability were also associated with the development of cerebral oedema of the vasogenic type [12]. Persistent functional deficits, both motor and cognitive, were also observed in the setting of neurogenic inflammation following TBI [12, 32]. More recently, a role for neurogenic inflammation in BBB dysfunction, cerebral oedema, and functional deficits has been described in stroke [34-36, 38].

Having established the presence of neurogenic inflammation and the role of SP in brain injury following trauma, subsequent experimental studies have examined the efficacy of blocking the effect of SP. An NK1 tachykinin receptor antagonist administered at 30mins following trauma conferred protection from injury-induced BBB permeability alterations, cerebral oedema (Figure 2) and functional deficits [12] in rodent models. Moreover, the therapeutic window was shown to be at least 12h following trauma with improvements in neurological

http://dx.doi.org/10.5772/57329

outcome and a reduction in neuronal injury still observed with such delayed treatment [32]. Studies using capsaicin pre-treatment to deplete the neuropeptides before injury have produced comparable improvements in BBB status, cerebral oedema and neurological outcome [33]. Taken together, such studies illustrate the involvement of neuropeptides in the genesis of cerebral oedema following acute injury to the brain [33] and demonstrate that the neuropeptide SP is primarily responsible for the development of neurogenic inflammation and subsequent alterations in BBB permeability and cerebral oedema which are observed in the setting of experimental acute brain injury [12, 32].

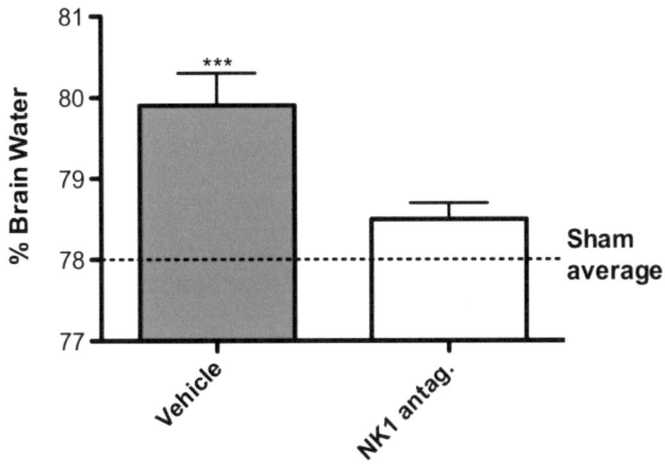

Figure 2. Cerebral oedema, as measured by wet weight dry weight, at 5h following diffuse traumatic brain injury in rats. Treatment with an NK1 antagonist significantly reduced cerebral oedema following trauma.

The efficacy of NK1 tachykinin receptor antagonists in treating BBB dysfunction and cerebral oedema in rodent TBI models is encouraging. However, given the disappointing lack of clinical translation of treatments shown to be neuroprotective in rodent models, it is becoming increasingly important to validate agents of promise in large animal models before any progression to clinical studies. Furthermore, we have recently reported that rodent TBI models do not produce consistent elevations in ICP in the absence of mass lesions, making them inappropriate for studying the evolution of increased ICP [48]. Accordingly, we have recently evaluated the efficacy of NK1 tachykinin receptor antagonists in an ovine model of TBI. This animal model incorporates a large gyrencephalic brain with large white matter domains and a significant tentorium cerebelli, features that are comparable to the human brain and essential in order to effectively study cerebral oedema and ICP dynamics. Administration of an NK1 tachykinin receptor antagonist at 30 mins following TBI produced a profound reduction in ICP by 4 h after injury (Figure 3) as compared to vehicle treated controls [6].

Blocking the action of SP with an NK1 antagonist significantly reduce ICP following trauma.

The NK1 tachykinin receptor antagonists have now been shown to be efficacious in reducing the BBB permeability, cerebral oedema, rises in ICP and functional deficits associated with TBI

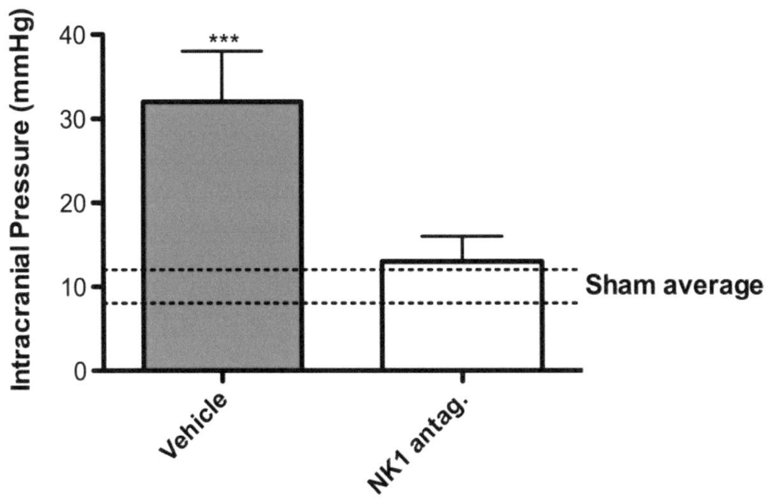

Figure 3. Intracranial pressure as measured at 4h following ovine traumatic brain injury.

in experimental models. Such studies have validated the use of NK1 tachykinin receptor antagonists in multiple models of trauma and in multiple species, including large animal models with a gyrencephalic brain. These studies in rodent and ovine models of TBI consistently demonstrate that SP release is a ubiquitous feature of acute injury to the brain. In addition, by using inactive enantiomers of the active ligands as well as a number of different structural antagonists, they emphasise that the efficacy of NK1 tachykinin receptor antagonists is a class effect rather than simple drug-specific effect. One can only conclude that, at least in these animal studies, the improvements observed were dependent upon inactivation of SP and its NK1 receptor.

A limited number of studies have also investigated the presence of neurogenic inflammation in human patients with TBI. In a cohort of patients who had sustained a TBI and subsequently died and undergone post-mortem and detailed neuropathological examination, SP immunoreactivity was increased compared to control cases [49]. Specifically, increased SP immunoreactivity was observed in the perivascular tissue surrounding the microvessels, and in particular around the post-capillary venules. Increases in SP were also observed in the perivascular axons, cortical neurons and astrocytes. The authors concluded that mechanical activation of the perivascular neurons initiated SP release and that SP played a significant role in initiating neurogenic inflammation following human TBI.

7. Conclusions

SP, through the process of neurogenic inflammation, has long been known to cause plasma extravasation and swelling in peripheral tissues. However, it has only been in recent years that the concept of neurogenic inflammation has been extended to the CNS and its role in BBB dysfunction and cerebral oedema appreciated. Furthermore, therapeutic intervention studies

http://dx.doi.org/10.5772/57329

that have blocked the action of SP have demonstrated profound reductions in BBB permeability, cerebral oedema, ICP and functional deficits in multiple species and models of TBI. Clearly, modulation of neurogenic inflammation using tachykinin NK1 receptor antagonists provides a novel therapeutic target for the treatment of cerebral oedema and elevated ICP in the setting of TBI, and other acute injuries to the brain.

Acknowledgements

Supported, in part, by the Neurosurgical Research Foundation, Australia, and the National Health and Medical Research Council of Australia (RJT #519365).

Author details

Renée J. Turner and Robert Vink

Adelaide Centre for Neuroscience Research, School of Medical Sciences, University of Adelaide, Adelaide, Australia

References

[1] Helps Y. Henley G. Harrison. J. Hospital separations due to traumatic brain injury, Australia 2004-05 2008; ISBN 9781740248266.

[2] Loane DJ. Faden AI. Neuroprotection for traumatic brain injury: translational challenges and emerging therapeutic strategies. Trends Pharmacol Sci 2010; 31 596-604.

[3] Jain KK. Neuroprotection in traumatic brain injury. Drug Discovery Today 2008; 13 1082-1089.

[4] Vink R. Nimmo A. Multifunctional drugs for head injury. Neurotherapeutics 2004; 6:28-42.

[5] Thornton E. Ziebell JM. Leonard AV. Vink R. Kinin receptor antagonists as potential neuroprotective agents in central nervous system injury. Molecules 2010; 15 6598-6618.

[6] Gabrielian L. Helps SC. Thornton E. Turner RJ. Leonard A. Vink R. Substance P anatgonists as a novel intervention for brain edema and raised intracranial pressure. Acta Neurochir Suppl 2013; 118 201-214.

[7] Rosenberg G. Neurolgical diseases in relation to the blood-brain barrier. J Cereb Blood Flow Metab 2012; 32 1139-51.

[8] Nag S. Kapadoa A. Stewart DJ. Review: molecular pathogenesis of blood-brain barrier breakdown in acute brain injury. Neuropathol Appl Neurobiol 2011; 37 3-23.

[9] Baskaya MK. Rao AM., Dogan A. Donaldson D. Dempsey RJ. The biphasic opening of the blood-brain barrier in the cortex and hippocampus after traumatic brain injury in rats. Neurosci Lett 1997; 226 33-36.

[10] Shapira Y. Setton D. Artru AA. and Shohami E. Blood-brain barrier permeability, cerebral edema, and neurologic function after closed head injury in rats. Anesthesia and analgesia 1993; 77 141-148.

[11] Strbian D. Durukan A. Pitkonen M. Marinkovic I. Tatlisumak E. Pedrono E. Abo-Ramadan U. Tatlisumak T. The blood-brain barrier is continuously open for several weeks following transient focal cerebral ischemia. Neuroscience 2008; 153 175-181.

[12] Donkin JJ. Nimmo AJ. Cernak I. Blumbergs PC. Vink R. Substance P is associated with the development of brain edema and functional deficits after traumatic brain injury. J Cereb Blood Flow Metab 2009; 29 1388-1398.

[13] Shlosberg D. Benifla M. Kaufer D. Friedman A. Blood-brain barrier breakdown as a therapeutic target in traumatic brain injury. Nature reviews. Neurology 2010; 6 393-403.

[14] Beck KD. Nguyen HX. Galvan MD. Salazar DL. Woodruff TM. Anderson A.J. Quantitative analysis of cellular inflammation after traumatic spinal cord injury: evidence for a multiphasic inflammatory response in the acute to chronic environment. Brain 2010; 133 433-447.

[15] Marmarou A. A review of progress in understanding the pathophysiology and treatment of brain edema. Neurosurg Focus 2007; 22 E1.

[16] Feickert HJ. Drommer S. Heyer R. Severe head injury in children: impact of risk factors on outcome. J Trauma 1999; 47 33-38.

[17] Hacke W. Schwab S. Horn M. Spranger M. De Georgia M. von Kummer, R. 'Malignant' middle cerebral artery territory infarction: clinical course and prognostic signs. Arch Neurol 1996; 53 309-315.

[18] Bor-Seng-Shu E. Figueiredo EG. Fonoff ET. Fujimoto Y. Panerai RB. Teixeira MJ. Decompressive craniectomy and head injury: brain morphometry, ICP, cerebral hemodynamics, cerebral microvascular reactivity, and neurochemistry. Neurosurg Review 2013; 36 361-370.

[19] Jasielski P. Glowacki M Czernicki Z. Decompressive craniectomy in trauma: when to perform, what can be achieved. Acta Neurochir. Suppl 2013; 118 125-128.

[20] Santana-Cabrera L. Perez-Acosta G. Rodriguez-Escot C. Lorenzo-Torrent R. Sanchez-Palacios M. Complications of post-injury decompressive craniectomy. Int J Critical Illness Injury Sci 2012; 2 186-188.

[21] Leeman SE. Ferguson SL. Substance P: an historical perspective. Neuropeptides 2000; 34 249-254.

[22] Maggi CA. The mammalian tachykinin receptors. Gen Pharmacol 1995; 26 911-944.

[23] Hokfelt T. Pernow B. Wahren J. Substance P: a pioneer amongst neuropeptides. J Internal Med 2001; 249 27-40.

[24] Harrison S. and Geppetti P. Substance P. Int J Biochem Cell Biol 2001; 33 555-576.

[25] Lundy FT. Linden GJ. Neuropeptides and Neurogenic Mechanisms in Oral and Periodontal Inflammation. Crit Rev Oral Biol Med 2004; 15 82-98.

[26] Pinto FM. Almeida TA. Hernandez M. Devillier P. Advenier C. Candenas ML. mRNA expression of tachykinins and tachykinin receptors in different human tissues. Eur J Pharmacol 2004 494 233-239.

[27] Black PH. Stress and the inflammatory response: a review of neurogenic inflammation. Brain Behav Immun 2002; 16 622-653.

[28] Seybold VS. McCarson KE. Mermelstein PG. Groth RD. Abrahams LG. Calcitonin gene-related peptide regulates expression of neurokinin1 receptors by rat spinal neurons. J Neurosci 2003; 23 1816-1824.

[29] Campos MM. Calixto JB. Neurokinin mediation of edema and inflammation. Neuropeptides 2000; 34 314-322.

[30] Alves RV. Campos MM. Santos AR. Calixto JB. Receptor subtypes involved in tachykinin-mediated edema formation. Peptides 1999; 20 921-927.

[31] Cao T. Gerard NP. Brain SD. Use of NK(1) knockout mice to analyze substance P-induced edema formation. Am J Physiol 1999; 277 R476-481.

[32] Donkin JJ. Cernak I. Blumbergs PC. Vink R. A substance P antagonist reduces axonal injury and improves neurologic outcome when administered up to 12 hours after traumatic brain injury. J Neurotrauma 2011; 28 217-224.

[33] Nimmo AJ. Cernak I. Heath DL. Hu X. Bennett CJ. Vink R. Neurogenic inflammation is associated with development of edema and functional deficits following traumatic brain injury in rats. Neuropeptides 2004; 38 40-47.

[34] Turner RJ. Blumbergs PC. Sims NR. Helps SC. Rodgers KM. Vink R. Increased substance P immunoreactivity and edema formation following reversible ischemic stroke. Acta Neurochir Suppl 2006; 96 263-266.

[35] Turner RJ. Helps SC. Thornton E. Vink R. A substance P antagonist improves outcome when administered 4 h after onset of ischaemic stroke. Brain Res 2011; 1393 84-90.

[36] Turner RJ. Vink R. Combined tissue plasminogen activator and an NK1 tachykinin receptor antagonist: An effective treatment for reperfusion injury following acute ischemic stroke in rats. Neuroscience 2012; 220 1-10.

[37] Vink R. Young A. Bennett CJ. Hu X. Connor CO. Cernak I. Nimmo AJ. Neuropeptide release influences brain edema formation after diffuse traumatic brain injury. Acta Neurochir Suppl 2003; 86 257-260.

[38] Turner RJ. Vink R. The role of substance P in ischaemic brain injury. Brain Sciences 2012; 3 123-142.

[39] Stumm R. Culmsee C. Schafer MK. Krieglstein J. Weihe E. Adaptive plasticity in tachykinin and tachykinin receptor expression after focal cerebral ischemia is differentially linked to gabaergic and glutamatergic cerebrocortical circuits and cerebrovenular endothelium. J Neurosci 2001; 21 798-811.

[40] Markowitz S. Saito K. Moskowitz MA. Neurogenically mediated leakage of plasma protein occurs from blood vessels in dura mater but not brain. J Neurosci 1987; 7 4129-4136.

[41] Cyrino LA. Cardoso RC. Hackl LP. Nicolau M. Effect of quercetin on plasma extravasation in rat CNS and dura mater by ACE and NEP inhibition. Phytother Res 2002; 16 545-549.

[42] Vink R. Donkin JJ. Cruz MI. Nimmo AJ. and Cernak I. A substance P antagonist increases brain intracellular free magnesium concentration after diffuse traumatic brain injury in rats. J Am Coll Nutr 2004; 23 538S-540S.

[43] Leonard AV. Thornton E. Vink R. Substance P as a mediator of neurogenic inflammation following balloon compression induced spinal cord injury. J Neurotrauma 2013; Sep 28.

[44] Bae YC. Oh JM. Hwang SJ. Shigenaga Y. Valtschanoff JG. Expression of vanilloid receptor TRPV1 in the rat trigeminal sensory nuclei. J Comp Neurol 2004; 478 62-71.

[45] Hu DE. Easton AS. Fraser PA. TRPV1 activation results in disruption of the blood-brain barrier in the rat. Br J Pharmacol 2005; 146 576-584.

[46] Cook NL. Vink R. Donkin JJ. van den Heuvel, C. Validation of reference genes for normalization of real-time quantitative RT-PCR data in traumatic brain injury. Journal of neuroscience research 2009; 87, 34-41.

[47] Harford-Wright E. Thornton E. Vink R. Angiotensin-converting enzyme (ACE) inhibitors exacerbate histological damage and motor deficits after experimental traumatic brain injury. Neurosci Lett 2010; 481 26-29.

[48] Gabrielian L. Willshire LW. Helps SC. van den Heuvel C. Mathias J. Vink R. Intracranial pressure changes following traumatic brain injury in rats: lack of significant change in the absence of mass lesions or hypoxia. J Neurotrauma 2011; 28 2103-2111.

[49] Zacest AC. Vink R. Manavis J. Sarvestani GT. Blumbergs PC. Substance P immunoreactivity increases following human traumatic brain injury. Acta Neurochir Suppl 2010; 106 211-216.

Chapter 6

The Non Invasive Brain Injury Evaluation, NIBIE – A New Image Technology for Studying the Mechanical Consequences of Traumatic Brain Injury

Hans von Holst and Svein Kleiven

Additional information is available at the end of the chapter

http://dx.doi.org/10.5772/57344

1. Introduction

Traumatic brain injury (TBI) is an epidemiologically well-known disorder that ranges from minor to severe conditions (Kleiven, Peloso, von Holst, 2003). The aetiology of TBI is due to external violence and similar all over the world. About 80% is defined as mild and in this category, most patients recover completely after a certain time, ranging from days to months. Another 10% is defined as moderate with a prolonged recovery period, depending on a number of causes such as type and localisation of injury. The final 10% covers severe injuries, where the recovery period usually remains for life. In contrast to most other diseases, TBI has a sudden onset with a substantial impact on the patient´s close relatives, as complications are obviously not only of a physiologically handicapped nature but also psychological, due to the presence of personality changes. With an increase in worldwide social standard, TBI may unfortunately increase further. Thus, the World Health Organization has predicted TBI to surpass many diseases as one of the major causes of disability and death during the next decade unless external causes are reduced (World Health Organization, 2003).

As the external causes of TBI in most cases are known, the focus on primary prevention has already reduced the number of TBI successfully. One of the reasons for this reduction is due to the reconstruction of accidents by means of simulation. By using simulation methods and models, it is easier to discover innovations for prevention of the most severe accidents. The use of simulations will be of even greater importance in the future. Of equal importance is secondary prevention at the scene of the accident, in hospital care and the subsequent tertiary prevention during neurological rehabilitation. The overall mechanisms associated with TBI have been extensively investigated during the last decades by focusing on the biochemical and biomechanical consequences (Narayan, R.K, Wilberger, J.E, Povlischock, J.T, 1995). The introduction of new and advanced image technologies such as computerized tomography (CT) and magnetic resonance tomography (MRI) have, together with the introduction of neurological intensive care units, further improved the knowledge and consequences of TBI (von Holst, 2007). Today the recommended treatment is based on clinical practice and personal experience, with an increased interest in evidenced-based medicine an even more comprehensive and generally accepted treatment of TBI resulting in an expected improved outcome. Utilisation of simulation models and methods in clinical practice will further improve future intensive care treatment after TBI. The same scenario holds true also for tertiary prevention in neurological rehabilitation, as simulation methods and models will be introduced. Thus, the combined increased activities of primary prevention, pre-hospital care, neurosurgical interventions and neurological rehabilitation have certainly improved outcomes following TBI.

In this chapter we present the concept of a new image technology defined as non-invasive brain injury evaluation technology focusing on reconstructive potentials of TBI and how this can be implemented into clinical practice aimed at improvement of outcomes.

2. Biomechanics in TBI

Clinically, brain injuries can be classified into two broad categories: focal injuries and diffuse injuries. The focal brain injury is a lesion causing local damage, which can be seen by the naked eye. The diffuse brain injury is associated with global disruption of brain tissue and is usually invisible. Focal injuries consist of epidural haematomas (EDH), subdural haematomas (SDH), intracerebral haematomas (ICH) and contusions (coup and contrecoup). Diffuse injuries consist of brain swelling, concussion and diffuse axonal injury (DAI) (Melvin, 1993).

The human brain is sensitive to rotational motion (Holbourn, 1943; Gennarelli et al., 1987). In pioneering work, Holbourn (1943) observed shear strain patterns in 2D gel models and claimed that translation is not injurious; on the other hand, rotation could explain the majority of traumatic brain injuries due to the nearly incompressible properties of brain tissue. The bulk modulus of brain tissue is roughly five to six orders of magnitude larger than the shear modulus (McElhaney et al., 1976), so that for a given impact, it tends to deform only in shear. This gives significant sensitivity of the strain in the brain to rotational loading and small sensitivity to linear kinematics (Kleiven, 2006). Therefore, rotational kinematics should be a better indicator of traumatic brain injury risk than linear acceleration. Additionally, it has been shown that the most common severe injuries, such as subdural haemorrhage and diffuse axonal injury are more easily caused by rotational head motion (Gennarelli et al., 1972, 1987). Gurdjian and Gurdjian (1975) suggested that a combination of skull deformation, pressures and inertial brain lag could present a clearer picture of head injury (Gennarelli et al., 1982) stated that all types of brain injury can be produced by angular acceleration. According to Ommaya (1985), rotation can produce both focal and diffuse brain injuries, while translation is limited to focal effects.

http://dx.doi.org/10.5772/57344

2.1. Focal injuries

2.1.1. Epidural haematoma (EDH) and subdural haematoma (SDH)

Epidural haematoma is a relatively infrequently occurring sequel to head trauma (0.2-6%, Cooper, 1982). It occurs as a result of trauma to the skull and the underlying meningeal vessels and is not due to brain injury (Melvin et al., 1993). The mortality rate of SDH in most studies is greater than 30%. Acute subdural haematoma, together with diffuse axonal injury account for more head injury deaths than all other lesions combined (Gennarelli, 1981). SDH is the most common of the severe traumatic brain injuries, accounting for around 50% of the total number of injuries in this category in Sweden (Kleiven et al., 2003). The most common mechanism of subdural haematoma is tearing of the veins that bridge the subdural space as they go from the brain surface to the various dural sinuses (Gennarelli and Thibault, 1982). Based on previous primate experiments, Gennarelli (1983) suggested that SDH was produced by short duration and high amplitude of angular accelerations. Lee and Haut (1989) studied the effects of strain rate on tensile failure properties of human bridging veins and determined the ultimate strain to be about εf=0.5, which they found to be independent of the strain rate (=0.1-250 s-1). Earlier research done by Löwenhielm (1974) showed that the failure strain was markedly reduced from about 0.8 to 0.2 as the strain rate was increased. Lee et al. (1987) used a 2D sagittal model and Huang et al. (1999) used a 3D model (Shugar, 1977) to study the mechanisms of SDH. They found that the contribution of angular acceleration to tearing of bridging veins, measured as observed change in distance between a node in the interior of the skull and a node in the brain, was greater than the translational acceleration. Substantially larger relative motions between the skull and the brain, as well as higher strain in the bridging veins have been found when switching from a translational to a rotational mode of motion using a detailed 3D head model including 11 pairs of the largest bridging veins (Kleiven, 2003).

2.1.2. Contusion

Cerebral contusion is one of the most frequently found lesions following head injury. It consists of heterogeneous areas of necrosis, pulping, infarction, haemorrhage and oedema (Melvin et al., 1993). Contusions generally occur at the site of impact (coup contusions) and at remote sites from the impact defined as contrecoup contusions. Contrecoup lesions are more significant than coup lesions. Cerebral contusion at the site of impact in the presence of skull fracture is likely to be induced by the direct impression of the skull against the underlying brain tissue and therefore, in terms of skull fracture, caused by contact force and predicted by linear acceleration. In the absence of skull fracture,cerebral contusion is likely to be induced by shearing and scratching of the brain tissue against edges and sharper ridges in the dura/skull, and therefore caused by excessive head rotational loading (Löwenhielm, 1974). Moreover, Shreiber et al. (1997) derived a threshold of 0.19 in principal logarithmic strain in the cortex for a 50% risk of cerebral contusions. As previously mentioned, this strain is sensitive only to the rotational kinematics and not the translational motion (Ueno, Melvin, 1995; Kleiven, 2007).

2.1.3. Intracerebral haematomas (ICH)

Intracerebral haematomas are well defined homogeneous collections of blood within the cerebral parenchyma. They are most commonly caused by sudden acceleration/deceleration of the head. Other causes are penetrating wounds and blows to the head. Through reconstruction of a motocross accident, it was possible to re-create the injury pattern in the brain of the injured rider using the maximal principal strain (Kleiven, 2007b). The strain levels at maximum for two locations of intra-cerebral haematomas were around 0.4-0.5, which is close to known thresholds for rupture of cerebral veins and arteries (Monson, 2003; Löwenhielm, 1974; Lee, Haut, 1989), indicating that the risk of intra-cerebral haematomas can be predicted by the pattern and magnitude of maximum principal strain.

2.2. Diffuse Injuries

2.2.1. Concussion

The classical cerebral concussion involves immediate loss of consciousness following injury (Melvin et al., 1993). This is the most commonly occurring head injury, accounting for around 70% of total injuries, with more than 99% of patients leaving the hospital within 14 days (Kleiven et al., 2003). Studies on giant squid axons (Thibault, 1993) suggest a maximal principal strain of around 0.10 to cause reversible injury to the axons, which could be used as an approximate axonal strain threshold for concussion. During simulations of concussions in the National Football League (NFL), the strain magnitude in the brain was found to be sensitive to only rotational kinematics and not translational motion (Kleiven, 2007b).

2.2.2. Diffuse axonal injury (DAI)

Diffuse axonal injury is associated with mechanical disruption of many axons in the cerebral hemispheres and subcortical white matter (see Fig. 3). Microscopic examination of the brain discloses axonal tearing throughout the white matter of both cerebral hemispheres. It also involves degeneration of long white matter tracts extending into the brain stem. High-resolution CT scans may show small haemorrhages and axonal swelling (Fig. 3). DAI involves immediate loss of consciousness lasting for days to weeks. Severe memory and motor deficits are present and posttraumatic amnesia may last for weeks. At the end of one month, 55% of patients are likely to have died (Gennarelli, 1982).

The maximum strain that may cause damage to the axons has been estimated in previous publications. Studies have been performed with giant squid axons (Thibault et al., 1990) and a strain of 0.3 was suggested as threshold of diffuse axonal injury. Bain and Meaney (2000) proposed a threshold of 0.2 in maximal principal strain in the brain tissue for the onset of the malfunction of the neurons in the brain, which could be seen as a first stage of DAI. Maximum principal Green Lagrange strain of 0.2 has also been shown to correlate with cell death and neuronal dysfunction associated with DAI (Morrison et al., 2003). Gennarelli et al. (1972) subjected squirrel monkeys to controlled sagittal plane head motions. It was found that in the animals subjected to pure translation of the head, cerebral concussion was not obtainable. In

http://dx.doi.org/10.5772/57344

contrast, the animals who were subjected to head rotations were all concussed. Visible brain lesions were noted in both translated and rotated groups but with a greater frequency and severity after rotation. Ueno and Melvin (1995) found, when applying kinematics to a 2D head model, that rotational acceleration has a dominant effect on shear deformation, while linear acceleration is related to pressure.

3. The concept of NIBIE

NIBIE, Non-Invasive Brain Injury Evaluation, is a technology for measuring and evaluating parameters such as intracranial pressure, strain and stress from CT, MRI and Positron Emission Tomography (PET) scans (Fig. 1). The primary purpose of NIBIE is as a diagnostic tool for screening of patients with TBI and observation of intensive care patients with neurosurgical disorders such as haematoma, oedema or tumours. In addition, stroke patients may benefit from the method. NIBIE was created as a result of an interdisciplinary collaborative research project between engineers at the Royal Institute of Technology and neurosurgeons at the Karolinska University Hospital in Stockholm, Sweden.

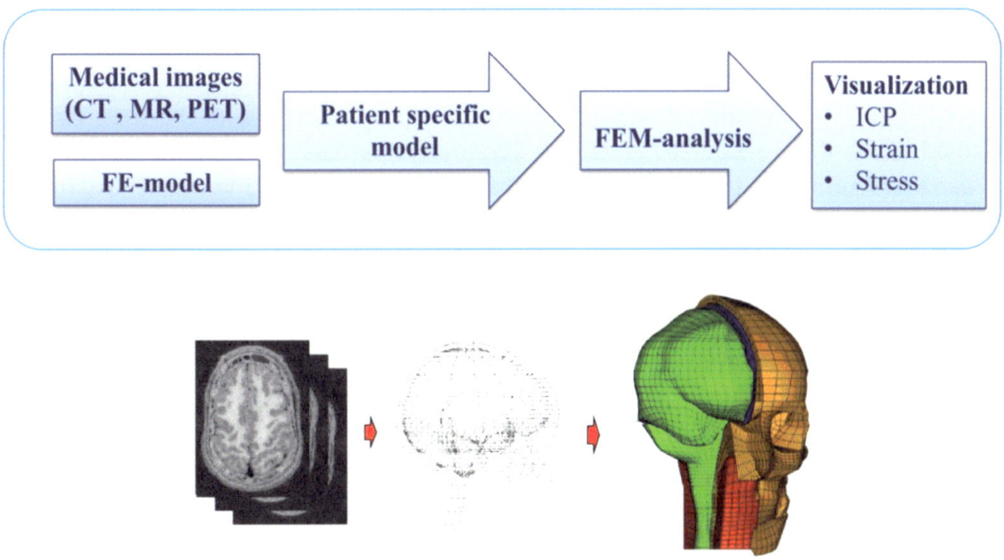

Figure 1. From medical image to FE modelling for visualization of ICP, strain and stress.

Research on numerical modelling of the head, which resulted in one of the most accurate finite-element models of the head available, made possible the use of the numerical model for medical applications in combination with existing medical diagnostic imaging data. The new image technology is well integrated with existing diagnostic tools and adjusted for each individual patient. NIBIE is unique in the sense that it can measure and define intracranial pressure without any neurosurgical procedure. It is the first non-invasive simulation method that is able to define biomechanic parameters in brain tissue on healthy and diseased people. Additionally, NIBIE has the capacity to define the strain level in the brain in order to evaluate the impact on the fibre tracts and classify what local and regional areas need to be focused on following TBI and stroke. Moreover, in neurosurgical procedures, NIBIE has proven to be safe for calculating the optimal neurosurgical approach into the brain with regard to the skull bone opening. Calculation of the most optimal head position with regard to gravitation is another possibility for reducing the pressure and strain level in the injured area. Further possibilities include the calculation of strain levels in fibre tracts following hydrocephalus and brain tumours to potentially explain gait apraxia, incontinence and cognitive dysfunctions. Thus, it is clear that the NIBIE software program is superior for the identification of biomechanical influence on brain tissue when compared to that found in CT, MRI and PET images.

4. Technology of NIBIE

The introduction of CT, MRI, PET and other images for medical use has substantially improved clinical knowledge and practice. Further improvement is to be expected parallel with developments in computer capacity. The next generation of innovative images supporting clinical neuroscience is the introduction of simulation technology. Among those that should be mentioned is the finite element (FE) method, or modelling, which dates back more than a hundred years. However, the phrase 'finite element' was initiated about 60 years ago when Ray Clough (1980) introduced and published what today is defined as the finite element method. Since then, a number of further improvements have been developed by other researchers worldwide. The advantage of FE modelling is that the numerical method has the capacity to deal and analyse larger and more complex geometrical structures by dividing, or discretizing them into smaller and simple geometrical structures defined as elements (Fig. 2). The elements, which are of one, two or three dimensions, are connected to each other by nodes and the system is usually defined as a mesh. Dividing the human head and neck using the FE technique made it possible for the first time to further study the biomechanical consequences of TBI and stroke.

Depending on the size of the structure to be analysed, the number of elements differ from a few to several hundred thousand. In general, the number of elements and nodes in the mesh is dependent on the accuracy needed for analysing a problem. Depending on the purpose of the analysis, a number of mathematical equations are connected to each of the elements, thereby giving the mesh a relationship between different mechanical loadings of, for instance, forces and displacements in the mesh. Parallel with the introduction of CT, MR and PET images, the FE method was further developed for engineering purposes such as simulations, of which the transportation industry should be mentioned in particular. Of special interest was the introduction of colour graphics in combination with powerful computers, which gave the FE method better understanding and capacity for clinical implementation.

In the last twenty years, an increased interest in the FE method has been highlighted in the health care sector mainly due to its capacity for performing reconstructive biomechanical

http://dx.doi.org/10.5772/57344

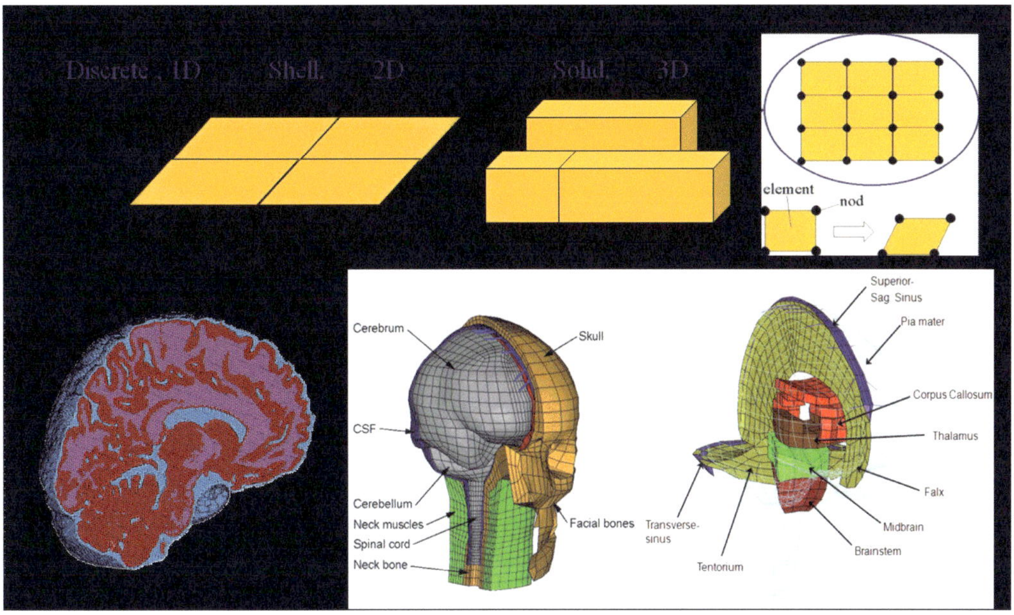

Figure 2. Schematic illustration of FE model development process from 1D, 2D and 3D elements to the construction of human head and neck models.

simulations of accidents (Kleiven, 2007). Although existing image technologies have substantially increased knowledge, there are still areas in need of new innovative image methods. This holds true especially in the field of clinical neuroscience. For instance, there is no image technology today that has the capacity to analyse a number of important biomechanical issues, which in turn will improve clinical assessment and clinical practice. Some of these problems include analysis of intracranial pressure in the brain tissue, the strain level in nerve fibres and stress in bone materials.

4.1. A computational 3D FE model

The development of advanced computational technology that started roughly three decades ago has increased interest in introducing human FE head models today (Ward and Thompson, 1975; Kumarezan, Radhakrishnan and Ganesan, 1995; Zhou, Khalif and King, 1995; Krabbel and Muller, 1996). With increased knowledge and interdisciplinary collaboration, more advanced 3D FE models have seen the light (Kleiven and von Holst, 2002; Johnson, 2008; Li, 2012). The technology has also become more accessible through easy-to-use interfaces, provided by most commercial FE-codes. The latest 3D versions include all anatomical structures of the human brain. As such, today a modern 3D FE head model includes the skin, all three bone layers, meninges, cerebrospinal fluid, brain tissue and aqueduct – including the foramina located in the skull base of the brain tissue. Depending on the purpose and need for analysis, the elements may consist of shell and solid tetrahedral or hexahedral configurations and can be used where they are most appropriate.

A prerequisite for using a modern FE method for TBI in clinical practice is access to a realistic and reliable 3D FE model of the human head. First, a FE model of the human head and brain includes the development of the outer and inner skull bone with the porous bone in between. Second, the model must further include the meninges including the dura, falx and tentorium and pia mater. Third, the construction of cerebrospinal fluid circulation requires the two lateral ventricles, the aqueduct connecting the third and fourth ventricles where it ends in the skull base and where the brain passes into the spinal cord. Fourth, the complete 3D FE model contains the grey and white matter of the brain tissue. Following complete development, it is necessary for the proposed 3D FE model to be validated via a significant amount of patient data before it can be accepted for use as a simulation tool in clinical practice. The 3D FE model can thus be extensively individualized and might tentatively have a substantial impact on clinical practice in the near future.

Increased knowledge from clinicians about the potential of using 3D FE modelling and what the FE method can add to existing knowledge will hasten the method to be implemented into clinical practice. For this purpose, an advanced and validated numerical model of the human brain was developed with the FE method; some of the results from such simulations are presented here, aiming at its introduction to neurosurgical practice of TBI and other neurological disorders. It can be tentatively suggested that the advanced FE model, including geometrically detailed descriptions, will predict injuries with good accuracy (Kumarezan, Radhakrishnan and Ganesan, 1995). It should be stressed that the FE method provides a means for studying how complex structures are affected by external loading and has long been used for applications in civil engineering and in the manufacturing industry. Thus, the method is well-suited for analysis of complex biological structures in the head and neck, as it will likely improve clinical practice in neurosurgery in the future.

5. Applied neuroscience

One of the most important neurological complications of moderate and severe traumatic brain injury (TBI) is the development of brain tissue oedema, which consists of an abnormal accumulation of fluid within the brain parenchyma. Oedema and its associated complications account for approximately 50% of deaths in patients with TBI (Marmarou, 2003). Raised intracranial pressure (ICP) is found in the majority of severe traumatic brain injuries caused by intracranial haematomas and brain swelling, and may be deleterious for the patient unless treated effectively (von Holst, 2007). Vasogenic and cytotoxic oedema are the two major types of oedema after TBI. Vasogenic oedema is due to blood-brain barrier disruption, resulting in increased extracellular water accumulation, while cytotoxic oedema is defined as increased intracellular water collection (Marmarou, 2003). As the aetiology of vasogenic oedema is relatively well understood, the treatment thereof is fairly effective. However, the mechanisms of cytotoxic brain oedema are still unclear, which renders a treatment of choice inadequate. Research has shown that the brain swelling observed in patients with TBI appears to be predominantly cellular, while vasogenic oedema is present at a minor degree (Marmarou,

Singoretti and Fatouros et al., 2006). This makes the clinical treatment of TBI oedema more complicated.

Although improvements in outcome have been made in recent decades, much remains to be done, especially since the recommended treatment is predominately based on clinical practice and personal experience. Clinically, local and minor oedema may be treated conservatively only by observation, while more extensive oedematous areas demand intensive care and where the treatment of choice usually follows evidence-based practice (Rabinstein, 2006; von Holst, 2007). Monitoring the ICP is an integral part of intensive care treatment following moderate and severe TBI. The longer the ICP exceeds 20–25 mm Hg, the poorer the outcome for the patient is expected (von Holst, 2007). On admittance to hospital, the patient with moderate and severe TBI is placed with the head at a 30-degree elevation aimed at optimizing the ICP, the cerebral perfusion pressure, the venous drainage from the head as well as the pulmonary function (Rabinstein, 2006; von Holst, 2007). Anatomically detailed FE models have been developed in an effort to simulate consequences of various impacts to the human head in the same group (Ho and Kleiven, 2009), which gives both researchers as well as clinicians a better understanding of TBI. Poroelasticity has been used to simulate hydrocephalus and brain oedema in early studies (Nagashima, Tamaki and Matsumotu et al., 1987). In more recent studies, simplified geometry, such as cylindrical or spherical geometry, has been adopted to investigate the mechanisms behind hydrocephalus (Kaczmarek, Subramaniam and Neff, 1997; Levine, 1999; Smille, Sobey and Molnar, 2005). Three dimensional (3D) FE models with anatomically detailed brain structures have also been used (Dutta-Roy, Wittek and Miller, 2008; Li, von Holst and Kleiven, 2009). Gravity has been shown to play an important role in brain shift during craniotomy in a series of studies using FE models (Coffey, Garg and Miga et al., 2010; Miga, Roberts and Kennedy et al., 2001). However, to date, there has been no experimental or clinical study evaluating the patient's head position with regard to oedema localization considering the gravity aspects, which could potentially alter both the clinical treatment as well as the outcome following TBI.

Formation of brain oedema involves fluid movement from the vasculature directly into the intracellular space (cytotoxic brain oedema) or extracellular space defined as the vasogenic oedema (Marmarou, 2003). The circulation of CSF tends to be disturbed for oedema patients with high ICP. However, the detailed mechanisms behind oedema and disturbed CSF circulation fall outside the range of this study. We used a fluid source added to the oedema to simulate extra fluid accumulation. In this model, we considered the influence of gravity on a patient with oedema at the posterior part of the brain for the supine and prone positions. The model was based on the normal CSF circulation model with the addition of a focal oedema at the posterior part of the brain. A higher localized interstitial fluid pressure (IFP) at the oedema zone due to extra oedema fluid accumulation is seen for both positions. The average IFP at the oedema zone decreased around 15%, from 3,331 Pa to 2,824 Pa when changing from the supine to prone position. For the supine position, the IFP decreased from the oedema zone the whole way to the frontal part of the brain. For the prone position, the IFP at tissue adjacent to the oedema showed a similar tendency to the supine position; however, IFP started to increase at a certain distance from the oedema zone due to hydrostatic pressure induced by gravity. The tissue pressure gradient drove the interstitial fluid away from the oedematous zone to other parts of the brain. When IFP increased, the brain tissue swelled due to the pressure gradient acting on the tissue skeleton. Since brain cells and CSF were nearly incompressible (Kaczmarek, Subramaniam and Neff, 1997), the enlarged volume was expected to be filled with extra fluid. The predicted water content increment was about 10% at the centre of the oedema and decreased towards other areas of the brain. The value of water content increment was nearly identical for both positions. This was to be expected, since the swelling of oedematous tissue was caused by the IFP gradient acting on the solid skeleton rather than the IFP. Both IFP and water content increment distribution in the brain was similar to those reported in experiments on cold-injured oedema in cats (Reulen and Kreysch, 1973).

5.1. Epidural, subdural haemorrhages and intracranial pressure

On arrival at the hospital from the scene of an accident it is not always easy to clinically judge the patient´s condition with regard to potential injuries in the brain tissue that might be in need of immediate handling by emergency staff. A better clinical evaluation can be done by performing a CT scan, which can be done within a short period of time. Among the most important injuries are the presence of epidural haematomas (EDH) and any signs of increased ICP. As shown in Fig. 3, the EDH is clearly visible. As this EDH will cause a substantial midline shift, the ultimate choice of treatment is an acute neurosurgical evacuation of the haemorrhage, which aims to reduce the expanding volume to avoid further secondary injuries. Additionally, as the ICP is probably increased in this patient, it is important to judge whether it should be monitored continuously or not. These two types of injuries are of great interest when it comes to the application of the NIBIE software image. The CT image is digitized over to NIBIE (Fig. 1) for the evaluation of secondary complications such as potential increased strain and ICP levels, two criteria which may have a dramatic effect on brain tissue with regard to secondary injuries.

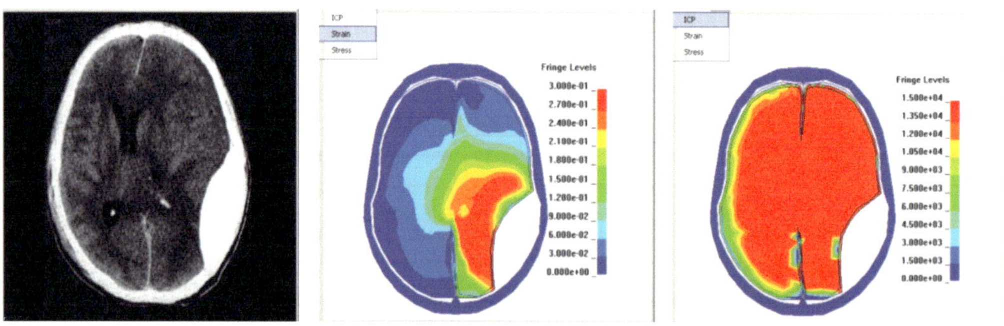

Figure 3. CT-verified epidural haematoma (left) analysed with NIBIE with regard to strain (middle) and ICP (right).

The increased strain level shown in Fig. 3 is highest in the nearest vicinity of the EDH and reduces with distance from the EDH. After the acute evacuation of the EDH, the strain level is simultaneously reduced, thereby preventing further secondary injuries. The strain level shown in Fig. 3 offers significant support to clinicians in neurological intensive care and for rehabilitation staff. Comparing the CT scan with the degree of strain level, NIBIE may also have the capacity to define the anatomical area of most importance for treatment and observation. This is also beneficial to the rehabilitation staff in terms of what region to focus on when the time is ready for more active neurological rehabilitation.

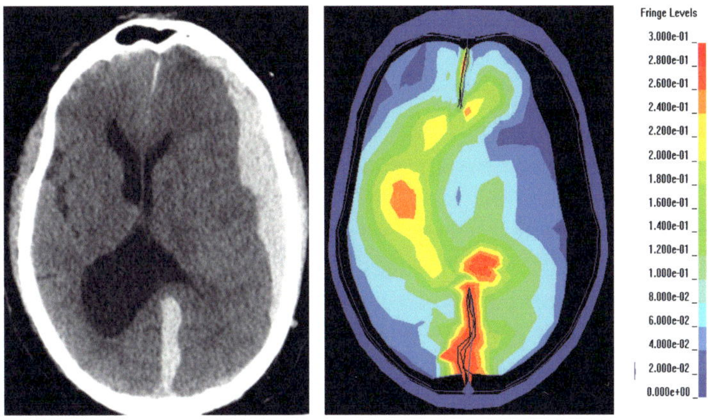

Figure 4. CT verified subdural haematoma (left) evaluated with NIBIE and strain level (right).

In contrast to the strain level, the FE model of NIBIE shows that the ICP is substantially increased throughout the brain. The evacuation of the EDH volume will not only reduce the increased strain level, but also the increased ICP. Usually, monitoring equipment is obligatory for a suspected increased ICP and is implanted right after the evacuation of the EDH during the same neurosurgical procedure. As NIBIE has proven reliable in defining normal or increased ICP levels, an implantation of monitoring equipment is not always necessary. In future, ICP can easily be defined by repeated CT and NIBIE evaluation during the neurological intensive care treatment period.

Not only patients suffering from EDH, but also those admitted with subdural haematomas may also benefit for the same reasons. While EDH is mostly acute and found in about two per cent of all types of injury, SDH is more frequent and divided into acute, sub-acute and chronic injury, depending on the age of the haematoma, which is not always easy to estimate. One possible and more precise age definition of the SDH could be the definition of strain level, which should be reduced when the haematoma has transferred from a clot to fluid (Fig. 4). In this case, NIBIE has the capacity to define the area of potential secondary injury and that may be found in nerve fibres showing an increased strain level.

With the introduction and use of NIBIE for clinical application, new injury criteria may be defined by focusing on strain and ICP levels. It is therefore quite possible that the new criteria will successively increase knowledge among health care staff. By using each slice from the CT and evaluating them in NIBIE, it is possible to gain a holistic view of the severity of the injury and hence better information on potential secondary injuries, and what clinicians should focus on when it comes to direct and more long term treatment. Of interest are patients with asymptomatic chronic SDH. Here, evaluation with NIBIE may give insight to whether the brain tissue shows any evidence of strain, which will be clinically helpful for clinicians in terms of their decisions for clinical treatment in some of the more questionable patients. This new insight will further encourage better clinical evaluation and treatment, resulting in better clinical outcomes for patients with EDH, SDH and increased ICP. This scenario also holds true for other cerebral injuries after TBI.

5.2. Influence of gravity

In general, gravity is an important factor for consideration in TBI, since it may have a dramatic influence on the pressure around an intracerebral lesion or haemorrhage. In the neurosurgical procedure, gravity is of paramount importance to facilitating an operation. When it comes to clinical treatment, this aspect is, however, not always considered in the intensive care phase. Here, the non-invasive brain injury evaluation, NIBIE, has proven promising in evaluating optimization of the head position. When the head of the patient with a cerebral lesion is positioned in the usual supine position, as is normally the case, the pressure is usually higher compared to if the same patient is instead considered for a more prone position (Fig. 5). Using NIBIE, it has been shown that by switching from the supine to prone position, the pressure in the injured area is reduced by about 15%, depending on the size and localization of the injury. Switching from a supine to a more prone position may thus cause a substantial reduction in secondary injury, thereby promoting shorter in-hospital treatment and a better outcome.

It has been shown that reducing an increased ICP after severe TBI can significantly reduce mortality rate. This is supported in clinical treatment where the patient's head is elevated to about 30 degrees to facilitate cerebral blood flow (Rabinstein, 2006). The pressure gradient in the central nervous system is not always equally distributed inside the cranium following TBI (Miindermann, 1999). Instead, it has been found that pressure gradients also exist between the two hemispheres or even between neighbouring areas of the same hemisphere, especially in the oedema zone, where local tissue pressure increases as a result of a pathological accumulation of extra fluid (Reulen and Kreysch, 1973). Additionally, it can be tentatively suggested that even a slightly increased interstitial fluid pressure IFP might trigger a series of pathological processes to the already injured and vulnerable tissue inside and surrounding the oedema zone (Li, von Holst and Kleiven, 2011). Following the biomechanical cascades of events in moderate and severe TBI, both necrosis and apoptosis was seen to be present, although they differed in some aspects. Necrosis was found as a response to tissue damage due to biomechanical or ischemic influences.

In contrast, neurons undergoing apoptosis are morphologically intact during the immediate post-traumatic period, with adequate biochemistry providing the existence of normal membrane potential (Werner and Engelhard, 2007). When the hydrostatic pressure reaches 30 mmHg the neurochemical cascade of events is initiated, thereby having the potential to trigger both neurodegenerative diseases and apoptosis in the brain tissue (Ju, Liu, Kim and Crowston et al., 2007). From a biomechanical point of view, the consequence of an increased interstitial fluid pressure may result in larger deformation of the cellular membranes, thus altering the

http://dx.doi.org/10.5772/57344

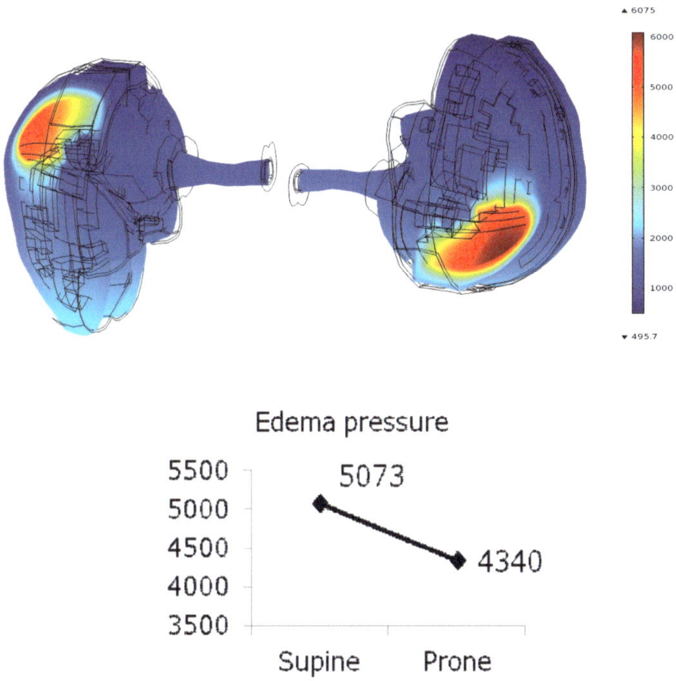

Figure 5. By changing the patient's head position from supine to a more prone position the pressure in the oedematous tissue is reduced by about 15 %.

pressure between the extracellular and intracellular environment. An elevated interstitial fluid pressure due to oedema compresses the vasculature, the consequence of which is reduced cerebral blood flow, potential ischemia and arteriolar dilatation. This makes the capillary pressure increase, which results in even more interstitial fluid accumulation (Werner and Engelhard, 2007).

Gravity causes a hydrostatic pressure gradient in compartments filled with fluid. The same pressure is present in the centre of gravity in the head and is thus set to zero. As is found in neurosurgical procedures, a postural change of the head position alters the distribution of hydrostatic pressure patterns according to the body's alignment to the gravity field (Hinghofer-Szalkay, 2011). However, at a certain location – referred to as hydrostatic indifferent point – the pressure remains constant during a given change of body position. When it comes to defining the most optimal head position in clinical practice between supine and prone positions, this remains to be evaluated when the indifferent point is presented. Thus, when positioning the TBI patient in the supine position as a routine, the treatment of choice is suboptimal, as it does not consider the anatomical localization of the injury. The potential verification of changing from supine to a more prone position has been shown by using an anatomically detailed 3D FE model in simulation research (Li, von Holst and Kleiven, 2011), where the effects of gravity on the oedema zone at the posterior part of the brain were investigated. It is hoped that these results may be implemented into clinical practice and may change the existing best evidence synthesis of intensive care for patients with TBI.

5.3. Decompressive hemicraniectomy

Decompressive procedures in neurosurgery are excellent methods for reducing the ICP in patients suffering from TBI or stroke. The number of such procedures has increased substantially worldwide with good effect on increased ICP. Of specific interest is decompression of the brain tissue by a larger hemicraniectomy. There are no real answers for these results. Thus, the debate on its overall effectiveness remains and complete consensus has not been achieved among clinicians (Cooper, Rosenfeld and Murray et al., 2011; Vashu and Sohail, 2011) From a global perspective, the number of TBI in need of decompressive hemicraniectomy will remain at an unacceptably high level for a long time to come. Presently, there is debate on how to interpret data from the decompressive craniotomy and apply it to clinical practice (Servadei, 2011).

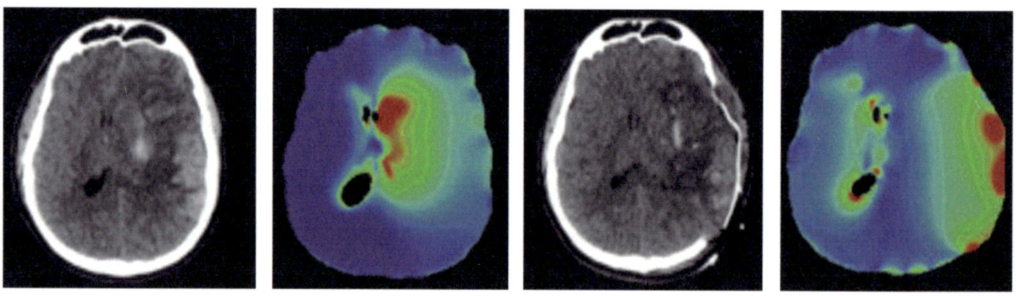

Figure 6. Evaluation of strain level in different anatomical areas in the brain tissue before (left black image with CT, left colour image with NIBIE) and after the decompressive hemicraniectomy (right black image with CT, right colour image with NIBIE).

The purpose of decompressive craniotomy is to reduce the ICP by allowing expansion of the brain tissue outside the skull bone. However, the treatment also results in stretching of the axonal fibres, which has been suggested to contribute to an unfavourable outcome for patients treated with decompressive craniotomy (Cooper, Rosenfeld and Murray et al., 2011; Stiver, 2009). Thus, there is a need for new clinical methods that may have the capacity to better consider the biomechanical consequences of swollen brain tissue after decompressive hemicraniectomy. With the new image technology of NIBIE, it is possible to evaluate the consequences by focusing on strain level before and after the neurosurgical procedure, which may have possibly improve our knowledge in this context. In fact, an increased strain may account for why we don´t see more significant improvement in these patients.

Prior to the neurosurgical procedure (Fig. 6, left), the CT showed increased strain levels of various degree in and around the injured brain tissue. When evaluating the consequences of decompressive hemicraniectomy, the strain level of the nervous tissue on the same side was also increased to over 60% and also more widespread (Fig. 6, right), and interfering with the

http://dx.doi.org/10.5772/57344

remote hemisphere. Increased strain level can also be found when evaluating several slices from NIBIE, where the decompressive hemicraniectomy causes the injured brain tissue to expand beyond the skull bone. This results in strain levels well over 60% (Fig. 7).

• Strain level in %

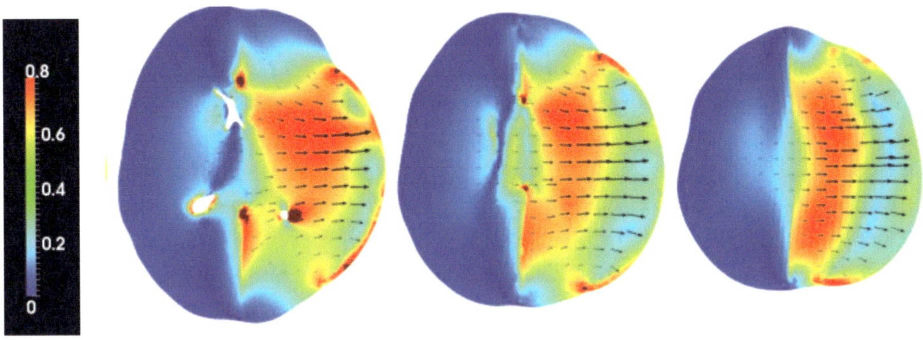

Figure 7. Three image sections from NIBIE showing increased strain levels of over 40% on the injured side (right hemisphere) after decompressive hemicraniectomy.

When biomechanical effects cause the nerve fibres to become stretched abnormally high during the neurosurgical procedure, the normal biochemical metabolism is also altered. Axons transfer electrical-chemical impulses between neurons, including their intact axons. These activities are critical for establishing normal clinical function in the nervous system. When parts of the axonal fibres are stretched, the capacity to transfer physiological impulses is altered and may even result in permanent loss of functional capability when stretched too much (Joseph, 1996). *In vitro* models for studying injury have shown that axonal stretch causes a number of neural derangements including neurofilament structure alterations (Chung, Staal and McCormack, 2005), immediate rise in intracellular calcium level after injury (Staal, Dickson and Gasperini, 2010), mechanical breaking of microtubules during stretch in axons (Tang-Schomer, M.D, Patel, A.R and Baas et al., 2010) and axonal swelling formation (Smith, Wolf and Lusardi et al., 1999).

An increase in strain levels as low as a 5% increase will alter neuronal function, while a strain level higher than 20% induces significant levels of cell injury in vitro (Morrison III, Cater, Wang et al., 2003). Bain et al. (2000) demonstrated that a strain level of approximately 21% will initiate electrophysiological changes, while a strain of approximately 34% results in morphological signs in the white matter. Combined, these studies have increased the knowledge in response to biomechanical stretch as found after the neurosurgical procedure. This may explain why decompressive craniotomy not only reduces ICP, but may also contribute to unfavourable outcomes for patients by increasing the strain level. Thus, with NIBIE, it is possible to quantify the axonal stretching to better understand the consequences of the neurosurgical procedure for many of the most severely injured patients, thereby having a better insight into the potential damages to the nervous tissue after TBI and stroke. In general, decompressive craniotomy results in complex axonal deformation; it is difficult to apply these cellular level thresholds to the tissue level, since the axons within the white matter do not necessarily lie in the same orientation as the stretching direction. Most of the previous axonal injury models use dynamic brain tissue during impact to study strain. This is different to the case of axonal stretch for the post-craniotomy stage, under which the axons are enduring slow dynamic events similar to a quasi-static stretching. The incorporation of fibre tracts in biomechanical simulation models is therefore necessary to obtain stretching along the axons, thus making it comparable with the threshold obtained from laboratory investigations.

The displacement and stretching of injured cerebral tissue after TBI and stroke is usually found for several days. This may further challenge the cerebral metabolism. Using a model of sciatic nerve stretch, it was reported that even minimal tension, if maintained for a significant amount of time, may result in loss of neuronal function (Fowler, Leonetti and Banich, 2001). Hence, it should be expected that the central nervous system will also sustain potential damage under long duration stretch, such as in the post-craniotomy stage, but with a different threshold level. The strain level representing the stretch of brain tissue has been quantified in a previous study (von Holst, Li and Kleiven, 2011). It was shown that following decompressive craniotomy, the strain level, as well as the water content in the brain tissue, was substantially increased. This may influence the axonal fibres in such a way that the neurochemical events are jeopardized. Axonal fibre tracts extracted from diffusion weighted (DW) images have been included in a biomechanical model simulating an impact event in order to study the axonal elongation occurring at the primary injury stage (Chatelin, Deck and Renard et al., 2011). Results from this study showed that stretching of axons correlates closely with diffuse axonal injury (DAI). However, axonal stretching during the post-craniotomy period, which may have prognostic value for the cognitive and neurological sequelae of patients treated with decompressive craniotomy, has not been previously studied.

6. Conclusion

Parallel with an increased theoretical knowledge of TBI, clinical neuroscience practice has witnessed a remarkable improvement with new technological equipment aimed at emergency care at the scene of accident, intensive care, as well as for neurosurgical procedures. Concurrently, the significant development of computer capacity during the last two decades has made it possible to implement various software programs having the capacity to better understand the mechanical consequences following an accident to the central nervous tissue. Thus, the integration of clinical practice with neuroengineering used for applied neurosurgery may further improve the future outcome of patients suffering from moderate and severe TBI. The integrated research between clinical neurosurgery and neuronic engineering has resulted in the development of a method for generating a numerical model based on 3D medical images, from either computer tomography (CT) or magnetic resonance imaging (MRI). We have found that simulation of TBI in numerical models can define the intracranial pressure without neurosurgical procedures. In addition, the strain level, which is related to the stretching of axons, can be used as a new predictive value in diagnoses, e.g., the effect of oedema and haemorrhage. Moreover, it is now possible to analyse the outcome of a treatment in, for

http://dx.doi.org/10.5772/57344

instance, decompressive hemicraniectomy. By using numerical models of the human brain, it is possible to further optimize present treatments of TBI. Finally, using NIBIE in education of health care staff in all categories in the new field of neuroengineering is of significant importance in order to better understand the consequences of diseases in the central nervous system.

Acknowledgements

The present chapter was sponsored by Vinnova, Karolinska University Hospital and the Royal Institute of Technology, Stockholm, Sweden.

Author details

Hans von Holst[1,2] and Svein Kleiven[2]

1 Dept. of Neurosurgery, Karolinska University Hospital, Stockholm, Sweden

2 Division of Neuronic Engineering, Royal Institute of Technology, Stockholm, Sweden

References

[1] Bain, B.C. and Meaney, D.F, (2000) Tissue-Level Thresholds for Axonal Damage in an Experimental Model of Central Nervous System White Matter Injury. J. Biomech. Engng.16, pps 615-622.

[2] Chatelin, S, Deck, C, Renard, F, Kremer, F, Heinrich, C, Armspach, J.P, Willinger, R, (2011). Computation of axonal elongation in head trauma finite element simulation. Journal of the Mechanical Behavior of Biomedical Materials, 4(8):1905–1919.

[3] Chung, R.S, Staal, J.A, McCormack, G.H, Dickson, T.C, Cozens, M.A, Chuckowree, J.A, Quilty, M.C, Vickers, J.C, (2005) Mild axonal stretch injury in vitro induces a progressive series of neurofilament alterations ultimately leading to delayed axotomy. Journal of neurotrauma, 22(10):1081–1091.

[4] Clough, R. W., (1980). The finite element method after twenty-five years: A personal review. Computers and Structures, Vol 12 (4), 361-370.

[5] Coffey A.M, Garg I, Miga M.I, Thompson R.C (2010). An evaluative tool for preoperative planning of brain tumor resection. Proc SPIE 7625:762531. doi:10.1117/12.844477

[6] Cooper, D.J, Rosenfeld, J.V, Murray, L, Arabi, Y.M, Davies, A.R, D'Urso, P, Kossmann, T, Ponsford, J, Seppelt, I, Reilly, P, Wolfe, R (2011). Decompressive craniecto-

my in diffuse traumatic brain injury. New England Journal of Medicine, 364(16): 1493–1502.

[7] Cooper, P.R, (1982). Post-traumatic intracranial mass lesions, in: Head injury, Williams and Wilkins, Baltimore/London. pages 185-232.

[8] Dutta-Roy T, Wittek A, Miller K (2008) Biomechanical modeling of normal pressure hydrocephalus. J Biomech 41:2263–2271.

[9] Fowler, S.S, Leonetti, J.P, Banich, J.C, Lee, J.M. Wurster, R, Young, M.R.I, 2001) Duration of neuronal stretch correlates with functional loss. Otolaryngology-Head and Neck Surgery, 124(6):641–644.

[10] Gennarelli, T.A, et al., (1972). Pathophysiological Responses to Rotational and Translational Accelerations of the Head, SAE Paper No. 720970, in: 16th Stapp Car Crash Conf., Society of Automotive Engineers, pp. 296-308.

[11] Gennarelli, T.A, et al., (1982) Diffuse Axonal Injury and Traumatic Coma in the Primate. Ann. Neurol. 12, 564-574.

[12] Gennarell, T.A, (1983). Head injuries in man and animals: clinical aspects. Acta neurochir suppl. 32, 1-13.

[13] Genarelli T.A, et al., (1987) Directional dependence of axonal injury and traumatic coma in the primate. Ann. Neurol. 12, 564-574.

[14] Gurdjian E.S, Lissner, H.R, (1944) Mechanisms of head injury as studied by the cathode ray oscilloscope: preliminary report. J Neurol Neurosurg Psychiatry 1:393-399.

[15] Hinghofer-Szalkay H, (2011) Gravity, the hydrostatic indifference concept and the cardiovascular system. Eur J Appl Physiol 111:163–174.

[16] Ho, J, (2008) Generation of patient specific finite element head models. Doctoral thesis report:7 ISBN: 978-91-7415-191-6.

[17] Ho, J, von Holst, H, Kleiven, S, (2009) Automatic generation and validation of patient specific finite element head models suitable for crashworthiness analysis. International Journal of Crashworthiness, 14(6): p. 555-563.

[18] Ho J, Kleiven, S, (2009) Can sulci protect the brain from traumatic injury? J Biomech 42: 2074–2080.

[19] von Holst, H, (2007) Traumatic brain injury. In: Feigin VL, Bennett DA (eds) Handbook of clinical neuroepidemiology. Nova Science, New York, pp. 197–232.

[20] von Holst, H, Li, X.G, Kleiven, S, (2011) Increased strain levels and water content in brain tissue after decompressive craniotomy evaluated with numerical analysis.

[21] Holburn, A.H.S, (1943) Mechanics of head injury. Lancet 2. October 9, 438-441.

[22] Huang H.M, Lee, M.C, Chiu, W.T, Chen, C.T, Lee, S.Y, (1999) Three-dimensional finite element analysis of subdural hematoma. The Journal of Trauma, Injury, Infection and Critical Care. 47 (3) 538-544.

[23] Joseph, R, (1996) Neuropsychiatry, neuropsychology, and clinical neuroscience: emotion, evolution, cognition, language, memory, brain damage, and abnormal behavior. Williams & Wilkins.

[24] Ju, W.K, Liu, Q, Kim, K.Y, Crowston, J.G, Lindsey, J.D, Agarwal, N, Ellisman, M.H, Perkins, G.A, Weinreb, R.N, (2007) Elevated hydrostatic pressure triggers mitochondrial fission and decreases cellular ATP in differentiated RGC-5 cells. Invest Ophthalmol Vis Sci 48:2145–2151.

[25] Kaczmarek, M, Subramaniam, R.P, Neff, S.R, (1997) The hydromechanics of hydrocephalus: steady-state solutions for cylindrical geometry. Bull Math Biol 59:295–323.

[26] Kleiven, S, von Holst, H,(2002) Consequences of head size following trauma to the human head. Journal of Biomechanics 35, 153-160.

[27] Kleiven, S., Peloso, P.M. von Holst, H, (2003) The epidemiology of head injuries in Sweden from 1987 to 2000. Inj Control Saf Promot, 10(3): p. 173-80.

[28] Kleiven, S, (2006) Evaluation of head injury criteria using an FE model validated against experiments on localized brain motion, intra-cerebral acceleration, and intracranial pressure, International Journal of Crashworthiness 11 (1), 65-79.

[29] Kleiven, S, (2007) Predictors for traumatic brain injuries evaluated through accident reconstructions. Stapp Car Crash J. 51: p. 81-114.

[30] Kleiven, S, (2007) Predictors for traumatic brain injuries evaluated through accident reconstructions. Stapp Car Crash J 51:1–35.

[31] Kumarezan, S, Radhakrishnan, S, Ganesan, N, (1995) Generation of geometry of closed human head and discretization for finite element analysis. Med. Biol Eng Comput 33, 349-353.

[32] Krabbel, G, Muller, R, (1996) Development of a finite element model of the human head using the visible human database. Abstract of the visible human project conference, 71-72.

[33] Lee, M.C, Haut, R.C, (1989). Insensitivity of tensile failure properties of human bridging veins to strain rate: Implications in biomechanics of subdural hematoma. J. Biomech. 22 (6/7), 537-542.

[34] Levine, D.N, (1999) The pathogenesis of normal pressure hydrocephalus: a theoretical analysis. Bull Math Biol 61:875–916.

[35] Li, X.G, (2012) Finite element and neuroimaging techniques to improve decision making in clinical neuroscience. Doctoral thesis report 2012:1 ISBN: 978-91-7501-240-7.

[36] Li, X.G, von Holst, H, Ho, J, Kleiven, S, (2009) Three dimensional poroelastic simulation of brain edema: initial studies on intracranial pressure. Proceedings of World Congress on Medical Physics and Biomedical Engineering, September 7–12, Munich. Springer, Heidelberg, pp. 1478–1481.

[37] Li, X, von Holst, H, Kleiven, S, (2011) Influence of gravity for optimal head positions in the treatment of head injury patients. Acta Neurochirurgica, 153(10): p. 2057-2064.

[38] Lövenhjelm, P, (1974) Dynamic Properties of the Parasagittal Bridging Veins. Z. Rechtsmedizin 74, pages 55-62.

[39] Marmarou, A, (2003) Pathophysiology of traumatic brain edema: current concepts. Acta Neurochir Suppl 86:7–10.

[40] Marmarou, A, Signoretti, S, Fatouros, P.P, Portella, G, Aygok, G.A, Bullock, M.R, (2006) Predominance of cellular edema in traumatic brain swelling in patients with severe head injuries. J Neurosurg 104:720–730.

[41] McElhaney, J, Roberts, V, Hilyard, J, (1976) Handbook of human tolerance, Japan Automobile Research Institute Inc., Tokyo, p. 143.

[42] Melwin, J.W, (1995). Injury assessment reference values for the CRABI 6-Month Infant Dummy in a Rear-Facing Infant Restraint with Air Bag Deployment. in: SAE International Congress and Exposition. , SAE Paper No. 950872, Society of Automotive Engineers.

[43] Miga, M.I, Roberts, D.W, Kennedy, F.E, Platenik, L.A, Hartov, A, Lunn, K.E, Paulsen, K.D, (2001) Modeling of retraction and resection for intraoperative updating of images. Neurosurgery 49:75–85.

[44] Morrison III, B.C, Cater, H.L, Wang, C.C-B, Thomas, F.C, Hung, C.T, Ateshian, G.A, Sundström, L.E, (2003) A tissue level tolerance criterion for living brain developed with an in vitro model of traumatic mechanical loading. 47TH Stapp Car Crash Conference Journal.

[45] Nagashima, T, Shirakuni, T, Rapoport, S.I, (1990) A two dimensional, finite element analysis of vasogenic brain edema. Neurol Med Chir 30:1–9.

[46] Narayab, R.K, Wilberger, J.E, Povlischock, J.T, (Editors) (1995) Neurotrauma. McGraw-Hill. ISBN 0-07-045662-3.

[47] Ommaya, A.K, Hirsch, A.E, (1971) Tolerances for Cerebral Concussion from Head Impact and Whiplash in Primates. J. Biomech, 4, 13-21.

[48] Rabinstein, A.A, (2006) Treatment of cerebral edema. Neurologist 12:59–73.

[49] Reulen, H.J, Kreysch, H.G, (1973) Measurement of brain tissue pressure in cold induced cerebral oedema. Acta Neurochir (Wein) 29:29–40.

[50] Schugar, T.A, (1977) A finite element head model Vol I: Theory, development and results. U.S. Depat. of Transportation Report No. DOT-HS-289-3-550-IA.

[51] Servadei, F, (2011) Clinical value of decompressive craniectomy. New England Journal of Medicine, 364(16):1558–1559.

[52] Shreiber, D.I, Bain, A.C, Meaney, D.F, Schreiber, D.I, Bain, A.C, (1997) In vivo thresholds for mechanical injury to the blood-brain barrier. 41st Stapp Car Crash Conf.

[53] Tang-Schomer, M.D, Patel, A.R, Baas, P.W, Smith, D.H, 2010 Mechanical breaking of microtubules in axons during dynamic stretch injury underlies delayed elasticity, microtubule disassembly, and axon degeneration. The FASEB Journal, 24(5):1401–1410.

[54] Thibault, L.E, (1993) Brain injury from the macro to the micro level and back again: What have we learned to date? in: IRCOBI Conf. 3-25.

[55] Smillie, A, Sobey, I, Molnar, Z, (2005) A hydroelastic model of hydrocephalus. J Fluid Mech 539:417–443.

[56] Smith, D.H, Wolf, J.A, Lusardi, T.A, Virginia, M.Y.L, Meaney, D.F, (1999) High tolerance and delayed elastic response of cultured axons to dynamic stretch injury. The Journal of neuroscience, 19(11):4263–4269.

[57] Staal, J.A, Dickson, T.C, Gasperini, R, Liu, Y, Foa, L, Vickers, J.C, (2010) Initial calcium release from intracellular stores followed by calcium dysregulation is linked to secondary axotomy following transient axonal stretch injury. Journal of neurochemistry, 112(5):1147–1155.

[58] Stiver, S.I, (2009) Complications of decompressive craniectomy for traumatic brain injury. Neurosurgical focus, 26(6):E7.

[59] Ueno, K, Melvin J.W,(1995) Finite element model study of head impact based on Hybrid III head acceleration; The effect of rotational and translational acceleration. Journal of Biomechanical Engineering. 117(3) 319-328.

[60] Vashu, R, Sohail, A, (2011) Decompressive craniectomy is indispensable in the management of severe traumatic brain injury. Acta neurochirurgica, 10:2065–2066.

[61] Ward, C.C, Thompson, R. B, (1975) The development of a detailed finite element brain model. SAE Paper No. 751163.

[62] Werner, C, Engelhard, K, (2007) Pathophysiology of traumatic brain injury. Br J Anaesth 99:4-9.

[63] World Health Organization (2003) World Health Report 2003: A vision for global health. Shaping the future. World Health Organization.

[64] Zhou, C., Khalif, T.B, King, A.I, (1995) New model comparing impact responses of the homogenous and inhomogeneous human brain. Stapp Car Crash Conference Proceedings, 121-137.

Section 2

Acute Management of Traumatic Brain Injury

Chapter 7

Management of Traumatic Brain Injury in the Intensive Care Unit

Farid Sadaka, Tanya M Quinn,
Rekha Lakshmanan and Ashok Palagiri

Additional information is available at the end of the chapter

http://dx.doi.org/10.5772/57298

1. Introduction

Traumatic brain injury (TBI) is a major source of death and severe disability worldwide. In the USA alone, this type of injury causes 290, 000 hospital admissions, 51, 000 deaths, and 80, 000 permanently disabled survivors [1, 2]. TBI occurs in two phases, primary and secondary brain injury. The primary injury results from the direct physical impact to the brain parenchyma resulting in structural and shearing injury of neurons, injury to vessels, and interruption of neurochemical processes. This leads to hemorrhage, edema, compression of intracranial structures. Primary injury is unalterable after the time of the trauma.The secondary injury, on the other hand, is characterized by a cascade of events that starts within minutes of the primary injury. As in ischemia –reperfusion injuries, the acute post-injury period in TBI is characterized by several pathophysiologic processes that start in the minutes to hours following injury and may last for hours to days. These result in further neuronal injury and are termed the secondary injury. The causes of the secondary injury can be evaluated by those that occur of the systemic or extracerebral level and those that occur on the cellular level. On the systemic level contributing factors include hypoxia, hypotension, hypercapnia, acidosis and hyperglycemia [3, 4].While the cellular mechanisms of secondary injury include all of the following: apoptosis, mitochondrial dysfunction, excitotoxicity, disruption in ATP metabolism, disruption in calcium homeostasis, increase in inflammatory mediators and cells, free radical formation, DNA damage, blood-brain barrier disruption, brain glucose utilization disruption, microcirculatory dysfunction and microvascular thrombosis [5-8]. All this leads to development of cerebral edema, blood-brain barrier disruption, vasospasm, increase in volume of bleeding and contusions, and intracranial hypertension.TBI patients, like other patients with brain injury, need multidisciplinary approach (neurointensivists, neurologists, neurosurgeons, and specialized nurses, respiratory therapists, physical therapists, nutritionists, etc.) for improved outcomes. Mortality and functional outcomes in all brain injured patients, including TBI patients improve when monitored and managed in neuroICUs and by neurointensivists [9-12]. This chapter will focus on the management of TBI patients in the intensive care unit.

2. Neurological assessment

The most widely used and most studied coma scale to date is the Glasgow coma scale (GCS) (Figure 1), first described by Teasdale and Jennett in 1974 and revised in 1976 with the addition of a sixth point in the motor response subscale for ''withdrawal from painful stimulus'' [13, 14]. The GCS was initially intended to assess level of consciousness after TBI in a Neurosurgical Intensive Care Unit (Neuro-ICU) [13].The GCS was broadly accepted as an instrument to classify the severity of TBI because it was easy to use and reproducible. It was used to classify the severity of TBI as mild (GCS 13-15), moderate (GCS 9-12), and severe (GCS 8 and below) [15, 16]. Since then it has become the gold standard against which newer scales are compared. As a result, the GCS was incorporated into several scoring systems, including the APACHE II [17], the Simplified Acute Physiology Score (SAPS) and SAPSII [18], the Revised Trauma Score (RTS) [19], the Circulation, Respiration, Abdomen, Motor, Speech scale (CRAMS) [20], the Traumatic Injury Scoring System (TRISS) [21], and A Severity Characterization of Trauma (ASCOT) scale [22], all of which are used to score the severity and predict outcome of TBI. However, the reliability of GCS in predicting patient outcomes is unsatisfactory, especially with regard to the verbal component. As a result, Widjicks et al. published a new scoring system in 2005, the Full Outline of UnResponsiveness (FOUR) score (Figure 2), a newer scale, developed to provide a more comprehensive assessment [23]. The FOUR score includes additional information not assessed by the GCS like brainstem reflexes, visual tracking, breathing patterns, and respiratory drive [23] (Figure 2). It is also more practical for evaluating critically ill intubated patients, as it does not depend on an evaluation of the verbal response. It has already been validated in various populations of comatose patients, including TBI patients [24-30]. While GCS lacks the ability to identify subtle changes in alteration of consciousness, the FOUR score assesses four variables: eye response, motor response, brainstem reflexes, and respiration pattern (Figure 2). The acronym also reflects the number of categories and the maximum number of potential points in each category, making it fairly simple to use and remember. In addition, the FOUR score can account for the intubated patient and can also differentiate between a locked-in state and a vegetative state, via the addition of testing eye tracking, thus incorporating midbrain and pontine functions, effectively allowing the examiner to localize lesions. Another advantage for the FOUR score is that is gives all components equal weight, making it linear which is ideal for a coma scale compared to the GCS score is weighted toward motor responses. The GCS scale does offer the benefit of rapid evaluation in the emergency department, it is readily reproducible by multiple personal, from nursing staff to trauma surgeons, and gives a rapid assessment of the severity of injury. This is likely attributable to its long standing application in the field, which makes it second nature for many to communicate the information. More studies are needed to favor one over the other. For now, using either GCS or FOUR score for initial neurological assessment as well as follow up neurologic checks is acceptable.

Best eye response (E)	Spontaneous--open with blinking at baseline	4
	Opens to verbal command, speech, or shout	3
	Opens to pain, not applied to face	2
	None	1
Best verbal response (V)	Oriented	5
	Confused conversation, but able to answer questions	4
	Inappropriate responses, words discernible	3
	Incomprehensible speech	2
	None	1
Best motor response (M)	Obeys commands for movement	6
	Purposeful movement to painful stimulus	5
	Withdraws from pain	4
	Abnormal (spastic) flexion, decorticate posture	3
	Extensor (rigid) response, decerebrate posture	2
	None	1

Figure 1. Glasgow Coma Scale

3. Intracranial pressure monitoring

Intracranial hypertension develops commonly in acute brain injury related to trauma [31, 32]. Raised Intracranial pressure (ICP) is an important predictor of mortality in patients with severe TBI, and aggressive treatment of elevated ICP has been shown to reduce mortality and improve outcome (32-39). Guidelines for the Management of Severe TBI, published in the Journal of Neurotrauma in 2007 [4] make a Level II recommendation that ICP should be monitored in all salvageable patients with a severe TBI (Glasgow Coma Scale [GCS] score of 3–8 after resuscitation) and an abnormal computed tomography (CT) scan. ICP monitoring is also recommended in patients with severe TBI and a normal CT scan if two or more of the following features are noted at admission: age over 40 years, unilateral or bilateral motor posturing, or systolic blood pressure < 90 mm Hg (Level III recommendation). In comatose TBI patients with an abnormal CT scan, the incidence of ICH was 53–63% [40]. Patients with a normal CT scan at admission, on the other hand, had a relatively low incidence of intracranial hypertension (13%). However, within the normal CT group, if patients demonstrated at least two of three adverse features (age over 40 years, unilateral or bilateral motor posturing,

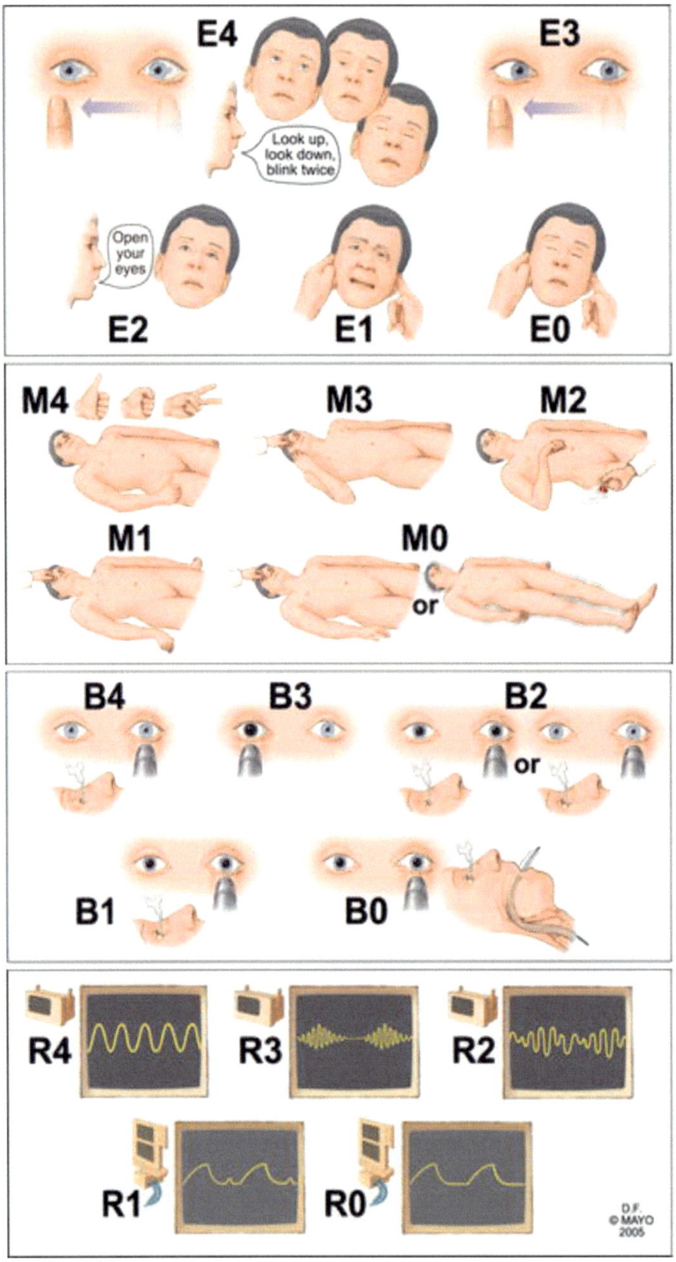

Figure 2. Full Outline of UnResponsiveness (FOUR) score. Eye response: E4 eyelids open or opened, tracking, or blinking to command; E3 eyelids open but not tracking; E2 eyelids closed but open to loud voice; E1 eyelids closed but open to pain; and E0 eyelids remain closed with pain. Motor response: M4 thumbs-up, fist, or peace sign; M3 localizing to pain; M2 flexion response to pain; M1 extension response to pain; and M0 no response to pain or generalized myoclonus status. Brainstem reflexes: B4 pupil and corneal reflexes present; B3 one pupil wide and fixed; B2 pupil or corneal reflexes absent; B1 pupil and corneal reflexes absent; and B0 absent pupil, corneal, and cough reflex. Respiration pattern: R4 not intubated, regular breathing pattern; R3 not intubated, Cheyne-Stokes breathing pattern; R2 not intubated, irregular breathing; R1 breathes above ventilatory rate; and R0 breathes at ventilator rate or apnea.

http://dx.doi.org/10.5772/57298

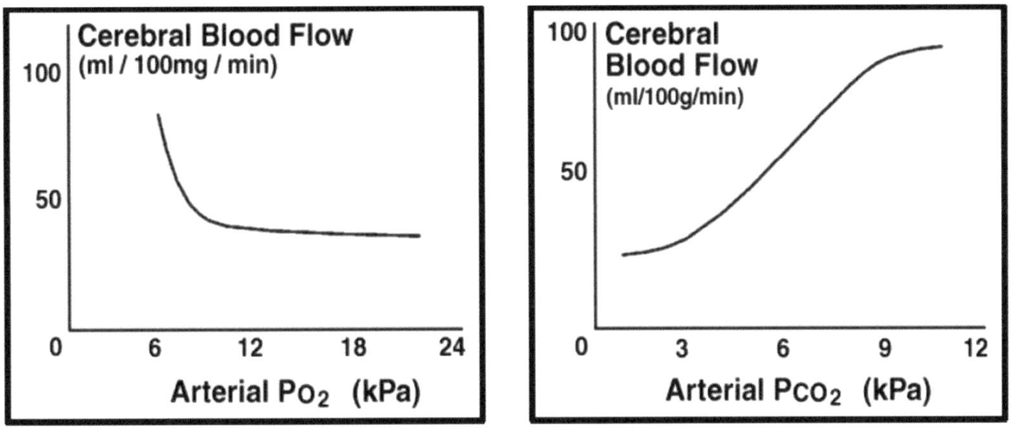

Figure 3. Oxygen tension and carbon dioxide effects on Cerebral Blood flow

or systolic BP < 90 mm Hg), their risk of intracranial hypertension was similar to that of patients with abnormal CT scans [4]. ICP is a strong predictor of outcome from severe TBI [33, 34, 36, 41-43]. Because of this, ethically a randomized trial of ICP monitoring with and without treatment is unlikely to be carried out. Similarly, a trial for treating or not treating systemic hypotension is not likely. Both hypotension and raised ICP are the leading causes of death in severe TBI. Furthermore, several studies have shown that patients who do not have intracranial hypertension or who respond to ICP-lowering therapies have a lower mortality than those who do not respond to therapy [5-12, 44-47]. As a result, Guidelines for the Management of Severe TBI recommend that treatment should be initiated with ICP thresholds above 20 mm Hg (level II) as well as target a cerebral perfusion pressure (CPP) within the range of 50-70 (level III) [4]. Prevention and/or treatment of intracranial hypertension is commonly accomplished by employing a progression of therapeutic approaches that are efficacious in controlling ICP and uniformly believed to be easily applied with minimal or rare negative side effects. These measures include(but are not limited to): elevation of the head of the bed, avoiding hypotension, hypoxia, and hypercapnea or prolonged hypocapnea, intravenous sedation and analgesia, administration of hyperosmolar agents (mannitol, hypertonic saline), and CSF drainage [4].

4. Oxygenation and ventilation

Hypoxia (PaO2 < 60 mmHg or O2 saturation < 90%) worsens secondary brain injury and thus significantly worsens outcome in patients with TBI [48, 49]. In addition, duration of hypoxemia (median duration ranging from 11.5 to 20 min) was found to be an independent predictor of mortality [50]. Elevated Carbon dioxide dilates the cerebral blood vessels, increasing the volume of blood in the intracranial vault and therefore increasing ICP [51]. On the other hand, hyperventilation leads to cerebral vasoconstriction, and thus can result in cerebral ischemia, despite possible improvements in CPP and ICP [52]. Thus, hyperventilation is recommended only as a temporizing measure to reduce an elevated ICP, preferably not below 30 mmHG unless absolutely necessary and only for few minutes while determining an etiology of the intracranial hypertension and initiating other treatment options or surgical intervention. The ventilator settings should be adjusted to maintain normoxia with a pulse oximetry (SpO2) around 95% or PaO2 around 80 mm Hg and eucapnia with PaCO2 of 35 to 40 mm Hg (in patients with chronic CO2 retention, such as COPD patients, CO2 should be maintained close to their baseline CO2 and normal pH). (Figure 3) At this point it is worth mentioning acute lung injury (ALI) and acute respiratory distress syndrome (ARDS). The incidence of ALI/ARDS in TBI is reported between 10 % and 30 % [53, 54]. ALI/ ARDS could develop secondary to aspiration, pneumonia, pulmonary contusion, massive blood transfusion, transfusion-related ALI (TRALI), sepsis, or neurogenic pulmonary edema. Management of ALI/ARDS entails low tidal volumes, higher positive end expiratory pressure, and permissive hypercapnea [55]. However, as mentioned above, hypercapnea (> 40 mmHG) is contraindicated in TBI patients with intracranial hypertension. One needs to balance need for low tidal volume and CO2 levels, and thus frequent ABG measurements is warranted.

5. Blood pressure and cerebral perfusion

Hypotension (SBP < 90 mm Hg) can worsen secondary injury in TBI and is associated with worsening mortality and neurologic outcomes [56, 57]. In TBI patients from the Trauma Coma Data Bank, early hypotension occurred in 34.6% of patients with severe traumatic brain injury and was shown to double the mortality rate (55% versus 27%). Late hypotension (in the ICU) occurred in 32% of patients. For patients whose only hypotensive episode occurred in the ICU, 66% died or were vegetative survivors compared with 17% of patients who never had a hypotensive episode [56]. We recommend IVF resuscitation to maintain euvolemia, using either invasive (e.g. CVP or pulmonary capillary wedge pressure) or noninvasive methods (e.g. Echocardiogram or NICOM-noninvasive cardiac output monitoring) to measure either static (CVP) or dynamic (Stroke volume index variation) surrogates of intravascular volume and hemodynamics. Hypotonic, hyponatremic, and sugar containing fluids should be avoided. If patient is euvolemic and remains hypotensive, then vasopressors should be started to maintain adequate blood pressure. Cerebral perfusion pressure is defined as mean arterial pressure (MAP) minus ICP (CPP = MAP – ICP). CPP < 60 mm HG should be avoided since it is associated with poor outcomes in patients with TBI [58]. Both 60 mmHg and 70 mm Hg are cited in the literature as the threshold above which CPP should be maintained. However, as reported earlier, the Guidelines for the Management of Severe Traumatic Brain injury recommend maintaining CPP between 50-70 mmHg [4]. In our patients, when the neurologic status is stable with a normal ICP, aggressive measures do not need to be taken as long as CPP >50 mmHg. Conversely, in patient's where exam is poor or ICP has been elevated/required treatment, then would recommend CPP >60 mmHg since the risk of secondary injury developing is more imminent in these patients. In the absence of cerebral ischemia, aggressive attempts to maintain CPP above 70 mmHg with fluids and vasopressors should be avoided because of the risk of ARDS [59].

6. Hyperosmolar therapy

Brain parenchyma is 80% water, and thus brain volume is very responsive to changes in water content. A hyperosmolar agent creates a gradient for water to move from brain parenchyma to the intravascular space across the blood-brain barrier (BBB). As a result, the effectiveness of a hyperosmolar agent depends on the extent it is prevented from crossing the BBB. Both mannitol and hypertonic saline posses this property and will be discussed in this section. Mannitol is given in a 20% or 25 % solution in boluses of 0.25 to 1.0 g per kilogram of body weight at intervals of 2, 4, 6, or more hours. In addition to the osmotic properties of mannitol, , it also lowers blood viscosity which leads to increase cerebral blood flow resulting in cerebral vasoconstriction (autoregulation), which in turn reduces cerebral blood volume and thus intracranial pressure [60]. In patients with impaired cerebral autoregulation, aggressive use of mannitol could result in increased ICP. As this implies, for mannitol to be fully effective, the blood brain barrier must be intact. In patients where the BBB is not intact, mannitol crosses the BBB and can draw fluid into brain resulting in an increase in ICP [61, 62], hence concentrations of mannitol should be assessed before each dose. Instead of measuring actual mannitol, an easier and more practical way is to measure serum osmolarity or osmolar gap (measured – calculated serum osmolarity) before infusing mannitol. Because of the above as well as to minimize risk of acute kidney injury, Mannitol should thus not be given if serum osmolarity is more than 320 mOsm/Kg H2O or osmolar gap > 10. Side effects of mannitol include hypotension, hypovolemia, hypokalemia, hyperkalemia, and acute kidney injury [63, 64]. Mannitol is contraindicated in patients with renal failure. Hypertonic saline increases serum osmolarity directly rather than by inducing osmotic dieresis, as well as by viscosity-related cerebral vasoconstriction [65]. It is used in a 3% solution infusion or in boluses of approximately 150 ml, in a 7.5% solution in 75-ml boluses, or in a 23.4% solution in 30-ml boluses every 2, 4, 6, or more hours. Serum osmolarity and serum sodium should be checked before dosing hypertonic saline. Hypertonic saline should not be given if serum Na is more than 160 mmol/liter. Side effects of hypertonic saline include fluid overload secondary to intravascular volume expansion, acidosis, hypokalemia, and hyperchloremia. There is a potential for development of central pontine myelinolysis with rapid increase in serum Na concentration, however, this phenomenon has not been documented in this scenario of management of intracranial hypertension. The question remains as to which one, mannitol or hypertonic saline, is superior? Recent evidence including two metanalysis, suggest that hypertonic saline may be more effective than mannitol in reducing ICP [66-69], however, high quality studies comparing the agents, while accounting for side effects and contraindications are lacking. Furthermore, it is very important to take into consideration patient characteristics, such as volume status, renal function, hemodynamic status, sodium levels, etc.., when choosing the appropriate hyperosmolar agent. Hyperosmolar therapy should be weaned gradually rather than stopped abruptly in the days to follow.

7. Refractory intracranial hypertension: Therapeutic hypothermia and barbiturate coma

Refractory intracranial hypertension (RICH) is defined as intracranial pressures that exceed 25 mm Hg for 30 minutes, 30 mm Hg for 15 minutes, or 40 mm Hg for 1 minute despite tier 1 and tier 2 therapies [70]. Tier 1 therapies include head of patient's bed at or more than 30 degrees, adequate sedation and analgesia, and adequate CSF draining if ICP is monitored by ventriculostomy. Tier 2 therapies include adequate hyperosmolar therapy (mannitol or hypertonic saline or both), mild hyperventilation (pCO2 goal of 30 – 35 mmHg), and neuromascular blockade. RICH occurs in approximately 15% of patients with traumatic brain injury [70]. If it is not aggressively treated, RICH can result in cerebral herniation and death.High-dose barbiturate administration is recommended to control elevated ICP refractory to maximum standard medical and surgical treatment (level II) [3]. High-dose barbiturates have been scarcely studied for this indication. In 2004, the Cochrane Injuries Group performed a systematic review of the barbiturate RCTs. In the only two studies examining the effect on ICP, the relative risk for refractory ICP with barbiturate therapy was 0.81 (95% CI 0.62–1.06). Concerning this indication, the Cochrane group concluded: "There is no evidence that barbiturate therapy in patients with acute severe head injury improves outcome. Barbiturate therapy results in a fall in blood pressure in one of four treated patients. The hypotensive effect of barbiturate therapy will offset any ICP lowering effect on cerebral perfusion pressure" [71]. Significant side effects of barbiturates include hypotension, arrhythmias, immunosuppression, hepatotoxicity, fever and injection site reactions. The Guidelines for the Management of Severe TBI recommend that more studies are needed to identify alternative agents for this indication - "elevated ICP refractory to standard therapy" [3]. Albeit small, there are more RCT evaluating the effect of therapeutic hypothermia on ICH in severe TBI (13 studies) than for barbiturates; all consistently demonstrating that hypothermia is effective in controlling ICP (This is reviewed in more detail elsewhere : Farid Sadaka, Christopher Veremakis, Rekha Lakshmanan and Ashok Palagiri (2013). Therapeutic Hypothermia in Traumatic Brain Injury, Therapeutic Hypothermia in Brain Injury, Dr. Farid Sadaka (Ed.), ISBN: 978-953-51-0960-0, InTech, DOI: 10.5772/48818. Available from: http://www.intechopen.com/books/therapeutic-hypothermia-in-brain-injury/therapeutic-hypothermia-in-traumatic-brain-injury). Complications from hypothermia include electrolyte imbalances, increase in incidence of infections, thrombocytopenia, coagulopathy, arrhythmias (especially bradycardia), pancreatitis, and rebound ICH (during re-warming). In one extensive review [72], Povlishock et al showed that posttraumatic hypothermia followed by slow rewarming appeared to provide maximal protection in terms of traumatically induced axonal damage, microvascular damage and dysfunction, contusional expansion, intracranial hypertension, and neurocognitive recovery. In contrast, hypothermia followed by rapid rewarming not only reversed the protective effects associated with hypothermic intervention, but exacerbated the traumatically induced pathology and its neurologic consequences. Povlishock's review concluded that the rate of posthypothermic rewarming is an important variable in assuring maximal efficacy following the use of hypothermic intervention. The most challenging issue appears to be rebound ICP during re-warming. We suggest that re-warming

http://dx.doi.org/10.5772/57298

only be considered if the patient's ICP is stable and <20mmHg for at least 48 hours, and, thereafter implemented at a rate not faster than 0.1 -0.25°C per hour. Surgical management of TBI patients such as Decompressive hemi-craniectomy or bilateral craniectomy are discussed in more detail in chapter http://www.intechopen.com/books/traumatic-brain-injury/surgical-treatment-of-severe-traumatic-brain-injury. Please refer to figure 4 for protocolized step-wise approach to ICP management.

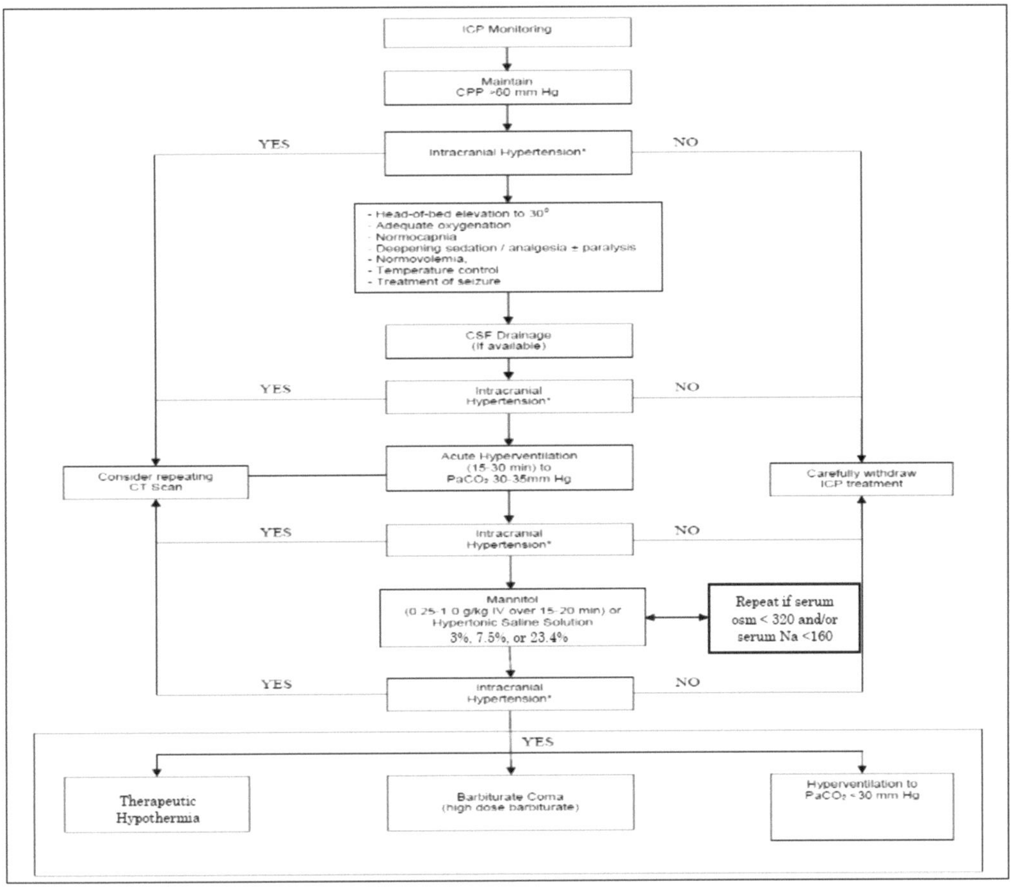

Figure 4. Stepwise approach to management of intracranial hypertension.

8. Temperature modulation and normothermia

Aside from role of hypothermia in ICP control in patients with refractory intracranial hypertension, therapeutic hypothermia has also been studied as a primary neuroprotectant in patients with severe TBI, based on the fact that early administration of TH could halt the secondary injury processes discussed above, and thus possibly improve outcome. This topic is reviewed in more detail elsewhere : Farid Sadaka, Christopher Veremakis, Rekha Lakshmanan and Ashok Palagiri (2013). Therapeutic Hypothermia in Traumatic Brain Injury, Therapeutic Hypothermia in Brain Injury, Dr. Farid Sadaka (Ed.), ISBN: 978-953-51-0960-0, InTech, DOI: 10.5772/48818. Available from: http://www.intechopen.com/books/therapeutic-hypothermia-in-brain-injury/therapeutic-hypothermia-in-traumatic-brain-injury. In short, although single-center studies were encouraging, multicenter trials with early administration of hypothermia for a defined period of time irrespective of ICP have almost uniformly been negative except maybe for patients undergoing craniotomy for hematoma evacuations. However, hypothermia was maintained for a fixed duration of only 48 hrs, and ICP elevations mainly occurred during and after rewarming. These results suggest that a period of 48 hours of hypothermia may be too short to have a beneficial effect on outcome. A standardized one size fit all may be inappropriate. The rate of rewarming plays an important role as well as pointed above.The rebound increase in ICP during and after rewarming in these studies and the encouraging outcomes from the randomized studies that induced hypothermia early and continued it throughout the period of intracranial hypertension point to the realization that individualizing the duration of hypothermia to fit a patient's ICP in future trials may be a better strategy than a predetermined period of hypothermia regardless of ICP. As for now, therapeutic hypothermia cannot be recommended for TBI patients aside from control of refractory ICP discussed above.All of the mechanisms of secondary brain injury in TBI discussed above (apoptosis, mitochondrial dysfunction, excitotoxicity, disruption in ATP metabolism, disruption in calcium homeostasis, increase in inflammatory mediators and cells, free radical formation, DNA damage, blood-brain barrier disruption, brain glucose utilization disruption, microcirculatory dysfunction and microvascular thrombosis) are temperature dependant. They are all stimulated and exacerbated by fever [73]. In addition, fever occurs with high frequency in this patient population, with up to 68% of patients experiencing at least one fever during their intensive care unit stay [74]. Fever in the TBI population may result from multiple causes and for reasons other than infection and has proven difficult to control. Disruption of the hypothalamic set point, tissue ischemia/infarction, surgery, medications, and blood product transfusions may all induce hyperthermia. Early hyperthermia following TBI is associated with a longer ICU length of stay and worsened neurologic outcomes [75-77]. Thereby, temperature should be controlled, fever should be aggressively treated, and normothermia should be maintained in patients with TBI.

9. Nutrition and glucose control

TBI, especially severe TBI, can cause increase metabolism and can create a hypercatabolic state that results in rapid depletion of nutrition reserves, as well as worsening immune function and morbidity [78, 79]. In TBI patients, adequate nutrition that is started early after injury is associated with enhanced immunity, decreased infectious morbidity, shortened length of hospitalization, improved neurological recovery and reduced mortality [80, 81]. Brain trauma foundation recommends that patients should be fed to attain full caloric replacement by day 7 post-injury [3]. American Society for Parenteral and Enteral Nutrition

(ASPEN) for nutrition support in critically-ill adult patients suggest that enteral nutrition should be started within the first 24–48 h following admission, as long as patients are hemodynamically stable [82]. The European Society for Parenteral and Enteral Nutrition (ESPEN) recommends initiating enteral nutrition within 24 h if possible [83]. Despite popular belief, recent evidence reiterates that bowel sounds and passing flatus or stool are not required for the initiation of enteral nutrition[82]. In circumstances when early enteral nutrition cannot be initiated, parenteral nutrition should be strongly considered.There is no question that hyperglycemia is associated with worse outcome in brain injured patients [84, 85]. Hypoglycemia leads to deprivation of the brain of its fuel which can lead to compromised brain energy metabolism and worsen the already existent brain injury, especially during the increase in glucose utilization and brain energy demand observed after TBI [86, 87]. However, the optimal target for systemic glucose control is not known. In patients with severe brain injury, tight systemic glucose control (80–120 mg/dL) was associated with reduced cerebral extracellular glucose availability and increased prevalence of brain energy crisis, which in turn correlates with increased mortality [88]. Intensive insulin therapy may thus impair cerebral glucose metabolism after severe brain injury. Based on the existing low quality evidence, the most recent guidelines from the Society of Critical Care Medicine (SCCM) suggest that blood glucose (BG) ≥ 150 mg/dL should trigger initiation of insulin therapy for most patients admitted to an ICU with the diagnosis of TBI, titrated to achieve BG values absolutely < 180 mg/dL, to minimize the adverse effects of hyperglycemia [89]. The guidelines also suggest that BG < 100 mg/dL be avoided during insulin infusion for patients with brain injury [89].

10. Anemia and transfusion

Hypoxia and hypotension worsen secondary brain injury and are important determinants of outcome in TBI patients. They both are associated with worse outcomes as discussed above. This could be the reason why many patients with TBI are still transfused to a hemoglobin threshold of 10 g/dl. Although red blood cells are an essential requirement for the transport of oxygen to the tissues, several problems are documented with red blood cells (RBC) transfusions such as infection, pulmonary complications such as transfusion-related acute lung injury (TRALI) and transfusion-associated circulatory overload (TACO), transfusion-related immunomodulation (TRIM) and multiorgan failure, and increased mortality [90]. Besides, there is no clear correlation between anemia and hypoxia or hypotension in TBI patients. In one retrospective study, linear regression showed that more days with hematocrit < 30% was associated with improved neurologic outcomes. In addition, transfusion of RBCs was significantly associated with worse outcomes [91]. In a subgroup analysis of a multicenter randomized controlled clinical trial involving 67 critically ill patients from the Transfusion Requirements in the Critical Care trial who sustained a closed head injury, patients were randomized to a restrictive RBC transfusion strategy (Hb 7.0 g/dL and maintained between 7.0 and 9.0 g/dL) or a liberal strategy (Hb 10.0 g/dL and maintained between 10.0 and 12.0 g/dL). This study was unable to detect significant improvements in mortality with a liberal as compared with a restrictive transfusion strategy in critically ill trauma patients with moderate-to severe TBI [92]. Guidelines for transfusion developed by EAST (Eastern Association for Surgery of Trauma) and the American College of Critical Care Medicine (ACCM) of the Society of Critical Care Medicine (SCCM) state that there is no benefit of a "liberal" transfusion strategy (transfusion when Hb is <10 g/dL) in patients with moderate-to-severe TBI [93]. Large multicenter prospective studies are needed to evaluate the effects of anemia and RBC transfusion in patients with TBI.

11. Deep venous thrombosis

The application of chemical venous thrombo-embolism (VTE) prophylaxis traumatic brain injury patients has been long been guided by the dogma of physicians practicing under assumptions rather than evidenced based guidelines. This has resulted in the blanket denial of the use of any chemical DVT prophylaxis up until more recent years, despite the common knowledge that trauma patient are at high risk for development of venous thromboembolic events. The overwhelming fear associated with propagating the intracranial injury has also limited the number of studies until recently. Most of the data that is currently applied to support the safety of chemical prophylaxis has been extrapolated from studies that were performed looking that the risks of post-operative hemorrhage in elective craniotomy patients. In 1998 Agnelli et. al. compared the use of enoxaparin combined with compression stockings to patients treated with compression stockings alone and found a significant reduction in the number of VTE without any increase in hemorrhage after elective neurosurgery [94]. This study and others like it opened the door for further application and research in the area of traumatic brain injury. The current formal recommendation by the Guidelines for the Management of Severe Traumatic Brain Injury 2007 state that the use of low-molecular weight heparin or low dose unfractionated heparin should be used in combination with mechanical prophylaxis, but there is an increased risk for expansion of intracranial hemorrhage. There is insufficient evidence to support recommendations regarding the preferred agent, dose, or timing [4].The lack of guidelines to follow has prompted many institutions to develop their own guidelines based on extrapolated data and to record and report their experiences. In 2010, clinicians at McGill University published their findings, which showed an acceptable VTE control rate without increased risk of expanding intracranial hemorrhage [95]. The general principles that predominate in the use of chemical VTE prophylaxis are the following: 1) patients not expected to go to the operating room in the next 24 hours for intracranial procedure 2) no evidence of systemic coagulopathy 3) 2 stable CT scans. All guidelines should be applied with the consideration of the injuries of the specific patient in question and altered as seen appropriate for the situation. The first phase of a randomized, double-blind study involving the early use of enoxaparin in trauma patients was just published in 2012 [96]. The study found at 2.3% higher rate of progression in the patients treated with enoxaparin over placebo, however, none were clinically significant. DEEP-II is intended to evaluate the efficacy of this VTE prevention and DEEP-III will apply to moderate-risk patients. A recently published systematic review and meta-analysis of the use of early chemical VTE prophylaxis in TBI patients found that it reduced the risk of VTE without progression of intracranial hemorrhage [97]. While there is certainly the need for more studies to quantify the risk associated with the early use of chemical VTE prophylaxis, there is evidence that supports the appropriate application in traumatic brain injury patients at high risk for developing DVT. Although we cannot not provide official recommendations, dosage, or timing of administration, it is used in our institution.

12. Conclusion

TBI is a devastating injury and often these patients would require monitoring and treatment in intensive care unit. Management of TBI patients requires multidisciplinary approach, frequent close monitoring and judicious use of multiple treatments to lessen secondary brain injury and improve outcomes. There is a lot of opportunity for further research in TBI, including but not limited to multimodal monitoring, and therapeutics to further improve outcomes in this very common mechanism of brain injury.

Author details

Farid Sadaka*, Tanya M Quinn, Rekha Lakshmanan and Ashok Palagiri

*Address all correspondence to: farid.sadaka@mercy.net

Mercy Hospital St. Louis/St. Louis University, Critical Care Medicine/Neurocritical Care, St. Louis, USA

The authors report no conflicts of interest. All authors declare that No competing financial interests exist. All authors report that no potential conflicts of interest exist with any companies/organizations whose products or services may be discussed in this article.

References

[1] Dombovy ML, Olek AC (1997) Recovery and rehabilitation following traumatic brain injury. Brain Inj 11(5):305-18.

[2] Rutland-Brown W, Langlois JA, Thomas KE, Xi YL (2006) Incidence of traumatic brain injury in the United States, 2003. J Head Trauma Rehabil 21: 544 - 48.

[3] Bader, MK. (2006) Recognizing and Treating Ischemic Insults to the Brain: The Role of Brain Tissue Oxygen Monitoring. Crit Care Nurs Clin N Am 18: 243-256.

[4] Brain Trauma Foundation; American Association of Neurological Surgeons; Congress of Neurological Surgeons; Joint Section on Neurotrauma and Critical Care,

AANS/CNS. Guidelines for the management of severe traumatic brain injury. J Neurotrauma 2007; 24 (Suppl 1):1-117.

[5] Small DL, Morley P, Buchan AM (1999) Biology of ischemic cerebral cell death. Prog Cardiovasc Dis 42:185–207.

[6] Dietrich WD, Atkins CM, Bramlett HM (2009) Protection in animal models of brain and spinal cord injury with mild to moderate hypothermia. J Neurotrauma 26(3): 301-12.

[7] Chesnut RM (1995) Secondary brain insults after head injury: clinical perspectives. New Horiz 3:366-75.

[8] Unterberg AW, Stover JF, Kress B, Kiening KL (2004) Edema and brain trauma. Neuroscience 129:1021-9.

[9] Mirski MA, Chang CW, Cowan R (2001) Impact of a neuroscience intensive care unit on neurosurgical patient outcomes and cost of care: evidence-based support for an intensivist-directed specialty ICU model of care. J Neurosurg Anesthesiol 13(2):83-92.

[10] Diringer MN, Edwards DF (2001) Admission to a neurologic/neurosurgical intensive care unit is associated with reduced mortality rate after intracerebral hemorrhage. Crit Care Med 29(3):635-40.

[11] Varelas PN, Conti MM, Spanaki MV, Potts E, Bradford D, Sunstrom C, Fedder W, Hacein Bey L, Jaradeh S, Gennarelli TA (2004) The impact of a neurointensivist-led team on a semiclosed neurosciences intensive care unit. Crit Care Med 32(11):2191-8.

[12] Varelas PN, Eastwood D, Yun HJ, Spanaki MV, Hacein Bey L, Kessaris C, Gennarelli TA (2006) Impact of a neurointensivist on outcomes in patients with head trauma treated in a neurosciences intensive care unit. J Neurosurg 104(5):713-9.

[13] Teasdale G, Jennett B (1974) Assessment of coma and impaired consciousness. A practical scale. Lancet 2:81–4.

[14] Teasdale GM, Jennett B (1976) Assessment and prognosis of coma after head injury. Acta Neurochir 34:45–55.

[15] Rimel RW, Giordani NP, Barth JT, Jane JJ (1982) Moderate head injury: completing the spectrum of brain trauma. Neurosurgery 11: 344–351.

[16] Rimel RW, Jane JA, Edlich RF (1979) An injury severity scale for comprehensive management of central nervous system trauma. JACEP 8: 64–67.

[17] Knaus WA, Draper EA, Wagner DP, Zimmerman JE (1985) APACHE II: a severity of disease classification system. Crit Care Med 13:818–29.

[18] Le Gall JR, Lemeshow S, Saulnier F (1993) A new Simplified Acute Physiology Score (SAPS II) based on a European/North American multicenter study. JAMA 270:2957–63.

[19] Gormican SP (1989) CRAMS scale: field triage of trauma Champion HR, Sacco WJ, Copes WS, Gann DS, Gennarelli TA, Flannagan ME. A revision of the trauma score. J Trauma 29:623–9.

[20] Gormican SP (1982) CRAMS scale: field triage of trauma victims. Ann Emerg Med 11:132–5.

[21] Boyd CR, Tolson MA, Copes WS (1987) Evaluating trauma care: the TRISS method. Trauma Score and the Injury Severity Score. J Trauma 27:370–8.

[22] Champion HR, Copes WS, Sacco WJ, Frey CF, Holcroft JW, Hoyt DB, Weigelt JA (1996) Improved predictions from a severity characterization of trauma (ASCOT) over Trauma and Injury Severity Score (TRISS): results of an independent evaluation. J Trauma 40:42–8. discussion 48–9.

[23] Wijdicks EF, Bamlet WR, Maramattom BV, Manno EM, McClelland RL (2005) Validation of a new coma scale: the FOUR score. Ann Neurol 58(4):585–93.

[24] Iyer VN, Mandrekar JN, Danielson RD, Zubkov AY, Elmer JL, Wijdicks EF (2009) Validity of the FOUR score coma scale in the medical intensive care unit. Mayo Clin Proc 84(8): 694–701.

[25] Stead LG, Wijdicks EF, Bhagra A, Kashyap R, Bellolio MF, Nash DL, Enduri S, Schears R, William B. (2009) Validation of a new coma scale, the FOUR score, in the emergency department. Neurocrit Care 10(1):50–4.

[26] Wolf CA, Wijdicks EF, Bamlet WR, McClelland RL (2007) Further validation of the FOUR score coma scale by intensive care nurses. Mayo Clin Proc 82(4):435–8.

[27] Fugate JE, Rabinstein AA, Claassen DO, White RD, Wijdicks EF (2010) The FOUR score predicts outcome in patients after cardiac arrest. Neurocrit Care 13(2):205-10.

[28] Weiss N, Mutlu G, Essardy F, Nacabal C, Sauves C, Bally C, Houbert M, Lecorre C, Germack V, Demeret S, Pierrot-Deseilligny C, Bolgert F (2009) The French version of the FOUR score: a new coma score. Rev Neurol 165:796–802.

[29] Akavipat P (2009) Endorsement of the FOUR score for consciousness assessment in neurosurgical patients. Neurol Med Chir 49:565–71.

[30] Sadaka F, Patel D, Lakshmanan R (2012) The FOUR score predicts outcome in patients after traumatic brain injury. Neurocrit Care 2012 16(1):95-101.

[31] Miller JD, Becker DP, Ward JD, Sullivan HG, Adams WE, Rosner MJ. Significance of intracranial hypertension in severe head injury. J Neurosurg 1977; 47:503–516.

[32] Miller JD, Dearden NM, Piper IR, Chan KH. Control of intracranial pressure in patients with severe head injury. J Neurotrauma 1992; 9(suppl 1):S317–S326.

[33] Marmarou A, Anderson PL, Ward JD, Choi SC, Young HF. Impact of ICP instability and hypotension on outcome in patients with severe head trauma. J Neurosurg 1991; 75(suppl):59–66.

[34] Ghajar J, Hariri RJ, Patterson RH. Improved outcome from traumatic coma using only ventricular cerebrospinal fluid drainage for intracranial pressure control. Adv Neurosurg 1993; 21:173–177.

[35] Juul N, Morris GF, Marshall SB, Marshall LF. Intracranial hypertension and cerebral perfusion pressure: influence on neurological deterioration and outcome in severe head injury. The Executive Committee of the International Selfotel Trial. J Neurosurg 2000; 92:1– 6.

[36] Becker DP, Miller JD, Ward JD, Greenberg RP, Young HF, Sakalas R. The outcome from severe head injury with early diagnosis and intensive management. J Neurosurg 1977; 47:491– 502.

[37] Patel HC, Menon DK, Tebbs S, Hawker R, Hutchinson PJ, Kirkpatrick PJ. Specialist neurocritical care and outcome from head injury. Intensive Care Med 2002; 28:547–553.

[38] Steiner T, Ringleb P, Hacke W. Treatment options for large hemispheric stroke. Neurology 2001; 57:S61–S68.

[39] Qureshi AI, Geocadin RG, Suarez JI, Ulatowski JA. Long-term outcome after medical reversal of transtentorial herniation in patients with supratentorial mass lesions. Crit Care Med 2000; 28:1556 –1564.

[40] Narayan RK, Kishore PR, Becker DP, Ward JD, Enas GG, Greenberg RP, Domingues Da Silva A, Lipper MH, Choi SC, Mayhall CG, Lutz HA 3rd, Young HF (1982) Intracranial pressure: to monitor or not to monitor? A review of our experience with severe head injury. J Neurosurg 56: 650–659.

[41] Lundberg N, Troupp H, Lorin H (1965) Continuous recording of the ventricular-fluid pressure in patients with severe acute traumatic brain injury. A preliminary report. J Neurosurg 22:581–590.

[42] Marshall LF, Smith RW, Shapiro HM (1979) The outcome with aggressive treatment in severe head injuries. Part I: the significance of intracranial pressure monitoring. J Neurosurg 50:20–25.

[43] Narayan RK, Greenberg RP, Miller JD, Enas GG, Choi SC, Kishore PR, Selhorst JB, Lutz HA 3rd, Becker DP (1981) Improved confidence of outcome prediction in severe head injury. A comparative analysis of the clinical examination, multimodality evoked potentials, CT scanning, and intracranial pressure. J Neurosurg 54:751–762.

[44] Eisenberg HM, Frankowski RF, Contant CF, Marshall LF, Walker MD (1988) High-dose barbiturate control of elevated intracranial pressure in patients with severe head injury. J Neurosurg 69: 15–23.

[45] Howells T, Elf K, Jones P, Ronne-Engström E, Piper I, Nilsson P, Andrews P, Enblad P (2005) Pressure reactivity as a guide in the treatment of cerebral perfusion pressure in patients with brain trauma. J Neurosurg 102:311–317.

[46] Saul TG, Ducker TB (1982) Effect of intracranial pressure monitoring and aggressive treatment on mortality in severe head injury. J Neurosurg 56:498–503.

[47] Timofeev I, Kirkpatrick P, Corteen E, Hiler M, Czosnyka M, Menon DK, Pickard JD, Hutchinson PJ (2006) Decompressive craniectomy in traumatic brain injury: outcome following protocol-driven therapy. Acta Neurochir Suppl 96:11–16.

[48] Stochetti N, Furlan A, Volta F (1996) Hypoxemia and arterial hypotension at the accident scene in head injury. J Trauma 40:764–767.

[49] Auer RN (2001) Non-pharmacologic (physiologic) neuroprotection in the treatment of brain ischemia. Ann N Y Acad Sci 939:271-82.

[50] Jones PA, Andrews PJ, Midgley S, Anderson SI, Piper IR, Tocher JL, Housley AM, Corrie JA, Slattery J, Dearden NM (1994) Measuring the burden of secondary insults in head injured patients during intensive care. J Neurosurg Anesthesiol 6:4–14.

[51] Paul RL, Polanco O, Turney SZ McAslan TC, Cowley RA. (1972) Intracranial pressure responses to alterations in arterial carbon dioxide pressure in patients with head injuries. J Neurosurg 36:714-720.

[52] Coles JP, Minhas PS, Fryer TD, Smielewski P, Aigbirihio F, Donovan T, Downey SP, Williams G, Chatfield D, Matthews JC, Gupta AK, Carpenter TA, Clark JC, Pickard JD, Menon DK (2002) Effect of hyperventilation on cerebral blood flow in traumatic head injury: clinical relevance and monitoring correlates. Crit Care Med 30(9): 1950-1959.

[53] Holland MC, Mackersie RC, Morabito D, Campbell AR, Kivett VA, Patel R, Erickson VR, Pittet JF (2003) The development of acute lung injury is associated with worse neurologic outcome in patients with severe traumatic brain injury. J Trauma 55:106-111.

[54] Bratton SL, Davis RL (1997) Acute lung injury in isolated traumatic brain injury.Neurosurgery 40:707-712.

[55] The Acute Respiratory Distress Syndrome Network (2000) Ventilation with lower tidal volumes as compared with traditional tidal volumes for acute lung injury and the acute respiratory distress syndrome. N Engl J Med 342:1301-1308.

[56] Chesnut RM, Marshall SB, Piek J, Blunt BA, Klauber MR, Marshall LF. (1993) Early and late systemic hypotension as a frequent and fundamental source of cerebral ischemia following severe brain injury in the Traumatic Coma Data Bank. Acta Neurochir Suppl (Wien) 59:121–5.

[57] Manley G, Knudson MM, Morabito D, Damron S, Erickson V, Pitts L (2001) Hypotension, hypoxia, and head injury: frequency, duration, and consequences. Arch Surg 136:1118-1123.

[58] Clifton GL, Miller ER, Choi SC, Levin HS (2002) Fluid thresholds and outcome from severe brain injury. Crit Care Med 30:739–745.

[59] Robertson CS, Valadka AB, Hannay HJ, Contant CF, Gopinath SP, Cormio M, Uzura M, Grossman RG (1999) Prevention of secondary ischemic insults after severe head injury. Crit Care Med 27:2086–2095.

[60] Muizelaar JP, Wei EP, Kontos HA, Becker DP (1983) Mannitol causes compensatory cerebral vasoconstriction in response to blood viscosity changes. J Neurosurg 59:822-8.

[61] Marshall LF, Smith RW, Rauscher LA, Shapiro HM (1978) Mannitol dose requirements in brain-injured patients. J Neurosurg 48:169-172.

[62] Wise BL, Perkins RK, Stevenson E, Scott KG (1964) Penetration of C14-labelled mannitol from serum into cerebrospinal fluid and brain. Exp Neurol 10:264-270.

[63] Better OS, Rubinstein I, Winaver JM, Knochel JP (1997) Mannitol therapy revisited (1940-1997). Kidney Int 52:886-94.

[64] Manninen PH, Lam AM, Gelb AW, Brown SC (1987) The effect of high dose mannitol on serum and urine electrolytes and osmolality inneurosurgical patients. Can J Anaesth 34:442-446.

[65] Hijiya N, Horiuchi K, Asakura T (1991) Morphology of sickle cells produced in solutions of varying osmolarities. J Lab Clin Med 117:60–66.

[66] Vialet R, Albanèse J, Thomachot L, Antonini F, Bourgouin A, Alliez B, Martin C (2003) Isovolume hypertonic solutes (sodium chloride or mannitol) in the treatment of refractory posttraumatic intracranial hypertension: 2 mL/kg 7.5% saline is more effective than 2 mL/kg 20% mannitol. Crit Care Med 31:1683–1687.

[67] Battison C, Andrews PJ, Graham C, Petty T (2005) Randomized, controlled trial on the effect of a 20% mannitol solution and a 7.5% saline/6% dextran solution on increased intracranial pressure after brain injury. Crit Care Med 33:196–202.

[68] Kamel H, Navi BB, Nakagawa K, Hemphill JC 3rd, Ko NU (2011) Hypertonic saline versus mannitol for the treatment of elevated intracranial pressure: A meta-analysis of randomized clinical trials. Crit Care Med 39:554–559.

[69] Mortazavi MM, Romeo AK, Deep A, Griessenauer CJ, Shoja MM, Tubbs RS, Fisher W (2012) Hypertonic saline for treating raised intracranial pressure: Literature review with meta-analysis. J Neurosurg 116:210–221.

[70] Censullo JL, Sebastian S (2003) Pentobarbital sodium coma for refractory intracranial hypertension. J Neurosci Nurs 35(5):252-62.

[71] Roberts I (2005) Barbiturates for acute traumatic brain injury. The Cochrane Library Volume 4.

[72] Povlishock JT, Wei EP (2009) Posthypothermic rewarming considerations following traumatic brain injury. J Neurotrauma 26:333–340.

[73] Polderman,Kees H (2008) Induced hypothermia and fever control for prevention and treatment of neurological injuries. Lancet 371: 1955–69.

[74] Albrecht RFII, Wass CT, Lanier WL (1998) Occurrence of potentially detrimental temperature alterations in hospitalized patients at risk for brain injury. Mayo Clin Proc 73(7):629–35.

[75] Geffroy A, Bronchard R, Merckx P, Seince PF, Faillot T, Albaladejo P, Marty J (2004) Severe traumatic head injury in adults: which patients are at risk of early hyperthermia? Intensive Care Med 30(5):785–90.

[76] Suz P, Vavilala MS, Souter M, Muangman S, Lam AM (2006) Clinical features of fever associated with poor outcome in severe pediatric traumatic brain injury. J Neurosurg Anesthesiol 18(1):5– 10.

[77] Diringer MN, Reaven NL, Funk SE, Uman GC (2004) Elevated body temperature independently contributes to increased length of stay in neurologic intensive care unit patients. Crit Care Med 32(7):1489–95.

[78] Charrueau C, Belabed L, Besson V, Chaumeil JC, Cynober L, Moinard C (2009) Metabolic response and nutritional support in traumatic brain injury: evidence for resistance to renutrition. J Neurotrauma 26(11):1911–20.

[79] Krakau K, Omne-Ponten M, Karlsson T, Borg J (2006) Metabolism and nutrition in patients with moderate and severe traumatic brain injury: a systematic review. Brain Inj 20(4):345–67.

[80] Sacks GS, Brown RO, Teague D, Dickerson RN, Tolley EA, Kudsk KA (1995) Early nutrition support modifies immune function in patients sustaining severe head injury. JPEN J Parenter Enteral Nutr 19(5):387–92.

[81] Ha°rtl R, Gerber LM, Ni Q, Ghajar J (2008) Effect of early nutrition on deaths due to severe traumatic brain injury. J Neurosurg 109(1):50–6.

[82] McClave SA, Martindale RG, Vanek VW, McCarthy M, Roberts P, Taylor B, Ochoa JB, Napolitano L, Cresci G; A.S.P.E.N. Board of Directors; American College of Critical Care Medicine; Society of Critical Care Medicine (2009) Guidelines for the provision and assessment of nutrition support therapy in the adult critically ill patient: Society of Critical Care Medicine (SCCM) and American Society for Parenteral and Enteral Nutrition (A.S.P.E.N.). JPEN J Parenter Enteral Nutr 33(3):277–316.

[83] Kreymann KG, Berger MM, Deutz NEP, Deutz NE, Hiesmayr M, Jolliet P, Kazandjiev G, Nitenberg G, van den Berghe G, Wernerman J; DGEM (German Society for Nutritional Medicine), Ebner C, Hartl W, Heymann C, Spies C; ESPEN (European So-

ciety for Parenteral and Enteral Nutrition) (2006) ESPEN guidelines on enteral nutrition: intensive care. Clin Nutr 25 (2):210–23.

[84] Rovlias A, Kotsou S (2000) The influence of hyperglycemia on neurological outcome in patients with severe head injury. Neurosurgery 46:335–342; discussion 42–43.

[85] Jeremitsky E, Omert LA, Dunham CM, Wilberger J, Rodriguez A (2005) The impact of hyperglycemia on patients with severe brain injury. J Trauma 58:47–50.

[86] Hayes RL, Katayama Y, Jenkins LW, Lyeth BG, Clifton GL, Gunter J, Povlishock JT, Young HF (1988) Regional rates of glucose utilization in the cat following concussive head injury. J Neurotrauma; 5:121–137.

[87] Bergsneider M, Hovda DA, Shalmon E, Kelly DF, Vespa PM, Martin NA, Phelps ME, McArthur DL, Caron MJ, Kraus JF, Becker DP (1997) Cerebral hyperglycolysis following severe traumatic brain injury in humans: A positron emission tomography study. J Neurosurg; 86:241–251.

[88] Oddo M, Schmidt JM, Carrera E, Badjatia N, Connolly ES, Presciutti M, Ostapkovich ND, Levine JM, Le Roux P, Mayer SA (2008) Impact of tight glycemic control on cerebral glucose metabolism after severe brain injury: a microdialysis study. Crit Care Med 36(12):3233-8.

[89] Jacobi J, Bircher N, Krinsley J, Agus M, Braithwaite SS, Deutschman C, Freire AX, Geehan D, Kohl B, Nasraway SA, Rigby M, Sands K, Schallom L, Taylor B, Umpierrez G, Mazuski J, Schunemann H (2012) Guidelines for the use of an insulin infusion for the management of hyperglycemia in critically ill patients. Crit Care Med 40(12): 3251-76.

[90] Sadaka F (2012) Red Blood Cell Transfusion in Sepsis: A Review. J Blood Disord Transfus S4:001.

[91] Carlson AP, Schermer CR, Lu SW (2006) Retrospective evaluation of anemia and transfusion in traumatic brain injury. J Trauma 61:567–571.

[92] McIntyre LA, Fergusson DA, Hutchison JS, Pagliarello G, Marshall JC, Yetisir E, Hare GM, Hébert PC (2006) Effect of a liberal versus restrictive transfusion strategy on mortality in patients with moderate to severe head injury. Neurocrit Care 5:4–9.

[93] Napolitano LM, Kurek S, Luchette FA, Anderson GL, Bard MR, Bromberg W, Chiu WC, Cipolle MD, Clancy KD, Diebel L, Hoff WS, Hughes KM, Munshi I, Nayduch D, Sandhu R, Yelon JA, Corwin HL, Barie PS, Tisherman SA, Hebert PC; EAST Practice Management Workgroup; American College of Critical Care Medicine (ACCM) Taskforce of the Society of Critical Care Medicine (SCCM) (2009) Clinical practice guideline: red blood cell transfusion in adult trauma and critical care. J Trauma 67(6): 1439-42.

[94] Agnelli G, Piovella F, Buoncristiani P, Severi P, Pini M, D'Angelo A, Beltrametti C, Damiani M, Andrioli GC, Pugliese R, Iorio A, Brambilla G (1998) Enoxaparin plus

compression stokings compared with compression stockings alone in the prevention of venous thromboembolism after elective neurosurgery. New Engl. J. Med 339: 80-85.

[95] Dudley RR, Aziz I, Bonnici A, Saluja RS, Lamoureux J, Kalmovitch B, Gursahaney A, Razek T, Maleki M, Marcoux J (2010) Early Venous Thromboembolic Event Prophylaxis in Traumatic Brain Injury with Low-Molecular-Weight Heparin: Risks and Benefits. J of Neurotrauma 27: 2165-2172.

[96] Phelan HA, Wolf SE, Norwood SH, Aldy K, Brakenridge SC, Eastman AL, Madden CJ, Nakonezny PA, Yang L, Chason DP, Arbique GM, Berne J, Minei JP (2012) A randomized, double-blinded, placebo-controlled pilot trial of anticoagulation in low-risk traumatic brain injury: The Delayed Versus Early Enoxaparin Prophylaxis I (DEEP I) study. J Trauma Acute Care Surg 73(6):1434-1441.

[97] Jamjoom, A. A B., Jamjoom, A. B. (2013) Safety and Efficacy of Early Pharmacological Thromboprophylaxis in Traumatic Brain Injury: Systematic Review and Meta-Analysis. J of Neurotrauma 30:1-9.

Chapter 8

Prevention of Seizures Following Traumatic Brain Injury

Matthew J. Korobey

Additional information is available at the end of the chapter

http://dx.doi.org/10.5772/57457

1. Introduction

Traumatic brain injury (TBI) is a devastating and complex disease state. Of the observed complications associated with TBI, the development of seizures following injury can be among the most challenging to manage. In addition to the direct effects on the brain, seizures following TBI lead to significant patient morbidity, with potential limitations on overall independence. Quality-of-life may also be adversely affected as the diagnosis of seizure carries the burden of extended exposure to anti-epileptic drugs. The use of these therapies often entails unfavorable medication related side-effects, frequent laboratory monitoring and dosage adjustments, as well as frequent physician visits. With these factors in mind, identification and prophylaxis of those at risk for post-traumatic seizures may provide significant improvement on patient outcomes.

2. Historical perspective

The association between head injury and seizures has long been observed. In the Hippocratic text, "On *Injuries of the Head*", wounds to the left temporal region were described as being associated with convulsion on the right side of the body. Later, Hippocratic surgeons noted that convulsions following head trauma were often a sign of impending fatality. [1] Interestingly, the reference to seizures was only in the immediate time frame surrounding the injury; chronic seizures were not noted to have been documented in those writings. [2] As medicine and neuroscience progressed, the recognition of the association between head trauma and the development of seizure disorder gradually began to grow. As an example, in the 14th century, the Italian physician Valescus de Tharanta described a patient who suffered a penetrating head wound which infiltrated into the pia mater and subsequently experienced seizures 7 to 8 times a day until death. [3] Approximately 150 years later, Berengarius da Carpi, an Italian physician and surgeon, documented that epileptic seizures developed approximately 60 days following a severe head wound which likely is the first documentation of post-traumatic epilepsy, or seizures occurring outside of the immediate post-injury period. This association was further strengthened by the writing of Duretus who reports his observations of a patient suffering a depressed skull fracture who developed seizures 6 years following injury. [1]

Despite the above observations, head injury failed to gain recognition as a *common* cause of epilepsy. Perhaps the first major medical publication noting this connection was book "*Epilepsy and Its Treatment*" in 1904 in which its author William Spratling specifically mentions trauma as an etiology for epilepsy. Spratling writes, "Traumatic lesions of the cranium and the cerebrum cause epilepsy without any predisposition to the disease", and he goes on to describe observed epileptic patterns in significant detail. [4]

The majority of our modern knowledge pertaining to seizures following TBI is derived from military literature. The Second World War, the Korean conflict, and the Vietnam War offered an extensive cohort from which to observe the effects of head trauma. While this data has provided tremendous insight, it does not translate well to head injuries in the civilian sector in regard to mechanism and potential complications. Recently, epidemiologic investigations have examined the impact of TBI and the frequency of post-traumatic seizures in civilians, yielding valuable information for this very different patient population.

3. Definitions

Post-traumatic seizure disorder is a broad term for new-onset seizures following TBI. This terminology fails to capture the distinct differences in etiology, incidence, outcome, and management of these seizures. It is these differences that necessitated the development of a classification system, separating the observed seizures into distinct diagnostic entities based on time of development. The first formal classification of post-traumatic seizures was suggested in 1939 when Elvidge proposed the observed seizures be separated into 3 distinct groups: immediate (within the first few hours following injury), delayed (occurring with 2 weeks of injury), and late (those occurring after a period of time). He further proposed that the immediate and delayed seizures be combined into a group of events termed early epilepsy which was felt to be distinctly different from late epilepsy. [5] This classification has been modified slightly, yet generally remains accepted to this day. The current accepted definitions adopted into practice today are:

- Immediate seizure: Seizures occurring less than 24 hours post injury.
- Early seizures: Seizures occurring less than 1 week post injury.
- Late seizure: Seizures occurring greater than 1 week post injury. [2] These seizures are considered "unprovoked" and are therefore considered epileptic in nature.
- Post-traumatic epilepsy: Recurrent late seizures, with no identifiable cause other then TBI

4. Epidemiology

The burden of post traumatic seizures (PTS) is often underestimated in relation to the overall incidence of epilepsy. A prevalence study by Hauser et al examined several characteristics of epileptic patients in Rochester, MN. Their results showed that approximately 20% of symptomatic disease and 5% of all epilepsies can be attributed to traumatic injury and that in patients less than the age of 65, brain trauma was the most frequently identified risk factor associated with the development of epilepsy. [6] In fact, trauma may be the primary cause of epilepsy in up to 30% of patients between the ages of 15 and 34. [2]

The incidence of seizures following TBI is commonly separated into two distinct categories; early seizures (occurring within 1 week of the index injury) and late (occurring greater than 1 week after the index injury). Each of these categories varies from the other in terms of incidence and risk factors.

The first major publication documenting the incidence of early PTS was presented in 1960. Jennett et al found that among 898 patients admitted with closed head injury, 4.2% experienced seizure within the first 7 days following the injury. [7] Approximately 20 years later, in a retrospective chart review of 2,747 head injury patients experiencing loss of consciousness, Annegers et all found the incidence of early seizure was 2.1%, and of those, 75% occurred in the first 24 hours following injury. [8] Desai et al reviewed the records of 702 patients with any degree of head trauma and found early seizure occurred in 4.1% of patients. [9] Annegers published a second retrospective chart review in 1998 after examining the records of 4,541 pediatric and adult head injury patients. This in-depth review reported the incidence of early PTS to be 2.6%. [10] Asikainen examined a group of 490 TBI patients (approximately 50% of patients studied were under the age of 16) admitted to a TBI rehabilitation facility. Early PTS were observed at an incidence of 16.4%, a number higher than what had been observed previously. The authors comment that there was a statistically significant relationship between the age at time of injury and occurrence of early PTS, as they reported that patients age 7 or younger experience early seizures at a rate of 30.8%, those ages 8 to 16 at a rate of 20%, and patients above the age of 16 at a rate of 8.4%. [11] These results mirrored findings in earlier studies in which pediatric head injury patients were more likely to experience early seizure, independent of the degree of head trauma as compared to adult patients with similar injuries. [7, 8]

Military data is the best source of information when assessing seizure risk following penetrating head injury. The Vietnam Head Injury Study (VHIS) provided valuable information regarding outcomes of patients with penetrating head wounds, including incidence of seizure. In this cohort, early PTS was observed at 4.3%. [12, 13]

While these studies offer useful insight into overall frequency of early PTS, there are limitations that must be addressed. The heterogeneity of the populations being assessed in the available literature makes generalization difficult. In the reports mentioned above, the characteristics of patients observed varies drastically based on severity of head injury, age distribution of the population observed, and mechanism of injury. In addition, the standard of care provided to the head injury patient has evolved and should be considered when assessing the occurrence of any disease sequelae. It must be noted that very little is mentioned in the epidemiologic literature about therapies provided to these patients. When evaluating the overall incidence of early PTS, these limitations must be taken into consideration.

Despite the above mentioned limitations, some of the most useful data gained from these epidemiologic studies is the identification of risk factors for early PTS. When looking at these risk factors as a whole and considering the various populations evaluated, several factors stand out. Age at the time of injury is associated with risk of early PTS, as younger patients are more likely to experience this complication than older patients. Jennett, as stated above, found that patients < 5 years of age were at increased risk of early PTS at any degree of head injury. Of the 75 patients in this age demographic, 9.4% developed early PTS while patients older than the age of 5 had in incidence rate of 3.8%. [7] Annegers et al. also identified children as being at higher risk of early PTS. The incidence of early PTS in patients <15 years of age was 30% while the rate was 10% in adult patients. [8] Similarly, Desai and Asikainen also found that younger patients had a greater risk of early PTS compared with older patients. [9, 11]

Severity of injury at the time of presentation has also been repeatedly associated with increased risk of early PTS. Early on, duration of post-traumatic amnesia (PTA), a guide to assessing the degree of brain injury as reported by Jennett, was found to be associated with an increased risk of early PTS. Approximately 11.7% of patients with PTA lasting greater than 24 hours were seen to have early PTS while those with shorter durations experienced early seizures at a rate of 2.7%. [7] Desai reported that focal deficit on admission and loss of consciousness (LOC)/PTA of more than 30 minutes following injury greatly increased the likelihood of early PTS. [9] It has also been shown that patients presenting with Glascow Coma Scale (GCS) scores of <10 have an increased risk of early PTS with an incidence as high as 20%. [14] Lee et al and Gilad et al examined the rates of PTS following mild head injury and found no increased incidence of seizure development than in the general population experiencing non-trauma related seizures further supporting the theory that severe head injuries pose a higher risk for PTS development. [15, 16] Skull fracture and penetrating head injury have been repeatedly shown to increase the risk of early PTS. Basilar skull fractures, specifically those that are depressed or in the temporal-parietal region, appear to infer the highest risk. [7, 9]

The presence of intracranial blood on radiographic imaging has also been positively associated with early PTS. It is observed that intracranial hemorrhage (ICH) may precede early PTS in up to 23% of patients while epidural/subdural hematoma and cortical contusion are seen in approximately 16% of such cases. [9, 14] Jennett utilized the presence of bloodstained cerebral spinal fluid (CSF) to infer the presence of ICH as cranial imaging at the time was very limited. In those with bloodstained CSF, early PTS were observed at a rate of 9.7% which is noted to be approximately double the incidence of patients observed in the entire series. [7] Additionally, a survey of approximately 3000 patients with severe closed head injury revealed that ICH was present in 66% of all patients experiencing early PTS. [17]

The incidence and risk factors for late PTS are significantly different than that of early PTS and must be described separately. The lowest incidence of late PTS reported was by Jennett at 10.2% whereas Caveness et al, in the VHIS, reported a rate of 27.9%. [7, 13] Annengers, having collected data on 2,747 patients, was able to compare the incidence of late PTS to the expected rate of new onset epilepsy. Using the age and sex-specific incidence rates for Rochester, Minnesota as a comparator, there was found to be a 3.6 fold increase in the risk of developing seizure disorder following head trauma. [8]

Several authors have assessed the duration of risk for the onset of late PTS and a strong trend has emerged from that data. The time-frame for the development of late PTS potentially extends several years post-injury. Salazar et al, utilizing data from the VHIS, found that 7% of all patients with late PTS reported their first seizure 10 or more years post injury while Annegers reports that in patients with moderate to severe TBI, the risk for late PTS extends as far as 20 years post injury. [10, 12] With the awareness of the extended risk for late PTS, it must be noted that more than half of all seizures occur within the 1 year following TBI and approximately 80% occurring within 2 years. [7, 12, 13, 18, 19]

Early PTS are found to be an important predictor of late PTS as Jennett reports a four-fold increase in late PTS if the patient seizes within 7 days of injury. [7] Annengers also found that in patients with moderate or severe TBI, early PTS were associated with an increased likelihood of late PTS. This was found not to be true in pediatric patients, mirroring the findings of Asikainen, who reported increasing age as predisposing factor for late onset seizures. [8, 11] Temkin and Englander report that the presence of intracranial blood, be that SDH or ICH, increases the risk of late seizure development some 400 times the expected rate seen in the general population. [14] Several investigators also report the severity of head injury, as stratified by GCS or an extended duration (most commonly > 24 hours) of PTA, exhibits a positive correlation with late PTS. [7, 10, 11, 13, 14, 19, 21]

Study Author	Population	N	Early Seizure %	Risk Factors Identified	Late Seizure %	Risk Factors Identified
Jennett and Lewin [7]	All admitted patients	896	4.2	Age<5, PTA >24 hours, skull fracture, ICH	10.2	Early PTS, PTA >24 hours, Depressed Skull fracture
Annengers et al. 1980 [8]	General population	2747	2.1	Age <15, severe injury	1.9	Moderate/ Severe TBI
Desai et al. [9]	All admitted patients	702	4.1	Age <16, focal defecit, LOC/ PTA >30 minutes, skull fracture, ICH	Not assessed	Not assessed
Annegers et al. 1998 [10]	Gerneral population	4541	2.6	Not assessed	2.1	Early PTS, Cortical Contusion/ SDH, PTA >24hours
Asikainen et al. [11]	TBI Rehab	490	16.4	Age <8	24.8	Coma >24 hours
Englander et al. [20]	Severe TBI	647	3	Not Assessed	10.2	Early PTS, cortical contusion, dural penetration, multiple intracranial procedures, >5mm midline shift, evacuated SDH

Table 1. Incidence and Risk Factors for PTS

Early	Late
• Penetrating Head Injury	• Early PTS
• Age <16 years	• Advanced Age
• Severity of injury (GCS <10)	• Intracranial Blood
• Basilar Skull Fracture	• Severity of Injury
• Intracranial Blood	

Table 2. Risk Factors for PTS

5. Pathophysiology

The exact mechanisms surrounding the pathogenesis of PTS are still highly debated. Despite extensive animal modeling and basic science bench work, an accurate replication of human TBI has yet to be created. With the lack of an accurate model, the hypotheses found regarding PTS development must be taken in such context. [22]

The pathology surrounding the development of PTS is likely different based on the time of seizure onset, with different mechanisms playing a role in the development of early PTS and late PTS. Early PTS are often considered injury-induced or "provoked" and therefore cannot be considered epileptic. Following the initial insult of tissue deformation and compression, a cascade of events begins to take place almost immediately. Vascular damage and increased permeability of the blood brain barrier (BBB) are noted. [22, 23] Inflammatory cascades are drastically up-regulated leading to neuronal and glial swelling. Glial swelling may lead to impaired neuronal oxygen delivery and subsequent energy depletion leading to cellular death. Glutamate, an excitatory neurotransmitter, is released in high quantities. High concentrations of extracellular glutamate may stimulate ion channel activation and an intracellular flooding of calcium, followed by eventual cell death. [22, 23, 24] In any type of penetrating injury, direct neuronal damage due to cortical laceration, contusion, imbedded bone fragments, or retained foreign bodies becomes extraordinarily irritating to the injured brain and a potential focus for seizure activity. Also, the presence of intracranial blood has been found to predispose the TBI patient to early PTS. When extravasated blood undergoes hemolysis, hemoglobin is engulfed by macrophages which degrade the hemoglobin into hemosiderin and biliverdin. Hemosiderin is deposited into neuronal tissue. These pathogenic deposits initiate lipid peroxidation, damaging cell membranes, and inhibiting Na-K ATPase. [25, 26] The combination of these factors is profoundly pro-excitatory and likely predisposes the post TBI patient to seizures proximal to the time of injury.

The pathology of late PTS is considerably different and more controversial. There are three core theories of epileptogenesis in this setting, the kindling model, the fluid-percussion model, and the synaptic plasticity model. Each of these models attempts to describe the changes taking place in the brain during the latent period between the time of injury and onset of first late seizure.

The kindling theory of epileptogenesis is well validated in several animal models. Briefly, weak electrical stimulation is applied to specific, susceptible regions of the brain in rodents until a seizure occurrs. It has been found that over time, less stimulation was needed to initiate a seizure and this reduced threshold is a change that appears permanent over the remaining life of the animal. In the post-TBI human brain, it is thought that subclinical seizures due to structure changes or brain lesions will reduce the seizure threshold to a point in which epileptogenesis occurs. [27, 28] While this is well described in several animal models, the findings do not completely translate into human patients. Additionally, compounds traditionally assumed effective in controlling kindling type epilepsy (phenytoin, carbamazapime) have been shown to be ineffective preventing late PTS. [29]

The fluid-percussion model may more accurately help depict epileptogensis following TBI in humans. In animal-based experiments, a small hole is drilled into the skull of the animal, followed by the delivery of a fluid-wave percussion against the intact dura initiated by a pendulum impact on a fluid column. The seizures following this type of injury appear after a latent period, much like is seen in late PTS. Sustained hyperexcitablity was also observed in the region surrounding the injured neocortex and this excitability was associated with intense glial reactivity. [30] This reactivity may impair glutamate metabolism, exposing injured neurons to potentially toxic concentrations of this excitatory neurotransmitter. [31] This also mirrors late PTS in humans, as epileptiform activity seems to generate from the injured, hyperexcitable region of brain. [30]

The brain has the ability to adapt to injury through both changes in function and structural reorganization. [32, 33, 34] These changes are termed synaptic plasticity and are found to be an adaptive mechanism aiding in neurologic recovery. [35] During the recovery process, compensatory axons sprout, forming new adaptive neural pathways. While generally beneficial to recovery, these sprouts may, in some settings, become a focus for epileptic activity and may allow a seizure to impact relatively distant regions of the brain. [36] This "maladaptive plasticity" after TBI makes management of this potential pathology very challenging as not all plasticity is harmful. In using therapies that target the sprouting axons, there may be significant impairment in neurological recovery as "adaptive plasticity" would be targeted as well. [37, 38]

6. Impact

Post-traumatic seizures have a broad, sweeping impact on the lives of post-TBI patients. In addition to loss of independence from seizure restrictions (restricted driving privileges); psychological health, employment, quality of life, and mortality may also suffer.

Recurrent seizures are identified as an important cause of hospital readmission among patients with severe TBI. [39, 40] These readmissions are not always directly related to PTS as psychologically related admissions are seen to increase as well. It is understood that TBI patients often suffer from mood disturbances and behavioral abnormalities. Patients with PTS are increasingly sensitive to these disorders and there is a significant increase in the number of psychiatric related hospitalizations among patients with PTS as compared to those without. [41] Additionally, it has been found that irritability and aggressive, disinhibited behavior were observed to be more frequent and severe in rehabilitation patients with PTS as opposed to those not experiencing seizures. [41]

Post-traumatic seizures have been found to correlate with poorer outcomes and quality of life. In the *VHIS*, seizures following penetrating head injury were 1 of 7 impairments that were independently predictive of poor employment status. [42] Also, Asikainen reports that patients with PTS have inferior outcomes per Glasgow Outcome Scale (GOS) scores when compared to those without PTS. These results were mirrored by the findings of Mazzini, who also notes that PTS correlated with poorer GOS scores as well as several other neurobehavioral scales in TBI patients 1 year post injury. [11, 43]

Mortality is seen to be higher in patients with PTS in multiple reports, although it is unknown if there is a direct correlation. [44, 45, 46, 47] It is highly possible that the increase in the incidence of death is more likely related to the severity of the injury as opposed to the development of seizures. [48, 49]

7. Prevention of post-traumatic seizures

With the evolution of the association between head injury and seizure activity, as well as the recognition of the above discussed repercussions, came considerable efforts to develop preventative strategies against epileptic activity post TBI. Early studies, some dating back to the late 1940's, examined the use of anti-epileptic medication in patients with a variety of head injuries. Rapport and Penry noted that in three distinct series of head injured patients, the use of anti-epileptic medication seemed to reduce the incidence of seizure activity when compared to those not receiving such treatment. Their survey of American board certified neurosurgeons found that, at the time, only 14% of those polled typically used PTS prophylaxis. There were no specific professional recommendations for prophylactic treatment and most physicians at the time felt the evidence was considerably sparse. In fact, the most common reasoning for not treating prophylactically was uncertainty of appropriate indication for treatment. When a PTS prophylaxis regimen was initiated, the use of diphenylhydantoin was by far the most commonly used medication, followed the combination of phenobarbitol + diphenylhydantoin. [50]

Phenytoin, the most commonly utilized agent in the hydantoin class, is the most extensively studied PTS prophylaxis medication. Phenytoin is available intravenously (IV) and because of its high (80-90%) bioavailability, can be given orally (PO) as well. [51] The drug enters the brain rapidly and is redistributed throughout the body, being disbursed well throughout all tissues. [52] Phenytoin has a complex, and not yet fully understood, mechanism of action. It appears that the primary site of action for phenytoin is the motor cortex, where it inhibits the spread of seizure activity. Phenytoin has been observed to promote sodium efflux from neurons, stabilizing the neuron and increasing the cell's threshold against hyperexcitability caused by excessive stimulation or environmental changes capable of reducing membrane sodium gradient. [53] Phenytoin is complex kinetically, saturating its metabolism at clinically utilized dosing ranges. Consequently, small changes in dose can translate to large changes in plasma concentration, potentially leading to toxicity or loss of efficacy. Extensively bound to albumin, phenytoin plasma levels must be interpreted in the context of the patient's serum albumin concentration and if necessary, corrected for. Phenytoin levels are appropriately drawn as a "trough", approximately 30 minutes prior administering a dose and the optimal concentration maximizing efficacy and minimizing toxicity is between 10-20mcg/mL. [53]

The impact of phenytoin on the occurrence of PTS has been investigated repeatedly with variable results depending on trial design and outcome assessed. Wohns and Wyler investigated the effect of phenytoin for PTS prophylaxis as compared to no treatment in the seriesz in 62 patients with severe head injury. The dosing of phenytoin was not standardized, but most patients received a 1g load with a maintenance dose of 400mg/day. Therapy was continued for 1 year in the 50 patients managed with phenytoin and plasma concentrations were maintained between 10-20mcg/mL. There were no early PTS observed in this series but a marked difference in the incidence of late PTS was observed. Of the 50 patients treated with phenytoin for seizure prophylaxis, 10% had EEG confirmed late PTS, while those without prophylaxis experience late PTS at a rate of 50%. There was no statistical assessment of this data, but the authors hypothesized that phenytoin may offer some degree of protection from late PTS. [54]

Young et al was not able to confirm the findings of Wohns and Wyler. In a study assessing the effectiveness of phenytoin prophylaxis in preventing late PTS, patients at high risk (>15%) of developing late PTS were assigned to receive either PTS prophylaxis with phenytoin or matching placebo for a minimum of 18 months. There was no measurable difference in the rate of late PTS between groups, however; all patients in the phenytoin arm whom developed PTS had plasma concentrations <12mcg/mL, potentially explaining the lack of efficacy. [55]

The landmark trial demonstrating the efficacy of phenytoin in PTS prophylaxis was published by Temkin et al in 1999. Patients with severe TBI were randomized to receive phenytoin or placebo for 12 months unless adverse drug reactions necessitated early discontinuation. Those receiving phenytoin were loaded with 20mg/kg via IV infusion within 24 hours of injury. The follow-up dosing was not standardized, but an unblinded study-staff member followed frequent blood levels and adjusted phenytoin dosing as need to maintain high therapeutic serum concentrations. The effect of phenytoin on both early and late PTS was evaluated as patients were followed for a total of 24 months (12 months post discontinuation of study medication). Early PTS occurred at a rate of 3.6% ± 1.3 in those receiving phenytoin versus 14.6% ± 2.6 in the placebo group ($p<0.001$). This effect was seen in spite of early serum phenytoin concentrations being less than desired. Unfortunately, the rate of late PTS was unaffected by phenytoin therapy with no significant difference in the number of PTS events when compared with placebo after both 1 year (21.2% ± 3.6 vs. 15.7% ± 3.2; $p>0.2$) and 2 years (27.5% ± 4 vs. 21.%1 ± 3.7; $p>0.2$) of treatment. No difference in mortality was observed and treatment was generally well tolerated, although more patients in the phenytoin group needed to therapy discontinued secondary to medication related adverse events. [56]

A follow-up study, looking at the same data set, attempted to determine if PTS prophylaxis with phenytoin following TBI had any effect on neurocognitive and psychosocial recovery. All patients, independent of treatment arm, (phenytoin or placebo for 1y post TBI) received neuropsychological and psychosocial testing 1, 12, and 24 months post injury. Interestingly, at 1 month post injury, those treated with phenytoin performed significantly worse than those receiving placebo across most neuropsychological parameters. The difference seemed to vanish at the 1 year assessment, but it was noted that at the 2 year assessment, there was a small but widespread negative effect seen in patients treated with phenytoin. Additionally, the negative trend was also seen in patients receiving placebo that developed PTS and were subsequently started on phenytoin for the remainder of the study period. [57]

The above mentioned data set was utilized a third time in order to assess if treatment with phenytoin for 1 or 2 weeks as prophylaxis of early PTS was associated with changes in mortality or morbidity due to adverse drug events. Rash was the most commonly observed drug reaction. Hypersensitivity was rarely seen, occurring only in 0.6% of patients during the first week of therapy and increasing to 2.5% at the end of the second week. Of note, very few adverse reactions were reported in the first week of therapy, with no reports of fever and only one report of leukocytosis. No difference in mortality was observed between patients receiving phenytoin and those receiving placebo, but it was noted that if the patient experienced early PTS, the risk of death increased significantly independent of the treatment arm ($p=0.03$ RR 2; 95% CI 1.1-3.7). [58] While this may again encourage the use of phenytoin as prophylaxis against early PTS, it is important to note that the analysis was not powered to assess a mortality benefit and when corrected for severity of injury, the mortality benefit was no longer observed. Considering the risk-benefit ratio for 1 week of therapy and considering the two previous investigations, a recommendation limiting PTS prophylaxis to 7 days post injury was put forth.

Phenytoin, while potentially efficacious as prophylaxis against early PTS, is a potentially toxic medication. The side-effect profile of this drug is well documented and while these effects are rarely seen in a 7 day treatment period, they must be none-the-less considered. Severe, cutaneous hypersensitivity reactions associated with phenytoin are well documented and potentially life-threatening. [52, 59, 60] Fever, a significant confounder in the critically ill, and potential cause of secondary neurological injury, is reported frequently as well. [52, 53, 61] Phenytoin is an inducer of CYP450 and the UGT isozyme system potentially leading to multiple drug interactions involving absorption, metabolism, and protein binding which may drastically impact the effect of the drug on physiology. [52, 53] Phenytoin's enteric absorption and bioavailability can also be altered when exposed to a food bolus or continuous enteral tube feeds. [52] Bauer evaluated the impact of continuous tube feeds on phenytoin absorption in 53 neurosurgical patients when administered via enteric route. All patients were found to have subtherapeutic serum phenytoin concentrations and approximately 60% continued to be subtherapeutic even after dosage increase. He notes that hypermetabolism and binding of drug to tubing were unlikely as symptoms of phenytoin toxicity were observed following discontinuation of tube feed. [62] In response to these findings, the investigator held tube feedings for 2 hours prior to and following phenytoin administration and flushed the feeding tube with

60mL of water before restarting. With this intervention, serum phenytoin concentrations became therapeutic with only slight dosage increases. [62] Currently, this practice is widely implemented but also leads to significant nutritional issues. [63] With phenytoin typically requiring thrice daily dosing, continuous tube feeds could potentially be held for 12 hours daily. This strategy is met with much resistance as appropriate nutrition may be significantly hindered.

Fosphenytoin (Cerebyx) was approved by the Food and Drug Administration in 1996 as a potential replacement for phenytoin. Fosphenytoin is a pro-drug, fully converting via phosphatases in the liver, red blood cells, and tissue, to phenytoin within 20 minutes of administration. The most clinically significant benefit of fosphenytoin is the ability to infuse the medication as a more rapid rate. Phenytoin loading is often limited by the risk of extravasation and cardiovascular collapse (hypotenion, bradycardia) and due to these concerns, can only be infused at a maximum rate of 50mg/ min. Dosed in phenytoin equivalents (PE), the use of fosphenytoin allows for faster infusion (100-150mg PE/min) and IM administration when there is a lack of effective IV access. [79]

Due to the above mentioned concerns, much effort has been invested in finding an efficacious treatment alternative to phenytoin. The drug most commonly utilized in place of phenytoin is levetiracetam. Levetiracetam is available in both IV and PO formulations with equal bioavailability. Enteral absorption is rapid and predictable, with peak plasma concentrations occurring less than 1 hour following administration. Levetiracetam does not undergo any significant metabolism, avoiding the CYP system which drastically reduces the incidence of clinically relevant drug interactions. Levetiracetam is excreted primarily by the kidney and dosage reductions are recommended in patients with significant renal impairment. [64] Levetiracetam's mechanism of action is poorly understood, but it is thought that it may be novel in its activity. Animal data seems to suggest that levetiracetam may be protective against the development of epilepsy in certain clinical scenarios, yet this is unproven in human subjects. [52, 64, 65] Levetiracetam is very well tolerated and is associated with very few adverse drug events; however, in several patient populations, levetiracetam has been noted to induce non-psychotic behavioral disorders and mood instability. [64] Serum levetiracetam concentrations are not affected by enteric feeds and serum drug level monitoring is not recommended although there are some reports of increased clearance in TBI patients. [66, 67]

The first published trial assessing the use of levetiracetam for PTS was published by Jones et al in 2008. In this analysis, 32 patients with severe TBI (GCS 3-8) were given levetiracetam as PTS prophylaxis at a dosage of 500mg IV every 12 hours for the first 7 days following injury. These patients were compared with a historical control of 41 patients with similar injuries. If a clinical seizure was suspected, an EEG was performed and was subsequently interpreted by a blinded neurologist. The EEG results were classified as normal or abnormal, with abnormal being further stratified into status epilepticus, seizure, or seizure tendency. The patients receiving levetiracetam for prophylaxis had 15 EEG examinations with 53% being abnormal. Seizure activity was only seen in 1 patient, while seizure tendency was observed in the 7 other abnormal EEG's. Of the 41 patients receiving phenytoin for prophylaxis, 12 required EEG and of those 12, none were considered abnormal. The results of this small, non-randomized investigation led the authors to conclude that levetiracetam was as effective as phenytoin in prevention of early PTS. They noted that the observed trend of an increase in seizure tendency required further investigation. [61]

Szalflarski at al completed a prospective, randomized, placebo controlled trial investigating the use of levetiracetam versus phenytoin for PTS prophylaxis. Patients with subarachnoid hemorrhage (SAH) or TBI were randomized in a 2:1 ratio to receive levetiracetam or phenytoin for 7 days following injury. Levetiracetam was loaded at 20mg/kg followed by 1000mg IV q12 hours. Phenytoin was also loaded, followed by twice daily maintenance dosing but no mention of desired blood level or strategy of dose adjustment was noted. Patients were placed on continuous EEG monitoring until awake and following commands or for a maximum of 72 hours. Of the 52 patients enrolled, 34 received levetiracetam while 18 were managed with phenytoin. No difference was observed in the incidence of seizure or overall mortality. Phenytoin use was associated with an increase in the incidence of neurological status decline and gastrointestinal upset while those treated with levetiracetam were devoid of significant adverse drug reactions. In surviving patients, those treated with levetiracetam had significantly better functional outcomes via GOS and the Disability Rating Scale (DRS) at both 3 and 6 months. [68]

The largest investigation to date comparing levetiracetam to phenytoin for early PTS prophylaxis enrolled 813 patients with severe TBI secondary to blunt impact trauma. At the discretion of the providing physician, patients received levetiracetam at a dose of 1g every 12 hours or phenytoin titrated to a serum level of 10-20mcg/mL. No significant difference in the rate of clinical seizures was noted between groups, nor any difference in adverse drug reactions, complications of therapy, or mortality. More patients had therapy discontinue due to medication side effects in the phenytoin group; however, hospital length of stay was increased in the levetiracetam group. [69]

With neither therapy providing a distinct advantage in several individual clinical trials, Zafar at al conducted a meta-analysis attempting to identify any difference in early PTS incidence when levetiracetam or phenytoin were used for prophylaxis. Eight studies (6 observational/ 2 randomized controlled trials) were included, leading to a total patient population of 990. There was a trend toward increased efficacy with levetiracetam, but this difference did not reach statistical significance even when corrected for heterogeneity. It was noted that levetiracetam may have a more favorable side effect profile, but this meta-analysis was not designed to formally assess adverse drug events. As no difference in efficacy was noted, the authors conclude that cost should be the driving factor in the choice of agent for PTS prophylaxis. [70]

A cost-utility analysis by Cotton et al attempted to determine whether levetiracetam or phenytoin was more cost effective as an early PTS prophylaxis strategy. Using quality-adjusted life years (QALY) as the primary cost-utility determinant, phenytoin was found to be the more cost- effective management strategy. Phenytoin was reported to have a ratio of $1.58/ QALY versus levetiracetam which had a ratio of $20.72/QALY. It must be noted that at the time of this analysis, levetiracetam was not available as a generic product. Levetiracetam has been available via several generic manufactures since May 2010 and the subsequent cost reduction must be taken in consideration when interpreting these results. [71]

Several other treatment modalities have been investigated as preventive strategies for PTS with little benefit derived. Valproate was compared with phenytoin for prevention of PTS in high risk TBI patients. No difference in the incidence of either early or late PTS was noted, but mortality in the valproate group (13.4%) was almost double that seen in those managed with phenytoin (7.2%) leading to the early closure of the investigation. [72] Phenobarbital has been assessed both alone and in combination with phenytoin as a PTS prophylaxis strategy. When used as monotherapy for PTS prophylaxis in severe head injury, no benefit was observed and, due to adverse drug reactions, medication compliance was poor. [73] In combination with phenytoin, there was an observed reduction in seizures, but this difference could have been influenced by the use of phenytoin, and significant concern of toxicity was noted. [74] Magnesium was thought to be potentially neuroprotective, but when higher serum concentrations were targeted with continuous infusion there was an increase in the incidence of hypotension, reduced cerebral perfusion pressure, and a doubling of mortality. Lower magnesium target ranges also failed to provide any benefit. [75] In an observational series, corticosteroids were found to provide no PTS protection and were found to worsen excitotoxicity and oxidative neuronal damage in animal models. [76] There is a theoretic benefit with targeted temperature management as this therapy has several neuroprotective effects such as reduced metabolic rate, reduced inflammatory response, fewer epileptic discharges, and a reduction in the production of reactive oxygen species. This modality has not been formally evaluated for its impact on PTS, but offers a potential area of future research and potential benefit. [77]

As of 2007, the Brain Trauma Foundation guidelines for the management of severe traumatic brain injury consider the use of anti-epileptic medications for early PTS prophylaxis a Level II recommendation and note that prolonged prophylactic therapy (>7days) is not recommended. At the time of publication, phenytoin was the preferred agent for early PTS prophylaxis, but in the time since, a significant amount data surrounding the use of levetiracetam has been published. At this point, the two therapies should be considered equivalent from an efficacy standpoint, but consideration needs to be given to the individual characteristics of each medication and institutional costs associated with their administration. [78]

Traumatic brain injury is a devastating disorder having significant impact on patient morbidity and mortality. Seizures following TBI are a potentially preventable complication if the patients at highest risk are identified quickly and are initiated on a prophylactic regimen. The highest risk patients (penetrating injury, age <16, GCS <10, basilar skull fracture, intracerebral blood) should be started on prophylactic anticonvulsant therapy as soon as possible. Based on the latest evidence, there are two potentially efficacious strategies; phenytoin (or substituted fosphenytoin) and levetiracetam. Phenytoin appears most effective when loaded at a dosage of 10-15mg/kg followed by a regimen of 100mg every 8 hours and should be titrated to a trough total phenytoin concentration of 10-20 mcg/mL. Levetiracetam is an acceptable alternative regimen found to be effective when dosed at 1000mg every 12 hours with no serum drug monitoring being currently recommended. Both phenytoin and levetiracetam should be administered for 7 days following initial injury and when utilized in this manner a significant reduction in early PTS is observed, potentially improving patient outcomes.

Drug Name (Brand)	Phenytoin (Dilantin)	Levetiracetam (Keppra)
Mechanism of Action	Enhanced sodium efflux from neurons of the motor cortex, stabilizing the threshold against hyperexcitability. Posttetanic potentiation at synapses is then reduced preventing cortical seizure foci from detonating adjacent cortical areas	Inhibition of voltage-dependent N-type calcium channels; facilitation of GABA-ergic inhibitory transmission; reduction of delayed rectifier potassium current; and/or binding to synaptic proteins which modulate neurotransmitter release
Route of Administration	IV/ PO	IV/ PO
Typical Loading Dose	10-15mg/kg x1 ***Fosphenytoin:*** 10-20mg PE/kg	None recommended
Typical Maintenance Dose	100mg q8hr ***Fosphenytoin:*** 4-6mg PE/kg	500-1000mg q12hr
Elimination Half-Life	7-42 hours	6-8 hours (Metabolite =8.4 hr)
Metabolism/ Excretion	Metabolism: Hepatic via hydroxylation Excretion: Extensively excreted in bile as well as in urine following gastrointestinal reabsorption	Metabolism: enzymatic hydrolysis via liver Excretion: 66% unchanged in urine
Dosing Considerations	Metabolism and excretion may be impaired in renal and hepatic disease. Consider following unbound phenytoin levels for dose adjustment	Consider dosage reduction in patients with renal failure
Major Adverse Effects	Rash Nausea/ vomiting Ataxia Confusion Fever Stevens-Johnson Syndrome Myelosuppression Nephrotoxicity Hypotension Bradycardia Venous irritation	Pancytopenia Thrombocytopenia Liver Failure Personality disturbances
Monitoring/ Laboratory Considerations	Monitor trough concentration for efficacy. Toxicity is a clinical diagnosis and must be assess based on patient	No routine monitoring currently recommended

http://dx.doi.org/10.5772/57457

Drug Name (Brand)	Phenytoin (Dilantin)	Levetiracetam (Keppra)
	presentation. At times, patients may require blood concentrations outside therapeutic range	
	If rapid therapeutic levels are needed, a level may be drawn 2 hrs after completion of IV load or 24 hrs after administration of an oral loading dose	
	Draw within 5-8 days of therapy initiation with subsequent doses of phenytoin adjusted accordingly If complex patient free phenytoin concentration is desirable	
	Goal concentration 10-20 mcg/ml (Total)	
	1-2.5mcg/mL (Free)	
	In patients with hypoalbuminemia, phenytoin concentration needs to be corrected mathematically or free level should be drawn	
	CORRECTED PHENYTOIN= (Measured Phenytoin) 0.1 x Albumin +0.1	

Table 3. makinji caption

Author details

Matthew J. Korobey

Mercy Hospital St. Louis, St. Louis MO, USA

References

[1] Temkin O. The Falling Sickness: A History of Epilepsy from the Greeks to the Beginnings of Modern Neurology. The Johns Hopkins Press. Baltimore, MA. 1945

[2] Lowenstein DH. Epilepsy after Head Injury: An Overview. Epilepsia 2009; 50(Suppl. 2): 4-9

[3] Henry FP. A Notice of the Life and Writings of Valescus de Tarenta. Maryland Medical Journal 1901; 44(6): 237-251

[4] Spratling WP. Epilepsy and Its Treatment. WB Saunders and Company. Philadelphia, PA. 1904

[5] Elvidge AR. Remarks on Post-Traumatic Convulsive State. Transactions of the American Neurological Association 1939; 65: 125-129

[6] Hauser WA, Annegers JF, Kurland LT. Prevalence of Epilepsy in Rochester, Minnestoa: 1940-1980. Epilepsia 1991; 32(4): 429-445

[7] Jennett WB, Lewin W. Traumatic Epilepsy after Closed Head Injury. J Neurol. Neurosurg. Psychiat. 1960; 23:295-301

[8] Annegers JF, Grabow JS, Groover RV et al. Seizures after Head Trauma: A Population Study. Neurology 1980; 30:683-689

[9] Desai BT, Whitman S, Coonley-Hoganson R et al. Seizures and Civilian Head Injuries. Epilepsia 1983; 24: 289-296

[10] Annegers JF, Hauser WA, Coan SP, Rocca WA. A Population-Based Study of Seizures after Traumatic Brain Injuries. NEJM 1998; 338: 20-24

[11] Asikainen I, Kaste M, Sarna S. Early and Late Posttraumatic Seizures in Traumatic Brain Injury Rehabilitation Patients: Brain Injury Factors Causing Late Seizures and Influence of Seizures on Long-Term Outcome. Epilepsia 1999; 40(5): 584-589

[12] Salazar AM; Jabbari B, Vance S et al. Epilepsy After Penetrating Head Injury. I. Clinicla Correlates: a Report of the Vietnam Head Injury Study. Neurology 1985; 35: 1406-1414

[13] Caveness WF, Meirowsky AM, Rish BL et al. The Nature of Posttraumatic Epilepsy. J Neurosurg 1979; 50: 545-553

[14] Temkin NR. Risk Factors for Posttraumatic Seizures in Adults. Epilepsia 2003; 44(Suppl 10): 18-20

[15] Lee ST, Lui TN. Early Seizures after Mild Closed Head Injury. J Neurosurg 1992; 76: 435-439

[16] Gilad R, Boaz M, Sadeh M et al. Seizures After Very Mild Head or Spine Trauma. J Neurotrauma 2013; 30: 469-472

[17] Lee TS, Luie TN, Wong CW et al. Early Seizures After Severe Closed Head Injury. Can J Neurol Sci 1997; 24: 40-43

[18] Weiss GH, Caveness WF. Prognostic Factors in the Persistence of Posttraumatic Epilepsy. J Neurosurg 1972; 37: 164-169

[19] Haltiner AM, Temkin NR, Dikmen SS. Risk of Seizure Recurrence After the First Late Posttraumatic Seizure. Arch Phys Med Rehabil 1997; 78: 835-840

[20] Endlander J, Bushnik T, Duong TT et al. Analyzing Risk Factors for Late Posttraumatic Seizures: A Prospective, Multicenter Investigation. Arch Phys Med Rehabil 2003; 84: 365

[21] Christensen J, Pedersen MG, Pedersen CB,Sidenius P, Olsen J, Vastergaard M. Long-Term Risk of Epilepsy After Traumatic Brain Injury IN Children and Young Adults: A Population-Based Cohort Study. Lancet 2009; 373: 1105-1110

[22] Garga N, Lowenstein SH. Posttraumatic Epilepsy: A Major Problem in Desperate need of Major Advances. Epilepsy Currents 2006; 6(1): 1-5

[23] Hunt RF, Boychuck JA, Smith BN. Neural Circuit Mechanisms of Post-traumatic Epilepsy. Frontiers in Cellular Neuroscience 2013; 7 (Article 89): 1-14

[24] Bao, Y, Bramlett HM, Atkins CM et al. Post-Traumatic Seizures Exacerbate Histopathological Damage after Fluid-Percussion Brain Injury. J Neurotrauma 2011; 23: 35-42

[25] Willmore LJ. Post-Traumatic Epilepsy: Cellular Mechanism and Implications for Treatment. Epilepsia 1990; 31(Suppl 3): S67-73

[26] Zasler N, Katz D, Zafonte RD. Brain Injury Medicine: Principles and Practice. New York, NY: Demos Medical Publishing, LLC; 2012

[27] Goddard BS. Developmnet of Epileptic Seizures Through Brain Stimulation at Low Intensity. Nature 1967; 214: 1020-1021

[28] Cavalheiro EA, Leite JP, Bortolotto ZA et al. Long-Term Effects of Pilocarpine in Rats: Structural Damage of the Brain Triggers Kindling and Spontaneous Recurrent Seizures. Epilepsia 1991; 32: 778-782

[29] Temkin NR, Jarell AD, Anderson GD. Antiepileptic Agents: How Close are We?. Drugs 2001; 61: 1045-1055

[30] D'Ambrosio R, Fairbanks JP, Fender JS et al. Post-Traumatic Epilepsy Following Fluid Percussion Injury in the Rat. Brain 2004; 127: 304-314

[31] Lieth E, Barber AJ, Xu B et al. Glial reactivity and Impaired Glutamate Metabolism in Short-Term Experimental Diabetic Retinophaty. Penn State Research Group. Diabetes 1998; 47 (5): 815-820

[32] Lowenstein DH, Thomas MJ, Smith DH, McIntosh TK. Selective Vulnerability of Dentate Hilar Neurons Following Traumatic Brain Injury: A Potential Mechanistic Link Between Head Trauma and Disorders of the Hippocampus. J Neurosci 1992; 12: 4846-4853

[33] Levin HA. Neuroplasticity Following Non-Penetrating Traumatic Brain Injury. Brain Inj 2003; 8: 665-674

[34] Seitz RJ, Huang Y, Knorr U et al. Large-Scale Plasticity of the Human Motor Cortex. Neuroreport 1995; 6: 742-744

[35] Albensi B, Janigro D. Traumatic Brain Injury and its Effects on Synaptic Plasticity. Brain Inj 2003; 17: 653-656

[36] Sankar R, Shin D, Liu H, Wasterlain C, Mazarati A. Epileptogenesis During Development: Injury, Circuit Recruitment, and Plasticity. Epilepsia 2002; 43 Suppl 5: 47-53

[37] Prince DA, Connors BW. Mechanisms of Epileptogenesis in Cortical Structures. Ann Neurol 1984; 16 (Suppl): S59-S64

[38] Montanez S, Kline AE, Gasser TA, Hernandez TD. Phenobarbital Administration Directed Against Kindled Seizures Delays Functional Recovary Following Brain Insult. Brain Res 2000; 860: 29-40

[39] Cifu DX, Kreutzer JS, Marwitz JH et al. Etiology and Incidence of Rehospitalization after Traumatic Brain Injury: A Multicenter Analysis. Arch Phys Med Rehabil 1999; 80: 85-90

[40] Shavelle RM, Strauss D, Whyte J, Day SM, Yu YL. Long-term Causes of Death After Traumatic Brain Injury. Am J Phys Med Rehabil 2001; 80: 510-6

[41] Swanson SJ, Rao SM, Grafman J, Salazar AM, Kraft J. The Relationship Between Seizure Subtype and Interictal Personality. Results from the Vietnam Head Injury Study. 1995; 118: 91-103

[42] Schwab K, Grafman J, Salazar Am, Kraft J. Residual Impairments and Work Status 15 Years After Penetrating Head Injury: Report from the Vietnam Head Injury Study. Neurology 1993; 43: 95-103

[43] Mazzini L, Cossa FM, Angelino E et al. Neuroradiologic and Neuropsychological Assessmetn of Long-Term Outcome. Epilepsia 2003; 44:569-74

[44] Corkin S, Sullivan EV, Carr FA. Prognostic Factors for Life Expectancy After Head Injury. Arch Neurol 1984; 41: 975-977

[45] Walker AE, BLumer D. The Fate of World War II Veterans with Post-traumatic Seizures. Arch Neurol 1989; 46: 23-26

[46] Weiss GH, Caveness WF, Eisiedel-Lechtape H, McNeel M. Life Expectancy and Causes of Death in a Group of Head Injured Veterans of World War I. Arch Neurol 1982; 39: 741-743

[47] Walker AE, Erculei F. Post-traumatic Epilepsy 15 Years Later. Epilepsia 1970; 11:17-26

[48] Rish B, Caveness W. Relation of Prophylactic Medication to the Occurrence of Early Seizures Following Craniocerbral Trauma. J Neurosurg 1973;38: 155-158

[49] Rish BL, Dillon JD, Weiss GH. Mortality Following Penetrating Craniocerebral Injuries. J Neurosurg 1983: 59;775-80

[50] Rapport RL, Penry JK. A Survey of Attitudes Toward the Pharmacological Prophyaxis of Posttraumatic Epilepsy. J Neurosurg 1973; 38: 159-166

[51] Rang HP, Dale MM, Ritter JM, Moore PK. Pharmacology. 5th Ed. St. Louis: Churchill Livingstone. 2003

[52] DiPiro JT, Talber RL, Yee GL. Pharmacotherapy, A Pathophysiological Approach, 8th Ed. McGraw-Hill. 2011

[53] Parenteral Dilantin [Package Insert]. New York, NY: Pfizer; 2011

[54] Wohns RNW, Wyler AR. Prophylactic Phenytoin in Severe Head Injuries. J Neurosurg 1979; 51: 507-509

[55] Young B, Rapp RP, Norton JA et al. Failure of Prophylactically Administered Phenytoin to Prevent Late Posttraumatic Seizures. J Neurosurg 1983; 58: 236-241

[56] Temkin NR, Dikmen SS, Wilensky AJ et al. A Randomized, Double-Blind Study of Phenytoin for the Prevention for Post-Traumatic Seizures. NEJM 1990; 323(8): 497-502

[57] Dikmen SS, Temkin NR, Miller B, Machamer J, Winn HR. Neurobehavioral Effects of Phenytoin Prophylaxis of Posttraumatic Seizrues. JAMA 1991; 265:1271-1277

[58] Haltiner AM, Newell DW, Temkin NR, Kikmen SS, Winn HR. Side Effects and Mortality Associated with Use of Phenytoin for Early Posttraumatic Seizure Prophylaxis. J Neurosurg 1999; 91: 588-592

[59] Jones GL, Wimbish GH, McIntosh WE. Phenytoin: Basic and Clinical Pharmacology. Med Res Rev 1983; 3: 383-434

[60] Stern RS. Exanthematous Drug Eruptions. N Engl J Med 2012; 366:2492-2501

[61] Jones KE, Puccio AM, Harshman KJ et al. Levetiracetam Versus Phenytoin for Seizure Prophylaxis in Severe Traumatic Brain Injury. Neurosurg Focus 2008; 25(4): E3 15

[62] Bauer LA. Interference of oral phenytoin absorption by continuous nasogastric feedings. Neurology 1982;32: 570-2

[63] Chan LN. To Hold (Enteral Feeding) or Not to Hold: That IS the Question; A Commentary and Tutorial. Practical Gastroenterology 2012; January: 13-21

[64] Keppra [Package Insert]. Symrna, GA: UCB; 2011

[65] Loscher W, Honack D, Rundfeldt C. Antiepileptogenic Effects of the Novel Anticonvulsant Levetiracetam (ucb LO59) in the Kindling Model of Temporal Lobe Epilepsy. J Pharmacol Exp Ther 1998;284:474–479.

[66] Cook AM, Arora S, Davis J, Pittman T. Augmented Renal Clearance of Vancomycin and Levetiracetam in a Traumatic Brain Injury Patient. Neurocrit Care 2013; 19: 210-214

[67] Klein P, Herr D, Pearl PL et al. Results of Phase II Pharmacokinetic Study of Levetiracetam for Prevention of Post-Traumatic Epilepsy. Epilepsy & Behavior 2012; 24: 457-461

[68] Szaflarski JP, Sangha KS, Lidsell CJ, Shutter LA. Prospective, Randomized, Single-Blinded Comparative Trial of Intravenous Levetiracetam Versus Phenytoin for Seizure Prohylaxis. Neurocrit Care 2010; 12: 165-172

[69] Inaba K, Menaker J, Branco DC et al. A Prospective Multicenter Comparison of Levetiracetam Versus Phenytoin for Early Posttraumatic Seizure Prophylaxis. J Trauma Acute Care Surg 2013; 74(3): 766-773

[70] Zafar SN, Kahn AA, Ghuri AA, Shamim MS. Phenytoin Versus Leviteracetam for Seizure Prophylaxis after Brain Injury – A Meta Analysis. BMC Neurology 2012; 12: 30-38

[71] Cotton BA, Kao LS, Kozar R, Holcomb JB. Cost-Utility Analysis of Levetiracetam and Phenytoin for Posttraumatic Seizure Prophylaxis. J Trauma 2011; 71: 375-379

[72] Temkin NR, Kikmen SS, Anderson GD et al. Valproate Therapy for Prevention of Posttraumatic Seizures: A Randomized Trial. J Neurosurg 1999; 91: 593-600

[73] Manaka S. Cooperative Prospective Study on Posttraumatic Epilepsy: Risk Factors and the Effect of Prophylactic Anticonvulsant. The Japanese Journal of Psychiatry and Neurology 1992; 46(2): 311-315

[74] Servit Z, Musil F. Prophylactic Treatment of Posttraumatic Epilepsy: Results of a Lont-Term Follow-Up in Czechoslovakia. Epilepsia 1981; 22: 315-320

[75] Temkin NR, Winn HR, Ellenbogen RG et al. Magnesium Sulfate for Neuroprotection after Traumatic Brain Injury: A Randomised Controlled Trial. Lancet 2007; 6: 29-38

[76] Watson NF, Barber JK, Doherty MJ, Miller JW, Temkin NR. Does Glucocorticoid Administration Prevent Late Seizures after Head Injury?. Epilepsia 2004; 45(6): 690-694

[77] Christian E, Sada G, Sung G, Giannotta SL. A Review of Selective Hypothermia in the Management of Traumatic Brain Injury. Neurosurg Focus 2008; 25 (4): E9 1- 8

[78] Brain Trauma Foundation. Guidelines for the Management of Severe Traumatic Brain Injury. J Neurotrauma 2007; 42, Supplement 1: S1-S106

[79] Kirschbaum K, Gurk-Turner C. Phenytoin vs Fosphenytoin. BUMC Proceedings 1999; 12:168-172.

Chapter 9

Immediate Treatment of the Anticoagulated Patient with Traumatic Intracranial Hemorrhage

David E. Tannehill

Additional information is available at the end of the chapter

http://dx.doi.org/10.5772/57366

1. Introduction

Traumatic brain injury with associated intracranial bleeding is an emergency requiring diagnostic and supportive measures directed at limiting subsequent morbidity and mortality. If this intracranial bleeding is discovered in a patient that is anticoagulated, further aggressive steps are indicated to reverse the effects of the anticoagulant and limit further bleeding. Hematoma or bleeding volume & growth does correlate with various outcome measures [1-3]. Following spontaneous ICH, hematoma expansion on follow-up neuroimaging is noted in as much as 26% of cases [1]. Particularly in anticoagulated patients, such hematoma expansion is associated with poor outcomes [4]. Hemostatic therapy with agents directed at trying to reverse the effects of the anticoagulant have been shown to potentially be effective in reducing hematoma expansion in such patients [5,6]. Failing to correct a high INR, for example, may be associated with higher mortality [7]. Therefore, prompt recognition of this condition and taking immediate action is important in treating traumatic intracranial hemorrhage. There are more options than ever for providing systemic anticoagulation to patients. These different agents work in different ways and have different pharmacokinetics and pharmacodyanmics. Published data supporting specific strategies for reversing the effects of these agents is limited, particularly for novel anticoagulants. However, the pharmacology and mechanism of action of some anticoagulant therapies (outlined in Figure 1 & Table 1) have led to at least initial studies that can provide some guidance.

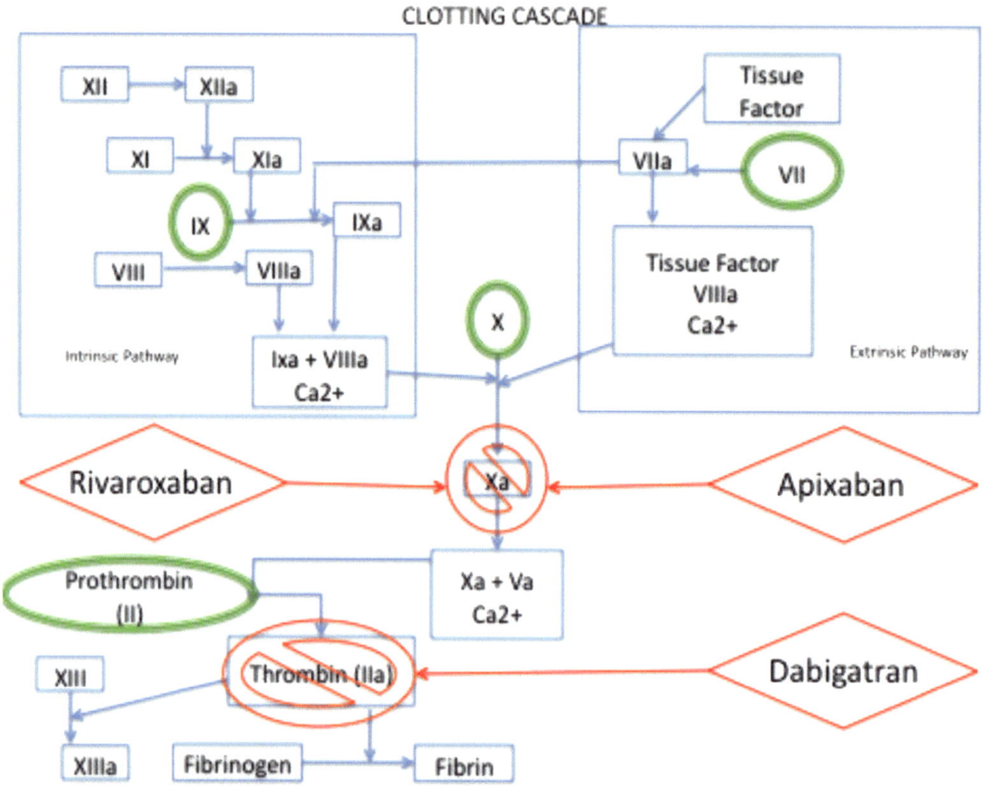

Figure 1. The clotting cascade. The vitamin K dependant factors (green circled factors) are inhibited by warfarin. Factor Xa is inhibited by rivaroxaban and apixaban. Dabigatran dirercty inhibits thrombin.

Medication	Mechanism of Action	T 1/2 (hrs)	Dose	Renal dosing	Excretion	Protein Bound
Warfarin	Vitamin K antagonist	20-60	2-10mg daily	None	92% urine bile	99%
Rivaroxaban	Direct factor Xa Inhibitor	5-13	10-30mg daily	Yes	66% urine 28% feces	92-95%
Apixaban	Direct factor Xa Inhibitor	8-12	2.5-5mg BID	Yes	27% urine feces	87%
Dabigatran	Direct thrombin inhibitor	12-17	110-150mg BID	Yes	80% urine	35%

Table 1. Pharmacology of warfarin and the novel anticoagulants.

2. Warfarin & strategies for warfarin-associated ICH

Studies suggest that hematoma expansion in patients anticoagulated with warfarin who suffer an intracranial hemorrhage may be limited by quickly normalizing INR [8]. Mortality rates for these patients may be as high as 66% for those with an INR > 3 [9]. Historically, clinicians have used vitamin K and fresh frozen plasma (FFP) to treat warfarin-associated bleeding, including ICH. Vitamin K is given to patients that have received warfarin so as to counteract the effects of warfarin inhibiting the vitamin-K dependant gamma-carboxylation of coagulation factors II, VII, IX and X and the anticoagulant factors protein C and S. Its time to maximum effectiveness can be measured in hours, so its use alone is not recommended [10]. FFP can be transfused to replenish the lower levels of vitamin K dependant clotting factors in patients that have been taking vitamin K antagonists. Recent data have led to a change in the American College of Chest Physicians guidelines to recommend the use of four-factor prothrombin complex concentrate to reverse the effects of warfarin in the treatment of major bleeding (Level 2C evidence) [11]. FFP has fallen out of favor for multiple reasons. Firstly, it must be thawed and this process can take anywhere from 20-60 minutes, causing the clinician to lose valuable time in trying to reverse the patient's coagulopathy. Additionally, FFP is given in a large weight-based volume (as much as two to four L total) that may be unadvisable in certain patients [12]. Lastly, FFP is a blood product and it is widely recognized that blood product administration in general and FFP specifically should be minimized as much as possible in the critically ill patient. This is because of the associated risks of acute lung injury, transfusion-associated circulatory overload, infection, etc [10, 68]. Additionally, it is quite common for FFP to be transfused inappropriately, or at least outside the recommendations of published guidelines [69].

Prothrombin complex concentrates (PCC) have been shown to normalize an elevated INR within 10-30 minutes of administration [13]. Prothrombin complex concentrates (PCC) are a group of products containing virus-reduced, concentrated, pooled plasma products made of a combination of three or four vitamin K-dependant clotting factors [10]. There are two versions commercially available, so-called three-factor and four-factor PCC. These agents have varying concentrations of coagulation factors II (prothrombin), VII, IX and X. Factor VIII Inhibitor Bypassing Activity, nonfiltered – or FEIBA NF – is an FDA approved four-factor PCC that is a nanofiltered, vapor-heated freeze-dried sterile human plasma fraction [14]. Until recently, it was the only four-factor complex available in the United States. As an activated product, it may hold more thrombogenic potential, limiting its use. Introduced in the summer of 2013, *Kcentra* (known as Beriplex P/N in Europe) is the only FDA-approved four-factor PCC available in the United States [15].

The preference for four-factor PCC over the three factor formulations is because of the increased concentration of factor VII in the four factor PCCs [11]. This may account for their possible superior effectiveness at reversing warfarin-associated anticoagulation [16]. Nonactivated four-factor PCC has only recently been available in the United States [15]. Prior to this, only FEIBA, an activated form of PCC was commercially available in the US. FEIBA has been suggested as an option for reversing the effects of VKA, but there have been concerns that this agent may increase thrombogenesis.[11]. Another option was to use one of the available three-factor PCCs, Profiline SD or Bebulin VH plus rVIIA, but this also comes with a similar thrombogenic risk, particularly arterial thromboses [18,19]. Four-factor PCC is the recommended agent for warfarin reversal in acute, severe bleeding [11]. While there are no randomized, controlled studies, there are multiple observational trials summarized by a systematic review of 18 studies in 654 patients. This review suggested that 4-factor prothrombin complex concentrates are more reliable for correcting the international normalized ratio (INR) compared with three factor formulations. This effect appears intact at least up until very elevated INRs (>4) where the effectiveness may be less reliable for both the three and four factor PCC [17]. While these agents are effective in reversing a high INR, it must be remembered that they may infer a risk of thromboembolism [18]. However, recent studies comparing four-factor PCC with FFP for urgent warfarin reversal seem to suggest that this adverse event rate - including thromboembolism - is at least the same, if not better than with FFP [71, 72].

3. Non-warfarin anticoagulants and strategies for treating ICH

An important point to remember when treating a patient that is anticoagulated with dabigatran, rivaroxaban or apixaban is that the physiologic effect of these drugs is much shorter in duration than that of warfarin. Warfarin can remain effective for up to 4 or 5 days as its half-life may be as long as 38-42 hours or longer. The direct-thrombin inhibitor, dabigatran, has a half-life of 12-17 hours. The factor Xa inhibitors, rivaroxaban (7-11hr) and apixaban (9-14hr) have similar, considerably shorter half-lives than warfarin and therefore are eliminated from the body more quickly. This elimination is somewhat dependant on renal function and therefore adjustments must be made in awaiting clearance of the drug's effect in patients with renal dysfunction. [19-22]

3.1. Dabigatran

Dabigatran is an oral direct thrombin inhibitor approved for the reduction of stroke risk in patients with nonvalvular atrial fibrillation. Maximum concentration is reached within about one hour of ingestion and the drug is cleared renally [23]. Dabigatran was shown in the RE-LY trial to be effective in the prevention of systemic embolism in non-valvular atrial fibrillation. The United States Food and Drug Administration therefore approved it in 2010 for this indication. The RE-LY study also demonstrated that dabigatran conferred a similar yearly rate of major bleeding with warfarin of about 3% and life-threatening bleeding of about 1.5%. [24]. Renal clearance of dabigatran suggests a potential role for hemodialysis in reversing the effects of this agent. [22].

FFP, however, will not reverse this agent, but rFVIIa may have a role [25]. In a murine model of ICH, FFP was shown to decrease ICH volume in animals that had received high dose, but not low dose, dabigatran. However, mortality remained higher in the mice that received high dose dabigatran and FFP did not reduce mortality. Beriplex, a four-factor PCC, decreased ICH hematoma size in both the high and low dose groups [26]. Other animal data in rabbits has suggested that PCC may be effective in reversing the anticoagulation effect of dabigatran [27]. There is a case report of FFP & PCC being used in combination to successfully treat dabigatran-induced gastrointestinal bleeding; however, full confirmation of this association was difficult because of patient comorbidities [28]. Similarly, FEIBA has been suggested in a case report as a potential therapy for dabigatran-associated bleeding [29].

However, a four factor PCC had no positive effect on the anticoagulant activity of dabigatran in a randomized, double-blind placebo controlled trial conducted by Eerenberg, et al. [30]. FEIBA at a lower than typical dose could have a role in reversing the anticoagulant effects of dabigatran, while the non-activated PCCs are likely less effective [31]. This same study also failed to show benefit of rVIIa for preventing hematoma expansion [31]. Conversely, a rat tail incision bleeding model suggested that rVIIa did reduce bleeding time [32]. However, there was no demonstrable positive effect on thrombin time, aPTT, or ecarin clotting time [33]. Similar results were found with both three and four factor PCC [19].

In summary, medical therapy for treatment of dabigatran associated traumatic ICH is somewhat limited by both a paucity of data and lack of consistent effectiveness demonstrated in what data does exist. Therefore, supportive therapies such as activated charcoal and dialysis may be the most prudent approach. Activated charcoal can be effective in reducing therapeutic levels of dabigatran if the ingestion was relatively recent [34] and dialysis is both pharmacologically plausible and has been shown to reduce dabigatran levels [22, 35].

3.2. Rivaroxaban

Rivaroxaban is an oral direct factor Xa inhibitor approved for the prophylaxis of deep vein thrombosis (DVT) post total knee or hip replacement [36], the treatment of DVT or pulmonary embolus (PE) [37, 38] as well as to reduce the risk of stroke in patients with nonvalvular atrial fibrillation [39]. The drug reaches maximum concentration within 2 to 4 hours of intake and has a 5 to 9 hour half-life [40]. Unlike dabigatran, rivaroxaban is highly protein-bound making it unlikely to be dialyzable [41].

Animal data suggest at most a moderate effect of rVIIa on reversing the anticoagulant effect of rivaroxaban. In a rat mesenteric bleeding model, this drug decreased the bleeding time, but had no effect on the inhibited factor Xa activity [42]. A primate model demonstrated a much less robust effect [43]. In another animal study, a four-factor PCC reversed the effect of rivaroxaban [44]. The same study by Eerenberg quoted above in regard to dabigatran suggested that the four factor PCC Cofact may have a positive effect on normalizing the inhibition of endogenous thrombin potential and the elevated PT induced by rivaroxaban [30]. In a randomized crossover *ex vivo* study in healthy volunteers, Marlu et al were able to demonstrate that the anticoagulant effect of rivaroxaban could be corrected with PCC or FEIBA and that rVIIa was much less effective [31]. This suggests that PCC may be effective for treating rivaroxaban-associated traumatic intracranial hemorrhage.

3.3. Apixaban

Apixaban is an oral factor Xa inhibitor indicated for the prevention of systemic embolism and stroke in patients with nonvalvular atrial fibrillation. It is the newest agent to be approved for use. In studies, the major bleeding event rate was approximately 2% compared with warfarin's 3% rate [49, 73]. The drug requires a dosage adjustment in renal impairment. It is also highly protein-bound and therefore not dialyzable. Activated charcoal is felt to be a potential intervention that could be employed early after ingestion, if possible [45]. There are no studies or case reports in the literature regarding treatment of apixaban-associated bleeding. Therefore, a similar approach to that used in treating rivaroxaban-associated bleeding is probably reasonable. However, in a rabbit model, bleeding induced by apixaban was not improved with either rFVIIa, PCC or fibrinogen [46]. Hopefully, with time and experience, we will learn more about treating life-threatening bleeding associated with this and the other novel anticoagulants.

4. Other agents to be considered for the treatment of traumatic ICH in the anticoagulated patient

4.1. rVIIa

Recombinant activated factor VII (rFVIIa) is indicated for hemophilia with inhibitors [47]. But it has also been used off-label for hemostasis in acute bleeding, surgical bleeding, trauma and ICH [48, 49, 50]. It has also been used as a medication to assist in hemostatic control of hemorrhage in trauma [51-56]. While this intervention has been shown to not impact mortality, it was shown to decrease overall blood product use [57].

rFVIIa works in the presence of tissue factor from injured or ischemic vascular subendothelium to promulgate clot formation as well as binding directly to platelets and enhancing thrombin burst to improve clot stability [58].

There is some concern that this agent may increase thromboembolic events [59, 60], particularly arterial thromboses [61] which tempers enthusiasm for using this agent. Its widespread use without compelling clinical evidence to do so as well as its substantial cost is concerning as well [62]. Combining these facts with the modest data available to suggest benefit in treating ICH in anticoagulated patients, there are likely better treatments except in unusual circumstances.

4.2. Summary

There is limited guidance in the literature to organize a clinical strategy for addressing the difficult problem of traumatic ICH in anticoagulated patients. However, what is available can be used to generate a reasonable strategy for treating this patient population, as some institutions have done (See Appendix 1). For warfarin-associated bleeding, the data and experience is much more robust. Unfortunately, clinical data is severely lacking on how to treat the patient population that has been taking one of the novel anticoagulants: dabigatran, rivaroxaban or apixaban. All of these patients should receive standard neurocritical care supportive measures

http://dx.doi.org/10.5772/57366

– airway protection, hemodynamic support with IVF and vasopressors, neurosurgical evaluation as well as discontinuing the patient's anticoagulant. Depending on the agent responsible for the patient's coagulopathy, some of the following may be indicated: activated charcoal, four-factor protein complex concentrate with or without vitamin K and hemodialysis/hemoperfusion (dabigatran only). FFP and rVIIa could be options, but these agents may potentially introduce unnecessary risk, namely transfusion-associated risks with FFP and thrombosis risk in the case of rVIIa [19]. It is possible that in the future, other agents such as aminocaproic acid [63, 64] may be discovered to have a role. For now, clinicians are left to rely on published guidelines from many professional organizations *(outlined below)* [11, 65-67, 70] to help the clinician sort through the multiple issues that need to be taken into consideration when attempting to reverse the effects of systemic anticoagulation in an attempt to limit the potentially substantial morbidity and mortality of traumatic brain injury in the anticoagulated patient.

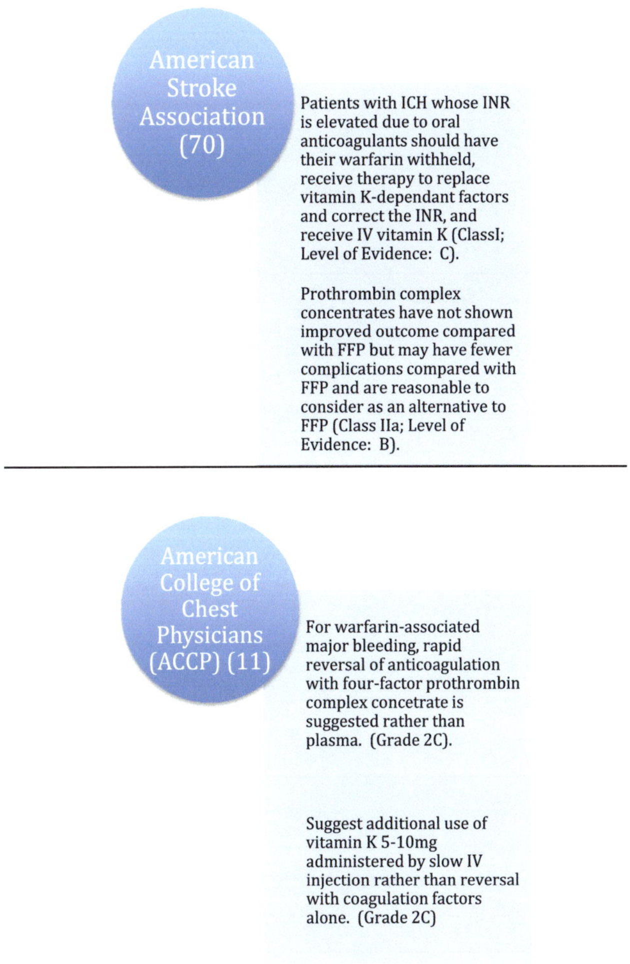

Italian Federation of Thrombosis Centers (66)

For life threatening bleeding associated with dabigatran, hemodialysis is a therapeutic option.

Direct factor Xa inhibitors could be partially antagonised by non-activated four-factor PCC at a dose of 50u/kg.

American Society of Hematology (65)

For urgent reversal of warfarin, 5-10mg of IV vitamin K and PCC or FFP should be given.

For urgent reversal of dabigatran in a patient with a prolonged aPTT, PCC, FEIBA, rFVIIa or hemodialysis should be considered.

Pharmaceutical Management Agency. The Government of New Zealand. (67)

For moderate to severe bleeding associated with dabigatran, consider:

- Tranexamic acid 15-30mg/kg IV x 1, possible continuous infusion 1mg/kg/hr.
- Charcoal if taken <2hr prior to presentation
- Prothrombinex-VF 25-50u/kg

For life-threatening bleeding associated with dabigatran, consider:

- interventions outlined for moderate/severe bleeding
- rVIIa 100mcg/kg IV

http://dx.doi.org/10.5772/57366

Appendix 1

Mercy Hospital St. Louis Trauma/Neuro ICU

Anticoagulant reversal protocol

This is the reversal protocol for a patient who has an intracranial hemorrhage and is taking one of the agents listed below. If the patient has had a cardiac ischemic event or other thromboembolic event within the last 30 days, the risks and benefits of PCCs, which have a higher rate of thrombotic events, over an FFP-based protocol should be weighed and clinical judgement should be applied.

If INR is 1.5-2.0 only FFP protocol should be utilized as Kcentra is not approved for reversal of warfarin in this INR range.

1. **WARFARIN- Related ICH- using PCCs (Kcentra):**

Warfarin with: INR < 1.2Recheck PT/INR in 12 hours

INR 1.2 – 1.4 Give Vitamin K 5 mg IV over 30 minutes if not already given and recheck PT/INR in 12 hours.

INR 1.5-1.9Use FFP protocol. Please dose based on weight and INR (See dosing Table 2]

INR ≥ 2.0 **Give Vitamin K 10 mg IV over 30 minutes** if not already given. Give Kcentra based on the table (see below). Recheck PT/INR 20 minutes after infusion complete. Treat again based on INR level.

Pre-treatment INR	2- <4	4-6	>6
Dose of Kcentra (Units of Factor IX)/kg body weight	25	35	50
Maxiumum dose (Units of Factor IX)	Not to exceed 2500	Not to exceed 3500	Not to exceed 5000

Table 2.

2. **WARFARIN- Related ICH- using FFP:**

If the patient has had a recent MI or other thromboembolic event and the decision is to not use PCCs, then consider using FFP according to the following calculations:

Warfarin with: INR < 1.2Recheck PT/INR in 12 hours

INR 1.2 – 1.4 Give Vitamin K 5 mg IV over 30 minutes if not already given and recheck PT/INR in 12 hours.

INR > 1.5Give Vitamin K 10 mg IV over 30 minutes if not already given. Give FFPs based on the table (see below). Recheck PT/INR 20 minutes after infusion complete. Treat again based on INR level.

INR	Fresh Frozen Plasma Dose by Weight (kg)		
	Less Than 75 kg	75-100 kg	Greater than 100 kg
1.5-2.5	2 units	4 units	6 units
2.6-3.5	3 units	5 units	7 units
3.6-5	4 units	6 units	8 units*
Greater than 5	6 units	8 units*	10 units*

Table 3.

This document is intended as a guide to the correct adult dose of FFP, it is not a directive, and should not be used in place of clinical assessment.

*Caution should be exercised if using this chart for calculating FFP volumes for overweight patients as the volume suggested may be an over estimation and may risk fluid overload.

HEPARIN- Related ICH:

Reversal of Intravenous Heparin-Related Intracerebral Hemorrhage	
Immediately stop heparin source	
Protamine dosing for reversal of intravenous heparin	
Time Elapsed	Protamine dose
Immediate	1 mg per 100 units of heparin
30-60 minutes	0.5 mg per 100 units of heparin
Greater than 2 hours but less than 4 hours	0.25 mg per 100 units of heparin
Protamine _____ mg (maximum dose 50 mg) intravenous over 10 minutes.	

Table 4.

3. ENOXAPARIN- Related ICH:

If taken within 24 hours (full dose only, no treatment is indicated for prophylactic doses of LMWH), give 1 mg protamine for every 1 mg of enoxaparin which was given up to a maximum dose of 50 mg. If another LMWH compound was given, please contact the pharmacy to discuss the protamine dosing.

4. tPA- Related ICH:

Immediately stop the thrombolytic infusion.

Send PT/PTT, platelet count, fibrinogen level, and type and cross.

Transfusions should begin while waiting for lab results.

Transfuse 6 pack of platelets.

Transfuse ______ units of cryoprecipitate (recommend 0.1-0.2 units/kg).

Consider giving aminocaproic acid (Amicar) 5 grams/250 cc NS IV over 60 minutes.

http://dx.doi.org/10.5772/57366

Once initial lab results return, calculate the amount of fibrinogen needed to achieve a level of > 150 mg/dl (one unit of cryoprecipitate increases fibrinogen by 7-10 mg/dl). This second dose should equal the total amount needed minus the initial dose.

Recheck CBC, PT/PTT, and fibrinogen one hour after transfusion therapy is completed.

Management of Bleeding in Patients

Taking Dabigatran/ Rivaroxaban/ Apixaban

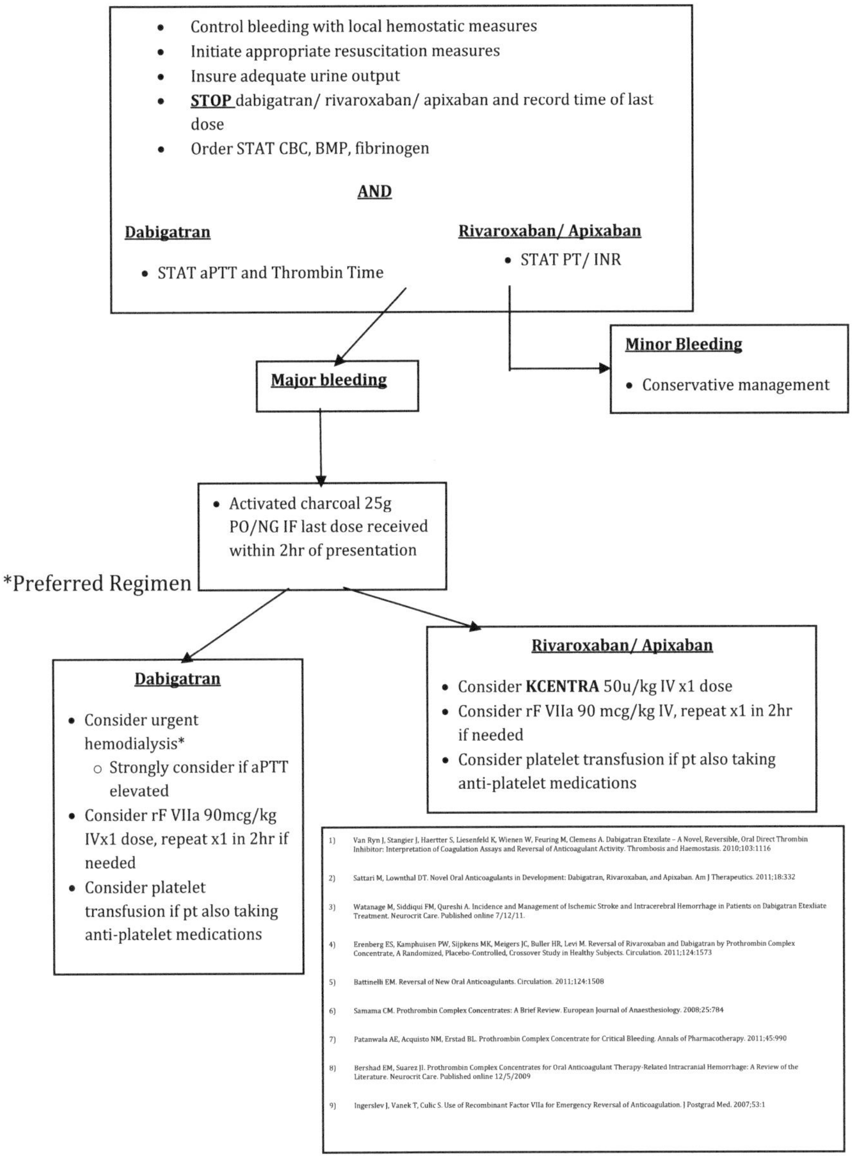

Author details

David E. Tannehill

Mercy Hospital St. Louis, USA

References

[1] Brott T, Broderick J, Kothari R, et al. Early hemorrhage growth in patients with intracerebral hemorrhage. Stroke. 1997 Jan;28(1):1-5.

[2] Davis SM, Broderick J, Hennerici M, et al. Hematoma growth is a determinant of mortality and poor outcome after intracerebral hemorrhage. Neurology. 2006 Apr 25;66(8):1175-81.

[3] Hemphill JC, Bonovich DC, Besmertis L, et al. The ICH score: a simple, reliable grading scale for intracerebral hemorrhage. Stroke. 2001;32:891-897.

[4] Cucchiara B, Messe S, Sansing L, et al. CHANT Investigators. Hematoma growth in oral anticoagulant related intracerebral hemorrhage. *Stroke.* Nov 2008;39(11):2993-6

[5] Mayer SA, Brun NC, Broderick J, et al; Europe/AustralAsia NovoSeven ICH Trial Investigators. Safety and feasibility of recombinant factor VIIa for acute intracerebral hemorrhage. *Stroke.* 2005;36(1):74-79

[6] Mayer SA, Brun NC, Begtrup K, et al; FAST Trial Investigators. Efficacy and safety of recombinant activated factor VII for acute intracerebral hemorrhage. *N Engl J Med.* 2008;358(20)2127-2137

[7] Menzin J, Hoesche J, Friedman M, Nichols C, et al. Failure to correct international normalzed ratio and mortality among patients with warfarin-related major bleeding: An analysis of electronic health records. *J Thromb Haemost.* April 2012;10(4):596-605

[8] Yasaka M, Minematsu K, Naritomi H, et al. Predisposing factors for enlargement of intracerebral hemorrhage in patients treated with warfarin. *Thromb Haemost.* Feb 2003;89(2):278-83

[9] Aguilar MI, Hart RG, Kase CS, et al. Treatment of warfarin-associated intracerebral hemorrhage: literature review and expert opinion. *Mayo Clinic Proc.* 2007;82:82-92

[10] Peacock WF, Gearhart MM, Mills RM. Emergency management of bleeding associated with old and new anticoagulants. *Clin Cardiol.* 2012

[11] Holbrook A, Schulman S, Witt D, et al. Evidence-based management of anticoagulant therapy: Antithrombotic therapy and prevention of thrombosis, 9th edition: American College of Chest Physicians Evidence-Based Clinical Practice Guidelines. *Chest.* 2012;141e152S-184S

[12] Makris M, van Veen JJ, Maclean R. Warfarin anticoagulation reversal: management of the asymptomatic and bleeding patient. *J Thromb Thrombolysis.* 2010;29:171-181

[13] Bershad EM, Suarez JL. Prothrombin complex concentrates for oral anticoagulant therapy-related intracranial hemorrhage: A review of the literature. *Neurocrit Care.* June 2010;12(3);403-13

[14] FEIBA NF package insert. http://www.fda.gov/downloads/biologicsbloodvaccines/bloodbloodproducts/approvedproducts/licensedproductsblas/fractionatedplasma-products/ucm221749.pdf

[15] Kcentra package insert. http://www.fda.gov/downloads/BiologicsBloodVaccines/BloodBloodProducts/ApprovedProducts/LicensedProductsBLAs/FractionatedPlas-maProducts/UCM350239.pdf

[16] Patanwala A, Acquisto N, Erstad B. Prothrombin complex concentrate for critical bleeding. *Ann Pharmacother.* 2011;45:990-999

[17] Voils SA, Baird B. Systematic review: 3-factor versus 4-factor prothrombin complex concentrate for warfarin reversal: Does it matter? *Thrombosis Research.* 130(2012): 833-840

[18] Dentali F, Marchesi C, Pierfranceschi M, et al. Safety of prothrombin complex concentrates for rapid anticoagulation reversal of vitamin K antagonists. A meta-analysis. *Thromb Haemost.* 2011;106:429-438

[19] Kaatz S, Kouides P, Garcia D, et al. Guidance on the emergent reversal of oral thrombin and factor Xa inhibitors. *Am J Hematol.* 2012;87:S141-S145

[20] Mahan CE, Fanikos J. New antithrombotics: The impact on global health care. Thromb Res. 2011;127:518-524

[21] Kubitza D, Becka m, Mueck W, et al. Effects of renal impairment on the pharmacokinetics, pharmacodynamics and safety of rivaroxaban, an oral, direct factor Xa inhibitor. *Br J Clin Pharmacol.* 2010;70:703-12

[22] Stangler J, Rathgen K, Stahle H, Mazur D. Influence of renal impairment on the pharmacokinetics and pharmacodynamics of oral dabigatran etexilate. An open-label, parallel-group, single-center study. *Clin Pharmacokinet.* 2010;49(4):259-268

[23] Pradaxa (dabigatran etexilate capsules) [package insert]. Ingelheim, Germany: Boehringer Ingelheim Pharma GmbH& Co

[24] Connolly S, Ezekowitz M, Yusuf S, et al. The RE-LY Steering Committee and Investigators. Dabigatran versus warfarin in patients with atrial fibrillation. *N Engl J Med.* 2009;361(12):1139-1151

[25] van Ryn J, Stangler J, Haertter S, et al. Dabigatran etexilate – a novel, reversible, oral direct thrombin inhibitor: Interpretation of coagulation assays and reversal of anticoagulant activity. *Thromb Haemost.* 2010;103(6):1116-1127

[26] Zhou W, Schwarting S, Illanes S, et al. Hemostatic therapy in experimental intracerebral hemorrhage associated with the direct thrombin inhibitor dabigatran. *Stroke.* 2011;42:3594-3599

[27] van Ryn J, Dorr B, Kaspereit F, et al. Beriplex P/N reverses bleeding in an acute renal injury model after dabigatran overdose in rabbits. *Pathophysiol Haemost Thromb.* 2010;37:A94-P486 (Abstract)

[28] Dumkow LE, Voss JR, Peters M, Jennings DL. Reversal of dabigatran-induced bleeding with a prothrombin complex concentrate and fresh frozen plasma. *Am J Health-Syst Pharm.* 2012;69:1646-50

[29] Dager WE, Gosselin CLS, Roberts AJ. *Crit Care Med.* 2013;41:e42-e46

[30] Eerenberg E, Kamphulsen P, Sijpkens M, et al. Reversal of rivaroxaban and dabigatran by prothrombin complex concentrate: A randomized, placebo-controlled, crossover study in healthy subjects. *Circulation.* 2011;124:1573-1579

[31] Marlu R, Hodaj E, Paris A, et al. Effect of non-specific reversal agents on anticoagulant activity of dabigatran and rivaroxaban. *Thromb Haemost.* 2012; 108:217-224

[32] van Ryn J, Ruehl D, Priepke H, et al. Reversibility of the anticoagulant effect of high doses of the direct thrombin inhibitor dagigatran, by recombinant factor VIIa or activated prothrombin complex concentrate (Abstract 0370). *Haematologica.* 2008;93(suppl 1):148

[33] van Ryn J, Kink-Eiband M, Clemens A. the successful reversal of dabigatran-induced bleeding by coagulation factor concentrates in a rat tail bleeding model do not correlate with ex vivo markers of anticoagulation. 53rd ASH Annual Meeting and Exposition, San Diego, CA; 2011

[34] van Ryn J, Sieger P, Kink-Elband M, et al. Adsorption of dabigatran etexilate in water or dabigatran in pooled human plasma by activated charcoal in vitro. 51st ASH Annual Meeting and Exposition. New Orleans, LA; 2009

[35] Warkentin T, Margetts P, Connolly S, et al. Recombinant factor VIIa (rFVIIa) and hemodialysis to manage massive dabigatran-associated postcardiac surgery bleeding. *Blood.* 2012 119;9:2172-2173

[36] Lassen MR, Ageno W, Borris LC, et al; for the RECORD3 Investigators. Rivaroxaban versus enoxaparin for thromboprophylaxis after total knee arthroplasty. *N Engl J Med.* 2008;358(26):2776-2786

[37] The EINSTEIN Investigators. Oral rivaroxaban for symptomatic venous thrombembolism. *N Engl J Med.* 2010; 363(26):2499-2510.

[38] The EINSTEIN-PE Investigators. Oral rivaroxaban for the treatment of symptomatic pulmonary embolism. *N Engl J Med.* 2012;366(14):1287-1297.

[39] Patel M, Mahaffey K, Garg J, et al. ROCKET AF Investigators Rivaroxaban versus warfarin in nonvalvular atrial fibrillation. *N Engl J Med.* 2011;365:883-891

[40] Xarelto (rivaroxaban) [package insert]. Titusville, NJ: Janssen Pharmaceuticals, 2013

[41] U.S. Food and Druge Administration. Xarelto: Highlights of prescribing information. 2011. http://www.accessdata.fda.gov/drugsatfda_docs/label/2011/202439s001lbl.pdf

[42] Tinel H, Huetter J, Perzborn E. Recombinant factor VIIa partially reversed the anticoagulant effect of high-dose rivaroxaban a novel, oral, direct factor Xa inhibitor in rats. *J Thromb Haemost.* 2007;5:Abstract P-W-652

[43] Gruber A, Marzec U, Buetehorn U, et al. Potential of activated prothrombin complex concentrate and activated factor VII to reverse the anticoagulant effects of rivaroxaban in primates. *Blood.* 2008;112:3825

[44] Pezborn E, Trabandt A, Selbach K, Tinel H. Prothrombin complex concentrate reverses the effects of high-dose rivaroxaban in rats (presented at the 21st International Congress on Thrombosis) [abstract]. *Pathophysiol Haemost Thromb.* 2010;37:A10-OC251

[45] Eliquis package insert. Bristol Meyers Squibb 2012. http://packageinserts.bms.com/pi/pi_eliquis.pdf

[46] Martin AC, Le Bonniec B, Fischer AM, et al. Evaluation of recombinant activated factor VII, prothrombin complex concentrate, and fibrinogen concentrate to reverse apixaban in a rabbit model of bleeding and thrombosis. *Int J Cardiol.* 2013 Aug 5 pii: S0167-5273(13)01363-6. doi: 10.1016/j.ijcard.2013.07.152. [Epub ahead of print]

[47] NovoSeven [package insert] Princeton, NJ 1999. http://factorviia.com/pi.pdf

[48] Marcel Levi, MD; Marjolein Peters, MD; Harry R. Bu®ller, MD. Efficacy and safety of recombinant factor VIIa for treatment of severe bleeding: A systematic review. *Crit Care Med.* 2005; 33:883–890

[49] Bauer K. Reversal of antithrombotic agents. *Am J Hematol.* 2012; 87:S119-S126

[50] Mayer SA, Brun NC, Broderick J, et al; Europe/AustralAsia NovoSeven ICH Trial Investigators. Safety and feasibility of recombinant factor VIIa for acute intracerebral hemorrhage. *Stroke.* 2005;36(1):74-79. 6

[51] Kenet G, Walden R, Eldad A, Martinowitz U. Treatment of traumatic bleeding with recombinant factor VIIa. *Lancet.* 1999;354:1879

[52] Martinowitz U, Kenet G, Segal E, et al. *J Trauma.* 2001;51:431-439

[53] Dutton RP, McCunn M, Hyder M, et al. Factor VIIa for correction of traumatic coagulopathy. *J Trauma.* 2004;57:709-718

[54] Spinella P, Perkins J, McLaughlin D, et al. The effect of recombinant activated factor VII on mortality in combat-related casualties with severe trauma and massive transfusion. *J Trauma.* 2008;64:286-293

[55] Boffard K, Riou B, Warren B, et al. NovoSeven Trauma Study Group. Recombinant factor VIIa as adjunctive therapy for bleeding control in severely injured trauma patients: two parallel randomized placebo-controlled, double-blind clinical trials. *J Trauma.* 2005;59:8-15

[56] Hsia C, Chin-Yee I, McAlister V. Use of recombinant activated factor VII in patients without hemophilia: a meta-analysis of randomized controlled trials. *Ann Surg.* 2008;248:61-68

[57] Hauser C, Boffard K, Dutton R, et al. *J Trauma.* 2010;69:489-500

[58] Hedner U. Mechanism of action, deelopment and clinical experience of recombinant fVIIa. *J Biotechnol.* 2006;124:747-757

[59] O'Connell K, Wood J, Wise R, et al. Thromboembolic adverse events after use of recombinant human coagulation factor VIIa. *JAMA.* 2006;295:293-298

[60] Thomas G, Dutton R, Hemlock B, et al. Thromboembolic complications associated with factor VIIa administration. *J Trauma.* 2007;62:564-569

[61] Levi M, Levy J, Andersen H, et al. Safety of recombinant activated factor VII in randomized clinical trials. *N Engl J Med.* 2010;363:1791-1800

[62] Yank V, Tuohy CV, Logan AC, et al. Systematic review: benefits and harms of in-hospital use of recombinant factor VIIa for off-label indications. *Ann Intern Med.* 2011;154:529-540

[63] Liu-DeRyke X, Rhoney D. Hemostatic therapy for the treatment of intracranial hemorrhage. *Pharmacotherapy.* 2008;28(4):485-495

[64] Piriyawat P, Morgenstern LB, Yawn DH, et al. Treatment of acute intracerebral hemorrhage with epsilon-aminocaproic acid: a pilot study. *Neurocrit Care.* 2004;1(1):47-51

[65] American Society of Hematology. 2011 Clinical practice guideline on the evaluation and management of immune thrombocytopenia (ITP) 2011. http://www.hematology.org/Practice/Guidelines/2934.aspx

[66] Pengo V, Crippa L, falanga A, et al. Questions and answers on the use of dabigatran and perspectives on the use of other new oral anticoagulants in patients with atrial fibrillation. A consensus document of the Italian Federation of Thrombosis Centers (FCSA). *Thromb Haemost.* 011;106:868-876

[67] Pharmaceutical Management Agency. 2011. Guidelines for management of bleeding with dabigatran – For possible inclusion into local management protocols. New Zealand Government. http://pharmac.govt.nz/2011/06/13/Dabigatran%20bleeding%20management.pdf

[68] O'Shaughnessy DF, Atterbury C, Bolton Maggs P, Murphy M, Thomas D, Yates S, Williamson LM, British Committee for Standards in Haematology, Blood Transfusion Task Force: Guidelines for the use of fresh-frozen plasma, cryoprecipitate and cryosupernatant. *Br J Haematol* 2004, 126:11-28

[69] Lauzier F, Cook D, Griffith L, Upton J, Crowther M: Fresh frozen plasma transfusion in critically ill patients. *Crit Care Med* 2007, 35:1655-1659.

[70] Morgenstern LB, Hemphill JC, Anderson C, et al. Guidelines for the Management of Spontaneous Intracerebral Hemorrhage: A Guideline for Healthcare Professionals from the American Heart Association/American Stroke Association. *Stroke.* 2010;41:2108-2129.

[71] Sarode R, Milling TJ Jr, Refaal MA, et al. Efficacy and safety of a 4-factor prothrombin complex concentrate in patients on vitamin K antagonists presenting with major bleeding: a randomized, plasma-controlled, phase IIIb study. *Circulation.* 2013 Sep 10;128(11):1234-43. doi: 10.1161/CIRCULATIONAHA.113.002283. Epub 2013 Aug 9.

[72] Hickey M, Gatien M, Taljaard M, et al. Outcomes of urgent warfarin reversal with frozen plasma versus prothrombin complex concentrate in the emergency department. *Circulation.* 2013 Jul 23; 128(4):360-4. doi: 10.1161/CIRCULATIONAHA.113.001875. Epub 2013 Jun 14.

[73] Granger CB, Alexander JH, McMurray JJ, et al. Apixaban versus warfarin in patients with atrial fibrillation. *N Engl J Med.* 2011;365:981-992.

Chapter 10

Surgical Treatment of Severe Traumatic Brain Injury

Matthew L. Dashnaw, Anthony L. Petraglia and
Jason H. Huang

Additional information is available at the end of the chapter

http://dx.doi.org/10.5772/57342

1. Introduction

Head injury is the number one cause of trauma-associated mortality, being directly associated with approximately half of all trauma-related deaths [1]. Every year in the United States, approximately 1.5 million head injuries occur, resulting in 250,000 hospitalizations and 52,000 deaths [2]. Traumatic brain injury (TBI) is the leading cause of death in persons less than 45 years of age [3]. Furthermore, by World Health Organization estimates, TBI will be the third leading cause of death and disability, across all age groups, by the year 2020 [4]. From a cost perspective, TBI results in an astounding $6 billion in direct costs and over $40 billion in indirect costs annually in the United States [5].

For neurosurgeons and intensivists involved, the management of TBI presents many challenges. Many patients with TBI also have traumatic injury to other organ systems, further complicating management. Centers treating a high volume of severe TBI may have better outcomes in terms of mortality and quality of life [6].

Current management of severe TBI consists of a host of surgical and non-surgical modalities. The majority of patients with severe TBI, defined by Glascow Coma Scale (GCS) 3-8, will be managed nonsurgically. Medical interventions are generally used to optimize intracranial pressure (ICP), maintain cerebral blood flow and oxygen delivery, minimize cerebral edema and maintain a healthy metabolic environment [7]. Surgical treatment in severe TBI is most commonly used for evacuation of intracranial hemorrhage (ICH), especially when there is decreased level of consciousness, focal neurologic signs and/or evidence of intracranial hypertension [7]. Prior to publication of the "Guidelines for the Surgical Management of Traumatic Brain Injury" in 2006, the role of surgery was often based on individual surgeon preference or subjective factors [8]. As noted by the guideline authors, there is a paucity of prospective, randomized controlled trials for surgical lesions in TBI, which precluded literature categorization with traditional "level of evidence" distinctions [8]. Instead, the guidelines offer "literature-based recommendations" for the following TBI lesion types warranting surgical consideration: acute epidural hematoma (aEDH), acute subdural hematoma (aSDH), traumatic parenchymal lesions, posterior fossa mass lesions, and depressed cranial fractures [8].

This chapter will review the aforementioned lesion types with respect to epidemiology, guidelines-based indications for surgery, timing, and technique. We will also review new data pertinent to each lesion type that has been published since the 2006 "Guidelines for the Surgical Management of Traumatic Brain Injury." We will pay particular attention to the surgical management of intractable intracranial hypertension, which has become the matter of intense debate in recent years. This debate relates to the publication of the DECRA (Decompressive Craniectomy in Diffuse Traumatic Brain Injury) trial in 2011 [9]. We will review this work in depth, as well as, the controversy surrounding its application to patient care.

2. Acute epidural hematoma

Epidural hematoma occurs in 2.7 – 4 % of TBI patients[10]. They most frequently occur in the temporal and temporoparietal regions as a result of linear skull fracture with subsequent damage to either the anterior or posterior divisions of the middle meningeal artery. Traumatic epidural hematoma of nonarterial origin may result from middle meningeal vein, diploic emissary vein, or venous sinus bleeding and accounts for roughly 25% of total cases [11]. Epidural hematoma in patients over 65 years of age is rare, due to the prominently adhered dura to the overlying skull. The classically described "lucid interval," with rapid deterioration after a period of post injury wakefulness occurs in about half of surgical cases[8]. Isolated EDH mortality is lower than that for ASDH and approximates 10%[12]. EDH represents one of the most urgent neurosurgical lesions, as severe brain compression can develop rapidly from high-pressure arterial bleeding often necessitating rapid evacuation. In the appropriate patient, surgical evacuation of posttraumatic epidural hematoma has been shown to be impressively cost-effective with regards to quality and duration of life preserved, when compared to other surgical procedures[13].

Guideline based indications for surgery [8]: aEDH larger than 30cc should be evacuated regardless of the patient's GCS score. In patients with GCS > 8 and no focal neurologic deficit, nonoperative management with serial CT scan and close observation in a center with available neurosurgical services can be considered if the hematoma meets the following criteria: volume < 30cc *and* less than 15mm thick *and* < 5mm of midline shift. It is important to mention that lesions in the temporal or posterior fossa may cause significant brainstem compression, associated with high mortality, in the absence of large size, significant midline shift, or elevated ICP [12,14,15]. These patients should have a much lower threshold for surgery.

Timing [8]: Patients meeting the above mentioned criteria should undergo surgical evacuation as soon as possible. Expeditious hematoma evacuation is particularly vital for comatose patients (GCS < 9) and/or with anisocoria.

Surgical technique [8]: There are insufficient data to support one surgical treatment method. Most authors would recommend craniotomy over simple burr hole evacuation in order to provide adequate access to and evacuation of the offending clot. Traditional teaching was for a large "question-mark" or "T" shaped incision and subsequent large trauma bone flap. With improved imaging quality and technology, the ability to localize the clot location prior to incision often allows for a smaller linear incision and more focused craniotomy which can be expanded if the need arises [7]. Also common practice is to make a small dural opening in order to inspect for a concomitant SDH which may sometimes develop with "reperfusion" or resuscitation of the trauma patient [7].

3. Acute subdural hematoma

Acute subdural hematoma is more common than aEDH, occurring in about 30% of severely head injured patients [16]. The most common causes of these lesions include motor vehicle accidents, falls (particularly in those > 75 years of age) and assaults. By definition, aSDH after trauma occurs within 14 days of injury and is associated with a higher mortality rate than EDH, with or without surgical intervention[12,16]. Compared to isolated EDH, the degree of underlying brain damage associated with aSDH is more severe [17]. Mortality rates traditionally quoted for those requiring surgery vary between 40 – 68% and is greatest in those with increased age, poor initial GCS, and other associated brain and systemic injuries [12]. A recent report observed 16% inpatient mortality for all comers with subdural hematoma, which did not vary significantly for those undergoing surgical intervention [18].

Guideline based indications for surgery [12]: Thickness greater than 10mm or midline shift greater than 5mm on CT should undergo surgical evacuation regardless of GCS score. Patients with GCS < 9 should undergo intracranial pressure monitoring. For comatose patients with less than a 10mm thick lesion or less than 5mm of midline shift, indications for surgical evacuation include a decrease of GCS by 2 or more points, ICP > 20mmHg, or asymmetric or fixed and dilated pupils.

Timing [12]: Most authors would agree surgical candidates should undergo evacuation as soon as possible. This principle was originally based on a study conducted over 30 years ago showing marked improvement in mortality if aSDH was evacuated within 4 hours of injury. Some have since questioned this notion claiming either no outcome difference, or a worsened outcome with more rapid time to evacuation [16,19,20]. However, a careful review of this data reveals that the patients who underwent rapid evacuatoin also had more severe neurologic injury prior to surgery, challenging the validity of the outcomes data [12,16,20].

Surgical technique: Craniotomy, generally via a large frontotemporoparietal approach, with or without bone flap removal and duraplasty is the preferred technique [12]. Multiple techniques for aSDH evacuation have been utilized in neurosurgery including trephination via twist drill or burr hole, craniotomy +/- duraplasty, subtemporal decompressive craniectomy, and large decompressive hemicraniectomy +/- duraplasty. For patients with poor GCS associated with aSDH, trephination and irrigation without craniotomy maybe associated with poorer outcomes compared to craniotomy or craniectomy [12,21]. Most studies comparing outcomes of those undergoing craniotomy vs. craniectomy suffer from selection bias as patients with more severe injury undergo craniectomy and worse outcomes based on initial presentation[12,22]. A subgroup of patients with a higher level of brain injury may benefit from decompressive craniectomy [23], and it is our practice to consider decompressive craniectomy in patients with midline shift that significantly exceeds the hematoma thickness suggesting a greater level of associated brain injury and swelling. A ventriculostomy is placed intraoperatively if the patient's preoperative GCS is less than 8 or if significant swelling is noted at the time of operation. Because brain shift is common in subdural hematoma and external landmarks can be difficult to palpate during surgery, ventriculostomy can be quite challenging in this setting. In addition to this, patients can also become coagulopathic in this setting making multiple passes less desirable. In this setting, consideration should be given to intraparenchymal ICP monitor placement if unable to successfully place a ventriculostomy.

4. Focal traumatic parenchymal lesions

Intraparenchymal hemorrhages, contusions or infarcts are associated with severe traumatic brain injury in up to 35% of cases, but only 20% of trauma craniotomy is undertaken for their removal [7,24-26]. These lesions occur most commonly in the frontal or temporal lobes due to the brain impacting against the frontal bone and sphenoid ridges, whereas parietal and occipital lobe hematoma is most often secondary to direct impact [7]. Parenchymal lesions tend to evolve over time and the resulting mass effect from larger lesions may lead to worsened secondary brain injury, neurological deterioration, herniation and death [27]. In addition to lesion blossoming, delayed traumatic intracerebral hematoma (DTICH), which occurs in areas of radiographically normal brain on initial CT scan, may lead to delayed neurological deterioration [24,28]. It is important that any patient with abnormal findings on initial CT scan be monitored closely for the aforementioned phenomena, which may develop subsequent to initial physical and radiographic examination. Much work has gone into prognosticating which patient and lesion characteristics may be prone to worse outcome. Risk factors for worse outcome include, but are not limited to increased age, lower GCS on admission, presence of skull fracture, absence of brainstem reflexes, status of basal cisterns on CT, ICH volume, severity of surrounding edema, preoperative neurological deterioration, and concomitant SDH[24]. Nonfocal lesions, specifically diffuse injury with intractable intracranial hypertension will be discussed in the next section with indications, timing and method for surgical management of focal parenchymal lesions reviewed now.

Guideline based indications for surgery[24]: Progressive neurological deterioration referable to the lesion, medically refractory intracranial hypertension (see section titled *Intractable Intracranial Hypertension* below for discussion) or signs of mass effect on CT should be considered for surgical evacuation. Also considered for surgical evacuation are patients with GCS 6-8 with frontal or temporal contusions greater than 20cm^3 in volume with at least 5mm of midline shift and/or cisternal compression on CT scan, as well as, patients with any lesion greater than 50 cm^3 in size. These indications have generally been derived by review of several studies which have focused on defining patient and lesion characteristics at high risk for subsequent neurological deterioration and assume that earlier operative intervention will improve likelihood for a more favorable outcome [12,24,29-31]. Candidates for nonoperative management with intensive monitoring and serial imaging, include those with lesions without significant mass effect on CT scan, no evidence for neurological compromise, and without intracranial hypertension. Such lesions are common and generally resorb in 4 to 6 weeks by macrophage phagocytosis and gliosis [7].

Timing and surgical technique [24]: For patients with focal lesions and the indications mentioned above, craniotomy with evacuation of the mass lesion as soon as possible is recommended. While stereotactic evacuation of focal posttraumatic lesions has been reported, we do not advocate this method as the majority of these patients also have a degree of diffuse injury with associated widespread secondary injury and a degree of cerebral edema with intracranial hypertension [24,25,32].

5. Intractable intracranial hypertension

Cerebral edema and subsequent intracranial hypertension are a major concern in combating the secondary injury of severe TBI [24,33]. At least 80% of severe TBI patients have elevated ICP and this is the major cause of death in those who die [7]. Intracranial pressure monitoring is currently recommended by clinical practice guidelines for patients with severe traumatic brain injury who have an abnormal CT scan of the head or those with a normal CT scan who meet other specified criteria [34]. The level II recommendation for treatment is for those with sustained intracranial pressure greater than 20mmHg [35]. Currently recommended nonoperative therapies for intracranial hypertension include head of bed elevation, hyperosmolar therapy (mannitol and/or hypertonic saline), intubation to ensure normocarbia with only short periods of hyperventilation as a temporizing measure if needed, analgesia, neuromuscular paralysis, ventricular drainage, hypothermia, and barbiturate or propofol induced burst suppression [7].

The role for surgery in the treatment of medically refractory intracranial hypertension has become the matter of intense debate in recent years. A variety of surgical procedures have been used for treatment of refractory intracranial hypertension, without a prominent mass lesion, including subtemporal decompression, temporal lobectomy, and circumferential craniotomy [24]. The two most utilized surgical procedures in this setting include the hemispheric decompressive craniectomy and the bifrontal decompressive craniectomy; the latter originally described by Kjellberg et al [24,36]. The 2006 "Guidelines for the Surgical Management of Traumatic Brain Injury" support bifrontal decompressive craniectomy within 48 hours of injury as a treatment option for patients with diffuse, medically refractory posttraumatic cerebral edema and resultant intracranial hypertension [24]. This is based on data associating intracranial hypertension with poor outcome and multiple studies showing that decompressive craniectomy can reliably manage intracranial hypertension [37-40]. The aforementioned data was noted to be less than ideal as it lacked data from prospective, randomized trials [24].

For example, particular attention was paid to the work of Polin et al, which retrospectively evaluated outcome in 35 patients undergoing bifrontal decompressive craniectomy for refractory posttraumatic cerebral edema, matched for age, admission GCS, sex, and maximal ICP with historical controls selected from the Traumatic Coma Data Bank [24,40]. Pertinent findings included favorable outcome association for surgery when performed less than 48 hours after injury compared to surgery performed longer than 48 hours after injury, especially in patients whose ICP had not yet been sustained above 40mmHg [40]. Also, medical management alone carried a 3.8 times relative risk of unfavorable outcome compared with decompressive craniectomy [40]. While this work and others argued in favor of bifrontal decompressive craniectomy as the potential intervention of choice in the proper patient, guideline authors also noted the lack of contemporaneous controls, and called for prospective, controlled trials to meaningfully compare outcome between surgical and nonsurgical groups in this clinical setting [24].

It was this goal that the DECRA (Decompressive Craniectomy in Diffuse Traumatic Brain Injury) trial was undertaken [9]. DECRA was a multicenter, randomized controlled trial conducted in 15 hospitals in Australia, New Zealand and Saudi Arabia designed to test the efficacy of bifrontotemporoparietal decompressive craniectomy in adults below 60 years of age with severe TBI in whom first-tier therapeutic measures failed to control ICP above 20 mmHg per Brain Trauma Foundation guidelines recommendation [9]. Randomization to either early decompressive craniectomy or standard medical management was undertaken for patients with an initial GCS <8 and when ICP was > 20 mmHg for > 15 minutes. At 6 months follow-up, 70% of patients in the craniectomy group had an unfavorable outcome vs. 51% of patients in the standard care group (odds ratio 2.21 [95% CI 1.14-4.26]; P = 0.002) [9]. This was despite the findings that decompressive craniectomy was associated with decreased intracranial pressure and shorter ICU stays [9]. Based on these results, the authors concluded that decompressive craniectomy was associated with more unfavorable outcomes and that the Australian health care system would save tens of millions of dollars by adhering to a medically-based treatment strategy rather than aggressive surgical decompression, and they predicted the trial would significantly alter clinical practice [38].

Many neurosurgeons and intensivists certainly did not share the aforementioned opinions of the DECRA trial authors, which provoked strong emotional reactions and has brought the methods and results of the trial under heavy scrutiny to explain the confounding lack of positive results [41-44]. Perhaps the most compelling criticism pertains to the baseline characteristics of the patients in each of the study's groups. The number of patients with bilateral unreactive pupils in the surgical group was nearly double that in the control group (27% vs. 12%) [9]. After controlling for this, the differences in outcomes became non-significant [9]. In addition to pupillary differences, radiologic findings were more severe in the surgical group (77% vs. 67% with Marshall grade III injury) and GCS scale was lower (5 vs. 6) in the surgical group; All of the aforementioned factors have prognostic significance [41]. The issue of patient crossover has also been noted as a potentially confounding factor in this trial [45,46]. A total of 19 patients (4 < 72 hours after randomization and 15 > 72 hours after randomization) in the standard care group had a decompressive craniectomy as a life saving procedure, which some involved in the trial believe may have eliminated equipoise for the involved neurosurgeons who likely felt that these patients had genuinely increased ICP [41,46]. These 19 patients were analyzed in the standard care group as there being an intention treat [45].

In addition to the baseline differences among groups in the DECRA trial, other issues with the trial have been raised. DECRA excluded patients with any traumatic mass lesions, greatly reducing the number of real-world patients the data may be applicable to [42]. Secondly, some feel that the procedure used may not be as efficacious as that originally described by Polin et al, as the DECRA procedure did not involve division of the sagittal sinus and falx cerebri [40, 42]. Furthermore, many believe the threshold ICP elevation for inclusion in the study (spontaneous increase in ICP > 20mmHg for > 15 min, continuous or intermittently, within a 1-hr period), may have been too low and that many neurosurgeons and intensivists would not normally consider decompressive craniectomy in patients who have an intracranial pressure around 20mmHg for such a short time [42,43,47]. Indeed, the median ICP for both groups during the 12 hours before randomization was 20mmHg, which is the upper limit of normal [9,42]. Based on many of the aforementioned points, the Section on Neurotrauma and Critical Care of the American Association of Neurological Surgeons and the Congress of Neurological Surgeons state "that no conclusions regarding management of the use of decompressive craniotomy in patients with traumatic brain injury should be drawn from this trial, and clinical practice should not be changed on the basis of these results [42]."

Despite its many criticisms, DECRA does represent the first randomized clinical trial of decompression to be completed in adult neurotrauma patients, and may "have placed us on the first rung of the evidence-based scientific ladder [41]." The protocol for the RESCUEicp (Randomized Evaluation of Surgery with Craniectomy for Uncontrollable Elevation of Intracranial Pressure) differs from that of the DECRA trial in terms of intracranial pressure threshold (> 25mmHg > 1-12 hours), decompressive craniectomy techniques allowed, timing of surgery (any time after injury vs. within 72 hours), acceptance of contusions and duration of follow-up (2 years) [48]. At the time of this writing, recruitment is near completion with anxiously awaited results, which may shed more light on the heavily debated topic of where the role of decompressive craniectomy lies in treatment of diffuse traumatic brain injury. In the meantime, decompressive craniectomy will likely continue to be used by many for refractory intracranial hypertension. Furthermore, the available data highlights the need for novel ways of treating patients with TBI, whether with neuroprotective agents or regenerative therapeutics, in addition to improved prevention initiatives [49].

6. Traumatic posterior fossa mass lesions

Compared with the aforementioned traumatic brain injuries, traumatic posterior fossa mass lesions are rare. In a recent retrospective review of 4315 patients of hospitalized TBI patients, only 41 (1%) were noted to have posterior fossa hematomas [50]. In these 41 patients, there were 18 patients with posterior fossa EDH, 10 with SDH, and 17 with intracerebellar hematoma [50]. EDH is the most common posterior fossa lesion reguiring surgery, followed by SDH and intracerebellar hemorrhage [7]. Though rare when compared with the incidence of supratentorial traumatic lesions, timely recognition and evacuation of surgical lesions is of the utmost importance for the patient [51]. Those caring for these patients must keep in mind that because of the limited volume of the posterior fossa and proximity of the neighboring brainstem, rapidly fatal deterioration can occur from obstructive hydrocephalus and brainstem compression from an expanding hematoma. Respiratory pattern changes and sudden increases in blood pressure may be a harbinger of impending crisis, while pupillary reflexes, ICP measurements, or altered sensorium are not reliable clues to impending herniation in this region [7]. With regards to data supporting surgery in this patient population, we must keep in mind as stated in the 2006 guidelines, "…surgery is generally viewed as required therapy in symptomatic patients with progressive dysfunction. Because of the potential adverse consequences of withholding or delaying surgery for such patients, studies depend on retrospective analyses. As a result, there is no Class I or Class II evidence to support recommendations for the surgical management of these injuries [51]."

Guideline based indications for surgery [51]: Patients with mass effect on computed tomographic (CT) scan or with neurological dysfunction or deterioration referable to the lesion should undergo operative intervention. Mass effect on CT scan is defined as distortion, dislocation, or obliteration of the fourth ventricle; compression or loss of visualization of the basal cisterns, or the presence of obstructive hydrocephalus. Management by close observation and serial imaging may be appropriate for patients with lesions with no significant mass effect on CT scan and without signs of neurologic dysfunction.

Timing [51]: In indicated cases, surgical evacuation should be performed as soon as possible because these patients can deteriorate rapidly, thus, worsening their prognosis.

Surgical technique: Suboccipital craniectomy is the predominant method for evacuation of posterior fossa mass lesions [51]. Generally, a ventriculostomy catheter should be placed before surgery for the purposes of CSF drainage and ICP reduction [7]. An important caveat to ventriculostomy in the setting of posterior fossa lesions is that many advocate slow CSF drainage to avoid the rare possibility of upward herniation [7,52]. Because optimal surgical positioning involves anterior flexion of the cervical spine, absence of cervical spine fracture must also be assured and documented. Careful attention must be paid to boney removal over the venous sinuses, which can be a major source of bleeding complicating surgery in this region, and should be prepared for [7]. With injuries that involve the subdural spaces and cerebellar parenchyma, a larger decompressive craniectomy including the rim of the foramen magnum inferior, up to the edge of the transverse sinus superiorly and laterally as far as the digastric groove should be undertaken to provide adequate decompression [7]. For lesions extending inferiorly with concomitant compression, the posterior arch of the atlas can also be removed [7].

7. Depressed cranial fractures

Depressed cranial fractures complicate up to 6% of head injuries in one series [53,54]. Compound depressed cranial fractures are depressed fractures with an overlying scalp laceration in continuity with the fracture site and with galeal disruption, while simple depressed cranial fractures have no galeal disruption. Besides sequalae from an associated hematoma with mass effect, the primary clinical concern for depressed skull fractures involve their association with infection and late seizure [54]. Depressed skull fractures overlying major venous sinuses are generally managed nonoperatively due to high associated risks of surgery, but have also been reported to be associated with delayed onset of intracranial hypertension [55]. As is common with the other injury types discussed previously, there is a lack of Class I literature evaluating indications, timing and surgical techniques which provide the best outcomes for these patients [54]. For all open cranial fractures, prophylactic antibiotics, specifically cefazolin or pipercillin/tazobactam for 5-7 days is generally recommended [56].

Guideline based indications for surgery [54]: Open (compound) cranial fractures depressed greater than the thickness of the cranium should undergo operative intervention to prevent infection. Open (compound) depressed cranial fractures may be treated nonoperatively if there is no clinical or radiographic evidence of dural penetration, significant intracranial hematoma, depression greater than 1 cm, frontal sinus involvement, gross cosmetic deformity, wound infection, pneumocephalus, or gross wound contamination. Nonoperative management of closed (simple) depressed cranial fractures is a treatment option.

With regards to frontal air sinus fractures, closed fractures, which only involve the posterior wall of the sinus, do not generally require surgical repair beyond scalp closure [7]. Compound frontal sinus fractures, which involve both anterior and posterior walls of the sinus, should be considered for surgical exploration and repair due to risk of delayed infection and/or CSF leak [7,57].

Timing [54]: Early operation is recommended to reduce the incidence of infection.

Surgical technique [54]: Elevation and debridement is recommended as the surgical method of choice. Primary bone fragment replacement is a surgical option in the absence of wound infection at the time surgery. All management strategies for open (compound) depressed fractures should include antibiotics.

8. Conclusions

Optimal outcome in severe TBI requires a coordinated effort between neurosurgeon, intensivist, nusrsing and rehabilitation to provide both surgical and nonsurgical interventions. With regards to surgery, prompt recognition and evacuation of surgical hematomas is vital. While surgical indications for many lesion types are based on retrospective data, the reviewed guidelines provide us with a framework from which we can build on to optimize treatment. The recent publication of the DECRA trial and presumed completion of the RESCUEicp trial provide hope that higher-level evidence may be gathered in this patient population. The role of decompressive craniectomy in intractable intracranial hypertension continues to evolve. The RESCUEicp study hopes to address the shortcomings of the DECRA study.

Author details

Matthew L. Dashnaw*, Anthony L. Petraglia and Jason H. Huang

*Address all correspondence to: matthew_dashnaw@urmc.rochester.edu

University of Rochester Medical Center, Department of Neurosurgery, Rochester, New York, USA

References

[1] MacKenzie EJ: Epidemiology of injuries: Current trends and future challenges. Epidemiol Rev 2000;22:112-119.

[2] Langlois JA, Rutland-Brown W, Wald MM: The epidemiology and impact of traumatic brain injury: A brief overview. J Head Trauma Rehabil 2006;21:375-378.

[3] Marshall LF: Head injury: Recent past, present, and future. Neurosurgery 2000;47:546-561.

[4] Murray CJ, Lopez AD: Global mortality, disability, and the contribution of risk factors: Global burden of disease study. Lancet 1997;349:1436-1442.

[5] Sharma S, de Mestral C, Hsiao M, Gomez D, Haas B, Rutka J, Nathens AB: Benchmarking trauma center performance in traumatic brain injury: The limitations of mortality outcomes. J Trauma Acute Care Surg 2013;74:890-894.

[6] Tepas JJ, 3rd, Pracht EE, Orban BL, Flint LM: High-volume trauma centers have better outcomes treating traumatic brain injury. J Trauma Acute Care Surg 2013;74:143-147; discussion 147-148.

[7] Winn HR, Youmans JR: Youmans neurological surgery. Philadelphia, Pa., W.B. Saunders,, 2011.

[8] Bullock MR, Chesnut R, Ghajar J, Gordon D, Hartl R, Newell DW, Servadei F, Walters BC, Wilberger JE, Surgical Management of Traumatic Brain Injury Author G: Surgical management of acute epidural hematomas. Neurosurgery 2006;58:S7-15; discussion Si-iv.

[9] Cooper DJ, Rosenfeld JV, Murray L, Arabi YM, Davies AR, D'Urso P, Kossmann T, Ponsford J, Seppelt I, Reilly P, Wolfe R, Investigators DT, Australian, New Zealand

Intensive Care Society Clinical Trials G: Decompressive craniectomy in diffuse traumatic brain injury. N Engl J Med 2011;364:1493-1502.

[10] Cordobes F, Lobato RD, Rivas JJ, Munoz MJ, Chillon D, Portillo JM, Lamas E: Observations on 82 patients with extradural hematoma. Comparison of results before and after the advent of computerized tomography. J Neurosurg 1981;54:179-186.

[11] Yilmazlar S, Kocaeli H, Dogan S, Abas F, Aksoy K, Korfali E, Doygun M: Traumatic epidural haematomas of nonarterial origin: Analysis of 30 consecutive cases. Acta Neurochir (Wien) 2005;147:1241-1248; discussion 1248.

[12] Bullock MR, Chesnut R, Ghajar J, Gordon D, Hartl R, Newell DW, Servadei F, Walters BC, Wilberger JE, Surgical Management of Traumatic Brain Injury Author G: Surgical management of acute subdural hematomas. Neurosurgery 2006;58:S16-24; discussion Si-iv.

[13] Pickard JD, Bailey S, Sanderson H, Rees M, Garfield JS: Steps towards cost-benefit analysis of regional neurosurgical care. BMJ 1990;301:629-635.

[14] Andrews BT, Chiles BW, 3rd, Olsen WL, Pitts LH: The effect of intracerebral hematoma location on the risk of brain-stem compression and on clinical outcome. J Neurosurg 1988;69:518-522.

[15] Marshall LF, Barba D, Toole BM, Bowers SA: The oval pupil: Clinical significance and relationship to intracranial hypertension. J Neurosurg 1983;58:566-568.

[16] Tallon JM, Ackroyd-Stolarz S, Karim SA, Clarke DB: The epidemiology of surgically treated acute subdural and epidural hematomas in patients with head injuries: A population-based study. Can J Surg 2008;51:339-345.

[17] Wilberger JE, Jr., Harris M, Diamond DL: Acute subdural hematoma: Morbidity, mortality, and operative timing. J Neurosurg 1991;74:212-218.

[18] Ryan CG, Thompson RE, Temkin NR, Crane PK, Ellenbogen RG, Elmore JG: Acute traumatic subdural hematoma: Current mortality and functional outcomes in adult patients at a level i trauma center. J Trauma Acute Care Surg 2012;73:1348-1354.

[19] Seelig JM, Becker DP, Miller JD, Greenberg RP, Ward JD, Choi SC: Traumatic acute subdural hematoma: Major mortality reduction in comatose patients treated within four hours. N Engl J Med 1981;304:1511-1518.

[20] Tien HC, Jung V, Pinto R, Mainprize T, Scales DC, Rizoli SB: Reducing time-to-treatment decreases mortality of trauma patients with acute subdural hematoma. Ann Surg 2011;253:1178-1183.

[21] Hatashita S, Koga N, Hosaka Y, Takagi S: Acute subdural hematoma: Severity of injury, surgical intervention, and mortality. Neurol Med Chir (Tokyo) 1993;33:13-18.

[22] Woertgen C, Rothoerl RD, Schebesch KM, Albert R: Comparison of craniotomy and craniectomy in patients with acute subdural haematoma. J Clin Neurosci 2006;13:718-721.

[23] Li LM, Kolias AG, Guilfoyle MR, Timofeev I, Corteen EA, Pickard JD, Menon DK, Kirkpatrick PJ, Hutchinson PJ: Outcome following evacuation of acute subdural haematomas: A comparison of craniotomy with decompressive craniectomy. Acta Neurochir (Wien) 2012;154:1555-1561.

[24] Bullock MR, Chesnut R, Ghajar J, Gordon D, Hartl R, Newell DW, Servadei F, Walters BC, Wilberger J, Surgical Management of Traumatic Brain Injury Author G: Surgical management of traumatic parenchymal lesions. Neurosurgery 2006;58:S25-46; discussion Si-iv.

[25] Miller JD, Butterworth JF, Gudeman SK, Faulkner JE, Choi SC, Selhorst JB, Harbison JW, Lutz HA, Young HF, Becker DP: Further experience in the management of severe head injury. J Neurosurg 1981;54:289-299.

[26] Wu JJ, Hsu CC, Liao SY, Wong YK: Surgical outcome of traumatic intracranial hematoma at a regional hospital in taiwan. J Trauma 1999;47:39-43.

[27] Bullock R, Golek J, Blake G: Traumatic intracerebral hematoma--which patients should undergo surgical evacuation? Ct scan features and icp monitoring as a basis for decision making. Surg Neurol 1989;32:181-187.

[28] Gentleman D, Nath F, Macpherson P: Diagnosis and management of delayed traumatic intracerebral haematomas. Br J Neurosurg 1989;3:367-372.

[29] Katayama Y, Tsubokawa T, Miyazaki S, Kawamata T, Yoshino A: Oedema fluid formation within contused brain tissue as a cause of medically uncontrollable elevation of intracranial pressure: The role of surgical therapy. Acta Neurochir Suppl (Wien) 1990;51:308-310.

[30] Marshall LF, Marshall SB, Klauber MR, Van Berkum Clark M, Eisenberg H, Jane JA, Luerssen TG, Marmarou A, Foulkes MA: The diagnosis of head injury requires a classification based on computed axial tomography. J Neurotrauma 1992;9 Suppl 1:S287-292.

[31] Mathiesen T, Kakarieka A, Edner G: Traumatic intracerebral lesions without extracerebral haematoma in 218 patients. Acta Neurochir (Wien) 1995;137:155-163, discussion 163.

[32] Coraddu M, Floris F, Nurchi G, Meleddu V, Lobina G, Marcucci M: Evacuation of traumatic intracerebral haematomas using a simplified stereotactic procedure. Acta Neurochir (Wien) 1994;129:6-10.

[33] Chesnut RM, Marshall LF, Klauber MR, Blunt BA, Baldwin N, Eisenberg HM, Jane JA, Marmarou A, Foulkes MA: The role of secondary brain injury in determining outcome from severe head injury. J Trauma 1993;34:216-222.

[34] Brain Trauma F, American Association of Neurological S, Congress of Neurological S, Joint Section on N, Critical Care AC, Bratton SL, Chestnut RM, Ghajar J, McConnell Hammond FF, Harris OA, Hartl R, Manley GT, Nemecek A, Newell DW, Rosenthal G, Schouten J, Shutter L, Timmons SD, Ullman JS, Videtta W, Wilberger JE, Wright DW: Guidelines for the management of severe traumatic brain injury. Vi. Indications for intracranial pressure monitoring. J Neurotrauma 2007;24 Suppl 1:S37-44.

[35] Brain Trauma F, American Association of Neurological S, Congress of Neurological S, Joint Section on N, Critical Care AC, Bratton SL, Chestnut RM, Ghajar J, McConnell Hammond FF, Harris OA, Hartl R, Manley GT, Nemecek A, Newell DW, Rosenthal G, Schouten J, Shutter L, Timmons SD, Ullman JS, Videtta W, Wilberger JE, Wright DW: Guidelines for the management of severe traumatic brain injury. Viii. Intracranial pressure thresholds. J Neurotrauma 2007;24 Suppl 1:S55-58.

[36] Kjellberg RN, Prieto A, Jr.: Bifrontal decompressive craniotomy for massive cerebral edema. J Neurosurg 1971;34:488-493.

[37] Aarabi B, Hesdorffer DC, Ahn ES, Aresco C, Scalea TM, Eisenberg HM: Outcome following decompressive craniectomy for malignant swelling due to severe head injury. J Neurosurg 2006;104:469-479.

[38] Cooper DJ, Rosenfeld JV: Does decompressive craniectomy improve outcomes in patients with diffuse traumatic brain injury? Med J Aust 2011;194:437-438.

[39] Honeybul S, Ho KM, Lind CR, Gillett GR: Observed versus predicted outcome for decompressive craniectomy: A population-based study. J Neurotrauma 2010;27:1225-1232.

[40] Polin RS, Shaffrey ME, Bogaev CA, Tisdale N, Germanson T, Bocchicchio B, Jane JA: Decompressive bifrontal craniectomy in the treatment of severe refractory posttraumatic cerebral edema. Neurosurgery 1997;41:84-92; discussion 92-84.

[41] Honeybul S, Ho KM, Lind CR: What can be learned from the decra study. World Neurosurg 2013;79:159-161.

[42] Timmons SD, Ullman JS, Eisenberg HM: Craniectomy in diffuse traumatic brain injury. N Engl J Med 2011;365:373; author reply 376.

[43] Sahuquillo J, Martinez-Ricarte F, Poca MA: Decompressive craniectomy in traumatic brain injury after the decra trial. Where do we stand? Curr Opin Crit Care 2013;19:101-106.

[44] Simard JM, Kahle KT, Walcott BP: Craniectomy in diffuse traumatic brain injury. N Engl J Med 2011;365:374; author reply 376.

[45] Honeybul S, Ho KM, Lind CR, Gillett GR: The future of decompressive craniectomy for diffuse traumatic brain injury. J Neurotrauma 2011;28:2199-2200.

[46] Marion DW: Decompressive craniectomy in diffuse traumatic brain injury. Lancet Neurol 2011;10:497-498.

[47] Servadei F: Clinical value of decompressive craniectomy. N Engl J Med 2011;364:1558-1559.

[48] Hutchinson PJ, Corteen E, Czosnyka M, Mendelow AD, Menon DK, Mitchell P, Murray G, Pickard JD, Rickels E, Sahuquillo J, Servadei F, Teasdale GM, Timofeev I, Unterberg A, Kirkpatrick PJ: Decompressive craniectomy in traumatic brain injury: The randomized multicenter rescueicp study (http://www.Rescueicp.Com). Acta Neurochir Suppl 2006;96:17-20.

[49] Chi JH: Craniectomy for traumatic brain injury: Results from the decra trial. Neurosurgery 2011;68:N19-20.

[50] Takeuchi S, Wada K, Takasato Y, Masaoka H, Hayakawa T, Yatsushige H, Shigeta K, Momose T, Otani N, Nawashiro H, Shima K: Traumatic hematoma of the posterior fossa. Acta Neurochir Suppl 2013;118:135-138.

[51] Bullock MR, Chesnut R, Ghajar J, Gordon D, Hartl R, Newell DW, Servadei F, Walters BC, Wilberger J, Surgical Management of Traumatic Brain Injury Author G: Surgical management of posterior fossa mass lesions. Neurosurgery 2006;58:S47-55; discussion Si-iv.

[52] Cuneo RA, Caronna JJ, Pitts L, Townsend J, Winestock DP: Upward transtentorial herniation: Seven cases and a literature review. Arch Neurol 1979;36:618-623.

[53] Heary RF, Hunt CD, Krieger AJ, Schulder M, Vaid C: Nonsurgical treatment of compound depressed skull fractures. J Trauma 1993;35:441-447.

[54] Bullock MR, Chesnut R, Ghajar J, Gordon D, Hartl R, Newell DW, Servadei F, Walters BC, Wilberger J, Surgical Management of Traumatic Brain Injury Author G: Surgical management of depressed cranial fractures. Neurosurgery 2006;58:S56-60; discussion Si-iv.

[55] Vender JR, Bierbrauer K: Delayed intracranial hypertension and cerebellar tonsillar necrosis associated with a depressed occipital skull fracture compressing the superior sagittal sinus. Case report. J Neurosurg 2005;103:458-461.

[56] Ali B, Ghosh A: Antibiotics in compound depressed skull fractures. Emerg Med J 2002;19:552-553.

[57] Sataloff RT, Sariego J, Myers DL, Richter HJ: Surgical management of the frontal sinus. Neurosurgery 1984;15:593-596.

Chapter 11

Nutrition in Traumatic Brain Injury: Focus on the Immune Modulating Supplements

T.M. Ayodele Adesanya, Rachael C. Sullivan,
Stanislaw P.A. Stawicki and David C. Evans

Additional information is available at the end of the chapter

http://dx.doi.org/10.5772/57305

1. Introduction

The hypermetabolic nature of post-traumatic brain injury (TBI) state makes adequate nutritional support critical. Maintenance of adequate nutritional intake has been shown to have a significant impact on outcomes after TBI. [1] While pre-injury and immediate post-injury malnutrition has been associated with lower survival after TBI, much remains to be learned about the role and optimization of nutritional support beyond the initial phases of recovery. One of the most rapidly evolving aspects of clinical investigation in this general area is focusing on the effects of immune-modulating nutrition on TBI outcomes. The secondary injury phase following brain trauma is characterized by neuroinflammation, free radical generation, excitatory toxicity, and oxidative stress. [2] In this chapter we will present our current state of understanding of immune-nutrition for TBI, highlighting modern clinical practices and emerging trends. Many nutritional supplements have shown promise in preclinical and animal trials, particularly in the area of neuroprotection prior to injury, but human clinical trials have been largely disappointing or nonexistent.

2. General overview of nutritional support following TBI

Trauma, including TBI, is associated with transient immune-suppression and high rates of nosocomial infection. Gastrointestinal mucosal health quickly deteriorates following trauma and stress.[3] Immune-modulating nutrition has been associated with lower complication and infection rates in surgical and critically ill patients and is recommended in SCCM and ASPEN guidelines for select patients including trauma patients. [4] These guidelines make broad recommendations for the initiation and management of enteral nutrition in critical illness and should serve as the evidence-based foundation for nutritional support programs. Early administration of enteral feeding, combined with immune-modulating nutrient supplementation, has been shown to promote both the structural integrity and immunological function of the gastrointestinal mucosa. Target caloric and protein intake goals should be calculated for each patient, accommodating fully for any baseline increases in nutritional needs due to the metabolic stress of injury. The general initial nutritional strategy should include the provision of more than 50 percent of the estimated total energy expenditure and 1–1.5 g/kg protein within 24 hours of injury. [5] The provision of these requirements by the enteral rather than the parenteral route is always preferred.

3. The immune-enhancing paradigm

Immune-enhancing nutritional ingredients will be the focus of the subsequent sections of this chapter. Specifically, we will discuss the use of omega-3 fatty acids, dietary nucleotides, arginine, glutamine, and various antioxidants in TBI. General principles of the immune-enhancing paradigm focus on aggressive supplementation of immune-modulating ingredients with the aim of promoting healing of injured brain tissue and minimizing loss of parenchyma in the area of penumbra—the threatened but still viable tissue around the periphery of acute brain injury[6]. Many immune-modulating strategies (including steroid administration in CRASH I) have been trialed, frequently without demonstrating benefit[7]. Protocols for the timing, dosage, and route for many of these immune-modulating elements are yet to be clearly defined, and the authors will focus on the most up-to-date evidence regarding the basic science and clinical research on this topic.

4. Omega-3 fatty acids

Omega-3 fatty acids (n-3FAs) such as eicosapentaenoic acid (EPA) and docosahexaenoic acid (DHA) have been shown to be of potential value in the management of patients with TBI. Present in dietary intake, n-3FAs are commonly found in fish oils and are associated with a wide range of possible health benefits. The human brain is composed of 60% lipid by dry weight, with DHA representing one of the most abundant fatty acids found in the brain. [2] N-3FAs contribute to membrane fluidity and thus affect many different aspects of neuronal development and physiology, including cell adhesion, axon guidance, synaptic integrity, and neurotransmission. [2] Additionally, n-3FAs may also play a role in defense against oxidative stress and inflammation.

Biological pathways

N-3FAs have been shown to mitigate the consequences of several key pathologic cellular pathways associated with TBI, including oxidative stress, apoptosis, inflammation, and neuronal excitotoxicity. [2]

- In response to disruption of neuronal membranes, arachidonic acid is released and converted to pro-inflammatory prostaglandins. N-3FA derivatives, on the other hand, inhibit activation and migration of inflammatory cells. Physiologically, n-3FAs also suppress T-cell activation and natural killer cell activity and decrease the total number of circulating leukocytes. [2]
- In the setting of acute oxidative stress, DHA has been demonstrated to lessen the burden of lipid and protein peroxidation. Its metabolites can also modulate expression of tissue repair signals, upregulate anti-apoptotic proteins, and downregulate pro-apoptotic proteins. [8]
- The excitotoxic neurotransmitter glutamate is released following TBI, leading to a disproportionate influx of calcium into neurons and subsequent cell death. [2] *In vitro*, DHA has been shown to decrease calcium influx and thus lessen the burden of glutamate cytotoxicity.

Cumulatively, the effects of n-3FAs on TBI-related cellular processes may promote cell survival and viability, highlighting the potential role of n-3FAs in improving neurological outcomes.

Animal studies

Animal studies investigating the role of n-3FAs in experimental TBI models have produced encouraging results.

- In rat fluid percussion injury (FPI) models, n-3FAs have been shown to decrease oxidative damage as measured by protein oxidation, improve post-traumatic cognitive disability as measured by water maze testing, and normalize dysregulated expression of genes linked to neuronal energy homeostasis. [9, 10]
- In a rat model of controlled cortical impact (CCI), n-3FAs were shown to restore TBI-induced deficits in neuronal dopamine release, important for maintenance of learning, attention, and other neurobehavioral phenomena. [2, 11]
- Finally, in response to impact acceleration TBI in rats, dietary DHA supplementation has been linked to decreased neuronal injury and decreased caspase-3 activity, a marker of apoptosis. [12-14]. Conversely, dietary deficiency of n-3FAs in animal models has been associated with impaired neurogenesis, decreased neuronal size, and neurobehavioral defects. [2] In rat models, n-3FA deficiency has been specifically connected with increased spinal cord vulnerability to neuronal damage, measured by reduced markers of synaptic plasticity and membrane homeostasis in the lumbar spinal cord. [15]

Clinical studies

Despite the laboratory and animal research showing potential benefits of n-3FAs in improving clinical outcomes following TBI, there have been no clinical trials to verify such benefits in human subjects. There have been promising case reports of n-3FA use in TBI and a pilot study which suggested a link between n-3FAs and prevention of post-traumatic psychiatric distress [16, 17], but more robust studies of clinical response will be necessary to ascertain therapeutic benefit. Challenges to conducting effective studies include inconsistent doses and sources of commercially-available n-3FA preparations. Various trials use different doses of n-3FA and the sources are inconsistent as suppliers vary the species of fish used to make fish oils. Low doses of n-3FA are incorporated in many commercial fish oils but experimental TBI studies typically have focused on higher dose supplementation. [18] Each product also has its own ratio of DHA and EPA. Two prescription n-3FA products are available in the United States, but both primarily consist of EPA.

Clinical Limitations

The clinical use of dietary n-3FAs for TBI has been historically limited by concerns for anti-thrombotic actions, but this concern has subsided with several studies in cardiology patients using combination regimens of fish oil and antiplatelet agents. [19, 20] Other threats to the n-3FA supply, including heavy metal toxicity in some fish oils, also confound the field.

5. Dietary oligonucleotides

Following traumatic brain injury, nucleotides are released into the extracellular space, acting in both an autocrine and paracrine fashion, via nucleotide receptors on neuronal cells. [21] From individual nucleosides to antisense strands and microRNAs, dietary oligonucleotides represent a possible therapeutic means through which the pathophysiological responses and functional outcomes of TBI can be modulated. [22]

Biological pathways

Much work has been done at the single nucleoside level, specifically focusing on the role of adenosine. Adenosine is a purine nucleoside, speculated to play a neuroprotective role in TBI, leading to lower neuronal metabolism and greater cerebral blood flow. [23] Adenosine and its metabolic derivatives have been shown to acutely upregulate after TBI in both animal models and human disease. [24-27] A product of ATP breakdown, cerebral adenosine acts via the purinergic signaling system and may reduce cellular death related to glutamate-mediated excitotoxicity. It also decreases free radical-related/oxidative damage. [28] Although different adenosine receptors have been observed to facilitate both beneficial and deleterious physiologic effects in post-TBI studies, adenosine and its downstream pathways remain clearly linked to the pathophysiology of TBI and represent possible targets for therapeutic modulation. [21, 29-33]

Antisense oligonucleotides, on the other hand, are short synthetic nucleotide strands that can bind to specific messenger RNA (mRNA) targets, making them susceptible to degradation and thus effectively blocking synthesis of corresponding proteins. [34] These nucleotides represent yet another promising avenue through which novel TBI management strategies can begin to utilize more recent biomedical research discoveries.

Animal studies

Animal TBI studies involving oligonucleotide-based therapies have produced some promising results. Oligomeric diets demonstrated potential benefit in rat models of TBI, preventing TBI-induced weight loss and thymus atrophy and, by extension, averting immune dysfunction. [35]

In terms of specific nucleosides, adenosine is increased in rat fluid percussion injury (FPI) and controlled cortical impact (CCI) models of TBI, and 2-chloroadenosine, an adenosine analogue, has been demonstrated to confer improved bioenergetic and functional outcomes after FPI. [27] Additionally, in a weight-drop closed head injury (CHI) TBI model, intraperitoneal injection of cytidine triphosphate (CTP) has been shown to decrease neuronal apoptosis and improve motor function post-TBI. [36]

Antisense oligonucleotides have also been studied in animal models of TBI.

- In a rat stab wound model of brain injury, antisense oligonucleotides against a monocyte chemoattractant protein (MCP-1) were able to inhibit inflammatory chemokine production. [37]
- In a rat FPI model, pretreatment with antisense oligonucleotides against a specific N-methyl-D-aspartate (NMDA) glutamate receptor (NMDA-R1) decreased mortality from 50% to 8% and improved behavioral recovery, both likely due to the prevention of glutamate excitotoxicity. [38]
- In a murine weight-drop CHI TBI model, injection of oligonucleotides against acetylcholinesterase (AChE) reduced mortality from 50% to 20% in trauma-sensitive mice, decreased post-TBI neuronal death, and improved neuromotor recovery as measured via a beam test for balance and coordination. [34]

Clinical studies

While no clinical studies have explored the use of oligonucleotide-based therapies for TBI, associations have been made in humans between TBI severity and increased CSF concentrations of adenosine. [23, 27] Clinical concerns of oligonucleotide treatments include the cardiovascular effects of purinergic modulation and the effects of oligonucleotides on other unrelated receptor targets. Certainly the array of cell-based and animal studies highlight clear potential for the translational relevance of dietary oligonucleotides. [33, 34]

6. Arginine

Arginine is a nonessential amino acid and is a component of both enteral and parenteral nutrition formulas. [39] Normally, arginine homeostasis is driven by dietary intake and metabolic degradation, and when its utilization increases during growth, development, or injury, arginine may be recognized as an essential amino acid. [39] Parenteral arginine supplementation in trauma patients has been demonstrated to confer improved wound healing and immune responses and has no known adverse effects as a nutritional supplement. [39] From this background, and given that serum levels of L-arginine and its metabolites have been shown to be significantly reduced in patients post-TBI, arginine represents a potential dietary adjuvant to enhance TBI therapy. [40]

Biological pathways

L-arginine is the immediate, endogenous precursor of nitric oxide (NO), an important physiological vasodilator. [41] Immediately post-TBI, there is an increase of NO, followed by a sustained decrease which can result in diminished cerebral blood flow (CBF) and consequent hypoperfusion. [42] Additionally, arginine is a precursor for proline and 4-hydroxyproline – both important for extracellular matrix (ECM) remodeling – and for creatine – an important energy source in both muscle and brain tissues that will be discussed independently later in this chapter. [40]

Animal studies

Administration of L-arginine has been shown in both rat and mouse controlled cortical impact (CCI) models to restore CBF and reduce contusion volume post-TBI in a dose-dependent manner. [40, 41, 43-48] It has been also been demonstrated in a rat fluid percussion injury (FPI) model to reduce immunoreactivity for nitrotyrosine, a marker of peroxynitrite ($ONOO^-$) super oxide radicals. [49, 50] While one study failed to confirm that hypertonic arginine produced significant cerebrovascular improvements over hypertonic saline in a rat FPI model, other dose and time studies in rats have even shown that L-arginine is most neuroprotective when 300mg/kg is given as soon as possible after injury. [39, 51] Rats treated with an arginase-specific inhibitor (Nω-hydroxy-nor-arginine) showed significantly reduced contusion volume post-TBI. [45]

Clinical studies and limitations

There have been no clinical studies exploring the potential therapeutic benefits of isolated arginine supplementation in post-TBI patients. Arginine is a common ingredient in commercially-available formulas and in low doses it appears to be safe. Trials of critical care formulas including arginine, fish oil, and various antioxidants appear to be safe and are effective at reducing infection rates in TBI and other critically-ill patients. [52] Although arginine seems to have strong translational promise, a number of potential risks exist before considering hyper-supplementation of arginine. For instance, while some studies have linked L-arginine to reduced neuronal damage, none have been able to demonstrate the same beneficial effects with regards to neurological function. [43] Secondly, the optimal dose of 300 mg/kg in rats is much larger than amounts of arginine found in typical nutritional formulations. [39] Finally, the roles of other arginine derivatives, such as arginine vasopressin, and nitric oxide (NO) signaling remain unclear in post-TBI pathophysiology. [46, 53-64]

7. Glutamine

Glutamine is a non-essential amino acid, widely distributed throughout the body. It is the most abundant free amino acid in circulation. [65, 66] Glutamine synthesis cannot keep up with

http://dx.doi.org/10.5772/57305

increased requirements such as those experienced during physiological stress, yet it is important for the immune response. Consequently, glutamine supplementation has been shown to decrease infectious complications in trauma patients. [65, 66] The brain serves prominently in glutamine metabolism and is a net producer of the amino acid. [67] In the brain, glutamine is involved in the glutamine-glutamate cycle which functions to conserve the carbon skeletons of neurotransmitters. [67] As part of this cycle, it is synthesized from glutamate and ammonia in astrocytes and also serves as the precursor for glutamate along with alpha-ketoglurate. [67-69]

While glutamate is recognized as an excitotoxic neurotransmitter released after TBI, patients with brain injury are also observed to experience profound hypoglutaminemia. [67, 68, 70-95] While the cause of this hypoglutaminemia is not known, this observed deficiency provides rationale for dietary supplementation.

Animal studies

While many animal studies assess post-TBI glutamatergic signaling, in a rat TBI model, glutamine administration was shown to decrease concentrations of pro-inflammatory cytokines and apoptotic cells in gastrointestinal tissue, thus reducing TBI-associated damage to gastrointestinal mucosa. [65, 66]

Clinical studies

Limited clinical studies have associated glutamine and alanine dietary supplementation with lower mortality rates, shorter hospital lengths of stay, decreased occurrences of pneumonia and stress ulcers, and higher lymphocyte counts in TBI patients. [96] While these results have been linked to an improved immunological response, future basic science and clinical studies are needed to advance our understanding of the translational potential of glutamine as a nutritional adjuvant in TBI therapy.

As a potential limitation of glutamine therapy, glutaminergic signaling has been implicated in basic science studies with post-traumatic epilepsy. [97] Additionally and perhaps of greater relevance, the recent REDOXS (REducing Deaths due to OXidative Stress) trial sought to investigate the effect of nutritional supplementation in critically-ill patients. A randomized trial, the study found that glutamine supplementation actually resulted in increased harm and mortality in critically ill patients and cautiously advocated that administration of glutamine be reserved for burn and trauma patients not in multiorgan failure. [98] Of note, much of the glutamine in that study was administered in parenteral form, and the body of literature using enteral glutamine has shown no such outcome.

8. Antioxidants

Biologic pathways

Reactive oxygen and nitrogen species (ROS/RNS, respectively) play an integral role in brain injury and posttraumatic neuronal degeneration. [99, 100] In the setting of acute traumatic stress, endogenous protective mechanisms such as glutathione (GSH) and superoxide dismutase (SOD) may become overwhelmed by increased production of free radicals. [99] This is driven in part by influx of excess of intracellular calcium into mitochondria. Lipid peroxidation mediated by oxygen radical species has been suggested as an important factor in posttraumatic neuronal degeneration. [100] In addition to disrupting the membrane phospholipid architecture, lipid peroxidation contributes to the formation of cytotoxic aldehyde-containing byproducts that bind to and impair the function of cellular proteins. [101] The oxidation of DNA and proteins then may trigger programmed cell death. This process is exacerbated during the reperfusion phase of injury, resulting in additional microvascular damage and neuronal cell death.

Clinical studies

Increasing amounts of evidence point to potential effectiveness of antioxidants in modulating the severity of TBI. [99, 100] Specifically, nutritional antioxidants may be critical in attenuating the deleterious effects of oxidative stress in ischemia and reperfusion type injuries. [102] Specific antioxidant agents that have been investigated in the setting of TBI include vitamin E (alpha-tocopherol), glucocorticoid methylprednisolone, tirilazad mesylate, 21-amino-steroids, green tea extract, ginkgo biloba extract, resveratrol, curcumin, and niacin. [100-102] In addition, evidence points to selenium as being an effective inhibitor of ROS-mediated apoptotic neural precursor cell death in TBI. [103] A full discussion of these antioxidants is beyond the scope of this chapter which focuses on immunonutrition, but several antioxidants are worthy of special mention.

Commercially available enteral formulas frequently tout "added antioxidants," but these typically are vitamin C, vitamin E, and beta-carotene. In the setting of TBI, enteral nutrition enriched with antioxidants and neuromodulatory agents seems to have some clinical benefit. [104] Although there were no mortality differences between the control and glutamine/probiotic enteral nutrition regimens, the glutamine/probiotic group demonstrated lower infection rates and infections per patient, as well as shorter intensive care stays and fewer ventilator days. [104]

The finding that plasma vitamin C levels are significantly lower in patients with brain trauma suggests that vitamin C plays a potential role in oxidative stress related to brain injury. [105] In addition to vitamin C, other nutritional factors may play a role in modulating oxidative damage associated with TBI, including vitamin E (alpha-tocopherol), beta-carotene, and coenzyme Q10. [106] Despite promising preliminary animal studies, data showing efficacy of specific or combined micronutrient supplementation in the setting of brain injury remains elusive. [106] A small study examining high-dose vitamin C and vitamin E showed some promise but should be interpreted as preliminary. [107]

It has to be noted that phase III clinical trials of neuroprotective agents in TBI have been somewhat disappointing. [108] In a multicenter trial of tirilazad mesylate in TBI, the experimental group was found to have similar mortality and neurologic recovery rates when compared to placebo. [109] However, a subgroup analysis suggested that tirilazad mesylate may contribute to reduced mortality in male patients with severe head injury accompanied by traumatic subarachnoid hemorrhage (34% tirilazad group mortality versus 43% placebo group mortality). [109]

One trial of polyethylene glycol (PEG)-conjugated SOD in TBI patients initiated within 8 hours of the injury showed a trend toward improved neurological outcomes. [110] Subsequent larger trials failed to reproduce any beneficial effect however. [111] Another agent, U-83836E, a second-generation lazaroid with non-steroidal structure, has been shown to decrease post-injury lipid peroxidation and protein nitration and enhance preservation of mitochondrial respiratory function and calcium buffering ability in a mouse model, and human studies using this agent may be warranted. [112] Melatonin is another antioxidant agent showing promise in providing neuroprotective benefits based on evidence from rat model of TBI. [111] A number of other promising agents have been investigated, but human evidence continues to be scarce.

Increasing amounts of evidence suggests that the most effective antioxidative approach to the brain-injured patient should involve combined treatment with mechanistically synergistic antioxidants. [101] Strategies within such a paradigm should include simultaneous scavenging of lipid peroxidation-initiating free radicals, inhibition of lipid peroxidation propagation, and removal of neurotoxic lipid peroxidation products. [101] Clinical trials with multidrug antioxidant regimens are needed before any recommendations can be made.

9. Branched-chain amino acids

Branched-chain amino acids (BCAAs) are essential amino acids that have important roles in energy metabolism and protein and neurotransmitter synthesis. [113] Valine, isoleucine, and leucine comprise the BCAAs, and these entities have important roles in regulating protein synthesis, gluconeogenesis, and energy metabolism as well as functioning as a major source of nitrogen for producing glutamine in the brain. [113] Because of the important baseline functions of these compounds, this would suggest that alterations in BCAA metabolism after TBI may actually play a role in decreased energy production and neurotransmitter synthesis, thereby contributing to TBI pathology. As such, the supplementation of BCAAs or their metabolites may have a role in the reduction of TBI pathology and possibly outcome.

Biologic Pathways

The metabolism of BCAAs is partially regulated by protein synthesis requirements and excess BCAAs are either catabolized or excreted. In terms of catabolism of excess BCAAs, the first step is catalyzed by the branched-chain aminotransferase isoenzymes, mitochondrial BCATm and cytosolic BCATc. The resulting product of this process is glutamate, which is a major excitatory neurotransmitter as well as a precursor of alpha-ketoglutarate. The second, irreversible step in BCAA catabolism is catalyzed by the mitochondrial branched-chain α-ketoacid dehydrogenase (BCKDC) enzyme complex. [114] BCKDC catalyzes oxidative decarboxylation of the BCKA products of the BCAT reaction, forming NADH and the respective branched-chain acyl CoA derivative of each BCAA. [114]

Animal studies

It is well-established that TBI causes cognitive impairment and altered net synaptic efficacy. In one study where brain injured mice or sham-injured mice either consumed water or water containing BCAAs, there was an overall cognitive improvement with a simultaneous restoration in net synaptic efficacy. [115] The major finding of this study was that dietary delivery of BCAAs ameliorates hippocampal-dependent cognitive dysfunction together with a restoration of net synaptic efficacy after concussive brain injury, and in every animal, cognitive improvement occurred only in conjunction with restored net synaptic efficacy. [115]

Clinical studies

Although the literature on this subject is rather sparse, there are some promising results. It has been reported that the levels of all three BCAAs in patients with mild TBI relative to healthy volunteers is decreased. BCAA levels are further reduced in patients with severe TBI compared with all groups. [113] In one study, it was shown that short-term intravenous supplementation of BCAAs in rehabilitation patients with TBI enhances recovery of cognitive function, induces a supraphysiologic plasma content of BCAAs, and increases tyrosine plasma concentration. [116] This study also revealed that plasma amino acid levels remained decreased in the posttraumatic rehabilitation phase (1-22 months). In this study, 40 patients with TBI were randomly assigned either intravenous BCAAs or placebo. Plasma tyrosine concentration improved in the group given BCAA supplementation and overall disability improvement was greater than that noted in the placebo group. The key conclusion of the study was simply that supplementation of BCAAs in TBI restores plasma levels to the normal range without having a negative effect on levels of precursors of brain catecholamines and serotonin. [116]

Another study revealed that BCAA supplementation may aid in recovery from a posttraumatic vegetative or minimally conscious state, thus reducing the risk of the vegetative state persisting over time. [117] This study, also performed by Aquilani et al., supplemented patients for 15 days by intravenous route with either BCAAs or placebo who were either in a posttraumatic vegetative or minimally conscious states. [117] The 15-day period of these trials is too short to draw any meaningful conclusions regarding that adaptation of BCAAs.

Another study sought to assess the impact of plasma BCAA and tyrosine levels following enterally-administered BCAAs; However, enteral administration failed to return plasma BCAA levels to the normal range. [118] In addition, it was found that elevated plasma phenylalanine was associated with decreased ICP and increased jugular venous oxygen saturation (SjvO2), while higher plasma isoleucine and leucine levels were associated with increased ICP and higher plasma leucine and valine were linked to decreased SjvO2. Therefore hyperalimentation with enteral nutrition should be carefully performed to avoid harmful side effects of amino acids while promoting improvements in brain metabolism. [118]

Summary

Although there are a small number of very preliminary but promising studies suggesting that BCAA supplementation may be beneficial to the TBI patient, further studies are needed to optimize the route and dosage of supplementation and to better elucidate the side effects of artificial supplementation such that supplementation produces no significant side effects.

10. Choline

Immediately following TBI, there is a transient period of excess cholinergic activity which may contribute to excitotoxicity via nicotinic and muscarinic receptor subtypes. However, the chronic phase of TBI is actually associated with decreased brain cholinergic function.

Acetylcholine acts on nicotinic and muscarinic acetylcholine receptors, and previous studies have suggested that TBI-related deficits in alpha-7 n-acetylcholine receptor (α7 nAChR) density may contribute to post-TBI cognitive deficits. [119] If this downregulation of α7 nACh receptors in fact contributes to the cognitive impairment seen as a result of TBI, a therapeutic option includes drugs or compounds that are selective agonists of α7 nAChRs and these may be helpful in ameliorating some measures of cognitive decline. [119]

One such compound that has been shown to bind α7 nAChR is choline. [119] Choline is an essential nutrient available from a wide variety of nutritional sources. It is an important molecule involved in synthesis of structural cell membrane phospholipids, other signaling molecules, and is also a precursor for acetylcholine. [120] As such, it is postulated that dietary choline supplementation may minimize cognitive deficits, reduce brain inflammation, and protect the penumbra.

Biologic pathways

Acetylcholine acts on nicotinic and muscarinic acetylcholine receptors, both of which are prominently located in brain regions that are involved with attention and cognition. [119] As previously stated, choline has been shown to be an agonist at α7 nAChRs, but not other nicotinic receptor subtypes. α7 nAChRs are known to be involved in both excitotoxicity and inflammatory pathways. Once TBI occurs, multiple biochemical pathways, including the aforementioned excitotoxicity and inflammatory pathways, are set into motion which leads to a chronic, neurodegenerative condition.

Animal studies

In one study, dietary choline supplementation was shown to significantly reduce brain injury-induced spatial learning deficits in a rat model. Additionally, the choline-supplemented diet helped reduce brain inflammation and spared cortical tissue. [119]
It is known that administration of cytidine-5'-diphosphate (CDP)-choline functions as a neurostimulant in neurological disorders of memory.[121] As such, its use in TBI was promising. Dixon et al. demonstrated that chronic CDP-choline treatment can attenuate neurological and cognitive performance deficits following TBI in rats. [122] CDP-choline treatment also increased post-injury resistance to the memory-disrupting effects of scopolamine. Exogenous administration of CDP-choline increased ACh release. [122] The mechanism of action is not definitively known, but CDP-choline may attenuate post-injury

functional deficits by several mechanisms, including providing the ACh precursor choline to drive up ACh synthesis, maintaining cell integrity by accelerating membrane formation, and/or stimulating brain metabolism. [122]

Clinical studies

Like with many immunonutrients, a number of large-scale studies have shown no benefit despite promising animal trials. Ruff et al. found that citicoline (an intermediate in the generation of phosphatidylcholine from choline) supplementation in TBI patients did not improve the extent or speed the recovery in patients following acute stroke. [123]

Similarly, Zafonte et al., completed the Citicoline Brain Injury Treatment Trial (COBRIT), a phase III, double-blind study comparing citicoline versus placebo. In this trial, 1213 study participants with complicated mild, moderate, or severe TBI were randomized to receive 2000 milligrams of citicoline or placebo daily for 90 days. The trial ran from 2007-2011 but was terminated early due to futility. The study did not demonstrate any benefits of citicoline treatment. [123]

Summary

Not all promising findings in the preclinical arena have been translated to success in patients. Choline supplementation in TBI rats holds promise. However, these have not held true in the patient models. This creates a need to understand the mechanism of how choline induces positive results in rats. Additionally, there may be other compounds or physiologic conditions that are necessary to allow for the beneficial effects of choline which are as of yet unknown.

11. Creatine

Creatine is a common dietary supplement, frequently used to increase strength and muscle mass. Creatine metabolism plays a key role in ATP turnover in the metabolically active brain. Endogenously expressed, cerebral creatine levels have been observed to decrease after TBI and recent studies have also shown that it provides significant neuroprotection against oxidative stress and ischemia. [124, 125] While investigations of creatine as a nutritional component of TBI therapy have been limited to animal models, much potential exists for clinical research to further define its translational relevance.

Biological pathways

The mechanisms of creatine-induced neuroprotection seem to be largely related to its effects on mitochondrial bioenergetics, binding to mitochondrial creatine kinase (CK) to exert structural protection allowing the enzyme to maintain its ability to inhibit free radical generation. [126, 127] Creatine supplementation lowers mitochondrial membrane potentials and reduces mitochondrial levels of reactive oxygen species (ROS) and calcium while maintaining the levels of adenosine triphosphate (ATP). [126] Physiologically, these effects result in inhibition of mitochondrial permeability and reduced neuronal loss. [126] Hybrid hydrophobic derivatives of creatine, creatinyl amino acids, have been synthesized with the aim to establish better penetration across the blood-brain barrier. *In vivo* these compounds maintain both their neuroprotective abilities and chemical stability. [128]

Animal studies

In experimental mouse and rat TBI models, chronic supplementation of creatine has been shown to decrease the extent of cortical damage by as much as 36% and 50%, respectively. [126] Compared to rats receiving a control diet, rats fed a creatine-enriched diet have also shown decreased levels of neurochemical markers of TBI-induced acute cellular injury.[127, 129] Many of these protective effects were demonstrated to follow a dose-dependent manner and cumulatively provide promising preclinical data to steer pilot clinical studies. [129]

12. Magnesium

Magnesium is essential for maintenance of vital cellular functions, including glycolysis, sustaining membrane structure and function, protein synthesis and DNA replication. [130] Magnesium also plays an important role in central nervous system following injury. It is known that after TBI, the normal homeostatic mechanisms of magnesium are deranged, resulting in a rapid decline in magnesium levels in the brain. [131] This disruption of normal magnesium homeostasis has actually been shown to correlate with the severity of neurologically-mediated behavioral deficits following injury. [132] As such, it has been postulated that magnesium pharmacotherapy may aid in the treatment of various CNS injuries, including ischemia and cortical lesions, and has been found to be effective in some of these arenas. Because of the critical function of magnesium, it is also postulated that manipulation of dietary magnesium may have an impact on the recovery of function following TBI.

Biological pathways

Magnesium plays an important role in homeostatic regulation of key pathways involved in the delayed secondary phase of brain injury. [133] During normal physiological processes, magnesium is a noncompetitive inhibitor of the NMDA receptors, thereby regulating calcium influx. [134] Following acute brain injury, tissue magnesium is depleted, leading to loss of homeostatic control of the NMDA receptors. The ensuing massive influx of calcium leads to neuronal degeneration and cell death. [133]

Animal studies

Previous research has shown that dietary magnesium deficiency prior to injury worsens recovery of function and that systemic administration of magnesium pre- or post-injury significantly improves functional recovery. A number of studies in rats have shown that treatment with magnesium after brain injury did offer neuroprotection. [133, 135-137] Bareyre et al. showed that in addition to beneficial effects on behavioral outcomes, magnesium supplementation in brain-injured rats attenuated cortical histological damage. [138] Magnesium therapy administered up to 24 hours after injury in rats significantly improved motor outcome and behavioral parameters in rats with severe diffuse traumatic axonal brain injury.

[139] Additionally, magnesium supplementation was shown to reduce long-term motor and cognitive deficits after TBI in rats which may result in decreased post-traumatic stress and anxiety. [140]

Clinical studies

Disruption of magnesium homeostasis has been observed in human traumatic brain injury. Despite a number of preclinical studies showing beneficial effects of magnesium supplementation in TBI, mostly in rat models, clinical studies in TBI patients have failed to show a consistent clinical benefit. Temkin et al. showed that continuous infusions of magnesium for 5 days given to patients within 8 hours of moderate or severe TBI were not neuroprotective and may even have a negative effect in the treatment of significant head injury. [141] However, in another prospective clinical trial by Dhandapani et al., magnesium sulfate administered to TBI patients within 12 hours of their injuries produced decreased mortality and improved neurologic patient outcome. [142] There have been a number of studies looking at the role of magnesium supplementation in combination with other pharmacological agents or physiological interventions, such as hypothermia and hyperoxia, again with varied results in both preclinical and clinical trials. A recent meta-analysis of all randomized controlled trials comparing magnesium supplementation in patients following acute TBI shows no evidence to support the use of magnesium beyond standard physiologic replacement. [143]

Summary

The success of magnesium in attenuating the process of neurodegeneration in animal models of brain injury has been widely studied with promising results. Unfortunately, these preclinical successes have not consistently translated into success in humans. Magnesium supplementation in TBI patients has produced varied results, requiring further investigation into not only magnesium supplementation but the secondary parameters that may affect clinical outcome in TBI patients.

13. Vitamin D

Vitamin D hormone (VDH; 1, 25-dihydroxyvitamin D3) is recognized as a neurosteroid with downstream implications in many different CNS signaling cascades. [144] VDH deficiency is associated with dysregulated neuronal physiology and has been demonstrated to both exacerbate TBI and reduce the efficacy of progesterone treatment for TBI. [145-147]. The relationship between VDH and TBI is perhaps most important in aging populations, within which the former is high in prevalence and the latter is rising in incidence. [148, 149]

Biological pathways

With regards to TBI, vitamin D generally acts in an anti-inflammatory manner, by regulating intracellular calcium levels (hence reducing the effects of glutamate excitotoxicity) and enhancing free radical scavenging. [144] Much TBI-related vitamin D research investigates it as a combined therapy with progesterone. The two hormones are proposed to act in a synergistic and perhaps compensatory manner, each of them having their own anti-inflammatory and oxidative damage-reducing properties. [144] Together, VDH and progesterone stimulate neural growth in cultured neurons following *in vitro* glutamate excitotoxicity. [145, 147]

Animal studies

In rat cortical contusion injury (CCI) models of TBI, combined therapy consisting of VDH and progesterone resulted in reduced expression of inflammatory genes; protection against cell death and DNA damage; and significant improvement in post-traumatic behavior in VDH-deficient rats. [148, 149]

Clinical studies

Limited clinical trials have shown promising results for VDH and progesterone combination therapy, improving outcomes and decreasing mortality rates after TBI. [144] VDH has a high safety profile and is inexpensive and easily administered. [147] Continued investigations will be critical to further elucidate its specific mechanisms of actions, differences in combination therapy and monotherapy, and potential for use in a therapeutic or preventative manner.

14. Zinc and other trace elements

Trace elements are known to be important modulators of cell physiology and growth, contributing to many key processes such as wound healing and the immune response. [150] Among trace elements, zinc is specifically critical for tissue repair and essential for the function of many enzymes and gene expression. [151, 152] The majority of zinc ions in the brain are bound to proteins while the remaining are sequestered in presynaptic neuronal vesicles. [151] Although neurotoxic at high levels, zinc mediates synaptic transmission and plasticity, and clinical studies have shown that after TBI patients lose excess zinc in urine in proportion to injury severity and are at increased risk for developing zinc deficiency. [152, 153] Dietary zinc regulates intestinal zinc absorption and plays an important role in zinc homeostasis, thus making zinc promising as a possible nutritional adjunct to TBI therapy. [154]

Biological pathways

To a large degree, there is some debate with regards to whether zinc is neuroprotective or neurotoxic. [155] Many studies have demonstrated zinc accumulation after brain injury, associating it with neurodegeneration and deposited aggregates of ubiquitinated proteins and thus linking altered zinc homeostasis to impaired protein degradation. [153, 156-160] Though zinc chelators were able to block these TBI-induced histological changes, they did not lead to improved post-TBI outcomes in rats. [152, 153]

Contrarily, the neuroprotective effects of zinc are also established at the basic science and animal model levels. After mechanical repetitive strain injury (RSI), neuronal-like cells have been shown to develop a cellular zinc deficiency, and zinc deficiency itself has been linked to impaired neuronal stem cell proliferation and compromised cellular repair. [161, 162]

Animal studies

In animal models, zinc reduces the development of behavioral deficits after TBI. [153, 163] Specifically, in a rat controlled cortical impact (CCI) model, zinc supplementation reduced anxiety and cognitive impairments. [153, 161, 163] This supplementation did not lead to increased neuronal cell death. [161] Further evidence of its potential therapeutic benefit comes from the fact that zinc deficiency has been demonstrated to result in increased cell death and altered glial immune responses in several different rat and mouse TBI models. [151, 153, 164, 165]

Clinical studies

Limited preclinical studies show that, after an initial period of total parenteral nutrition, dietary zinc supplementation of 22 milligrams per day using zinc gluconate significantly increases visceral protein mass in post-TBI patients, is associated with improved Glasgow Coma Scores, as well as mortality decrease from 26% to 12%. [153] With the recommended upper limit of dietary zinc being 40 milligrams per day, further clinical studies will clearly define the optimal doses and time windows to improve post-TBI deficits and prevent neurotoxicity and undesired effects to other organs. [153, 166]

Other trace elements

While zinc has been the most thoroughly studied trace element in the contexts of TBI therapy, few studies have investigated the potential roles of others. While most of these elements still present the same concerns of toxicity versus protection, preliminary results seem promising for continued research. [154]

- As described above for zinc, copper deficiency has also been linked to increased neuronal apoptosis in a rat model of TBI. [165]
- To prevent deposition of free iron from heme degradation, administration of heme oxygenase (HO) inhibitors such as tin protoporphyrin or iron chelators have shown to reduce pathophysiologic and neuromotor changes in post-TBI models. [167, 168]
- As mentioned previously in this chapter, selenium, acting as an antioxidant, reduces reactive oxygen species (ROS)-mediated apoptosis of neural precursor cells both *in vitro* and in a mouse model of TBI. [103]

15. Conclusion

TBI represents a heterogeneous pathophysiological process that is clearly a challenge to manage. Multiple clinical studies of nutritional strategies have not defined a specific pathway that can serve as a sole, standalone target in TBI nutritional therapy. Multidimensional treatment plans, perhaps incorporating some of the described nutritional adjuvants, will thus merit more investigations from both the bench and the bedside to elucidate effective strategies to best treat TBI patients. Unfortunately, many strategies that are promising in the lab or in animal models have not borne fruit in clinical trials to date.

Author details

T.M. Ayodele Adesanya, Rachael C. Sullivan, Stanislaw P.A. Stawicki and David C. Evans

The Ohio State University, Department of Surgery, Columbus, Ohio, USA

References

[1] Charrueau C, Belabed L, Besson V, Chaumeil JC, Cynober L, Moinard C. Metabolic response and nutritional support in traumatic brain injury: evidence for resistance to renutrition. J Neurotrauma. 2009;26(11):1911-20.

[2] Hasadsri L, Wang BH, Lee JV, Erdman JW, Llano DA, Barbey AK, et al. Omega-3 Fatty acids as a putative treatment for traumatic brain injury. J Neurotrauma. 2013;30(11):897-906.

[3] Ivatury RR, Simon RJ, Islam S, Fueg A, Rohman M, Stahl WM. A prospective randomized study of end points of resuscitation after major trauma: global oxygen transport indices versus organ-specific gastric mucosal pH. J Am Coll Surg. 1996;183(2):145-54.

[4] McClave SA, Martindale RG, Vanek VW, McCarthy M, Roberts P, Taylor B, et al. Guidelines for the Provision and Assessment of Nutrition Support Therapy in the Adult Critically Ill Patient: Society of Critical Care Medicine (SCCM) and American Society for Parenteral and Enteral Nutrition (A.S.P.E.N.). JPEN Journal of parenteral and enteral nutrition. 2009;33(3):277-316. Epub 2009/04/29.

[5] Härtl R, Gerber LM, Ni Q, Ghajar J. Effect of early nutrition on deaths due to severe traumatic brain injury. J Neurosurg. 2008;109(1):50-6.

[6] Liu S, Levine SR, Winn HR. Targeting ischemic penumbra: part I - from pathophysiology to therapeutic strategy. J Exp Stroke Transl Med. 2010;3(1):47-55.

[7] Roberts I, Yates D, Sandercock P, Farrell B, Wasserberg J, Lomas G, et al. Effect of intravenous corticosteroids on death within 14 days in 10008 adults with clinically significant head injury (MRC CRASH trial): randomised placebo-controlled trial. Lancet. 2004;364(9442):1321-8. Epub 2004/10/12.

[8] Mayurasakorn K, Williams JJ, Ten VS, Deckelbaum RJ. Docosahexaenoic acid: brain accretion and roles in neuroprotection after brain hypoxia and ischemia. Current opinion in clinical nutrition and metabolic care. 2011;14(2):158-67. Epub 2010/12/24.

[9] Wu A, Ying Z, Gomez-Pinilla F. Omega-3 fatty acids supplementation restores mechanisms that maintain brain homeostasis in traumatic brain injury. J Neurotrauma. 2007;24(10):1587-95.

[10] Wu A, Ying Z, Gomez-Pinilla F. Dietary omega-3 fatty acids normalize BDNF levels, reduce oxidative damage, and counteract learning disability after traumatic brain injury in rats. J Neurotrauma. 2004;21(10):1457-67.

[11] Shin SS, Dixon CE. Oral fish oil restores striatal dopamine release after traumatic brain injury. Neurosci Lett. 2011;496(3):168-71.

[12] Mills JD, Hadley K, Bailes JE. Dietary supplementation with the omega-3 fatty acid docosahexaenoic acid in traumatic brain injury. Neurosurgery. 2011;68(2):474-81; discussion 81.

[13] Mills JD, Bailes JE, Sedney CL, Hutchins H, Sears B. Omega-3 fatty acid supplementation and reduction of traumatic axonal injury in a rodent head injury model. J Neurosurg. 2011;114(1):77-84.

[14] Bailes JE, Mills JD. Docosahexaenoic acid reduces traumatic axonal injury in a rodent head injury model. J Neurotrauma. 2010;27(9):1617-24.

[15] Ying Z, Feng C, Agrawal R, Zhuang Y, Gomez-Pinilla F. Dietary omega-3 deficiency from gestation increases spinal cord vulnerability to traumatic brain injury-induced damage. PLoS One. 2012;7(12):e52998.

[16] Lewis M, Ghassemi P, Hibbeln J. Therapeutic use of omega-3 fatty acids in severe head trauma. Am J Emerg Med. 2013;31(1):273.e5-8.

[17] Matsuoka Y, Nishi D, Yonemoto N, Hamazaki K, Hamazaki T, Hashimoto K. Potential role of brain-derived neurotrophic factor in omega-3 Fatty Acid supplementation to prevent posttraumatic distress after accidental injury: an open-label pilot study. Psychother Psychosom. 2011;80(5):310-2.

[18] Zhou M, Martindale RG. Immune-modulating enteral formulations: optimum components, appropriate patients, and controversial use of arginine in sepsis. Current gastroenterology reports. 2007;9(4):329-37.

[19] Bays HE. Safety considerations with omega-3 fatty acid therapy. The American journal of cardiology. 2007;99(6a):35c-43c. Epub 2007/03/21.

[20] Watson PD, Joy PS, Nkonde C, Hessen SE, Karalis DG. Comparison of bleeding complications with omega-3 fatty acids + aspirin + clopidogrel--versus--aspirin + clopidogrel in patients with cardiovascular disease. The American journal of cardiology. 2009;104(8):1052-4. Epub 2009/10/06.

[21] Burgos M, Neary JT, González FA. P2Y2 nucleotide receptors inhibit trauma-induced death of astrocytic cells. J Neurochem. 2007;103(5):1785-800.

[22] Bhalala OG, Srikanth M, Kessler JA. The emerging roles of microRNAs in CNS injuries. Nat Rev Neurol. 2013;9(6):328-39.

[23] Bell MJ, Robertson CS, Kochanek PM, Goodman JC, Gopinath SP, Carcillo JA, et al. Interstitial brain adenosine and xanthine increase during jugular venous oxygen desaturations in humans after traumatic brain injury. Crit Care Med. 2001;29(2):399-404.

[24] Kochanek PM, Clark RS, Obrist WD, Carcillo JA, Jackson EK, Mi Z, et al. The role of adenosine during the period of delayed cerebral swelling after severe traumatic brain injury in humans. Acta Neurochir Suppl. 1997;70:109-11.

[25] Bell MJ, Kochanek PM, Carcillo JA, Mi Z, Schiding JK, Wisniewski SR, et al. Interstitial adenosine, inosine, and hypoxanthine are increased after experimental traumatic brain injury in the rat. J Neurotrauma. 1998;15(3):163-70.

[26] Schubert P, Kreutzberg GW. Cerebral protection by adenosine. Acta Neurochir Suppl (Wien). 1993;57:80-8.

[27] Robertson CL, Bell MJ, Kochanek PM, Adelson PD, Ruppel RA, Carcillo JA, et al. Increased adenosine in cerebrospinal fluid after severe traumatic brain injury in infants and children: association with severity of injury and excitotoxicity. Crit Care Med. 2001;29(12):2287-93.

[28] Mitchell HL, Frisella WA, Brooker RW, Yoon KW. Attenuation of traumatic cell death by an adenosine A1 agonist in rat hippocampal cells. Neurosurgery. 1995;36(5): 1003-7; discussion 7-8.

[29] Li W, Dai S, An J, Xiong R, Li P, Chen X, et al. Genetic inactivation of adenosine A2A receptors attenuates acute traumatic brain injury in the mouse cortical impact model. Exp Neurol. 2009;215(1):69-76.

[30] Dai SS, Xiong RP, Yang N, Li W, Zhu PF, Zhou YG. (Different effects of adenosine A2A receptors in the models of traumatic brain injury and peripheral tissue injury). Sheng Li Xue Bao. 2008;60(2):254-8.

[31] Dai SS, Zhou YG. Adenosine 2A receptor: a crucial neuromodulator with bidirectional effect in neuroinflammation and brain injury. Rev Neurosci. 2011;22(2):231-9.

[32] Dai SS, Li W, An JH, Wang H, Yang N, Chen XY, et al. Adenosine A2A receptors in both bone marrow cells and non-bone marrow cells contribute to traumatic brain injury. J Neurochem. 2010;113(6):1536-44.

[33] Choo AM, Miller WJ, Chen YC, Nibley P, Patel TP, Goletiani C, et al. Antagonism of purinergic signalling improves recovery from traumatic brain injury. Brain. 2013;136(Pt 1):65-80.

[34] Shohami E, Kaufer D, Chen Y, Seidman S, Cohen O, Ginzberg D, et al. Antisense prevention of neuronal damages following head injury in mice. J Mol Med (Berl). 2000;78(4):228-36.

[35] Moinard C, Delpierre E, Loï C, Neveux N, Butel MJ, Cynober L, et al. An oligomeric diet limits the response to injury in traumatic brain-injured rats. J Neurotrauma. 2013;30(11):975-80.

[36] Zhao J, Liu Q, Cui J, Hong J, Song Z. (Research on motor dysfunction and the role of CTP after traumatic brain injury in rats). Sichuan Da Xue Xue Bao Yi Xue Ban. 2003;34(3):559-61.

[37] Ghirnikar RS, Lee YL, Li JD, Eng LF. Chemokine inhibition in rat stab wound brain injury using antisense oligodeoxynucleotides. Neurosci Lett. 1998;247(1):21-4.

[38] Sun FY, Faden AI. Pretreatment with antisense oligodeoxynucleotides directed against the NMDA-R1 receptor enhances survival and behavioral recovery following traumatic brain injury in rats. Brain Res. 1995;693(1-2):163-8.

[39] Cherian L, Chacko G, Goodman C, Robertson CS. Neuroprotective effects of L-arginine administration after cortical impact injury in rats: dose response and time window. J Pharmacol Exp Ther. 2003;304(2):617-23.

[40] Jeter CB, Hergenroeder GW, Ward NH, Moore AN, Dash PK. Human traumatic brain injury alters circulating L-arginine and its metabolite levels: possible link to cerebral blood flow, extracellular matrix remodeling, and energy status. J Neurotrauma. 2012;29(1):119-27.

[41] Cherian L, Robertson CS. L-arginine and free radical scavengers increase cerebral blood flow and brain tissue nitric oxide concentrations after controlled cortical impact injury in rats. J Neurotrauma. 2003;20(1):77-85.

[42] Louin G, Neveux N, Cynober L, Plotkine M, Marchand-Leroux C, Jafarian-Tehrani M. Plasma concentrations of arginine and related amino acids following traumatic brain injury: Proline as a promising biomarker of brain damage severity. Nitric Oxide. 2007;17(2):91-7.

[43] Degeorge ML, Marlowe D, Werner E, Soderstrom KE, Stock M, Mueller A, et al. Combining glial cell line-derived neurotrophic factor gene delivery (AdGDNF) with L-arginine decreases contusion size but not behavioral deficits after traumatic brain injury. Brain Res. 2011;1403:45-56.

[44] Cherian L, Hlatky R, Robertson CS. Comparison of tetrahydrobiopterin and L-arginine on cerebral blood flow after controlled cortical impact injury in rats. J Neurotrauma. 2004;21(9):1196-203.

[45] Bitner BR, Brink DC, Mathew LC, Pautler RG, Robertson CS. Impact of arginase II on CBF in experimental cortical impact injury in mice using MRI. J Cereb Blood Flow Metab. 2010;30(6):1105-9.

[46] Cherian L, Chacko G, Goodman JC, Robertson CS. Cerebral hemodynamic effects of phenylephrine and L-arginine after cortical impact injury. Crit Care Med. 1999;27(11):2512-7.

[47] Mendez DR, Cherian L, Robertson CS. Laser Doppler flow and brain tissue PO2 after cortical impact injury complicated by secondary ischemia in rats treated with arginine. J Trauma. 2004;57(2):244-50.

[48] Liu H, Goodman JC, Robertson CS. The effects of L-arginine on cerebral hemodynamics after controlled cortical impact injury in the mouse. J Neurotrauma. 2002;19(3):327-34.

[49] Avila MA, Sell SL, Kadoi Y, Prough DS, Hellmich HL, Velasco M, et al. L-Arginine decreases fluid-percussion injury-induced neuronal nitrotyrosine immunoreactivity in rats. J Cereb Blood Flow Metab. 2008;28(10):1733-41.

[50] DeWitt DS, Smith TG, Deyo DJ, Miller KR, Uchida T, Prough DS. L-arginine and superoxide dismutase prevent or reverse cerebral hypoperfusion after fluid-percussion traumatic brain injury. J Neurotrauma. 1997;14(4):223-33.

[51] Prough DS, Kramer GC, Uchida T, Stephenson RT, Hellmich HL, Dewitt DS. Effects of hypertonic arginine on cerebral blood flow and intracranial pressure after traumatic brain injury combined with hemorrhagic hypotension. Shock. 2006;26(3):290-5.

[52] Bower RH, Cerra FB, Bershadsky B, Licari JJ, Hoyt DB, Jensen GL, et al. Early enteral administration of a formula (Impact) supplemented with arginine, nucleotides, and fish oil in intensive care unit patients: results of a multicenter, prospective, randomized, clinical trial. Crit Care Med. 1995;23(3):436-49. Epub 1995/03/01.

[53] Trabold R, Krieg S, Schöller K, Plesnila N. Role of vasopressin V(1a) and V2 receptors for the development of secondary brain damage after traumatic brain injury in mice. J Neurotrauma. 2008;25(12):1459-65.

[54] Rauen K, Trabold R, Brem C, Terpolilli NA, Plesnila N. Arginine vasopressin V1a receptor deficient mice have reduced brain edema and secondary brain damage following traumatic brain injury. J Neurotrauma. 2013.

[55] Szmydynger-Chodobska J, Fox LM, Lynch KM, Zink BJ, Chodobski A. Vasopressin amplifies the production of proinflammatory mediators in traumatic brain injury. J Neurotrauma. 2010;27(8):1449-61.

[56] Yuan ZH, Zhu JY, Huang WD, Jiang JK, Lu YQ, Xu M, et al. Early change of plasma and cerebrospinal fluid arginine vasopressin in traumatic subarachnoid hemorrhage. Chin J Traumatol. 2010;13(1):42-5.

[57] Kleindienst A, Brabant G, Morgenthaler NG, Dixit KC, Parsch H, Buchfelder M. Following brain trauma, copeptin, a stable peptide derived from the AVP precusor, does not reflect osmoregulation but correlates with injury severity. Acta Neurochir Suppl. 2010;106:221-4.

[58] Xu M, Su W, Huang WD, Lu YQ, Xu QP, Chen ZJ. Effect of AVP on brain edema following traumatic brain injury. Chin J Traumatol. 2007;10(2):90-3.

[59] Huang WD, Pan J, Xu M, Su W, Lu YQ, Chen ZJ, et al. Changes and effects of plasma arginine vasopressin in traumatic brain injury. J Endocrinol Invest. 2008;31(11): 996-1000.

[60] Huang WD, Yang YM, Wu SD. Changes of arginine vasopressin in elderly patients with acute traumatic cerebral injury. Chin J Traumatol. 2003;6(3):139-41.

[61] Sanui M, King DR, Feinstein AJ, Varon AJ, Cohn SM, Proctor KG. Effects of arginine vasopressin during resuscitation from hemorrhagic hypotension after traumatic brain injury. Crit Care Med. 2006;34(2):433-8.

[62] Mésenge C, Charriaut-Marlangue C, Verrecchia C, Allix M, Boulu RR, Plotkine M. Reduction of tyrosine nitration after N(omega)-nitro-L-arginine-methylester treatment of mice with traumatic brain injury. Eur J Pharmacol. 1998;353(1):53-7.

[63] Wada K, Chatzipanteli K, Busto R, Dietrich WD. Effects of L-NAME and 7-NI on NOS catalytic activity and behavioral outcome after traumatic brain injury in the rat. J Neurotrauma. 1999;16(3):203-12.

[64] Gahm C, Danilov A, Holmin S, Wiklund PN, Brundin L, Mathiesen T. Reduced neuronal injury after treatment with NG-nitro-L-arginine methyl ester (L-NAME) or 2-sulfo-phenyl-N-tert-butyl nitrone (S-PBN) following experimental brain contusion. Neurosurgery. 2005;57(6):1272-81; discussion -81.

[65] Chen G, Shi J, Qi M, Yin H, Hang C. Glutamine decreases intestinal nuclear factor kappa B activity and pro-inflammatory cytokine expression after traumatic brain injury in rats. Inflamm Res. 2008;57(2):57-64.

[66] Feng D, Xu W, Chen G, Hang C, Gao H, Yin H. Influence of glutamine on intestinal inflammatory response, mucosa structure alterations and apoptosis following traumatic brain injury in rats. J Int Med Res. 2007;35(5):644-56.

[67] Petersen SR, Jeevanandam M, Holaday NJ, Lubhan CL. Arterial-jugular vein free amino acid levels in patients with head injuries: important role of glutamine in cerebral nitrogen metabolism. J Trauma. 1996;41(4):687-94; discussion 94-5.

[68] Platt SR. The role of glutamate in central nervous system health and disease--a review. Vet J. 2007;173(2):278-86.

[69] Luo P, Fei F, Zhang L, Qu Y, Fei Z. The role of glutamate receptors in traumatic brain injury: implications for postsynaptic density in pathophysiology. Brain Res Bull. 2011;85(6):313-20.

[70] Baethmann A, Maier-Hauff K, Schürer L, Lange M, Guggenbichler C, Vogt W, et al. Release of glutamate and of free fatty acids in vasogenic brain edema. J Neurosurg. 1989;70(4):578-91.

[71] Globus MY, Alonso O, Dietrich WD, Busto R, Ginsberg MD. Glutamate release and free radical production following brain injury: effects of posttraumatic hypothermia. J Neurochem. 1995;65(4):1704-11.

[72] Yamamoto T, Rossi S, Stiefel M, Doppenberg E, Zauner A, Bullock R, et al. CSF and ECF glutamate concentrations in head injured patients. Acta Neurochir Suppl. 1999;75:17-9.

[73] Rao VL, Başkaya MK, Doğan A, Rothstein JD, Dempsey RJ. Traumatic brain injury down-regulates glial glutamate transporter (GLT-1 and GLAST) proteins in rat brain. J Neurochem. 1998;70(5):2020-7.

[74] Matsushita Y, Shima K, Nawashiro H, Wada K. Real-time monitoring of glutamate following fluid percussion brain injury with hypoxia in the rat. J Neurotrauma. 2000;17(2):143-53.

[75] Matsushita Y, Shima K, Nawashiro H, Wada K, Tsuzuki N, Miyazawa T. Real time monitoring of glutamate following fluid percussion brain injury with hypoxia in the rat. Acta Neurochir Suppl. 2000;76:207-12.

[76] Stover JF, Schöning B, Beyer TF, Woiciechowsky C, Unterberg AW. Temporal profile of cerebrospinal fluid glutamate, interleukin-6, and tumor necrosis factor-alpha in relation to brain edema and contusion following controlled cortical impact injury in rats. Neurosci Lett. 2000;288(1):25-8.

[77] Ros J, Jones D, Pecinska N, Alessandri B, Boutelle M, Landolt H, et al. Glutamate infusion coupled with hypoxia has a neuroprotective effect in the rat. J Neurosci Methods. 2002;119(2):129-33.

[78] van Landeghem FK, Stover JF, Bechmann I, Brück W, Unterberg A, Bührer C, et al. Early expression of glutamate transporter proteins in ramified microglia after controlled cortical impact injury in the rat. Glia. 2001;35(3):167-79.

[79] Stover JF, Unterberg AW. Increased cerebrospinal fluid glutamate and taurine concentrations are associated with traumatic brain edema formation in rats. Brain Res. 2000;875(1-2):51-5.

[80] Vespa P, Prins M, Ronne-Engstrom E, Caron M, Shalmon E, Hovda DA, et al. Increase in extracellular glutamate caused by reduced cerebral perfusion pressure and seizures after human traumatic brain injury: a microdialysis study. J Neurosurg. 1998;89(6):971-82.

[81] Palmer AM, Marion DW, Botscheller ML, Redd EE. Therapeutic hypothermia is cytoprotective without attenuating the traumatic brain injury-induced elevations in interstitial concentrations of aspartate and glutamate. J Neurotrauma. 1993;10(4):363-72.

[82] van Landeghem FK, Weiss T, Oehmichen M, von Deimling A. Decreased expression of glutamate transporters in astrocytes after human traumatic brain injury. J Neurotrauma. 2006;23(10):1518-28.

[83] Zlotnik A, Gurevich B, Tkachov S, Maoz I, Shapira Y, Teichberg VI. Brain neuroprotection by scavenging blood glutamate. Exp Neurol. 2007;203(1):213-20.

[84] Cao R, Hasuo H, Ooba S, Akasu T, Zhang X. Facilitation of glutamatergic synaptic transmission in hippocampal CA1 area of rats with traumatic brain injury. Neurosci Lett. 2006;401(1-2):136-41.

[85] Yi JH, Hazell AS. Excitotoxic mechanisms and the role of astrocytic glutamate transporters in traumatic brain injury. Neurochem Int. 2006;48(5):394-403.

[86] Muir KW. Glutamate-based therapeutic approaches: clinical trials with NMDA antagonists. Curr Opin Pharmacol. 2006;6(1):53-60.

[87] Mukhin A, Fan L, Faden AI. Activation of metabotropic glutamate receptor subtype mGluR1 contributes to post-traumatic neuronal injury. J Neurosci. 1996;16(19): 6012-20.

[88] Yi JH, Herrero R, Chen G, Hazell AS. Glutamate transporter EAAT4 is increased in hippocampal astrocytes following lateral fluid-percussion injury in the rat. Brain Res. 2007;1154:200-5.

[89] Hinzman JM, Thomas TC, Burmeister JJ, Quintero JE, Huettl P, Pomerleau F, et al. Diffuse brain injury elevates tonic glutamate levels and potassium-evoked glutamate release in discrete brain regions at two days post-injury: an enzyme-based microelectrode array study. J Neurotrauma. 2010;27(5):889-99.

[90] Dai SS, Zhou YG, Li W, An JH, Li P, Yang N, et al. Local glutamate level dictates adenosine A2A receptor regulation of neuroinflammation and traumatic brain injury. J Neurosci. 2010;30(16):5802-10.

[91] Chamoun R, Suki D, Gopinath SP, Goodman JC, Robertson C. Role of extracellular glutamate measured by cerebral microdialysis in severe traumatic brain injury. J Neurosurg. 2010;113(3):564-70.

[92] Allen JW, Ivanova SA, Fan L, Espey MG, Basile AS, Faden AI. Group II metabotropic glutamate receptor activation attenuates traumatic neuronal injury and improves neurological recovery after traumatic brain injury. J Pharmacol Exp Ther. 1999;290(1): 112-20.

[93] Zlotnik A, Sinelnikov I, Gruenbaum BF, Gruenbaum SE, Dubilet M, Dubilet E, et al. Effect of glutamate and blood glutamate scavengers oxaloacetate and pyruvate on neurological outcome and pathohistology of the hippocampus after traumatic brain injury in rats. Anesthesiology. 2012;116(1):73-83.

[94] Maxwell WL, Bullock R, Landholt H, Fujisawa H. Massive astrocytic swelling in response to extracellular glutamate--a possible mechanism for post-traumatic brain swelling? Acta Neurochir Suppl (Wien). 1994;60:465-7.

[95] Meldrum BS. The role of glutamate in epilepsy and other CNS disorders. Neurology. 1994;44(11 Suppl 8):S14-23.

[96] Yang DL, Xu JF. Effect of dipeptide of glutamine and alanine on severe traumatic brain injury. Chin J Traumatol. 2007;10(3):145-9.

[97] Tani H, Bandrowski AE, Parada I, Wynn M, Huguenard JR, Prince DA, et al. Modulation of epileptiform activity by glutamine and system A transport in a model of post-traumatic epilepsy. Neurobiol Dis. 2007;25(2):230-8.

[98] Heyland DK, Dhaliwal R. Role of Glutamine Supplementation in Critical Illness Given the Results of the REDOXS Study. JPEN Journal of parenteral and enteral nutrition. 2013;37(4):442-3.

[99] Slemmer JE, Shacka JJ, Sweeney MI, Weber JT. Antioxidants and free radical scavengers for the treatment of stroke, traumatic brain injury and aging. Curr Med Chem. 2008;15(4):404-14.

[100] Hall ED, Yonkers PA, Andrus PK, Cox JW, Anderson DK. Biochemistry and pharmacology of lipid antioxidants in acute brain and spinal cord injury. J Neurotrauma. 1992;9 Suppl 2:S425-42.

[101] Hall ED, Vaishnav RA, Mustafa AG. Antioxidant therapies for traumatic brain injury. Neurotherapeutics. 2010;7(1):51-61.

[102] Ikeda K, Negishi H, Yamori Y. Antioxidant nutrients and hypoxia/ischemia brain injury in rodents. Toxicology. 2003;189(1-2):55-61.

[103] Yeo JE, Kang SK. Selenium effectively inhibits ROS-mediated apoptotic neural precursor cell death in vitro and in vivo in traumatic brain injury. Biochim Biophys Acta. 2007;1772(11-12):1199-210.

[104] Falcão de Arruda IS, de Aguilar-Nascimento JE. Benefits of early enteral nutrition with glutamine and probiotics in brain injury patients. Clin Sci (Lond). 2004;106(3):287-92.

[105] Polidori MC, Mecocci P, Frei B. Plasma vitamin C levels are decreased and correlated with brain damage in patients with intracranial hemorrhage or head trauma. Stroke. 2001;32(4):898-902.

[106] Gilgun-Sherki Y, Rosenbaum Z, Melamed E, Offen D. Antioxidant therapy in acute central nervous system injury: current state. Pharmacol Rev. 2002;54(2):271-84.

[107] Razmkon A, Sadidi A, Sherafat-Kazemzadeh E, Mehrafshan A, Jamali M, Malekpour B, et al. Administration of vitamin C and vitamin E in severe head injury: a randomized double-blind controlled trial. Clin Neurosurg. 2011;58:133-7.

[108] Bullock MR, Lyeth BG, Muizelaar JP. Current status of neuroprotection trials for traumatic brain injury: lessons from animal models and clinical studies. Neurosurgery. 1999;45(2):207-17; discussion 17-20.

[109] Marshall LF, Maas AI, Marshall SB, Bricolo A, Fearnside M, Iannotti F, et al. A multicenter trial on the efficacy of using tirilazad mesylate in cases of head injury. J Neurosurg. 1998;89(4):519-25.

[110] Muizelaar JP, Marmarou A, Young HF, Choi SC, Wolf A, Schneider RL, et al. Improving the outcome of severe head injury with the oxygen radical scavenger polyethylene glycol-conjugated superoxide dismutase: a phase II trial. J Neurosurg. 1993;78(3):375-82.

[111] Bains M, Hall ED. Antioxidant therapies in traumatic brain and spinal cord injury. Biochim Biophys Acta. 2012;1822(5):675-84.

[112] Mustafa AG, Singh IN, Wang J, Carrico KM, Hall ED. Mitochondrial protection after traumatic brain injury by scavenging lipid peroxyl radicals. J Neurochem. 2010;114(1):271-80.

[113] Jeter CB, Hergenroeder GW, Ward NH, Moore AN, Dash PK. Human mild traumatic brain injury decreases circulating branched-chain amino acids and their metabolite levels. J Neurotrauma. 2013;30(8):671-9.

[114] Cole JT, Sweatt AJ, Hutson SM. Expression of mitochondrial branched-chain aminotransferase and α-keto-acid dehydrogenase in rat brain: implications for neurotransmitter metabolism. Front Neuroanat. 2012;6:18.

[115] Cole JT, Mitala CM, Kundu S, Verma A, Elkind JA, Nissim I, et al. Dietary branched chain amino acids ameliorate injury-induced cognitive impairment. Proc Natl Acad Sci U S A. 2010;107(1):366-71.

[116] Aquilani R, Iadarola P, Contardi A, Boselli M, Verri M, Pastoris O, et al. Branched-chain amino acids enhance the cognitive recovery of patients with severe traumatic brain injury. Arch Phys Med Rehabil. 2005;86(9):1729-35.

[117] Aquilani R, Boselli M, Boschi F, Viglio S, Iadarola P, Dossena M, et al. Branched-chain amino acids may improve recovery from a vegetative or minimally conscious state in patients with traumatic brain injury: a pilot study. Arch Phys Med Rehabil. 2008;89(9):1642-7.

[118] Vuille-Dit-Bille RN, Ha-Huy R, Stover JF. Changes in plasma phenylalanine, isoleucine, leucine, and valine are associated with significant changes in intracranial pressure and jugular venous oxygen saturation in patients with severe traumatic brain injury. Amino Acids. 2012;43(3):1287-96.

[119] Guseva MV, Hopkins DM, Scheff SW, Pauly JR. Dietary choline supplementation improves behavioral, histological, and neurochemical outcomes in a rat model of traumatic brain injury. J Neurotrauma. 2008;25(8):975-83.

[120] Blusztajn JK. Choline, a vital amine. Science. 1998;281(5378):794-5.

[121] Caamano J, Gomez MJ, Franco A, Cacabelos R. Effects of CDP-choline on cognition and cerebral hemodynamics in patients with Alzheimer's disease. Methods and findings in experimental and clinical pharmacology. 1994;16(3):211-8. Epub 1994/04/01.

[122] Dixon CE, Ma X, Marion DW. Effects of CDP-choline treatment on neurobehavioral deficits after TBI and on hippocampal and neocortical acetylcholine release. J Neurotrauma. 1997;14(3):161-9.

[123] Ruff RL, Riechers RG. Effective treatment of traumatic brain injury: learning from experience. JAMA. 2012;308(19):2032-3.

[124] Signoretti S, Di Pietro V, Vagnozzi R, Lazzarino G, Amorini AM, Belli A, et al. Transient alterations of creatine, creatine phosphate, N-acetylaspartate and high-energy phosphates after mild traumatic brain injury in the rat. Mol Cell Biochem. 2010;333(1-2):269-77.

[125] Zhu S, Li M, Figueroa BE, Liu A, Stavrovskaya IG, Pasinelli P, et al. Prophylactic creatine administration mediates neuroprotection in cerebral ischemia in mice. J Neurosci. 2004;24(26):5909-12.

[126] Sullivan PG, Geiger JD, Mattson MP, Scheff SW. Dietary supplement creatine protects against traumatic brain injury. Ann Neurol. 2000;48(5):723-9.

[127] Saraiva AL, Ferreira AP, Silva LF, Hoffmann MS, Dutra FD, Furian AF, et al. Creatine reduces oxidative stress markers but does not protect against seizure susceptibility after severe traumatic brain injury. Brain Res Bull. 2012;87(2-3):180-6.

[128] Burov S, Leko M, Dorosh M, Dobrodumov A, Veselkina O. Creatinyl amino acids: new hybrid compounds with neuroprotective activity. J Pept Sci. 2011;17(9):620-6.

[129] Scheff SW, Dhillon HS. Creatine-enhanced diet alters levels of lactate and free fatty acids after experimental brain injury. Neurochem Res. 2004;29(2):469-79.

[130] Hoane MR, Gilbert DR, Barbre AB, Harrison SA. Magnesium dietary manipulation and recovery of function following controlled cortical damage in the rat. Magnes Res. 2008;21(1):29-37.

[131] Vink R, McIntosh TK, Demediuk P, Faden AI. Decrease in total and free magnesium concentration following traumatic brain injury in rats. Biochem Biophys Res Commun. 1987;149(2):594-9.

[132] Vink R, McIntosh TK. Pharmacological and physiological effects of magnesium on experimental traumatic brain injury. Magnes Res. 1990;3(3):163-9.

[133] van den Heuvel C, Vink R. The role of magnesium in traumatic brain injury. Clin Calcium. 2004;14(8):9-14.

[134] Garfinkel L, Garfinkel D. Magnesium regulation of the glycolytic pathway and the enzymes involved. Magnesium. 1985;4(2-3):60-72.

[135] Feldman Z, Gurevitch B, Artru AA, Oppenheim A, Shohami E, Reichenthal E, et al. Effect of magnesium given 1 hour after head trauma on brain edema and neurological outcome. J Neurosurg. 1996;85(1):131-7.

[136] Hoane MR. Magnesium therapy and recovery of function in experimental models of brain injury and neurodegenerative disease. Clin Calcium. 2004;14(8):65-70.

[137] Heath DL, Vink R. Optimization of magnesium therapy after severe diffuse axonal brain injury in rats. J Pharmacol Exp Ther. 1999;288(3):1311-6.

[138] Bareyre FM, Saatman KE, Raghupathi R, McIntosh TK. Postinjury treatment with magnesium chloride attenuates cortical damage after traumatic brain injury in rats. J Neurotrauma. 2000;17(11):1029-39.

[139] Heath DL, Vink R. Improved motor outcome in response to magnesium therapy received up to 24 hours after traumatic diffuse axonal brain injury in rats. J Neurosurg. 1999;90(3):504-9.

[140] Vink R, O'Connor CA, Nimmo AJ, Heath DL. Magnesium attenuates persistent functional deficits following diffuse traumatic brain injury in rats. Neurosci Lett. 2003;336(1):41-4.

[141] Temkin NR, Anderson GD, Winn HR, Ellenbogen RG, Britz GW, Schuster J, et al. Magnesium sulfate for neuroprotection after traumatic brain injury: a randomised controlled trial. Lancet Neurol. 2007;6(1):29-38.

[142] Dhandapani S, Gupta A, Vivekanandhan S, Sharma B, Mahapatra A. Randomized controlled trial of magnesium sulphate in severe closed traumatic brain injury. The Indian Journal of Neurotrauma; 2008. p. 27-33.

[143] Arango MF, Bainbridge D. Magnesium for acute traumatic brain injury. Cochrane Database Syst Rev. 2008(4):CD005400.

[144] Aminmansour B, Nikbakht H, Ghorbani A, Rezvani M, Rahmani P, Torkashvand M, et al. Comparison of the administration of progesterone versus progesterone and vitamin D in improvement of outcomes in patients with traumatic brain injury: A randomized clinical trial with placebo group. Adv Biomed Res. 2012;1:58.

[145] Atif F, Sayeed I, Ishrat T, Stein DG. Progesterone with vitamin D affords better neuroprotection against excitotoxicity in cultured cortical neurons than progesterone alone. Mol Med. 2009;15(9-10):328-36.

[146] Cekic M, Cutler SM, VanLandingham JW, Stein DG. Vitamin D deficiency reduces the benefits of progesterone treatment after brain injury in aged rats. Neurobiol Aging. 2011;32(5):864-74.

[147] Cekic M, Sayeed I, Stein DG. Combination treatment with progesterone and vitamin D hormone may be more effective than monotherapy for nervous system injury and disease. Front Neuroendocrinol. 2009;30(2):158-72.

[148] Cekic M, Stein DG. Traumatic brain injury and aging: is a combination of progesterone and vitamin D hormone a simple solution to a complex problem? Neurotherapeutics. 2010;7(1):81-90.

[149] Stein DG, Cekic MM. Progesterone and vitamin d hormone as a biologic treatment of traumatic brain injury in the aged. PM R. 2011;3(6 Suppl 1):S100-10.

[150] Ekin S, Berber I, Kiymaz N. Effects of dexamethasone on trace elements and serum protein patterns following brain trauma in rats. Biol Trace Elem Res. 2005;107(1):53-60.

[151] Doering P, Stoltenberg M, Penkowa M, Rungby J, Larsen A, Danscher G. Chemical blocking of zinc ions in CNS increases neuronal damage following traumatic brain injury (TBI) in mice. PLoS One. 2010;5(4):e10131.

[152] Hellmich HL, Eidson K, Cowart J, Crookshanks J, Boone DK, Shah S, et al. Chelation of neurotoxic zinc levels does not improve neurobehavioral outcome after traumatic brain injury. Neurosci Lett. 2008;440(2):155-9.

[153] Cope EC, Morris DR, Levenson CW. Improving treatments and outcomes: an emerging role for zinc in traumatic brain injury. Nutr Rev. 2012;70(7):410-3.

[154] Levenson CW. Trace metal regulation of neuronal apoptosis: from genes to behavior. Physiol Behav. 2005;86(3):399-406.

[155] Levenson CW. Zinc supplementation: neuroprotective or neurotoxic? Nutr Rev. 2005;63(4):122-5.

[156] Hellmich HL, Eidson KA, Capra BA, Garcia JM, Boone DR, Hawkins BE, et al. Injured Fluoro-Jade-positive hippocampal neurons contain high levels of zinc after traumatic brain injury. Brain Res. 2007;1127(1):119-26.

[157] Suh SW, Chen JW, Motamedi M, Bell B, Listiak K, Pons NF, et al. Evidence that synaptically-released zinc contributes to neuronal injury after traumatic brain injury. Brain Res. 2000;852(2):268-73.

[158] Suh SW, Frederickson CJ, Danscher G. Neurotoxic zinc translocation into hippocampal neurons is inhibited by hypothermia and is aggravated by hyperthermia after traumatic brain injury in rats. J Cereb Blood Flow Metab. 2006;26(2):161-9.

[159] Sun KJ, Zhu L, Wang HD, Ji XJ, Pan H, Chen M, et al. Zinc as mediator of ubiquitin conjugation following traumatic brain injury. Brain Res. 2013;1506:132-41.

[160] Zhu L, Wang HD, Yu XG, Jin W, Qiao L, Lu TJ, et al. Erythropoietin prevents zinc accumulation and neuronal death after traumatic brain injury in rat hippocampus: in vitro and in vivo studies. Brain Res. 2009;1289:96-105.

[161] Cope EC, Morris DR, Scrimgeour AG, Levenson CW. Use of zinc as a treatment for traumatic brain injury in the rat: effects on cognitive and behavioral outcomes. Neurorehabil Neural Repair. 2012;26(7):907-13.

[162] Li Y, Hawkins BE, DeWitt DS, Prough DS, Maret W. The relationship between transient zinc ion fluctuations and redox signaling in the pathways of secondary cellular injury: relevance to traumatic brain injury. Brain Res. 2010;1330:131-41.

[163] Cope EC, Morris DR, Scrimgeour AG, VanLandingham JW, Levenson CW. Zinc supplementation provides behavioral resiliency in a rat model of traumatic brain injury. Physiol Behav. 2011;104(5):942-7.

[164] Yeiser EC, Vanlandingham JW, Levenson CW. Moderate zinc deficiency increases cell death after brain injury in the rat. Nutr Neurosci. 2002;5(5):345-52.

[165] Penkowa M, Giralt M, Thomsen PS, Carrasco J, Hidalgo J. Zinc or copper deficiency-induced impaired inflammatory response to brain trauma may be caused by the concomitant metallothionein changes. J Neurotrauma. 2001;18(4):447-63.

[166] Zhu L, Yan W, Qi M, Hu ZL, Lu TJ, Chen M, et al. Alterations of pulmonary zinc homeostasis and cytokine production following traumatic brain injury in rats. Ann Clin Lab Sci. 2007;37(4):356-61.

[167] Potts MB, Koh SE, Whetstone WD, Walker BA, Yoneyama T, Claus CP, et al. Traumatic injury to the immature brain: inflammation, oxidative injury, and iron-mediated damage as potential therapeutic targets. NeuroRx. 2006;3(2):143-53.

[168] Chang EF, Claus CP, Vreman HJ, Wong RJ, Noble-Haeusslein LJ. Heme regulation in traumatic brain injury: relevance to the adult and developing brain. J Cereb Blood Flow Metab. 2005;25(11):1401-17.

Chapter 12

Evidence-Based Treatment of Chronic Subdural Hematoma

Jehuda Soleman, Philipp Taussky,
Javier Fandino and Carl Muroi

Additional information is available at the end of the chapter

http://dx.doi.org/10.5772/57336

1. Introduction

Chronic subdural hematoma (cSDH) is one of the most frequent neurosurgical entities caused by head trauma. Since cSDH affects mainly elderly patients and the population continues to age, it has become a common neurosurgical disease seen by both general and specialized health-care practitioners. Despite the increasing prevalence of cSDH, class I studies, and evidence regarding the management of this disease is lacking. We provide an overview of the epidemiology, pathophysiology and etiology of cSDH and discuss several controversial aspects of its management; including indication and timing of surgery, steroid treatment, the effectiveness of anti-epileptic prophylaxis, comparative effectiveness of various techniques for surgical evacuation, the timing of postoperative resumption of anticoagulant medication, and protocols for mobilization following evacuation of cSDH. Complications of surgical evacuation such as recurrent hematoma, postoperative epilepsy, brain injury and/or iatrogenic intracerebral bleeding due to hematoma evacuation, drainage insertion or irrigation, and ways to avoid them are also discussed. As the incidence of cSDH is expected to increase and most treatment aspects lack clear consensus, further large prospective studies are needed. For this reason, a randomized, prospective study evaluating one aspect of the management of cSDH is currently in progress at our institution.

2. Definition

A chronic subdural hematoma (cSDH) is defined as chronic (≥3 weeks) intracranial bleeding between the dura mater (which adheres to the skull), and the arachnoid mater (which envelops the brain). The underlying cause of cSDH is usually traumatic tearing of the bridging veins which connect the brain surface with the dura mater.[1].

3. Epidemiology

The incidence of cSDH is estimated at 1.7-18 per 100’000 people, rising up to 58 per 100’000 people in patients above the age of 65 [1-4]. The average age of patients with cSDH is approximately 63 years old [5]. As the population continues to mature, incidence is expected to double by the year 2030 [6, 7]. A large demographic study found the prevalence of cSDH in patients older than 65 to be significantly higher (69% vs. 31%) [8]. In addition, men are more frequently affected than women (64% vs. 33%). In 77% of the cases, the patient has suffered a fall in the past and 41% of the patients were either treated with oral anticoagulants or platelet aggregation inhibitors. The reported recurrence rates range from 2.3% to 33% [8-11]. The most common risk factors are: advanced age, alcohol abuse, seizures, cerebrospinal fluid (CSF) shunts, coagulopathies, blood thinners, and patients at risk for falling (e.g. hemiplegia). In 20-25% of the cases, cSDHs are bilateral [5]. cSDH remains one the most frequent diagnoses in neurosurgical practice.

4. Pathophysiology

The entity of cSDH was first described by Virchow in 1857 [12]. He named it “pachymeningitis haemorrhagica interna”, recognizing its inflammatory and hemorrhagic elements [12]. Interestingly, the subdural space is a virtual space which does not exist in healthy individuals, as the dura and arachnoid are tethered by a layer of dural border cells (DBC)[1, 7, 13, 14]. The DBC is characterized by a paucity of tight junctions and an enlarged extracellular space containing amorphous material [7, 14] (Figure 1).

With increasing brain atrophy, the arachnoid is pulled away from the dural layer, which remains attached to the skull. The resultant force stretches the DBC layer and the veins traversing it (bridging veins). Any minor additional force can cause these veins to tear, causing a leakage of blood into the DBC and creating an acute SDH. Therefore, the majority of cSDHs are caused by an undiagnosed trivial head injury, primarily in patients with brain atrophy. This trauma leads to a minor acute SDH. Today, it is widely accepted that cSDHs are a result of the failure of small acute SDH to heal. Following the initial trauma and development of a cSDH, fibrin deposition occurs, followed by organization, enzymatic fibrinolysis and liquefaction of the hematoma. The blood in the subdural space triggers an inflammatory response. After approximately two weeks, an inner (cortical) and outer (dural) neomembrane is formed inside the DBC layer through dural collagen synthesis and fibroblast spread [1, 15, 16]. Ingrowth of fragile neocapillaries into the neomembranes of the hematoma can lead to further microbleeds within the subdural space [1]. Less commonly, the SDH may result from arterial ruptures (20-30%), hemorrhage into an existing subdural hygroma or spontaneously, mostly influenced by anticoagulants or antiplatelet therapy [1, 17, 18].

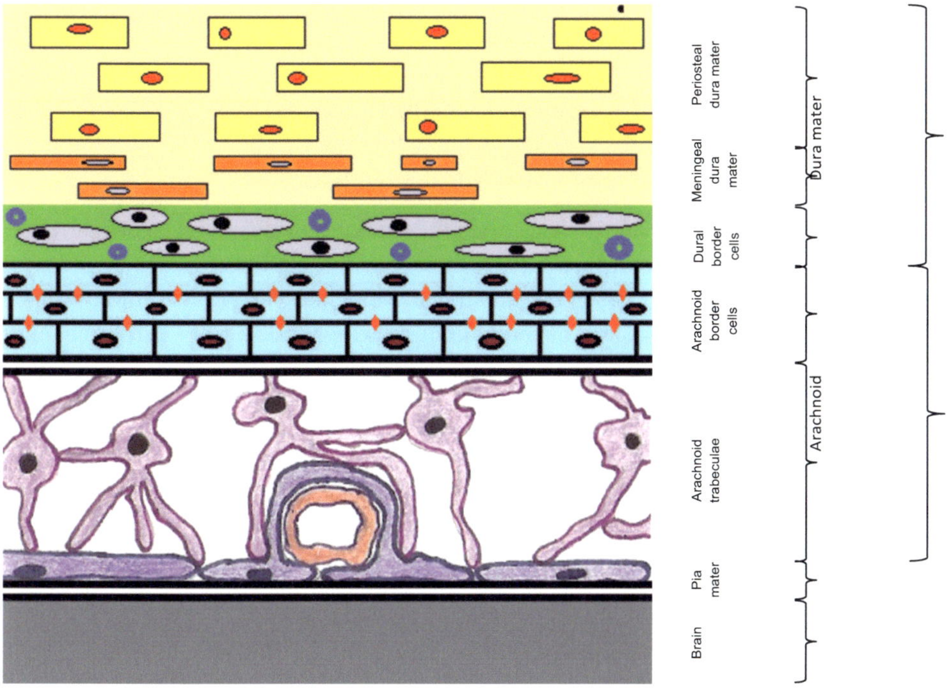

Figure 1. Schematic representation of the ultrastructure of the meninges (adapted from Haines and Santarius et al [7, 14]. The dura mater is composed of fibroblasts and large amounts of collagen. The arachnoid barrier cells are supported by a basement membrane and bound together by numerous tight junctions (red diamonds). The dural border cell layer (green) is formed by flattened fibroblasts with no tight junctions and no intercellular collagen. It is therefore a relatively loose layer positioned between the firm dura matter and arachnoid. The subdural space is a potential space that can form within the dural border cell layer. The bridging veins passing through the dural border cell layer are a potential source of bleeding.

The factors responsible for the maintenance or enlargement of cSDH over time are still ambiguous. It is most likely influenced by multiple factors, which vary from case to case. Over the years, several theories have been debated:

a. "Osmotic theory": The initial acute hematoma resorbs through fibrinolysis and the remaining resorption products within the subdural space lead to an elevated osmotic gradient. Due to the osmotic pressure, CSF follows the osmotic gradient and drifts into the subdural space, leading to an expansion of the hematoma [19, 20].

b. "Oncotic theory": Due to the low oncotic pressure inside the hematoma capsule, blood permeates from the dural vessels into the subdural space, leading to an expansion of the hematoma [16].

c. "Microbleeds theory": As they lack a muscle layer and periocytes, the neocapillaries forming inside the neomembrane are fragile. This leads to repeated microhemorrhaging into the subdural space and expansion of the hematoma [1, 7, 21].

d. "Anticoagulant and profibrinolytic theory": Under normal circumstances capillary leaks are stopped by blood clots. However, the neomembrane surrounding the SDH is saturated with profibrinolytic and anticoagulation factors. Different studies have shown acceleration of fibrinolysis [22], high levels of tissue plasminogen activator [23, 24], and high concentration of fibrin degradation products within the subdural fluid. All these factors obviate hemostasis within the subdural hematoma.

e. "Inflammatory and growth factors theory": Inflammation leads to high concentrations of vascular endothelial growth factor (VEGF) within the subdural fluid. VEGF receptor subtype I was seen in the cells of the neomembrane. These factors within the hematoma lead to an increased promotion of the ongoing angiogenesis and hyperpermeability in cSDHs. Since VEGF increases the permeability of capillaries, it contributes directly to the expansion of the hematoma [25-27].

As stated above, current evidence suggests that the maintenance or enlargement of cSDH is caused by multiple factors. The stimulus is probably a mixture of the "microbleed theory", the "anticoagulant and profibrinolytic theory", and the "inflammatory and growth factors theory". These theories are currently accepted, while the "osmotic-" and "oncotic theory" has been largely abandoned [1, 7, 28, 29].

5. Etiology/risk factors

Trauma

The most common cause of cSDH is a traumatic event, mostly a minor head trauma. In most large series, approximately two thirds of the patients have suffered one [2, 30]. Reports exist of cSDHs due to birth trauma in neonates [2, 31].

Advanced age

The elderly are at risk, firstly due to brain atrophy, whereby the bridging veins are stretched and have become more fragile. Secondly, older people tend to fall more often and suffer minor head traumas. Lastly, with increasing age, the incidence of blood thinner administration rises leading to increased risk for hemorrhage.

Chronic Alcoholism

Alcoholism is well known to be associated with cSDH. The reason for the greater propensity of patients with chronic alcohol addiction for hematoma formation is unknown. Yet, higher trauma rate, brain atrophy, and coagulopathy (secondary to liver malfunction), most definitely play a major role in cSDH development in these patients.

Gender

Men, from all age groups, suffer disproportionally higher rates of cSDH than women [2, 7, 32]. The underlying reasons for this discrepancy are not known but may be because they are more likely to suffer trauma [2].

Coagulopathy

Coagulopathies, including therapeutic anticoagulation and antiplatelet therapy, are known contributors to the pathogenesis of cSDH. Medical conditions include: sepsis, hepatic failure, all forms of hemophilia, disseminated intravascular coagulation, and renal dialysis [2, 5].

Intracranial hypotension

Intracranial hypotension, as a result of overshunting after placement of a ventriculoperitoneal (VP) shunt, or CSF leak (iatrogenic or spontaneous) are less common causes of cSDH. Overshunting results in increased retraction of the bridging veins, leading to a higher likelihood of hematoma formation. Subdural hematomas may result in up to 8% of the patients shunted for normal-pressure hydrocephalus [33]. Since the introduction of adjustable-pressure valves, successful management of shunt-related cSDHs has been reported [34, 35]. Even with adjustable-pressure valves, intracranial hypotension resulting from overshunting is still a significant problem. The most common causes of a CSF leak leading to cSDH are traumatic or postoperative CSF fistulas, lumbar puncture or drainage, iatrogenic or disease-induced dehydration or spontaneous events [36]. Overall, it is a very rare condition.

Other causes

Other rare causes of spontaneous subdural hematomas have been described: vascular malformation (e.g. cerebral aneurysms and arterial venous-malformation) [37, 38], benign (e.g. convexity meningiomas) and malignant tumors [39-41], carcinomatosis/sarcomatosis meningiosa, and infections (e.g. meningitis and tuberculosis) [5].

6. Classifications

A *clinical* classification of cSDH was presented by Markwalder in 1981 [42]:

- grade 0: asymptomatic
- grade 1: alert, oriented, mild symptoms (e.g. headache)
- grade 2: drowsy or disoriented, variable neurological deficits (e.g. hemiparesis)
- grade 3: stupor but responds to stimuli, severe focal signs (e.g. hemiplegia)
- grade 4: coma, flexes/extends to pain

This grading system is used to pre- and postoperatively evaluate the clinical course of the patient.

Other scores frequently used for follow-up evaluation of cSDH patients include the Glasgow Coma Scale (GCS) [43], the Glasgow Outcome Score [44], and the modified Rankin scale (mRS) [45].

A *radiological* classification of the internal architecture of the hematoma, corresponding to possible stages in the natural history of cSDH, was suggested by Nakaguchi and colleagues in 2001 [9, 46]:

- Homogeneous type was defined as a hematoma exhibiting homogeneous high-density.
- Laminar type was defined as a subtype of the homogeneous type with a thin high-density layer along the inner membrane.
- Separated type was defined as a hematoma containing two components with different densities with a clear boundary between them; that is, a lower density component located above a higher density component.
- Trabecular type was defined as a hematoma with inhomogeneous contents and a high-density septum running between the inner and outer membrane on a low-density to isodense background

This scale can be helpful in predicting the recurrence rate based on the internal architecture of the hematoma as seen on computerized tomography. Recurrence rate was shown to be lower in the homogeneous and trabecular type [46].

7. Clinical presentation

Injuries associated with cSDHs tend to be minor, without any accompanying severe brain injury. The accumulation of blood within the subdural space occurs slowly, over an extended period (weeks-months), and under comparatively low pressure [7]. The coexistence of brain atrophy, in most cases, and the slow formation and expansion of the hematoma, allows the hematoma to reach a large size without triggering neurologic symptoms. Symptoms arise when the pressure caused by the hematoma leads to a compression of the cortex that cannot be tolerated or compensated anymore. In selected cases these symptoms manifest in a dramatic and acute fashion (e.g. coma) and might even lead to death. Acute deterioration can occur with small changes in hematoma volume when significant mass effect is already present. It is also not unusual for acute deterioration to occur secondary to acute bleeding within the subdural space, with a preexisting cSDH.

Therefore patients with cSDH can be asymptomatic, have very mild symptoms such as headache, nausea, vomiting, vertigo, fatigue, confusion, gait disturbance, mental deterioration, limb weakness, incontinence, or language difficulties (e.g. word-finding difficulties), or present with acute and grave symptoms such as hemiplegia, seizures, or coma (Table 1).

8. Diagnostic work-up

After assessing the patient's history - including previous falls, minor head injuries, onset and course of clinical symptoms, cardiovascular disease, coagulopathies, medication, alcohol or

general symptoms (frequent)	mild symptoms (frequent)	severe/acute symptoms (rare)
headache	confusion	hemiplegia
fatigue	gait disturbance	aphasia
nausea/vomiting	mental deterioration	seizure
vertigo	language difficulties	coma
	limb weakness	
	incontinence	

Table 1. Clinical presentation of patients with cSDH

drug abuse - and completing the physical examination and blood work-up, brain imaging should be conducted to reach a diagnosis. Computed tomography (CT) is the most important imaging method in the initial evaluation of cSDHs [36, 47]. The best diagnostic signal is a crescent-shaped iso- to hypodens extra-axial collection on CT. Typically, the hematoma is spread over the whole effected hemisphere [48]. In addition to the enhancement of encapsulating membranes, those leading to septation within the collection may be seen [48]. The hematoma density varies depending on the stage of hematoma evolution. An acute SDH (<3 days old; hyperdens on native CT), progresses over ±3 weeks to a subacute SDH [3 days to 3 weeks old; isodens on native CT) and finally to a cSDH (≥3 weeks old, hypodens on native CT) (Figure 2) [48]. Mixed hematomas containing acute, subacute and chronic shares are often present. Special attention should be paid to isodens subacute SDH since it might be missed on the initial scan. Magnetic resonance imaging (MRI) is more accurate than CT; hematoma thickness can be measured reliably, as isodense and small cSDHs are easier to identify. In almost all cases, hematoma membranes are detected on an MRI, but only 27% are discovered on CT scans [47]. Nevertheless, CT is still preferred for the diagnosis of cSDH as it is cheaper, more accessible, and faster. MRI, when available, is useful in delineating the exact margins of the cSDH and determining the internal hematoma structures [47]. Table 2 summarizes CT and MRI findings in cSDH.

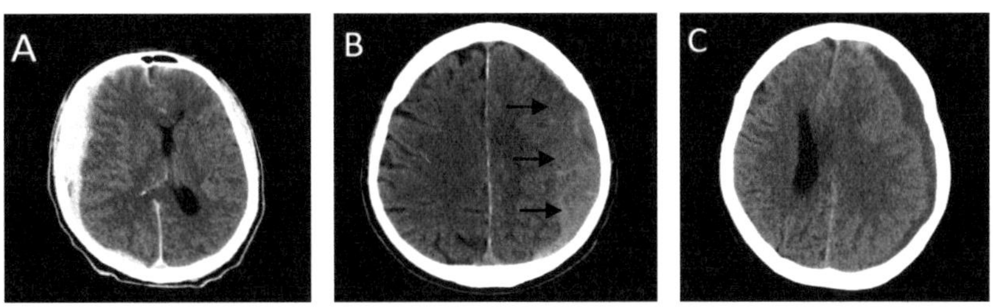

Figure 2. CT images with an acute SDH, <3 days old, hyperdens (A), subacute SDH, 3 days to 3 weeks old, isodens (B), and cSDH, >3 weeks old, hypodens (C)

	CT
Native CT	Iso - or hypodens
Contrast CT	Inward displacement of enhancing cortical vessels Enhancement of dura and membranes
	MRI
T1 WI	Isointense to CSF if chronic (no active/acute rebleeds) Hypointens with active/acute rebleeds or elevated proteins
Contrast TI WI	Peripheral and/or dural enhancement Delayed scans show contrast diffusion into cSDH
T2 WI	Variable, depending on evolution stage Isointense to CSF if chronic (no active/acute rebleeds) Hypointens with active/acute rebleeds
FLAIR	Hyperintens to CSF Most sensitive sequence to detect cSDH
DWI	Variable signal

T1 WI: T1 weighted imaging, T2 WI: T2 weighted imaging, FLAIR:, DWI: diffusion weighted imaging

Table 2. Characteristic findings of cSDH on CT and MRI, adapted from Osborn et al. [48]

9. Management

The decision to evacuate a cSDH is influenced by both the clinical presentation of the patient and the radiographic appearance of the lesion. An asymptomatic patient with a small cSDH is often best observed clinically and radiologically, in a carefully monitored setting. Although the size of a cSDH may play a role in the decision to perform surgery, absolute cutoffs sizes do not exist. Moreover, spontaneous resolution of cSDH with significant thickness has been reported - only in a small number of case series, in elderly patients (>70 years) with brain atrophy and without clinical or radiographic evidence of increased intracranial pressure [49, 50]. Other conservative treatments using ACE inhibitors or corticosteroids were reviewed, but justification for treatment using ACE inhibitors or steroids has largely been theoretical, and further research is clearly warranted. It is generally accepted that in the presence of neurologic symptoms and radiologic findings, patients should undergo surgical evacuation. The paucity of quality data from well-designed studies makes it difficult to identify the most effective surgical approach for cSDH. While surgical drainage is well-recognized as an effective treatment of cSDH [6, 7], multiple standard surgical techniques exist. These include burr hole craniostomy (BHC), twist drill craniostomy (TDC) and craniotomy. TDC produces a small opening of 10mm to the skull, while BHC enables a larger opening of 30mm [1]. During a craniotomy, a substantial piece of bone (>30mm) is removed and, following the hematoma evacuation, is replaced and fixed to the skull defect [1, 11].

a. Surgical Management

Indication for surgery

It is generally accepted that a patient presenting with neurologic symptoms and a radiologically proven cSDH, should undergo immediate surgical evacuation. Clearly, an asymptomatic patient showing no evidence of brain compression and/or midline shift on radiographic films is best managed conservatively and observed in a carefully monitored setting. A surgical approach is advised only if significant changes in neurologic status occur. Management of patients with cSDH leading to brain compression and/or midline shift, but lacking neurologic symptoms is very controversial. To our knowledge, no studies evaluating conservative vs. operative management in this group of patients exist. Widely used cutoffs for the indication of surgical evacuation (even in asymptomatic patients) are cSDH with maximum hematoma thickness exceeding that of the skull; or greater than 1cm [5]. An evidence-based hematoma cutoff size for the indication of operative treatment does not exist.

Craniotomy

Craniotomy was the treatment of choice for cSDH until the mid-1960s. Craniotomy exposes the largest portion of the brain and thus provides the surgeon with the most expansive operative space. It is however the most invasive of the treatment options, with the longest operating time, the largest amount of blood loss and the most postoperative complications. In 1964, Svien and Gelety published a series comparing craniotomy and BHC for the treatment of cSDH [51]. Patients treated with BHC showed lower recurrence rates and better functional outcomes than those who underwent a craniotomy. Two meta-analyzes comparing BHC, TDC and craniotomy showed similar results [11, 52]. Even though class I studies comparing these three surgical methods do not exist, the primary treatment for cSDH remains BHC, while craniotomy is considered a second-tier remedy [1, 11, 53]. Most surgeons nowadays would agree that craniotomy should only be considered if subdural collection reaccumulates, solid or calcified hematoma occur, the brain fails to expand and obliterate the subdural space, or numerous thick membranes are present [1, 7, 11, 36].

Twist drill craniostomy

TDC can be performed bedside under local anesthesia, making it an attractive treatment option, especially in polymorbid patients who are poor surgical candidates. A closed drainage system is placed at time of surgery to allow continuous drainage and promote postoperative brain expansion [1]. TDC is probably most effective in cases where the blood is almost completely liquefied and no membranes are present [1]. The morbidity and mortality of TDC seems to be similar or even superior to BHC [1, 11], however TDC is associated with significant higher recurrence rates than BHC [1, 7, 11]. In addition, there is a theoretical increased risk of contamination when performed at the bedside.

Burr hole Craniostomy

BHC is probably the treatment most frequently implemented for cSDH [1, 7, 54, 55]. Based on reviews by Weigel et al. and Laga et al., BHC seems to be the most efficient method as it balances a low recurrence rate against morbidity and mortality better than TDC and craniotomy [7, 11, 52].

Although BHC is the treatment of choice for cSDH in most neurosurgical departments and is performed frequently, many controversies and questions concerning the operational techniques and postoperative management still remain unanswered. In fact, it is quite astonishing that so few class I studies (Table 3) attempting to resolve these controversies and questions have been conducted over the past decades. The preferred operational technique (TDC, BHS vs. craniotomy), number of burr holes (one vs. two), role of intraoperative hematoma irrigation, localization of drainage (subdural vs. subperiosteal), and postoperative management have all been studied, yet studies with class I evidence are lacking, making evidence-based treatment and recommendations very difficult.

Other surgical methods

Various other surgical methods have been published, mostly within the limits of single retrospective studies or case reports. Among the methods described are: use of a tissue plasminogen activator in addition to TDC [56], minimally invasive hematoma evacuation using hollow screws [57], subduro-peritoneal shunt in infants [58], in older patients, and for recurrent cSDH [59], small craniotomy and endoscopic hematoma removal [60], replacement of the hematoma with oxygen via percutaneous subdural tapping [61], carbon dioxide insufflation in addition to BHC and closed-system drainage [62], embolization of middle meningeal artery in refractory cSDH [63-66], and implantation of an ommaya reservoir for repeated punctures and aspiration of subdural fluid [67, 68].

In order for these various techniques to be adopted as standard treatments for cSDH, further well-designed and comparative studies are necessary.

Comparison of the various operational techniques

The vast majority of studies comparing TDC, BHC and craniotomy have been small single-center retrospective studies (class II or III evidence) [1].

In their meta-analysis from 2003, Weigel et al. showed that TDC and BHC are safer and more efficient than craniotomy. Craniotomy and BHC have lower recurrence rates than TDC. They concluded that BHC has the best cure to complication ratio and is superior to TDC in the treatment of recurrences (type B recommendation) [11].

Ducruet et al. concluded in their meta-analysis from 2012, that TDC produces the best outcome and lowest complication rates as compared to BHC and craniotomy, while BHC results in lower mortality and recurrence rates than TDC or craniotomy (type C recommendation) [1]. Their recommendation is to observe small and asymptomatic cSDH, while large and symptomatic cSDH should be managed primarily with TCD or BHC. For high-risk surgical candidates with unseptated hematomas, the treatment of choice should be bedside TCD, while a craniotomy should be performed in cSDHs with significant membranes.

In conclusion, according to the current knowledge and based on the two stated meta-analyzes, BHC results in the best cure to complication ratio in most patients. In high-risk surgical patients, bedside TDC using local anesthesia might be the best treatment, while cSDH with significant membranes, acute shares, multiple recurrences, or calcification are best evacuated by craniotomy. As class I evidence-based studies and type A recommendations are lacking, further prospective randomized multi-center studies are needed.

Classes of Evidence	
Class I	Evidence provided by one or more well-designed randomized controlled clinical studies.
Class II	Evidence provided by one or more well-designed clinical studies such as prospective open, case–control studies, etc.
Class III	Evidence provided by expert opinion, non-randomized historical controls, or case reports of one or more patients.
Strength of recommendation	
Type A	Strong recommendation, based on class I evidence or overwhelming class II evidence when circumstances preclude randomized clinical trials.
Type B	Recommendation based on class II evidence.
Type C	Recommendation based on strong consensus of class III evidence.

(Classes of evidences and strength of recommendations adopted from the guidelines of the American Academy of Neurology and Weigel et al. 2012 [11, 69])

Table 3. Overview of evidence-based criteria.

Number of burr holes

While performing BHC, some surgeons prefer a single burr hole while others use two. There is no conclusive evidence to support either approach. Taussky et al. demonstrated that patients treated with a single burr hole have significantly higher recurrence rates, longer hospitalization, and more wound infections [10]. On the other hand, two different studies suggested no significant difference regarding recurrence, complications, mortality or outcome in patients treated with two burr holes as compared to one [70, 71]. A recent meta-analysis summarizing five retrospective cohort studies of 355 double BHC and 358 single BHC in 631 patients suggests that single BHC is as good as double BHC in evacuating chronic subdural hematoma and is not associated with a higher revision rate compared to double BHC (class III evidence) [72].

Irrigation

The role of irrigation after concluding the burr holes is still unclear. Four class III and one class II evidence publications have evaluated the role of irrigation in BHC, while one class III evidence study investigated the effect in TDC. Three publications (class III evidence) compared BHC, with and without irrigation; all studies found no significant difference in recurrence rates [73-75]. Two studies report on the use of continuous inflow and outflow irrigation. Ram et al. reported fewer recurrences in the irrigation group, yet significance was not achieved, due to low recurrence numbers [1/19 vs. 4/18] (class II evidence) [76]. In a retrospective study, Hennig et al. found significantly lower recurrence rates in patients treated with inflow outflow drainage compared to BHC with intraoperative irrigation and postoperative closed system drainage, BHC with intraoperative irrigation only, and craniotomy (class III evidence) [77]. In TDC, a significantly reduced rate was shown when using intraoperative irrigation (class III evidence) [78]. The use of irrigation had no impact on mortality or morbidity, both in BHC and TDC [77, 78].

Use of closed-system drainage

A survey conducted in 2008 in Great Britain showed that most surgeons did not insert closed-system drainage after operative treatment of chronic subdural hematoma [54]. However, a Canadian survey in 2005 showed that most surgeons in Canada utilize closed-system drainage [55]. Practices in many centers around the world changed after Santarius et al. published the results of their randomized controlled trial, which demonstrated a significant benefit in recurrence, mortality and discharge outcome for patients with subdural drain placement after BHC using two burr holes [6]. The placement of closed-system drainage is deemed to be standard in the operative treatment of cSDH with BHC and is considered a type A recommendation.

Drainage localization

Even though the insertion of a subdural drainage is considered safe, its proximity to the surface of the brain and the fact that it is inserted through a small burr hole make complications such as brain injury, intracranial bleeding, epilepsy, and subdural infection or empyema still possible. Consequently, a less invasive method - the insertion of a subperiosteal drainage - was proposed by some surgeons [3, 79-81]. Multiple studies have been published lately comparing the recurrence and complication rates of subperiosteal (or subgaleal) drainage with subdural drainage. Studies evaluating this novel method by Gazzeri et al. and Zumofen et al. showed similar recurrence and complication rates as with subdural drainage [3, 81]. Bellut et al. published results on the direct comparison of subdural and subperiosteal drainage. They found no statistical difference in recurrence or complication rates, although a tendency towards fewer complications in the subperiosteal group, and less recurrences in the subdural group was noted [79]. They recommend the usage of subperiosteal drainage in patients over 80 years of age or in those with predictable high risk for complications (type C recommendations) [79]. A recently published prospective randomized single-center study comparing BHC with subdural drainage and BHC with subgaleal drainage - 25 patients each - showed no recurrence at 6 months in either group, however, the overall outcome at 6 months was significantly better in the subperiosteal group (type B recommendation) [80]. Despite the prospective and randomized setting of this study, it was not sufficiently powered and the number of patients included was small (25 in each group). Definitive conclusions based on existing publications cannot be drawn and further large prospective studies are therefore warranted. In our institution, a prospective randomized trial has been initiated and this matter is presently being investigated. We aim to collect a sample size of 150 patients in each group (power of 80%) to demonstrate the difference in recurrence rates and overall outcome ($\alpha < 0.05$), (www.clinicaltrials.gov, Identifier: NTC01869855).

Mobilization of patients following BHC

Mobilization following surgical treatment of cSDH is an important aspect of postoperative care, especially since most patients are older and more susceptible to complications of immobility such as pneumonia, deep venous thrombosis and pulmonary embolism [1]. On the other hand, delayed mobilization might promote brain expansion and thus prevent recurrence of cSDH [82, 83]. Studies on this topic have reached mixed conclusions about the influence of patient mobilization on cSDH recurrence. Two prospective, randomized studies have concluded that recurrence rates after BHC are independent of patients' post-operative position (class I evidence) [84, 85]. In contrast, one prospective randomized study showed a statistically higher recurrence rate in patients mobilized immediately after surgery, although only one recurrence led to repeat surgery (class I evidence) [82]. In terms of complications due to postoperative immobilization, Kurabe et al. [84] showed significantly higher complication rates, while Abouzari et al. found no difference [82]. Due to the heterogeneity of these studies, making comparisons and conclusions are difficult. Even though Abouzari et al. [82] and Kurabe et al. [84] both employed similar surgical methods, the mean age of patients differed by approximately 20 years (77.3 years vs. 56.5 years), which may have influenced the differences in recurrence and medical complication rates observed.

Conclusion - surgical treatment of cSDH based on the available literature

In summary, defining the optimal and best known evidence-based treatment of cSDH is quite difficult, since most studies provide class III evidence leading to type C recommendations. Based on current knowledge, the following recommendations can be made (Table 4):

1. The preferred surgical method seems to be BHC since it produces the best cure to complication ratio in most patients (type C recommendation).

2. In high-risk surgical patients, bedside TDC under local anesthesia might be the best treatment (type C recommendation).

3. cSDH with significant membranes, acute shares, multiple recurrences, or calcification are best evacuated by craniotomy (type C recommendation).

4. Single BHC is as good as double BHC and is not associated with a higher revision rate compared to double BHC (type C recommendation).

5. The use of irrigation has no impact on mortality or morbidity, its role in avoiding recurrence is unclear (type C recommendation).

6. The placement of closed-system drainage after BHC leads to significantly lower recurrence rates (type A recommendation).

7. Subperiosteal (subgaleal) drainage is as good as subgaleal drainage (type C recommendation).

8. Immediate postoperative mobilization might lead to higher recurrence rates but might also prevent medical complications caused by immobilization (type A recommendation exists for both early and late mobilization).

b. Conservative Management

Although, surgical drainage is well recognized as an effective treatment of cSDH and is usually the treatment of choice, numerous reports have described spontaneous resorption of cSDH [7, 86]. In addition, therapy with corticosteroids and ACE-inhibitors was reviewed by different authors, showing good results [87-90]. Two surveys among neurosurgeons, one in Canada and one in the United Kingdom and the Republic of Ireland, found that conservative management is seldom practiced [54, 55], probably due to the poorer outcomes and prolonged hospital stay. Conservative management is reserved for asymptomatic cases, patients refusing surgical treatment or those with a high perceived operation risk [7, 91, 92]. Still, multi-centered prospective studies are needed to evaluate the efficacy of conservative treatment options in patients with cSDH.

Corticosteroids

Tissue plasminogen activator activity, interleukin-6 and -8, and VEGF expression were shown to be inhibited by corticosteroids [7, 93-95]. Since these inflammation and angiogenesis factors play a role in the pathophysiology of cSDH, corticosteroids have been proposed as a therapy for cSDH [7, 23, 24, 96]. Few systematic studies evaluating the role of corticosteroids in the treatment of cSDH have been published. Glover et al. showed that corticosteroids inhibit the growth of neomembranes in cSDH [97]. In 1970, Benders work group retrospectively analyzed 100 patients treated for cSDH without surgery. They concluded that since the introduction of corticosteroids, the incidence of successful treatment by medical means is higher and patients show excellent recovery [87]. Decaux et al. reported two cases successfully treated with corticosteroids [88]. A retrospective study comparing conservative treatment with corticosteroids and surgical treatment using TDC and closed drainage system was conducted by Delgado-Lopez et al. [89]. Patients with a Markwalder score of 1 or 2 (n=101) were treated with corticosteroids (Dexamethasone® 4mg TID for 48 to 72 hours), while patients with a Markwalder score of 3-4 (n=19) were treated surgically. Of the patients treated with corticosteroids, surgery was avoided in 2/3 of the cases and 96% showed a favorable outcome, compared to 93% in the surgical group. Duration of hospitalization was shorter in the corticosteroids group (6 vs. 8 days). Medical complications, mainly mild hyperglycemic impairments caused by corticosteroids, occurred in 27.8%. The authors concluded that corticosteroids are a feasible and safe option in the management of cSDH, and were able to cure or improve the condition of two thirds of the patients. A recently published review of the literature showed that secondary intervention rate lies between 3 and 28%, lethality rate ranged from 0 to 13%, and good outcome was seen in 83-100%. Hyperglycemia occurred more often in patients treated with corticosteroids, while in two studies one case of gastrointestinal bleeding was observed. All five observational studies suggest that corticosteroids might be beneficial in treatment of cSDH [98].

However, well designed studies that support or refute the use of corticosteroids in cSDH are lacking. Although the few existing studies show promising results, the rational for using corticosteroids is still based on theory, and more research into their treatment of cSDH is warranted [7, 89].

Angiotensin converting enzyme-inhibitors

Hypothesizing that hyperangiogenesis plays a role in the pathogenesis of cSDH, Weigel et al. analyzed the recurrence rates of surgically treated cSDH, in patients with and without concurrent treatment of hypertension with angiotensin converting enzyme (ACE) inhibitors [90]. The recurrence rate between the two was significant: 5% in patients taking, and 18% in those not taking ACE-inhibitors. Moreover, VEGF levels were significantly lower in the hematomas of patients taking ACE inhibitors. The authors conclude that ACE-inhibitor treatment for hypertension might lower the risk of recurrence in patients undergoing surgery and possibly even lower the risk of development of cSDH, through their antiangiogenic mechanism. To the best of our knowledge, there are no further publications studying the role of ACE-inhibitors in the treatment of cSDH, or comparing them to surgical treatment. Therefore, the impact and potential of ACE-inhibitors in the treatment of cSDH remains to be determined.

Other conservative methods

Several authors have described spontaneous resorption of cSDH, while conducting a "wait and watch" or "wait and scan" regime [49, 86]. One small study described successful treatment in 95% of cases using Mannitol 20%, leading to no sequelae, recurrence, or complications at follow-up [92].A recent study by Kageyama et al. showed promising results using tranexamic acid - an antifibrinolytic drug - for conservative treatment of cSDH. Treated solely with tranexamic acid, all 18 patients in their cohort displayed hematoma resolution without progression or recurrence at follow-up [99]. Further investigation of the natural healing course of cSDH and various conservative treatment options is needed.

10. Anticoagulation therapy in patients with cSDH

Neurosurgeons are faced increasingly with issues involving the treatment of anticoagulated patients [1, 100]. Especially in patients with cSDH - where up to 43% are anticoagulated and in whom anticoagulants were shown to increase the risk for development (up to 42.5 times) - the correct handling of anticoagulation therapy before, during and after surgery has become a major concern [8, 101].

Reversal of anticoagulation

Although class I evidence comparing outcome of patients undergoing surgical evacuation of cSDH with and without reversal of anticoagulant medication are not available, there is a consensus that patients presenting with cSDH while on anticoagulation therapy require a rapid reversal [1, 102]. The risk of hematoma expansion or complications during potential neurosurgical interventions would otherwise be too high.

In cases where immediate reversal is not critical, vitamin K can be used for a more gradual change of international normalized ratio (INR) [102]. Urgent reversal of oral anticoagulation (e.g. warfarin) is usually accomplished using fresh frozen plasma (FFP) transfusion, prothrombin complex concentrate (PCC) or recombinant Factor VIIa (rFVIIa) [103-106]. To avoid INR rebound, vitamin K should always be given adjuvant to FFP, PCC and rFVIIa [107]. In the case of reversal using FFP, the required volume can promote fluid overload in patients with cSDH, as these often present with cardiac or renal impairment [1, 103]. In addition, transfusion-related lung injury (TRALI) is an underestimated but feared complication caused by blood product transfusions such as FFP [108]. PCC exists as a 3-factor PCC (containing factor II, IX, and X, e.g. Uman Complex DI® and Profilnine SD®) or a 4-factor PCC (containing factor II, XII, IX, and X, e.g. Beriplex® and Kcentra®). While 4-factor PCC was shown to be affective in emergency reversal of oral anticoagulation irrespective of the starting INR, 3-factor PCC should be restricted to patients presenting with an INR< 4.0 (type C recommendation) [109]. The most feared complication of PCC is thrombosis and although it has been suggested that this risk might be less with the 3- than the 4-factor PCC, there is no study confirming this [109]. The role of rFVIIa remains unclear, due to the known and relative frequent side-effects, such as deep vein thrombosis and pulmonary embolism, and high costs of this new drug [104, 105, 110]. A comparison of FFP, rFVIIa and PCC treatment for intracranial hemorrhaging showed that rapid reversal of oral anticoagulants using rFVIIa and PCC is more effective than FFP. In addition, rFVIIa is considerably more expensive and might have greater risk of INR rebound than PCC [107].

We therefore recommend that patients presenting with cSDH while on anticoagulation should be rapidly reversed using PCC adjuvant to vitamin K (type C recommendation, class III evidence) (Table 4). Alternatively, in institutions where PCC is not available FFP should be applied.

Timing of resumption of anticoagulant therapy

The timing of resumption of anticoagulant therapy in patients treated surgically for cSDH must be chosen cautiously. The increased risk of thromboembolic complications due to prolonged discontinuation of anticoagulation must be balanced carefully against the increased risk of hemorrhage if oral anticoagulation is commenced soon after surgery. There is little empirical evidence to support a definitive decision on when to restart oral anticoagulation in these patients. Still, a few studies addressing this issue do exist [111-115]. Yeon et al. (in a prospective design) and Kawamata et al. (in a retrospective design), both showed that resumption of anticoagulation within 3 days is safe and does not lead to higher risk of recurrent cSDH or intracranial bleeding (type B recommendation) [113, 114]. On the other hand, Foster et al. showed in their retrospective series that early postoperative treatment with low-molecular-weight heparin or oral anticoagulation may affect reoperation rate and lead to a poorer outcome (type C recommendations) [112]. In 2013, Chari et al. conducted a meta-analysis; they reported a lower bleeding risk (11% vs. 22%) and paradoxically, a higher thromboembolism risk (2.2% vs. 0%) when anticoagulation was restarted, as opposed to a prolonged discontinuation [111]. They stated that no conclusions can be drawn from their data, due to the small cohort group (67 patients in 3 studies).

The indication for anticoagulation is an additional important factor when considering reinstating treatment. It is clear that patients with a mechanical heart valve require anticoagulation due to the high risk of thromboembolic complications. The decision is more complex for atrial fibrillation, requiring a balance between the risk of recurrence and a thromboembolic event. Chari et al. recommend comparing two validated scores: the HAS-BLED (hypertension, abnormal renal/liver function, stroke, bleeding history or predisposition, labile INR, elderly, drugs/alcohol concomitantly) score - assessing bleeding risk under anticoagulation - and the CHA_2DS_2-VASc (congestive heart failure, hypertension, age ≥75 years (doubled), diabetes mellitus, stroke (doubled), vascular disease, age 65–74 years, sex) score - evaluating the thromboembolic risk without anticoagulation - to determine risk on a patient-by-patient basis [104, 111, 116, 117]. The two scores may correlate in terms of risk/year and consideration of which score is higher might help in the decision making.

A definitive recommendation on when to restart oral anticoagulation after surgical evacuation of cSDH cannot be made. Based on the literature available, oral anticoagulation can be reinstated 72 hours after surgery, particularly in patients with a high thromboembolic risk (type B recommendation, class II evidence). A comparison of the HAS-BLED score and the CHA_2DS_2-VASc score might be helpful when deciding whether to restart anticoagulation in patients with atrial fibrillation (type C recommendation, class III evidence) (Table 4).

11. Antiplatelet therapy in patients with cSDH

As with oral anticoagulation, antiplatelet therapy in patients with cSDH presents a significant neurosurgical challenge. Patients seem to be at greater risk for development of cSDH while on these medications [101, 118]. However, it remains unclear if the recurrence rate is affected by antiplatelet therapy [1, 101, 112, 118].

Reversal of antiplatelet therapy

Even though there is no definitive evidence for the discontinuation of antiplatelet therapy prior to surgical evacuation of cSDH, most surgeons prefer to discontinue and convert antiplatelet therapy before surgery. The most effective way for reversing antiplatelet therapy is discontinuing the drug for 7 days. Aspirin inhibits the platelet enzyme cyclooxygenase irreversibly, meaning that as long as aspirin is administered, all thrombocytes in the blood system are non-functional and aggregation is inhibited for the entire platelet lifespan [119]. The period therefore required for total recovery of platelet function after the last administration of aspirin is their 7 day lifespan [119]. Platelet transfusion and desmopressin application have also been described as methods to reverse antiplatelet therapy, yet they are controversial and have not been analyzed rigorously with regards to cSDH [1, 119, 120].

Since evidence is still lacking, we recommend case-by-case management for patients on antiplatelet drugs presenting with cSDH. Generally, in patients with minor symptoms, antiplatelet therapy should be discontinued for 7 days, while meticulous clinical and radiologic follow-up is warranted. Platelets could be administered during surgery for patients undergoing emergency procedures (Table 4).

Timing of resumption of antiplatelet therapy

Little evidence exists to determine the optimal timing of postoperative resumption of antiplatelet therapy in patients surgically treated for cSDH. Several studies have produced controversial results on recurrence risk. Two studies showed no significant difference in recurrence of hematoma in patients who did or did not receive antiplatelet agents preoperatively [112, 121], while a further study found a significant difference in recurrence rate [101]. Therefore, evidence-based recommendations on when to reinstate antiplatelet agents cannot be made. Case-by-case management is advised until further prospective studies produce evidence-based recommendations (Table 4).

12. Anticonvulsant therapy in patients with cSDH

The efficacy and indication of antiepileptic drugs (AED) in patients presenting with symptomatic cSDH has been debated without reaching a consensus on its use [1, 122]. The reported seizure rate in patients treated surgically for cSDH varies from 2.3% to 17% [123, 124], and affects 1% to 23.4% of individuals postoperatively [123, 125]. Two studies showed no significant differences in seizure rate secondary to the prophylactic administration of AEDs [124, 126]. They concluded that AED morbidity outweighs the benefits except in patients at high risk for seizures, like alcoholics. Another study found a significant increase in morbidity and mortality in patients with cSDH and new-onset seizures. They therefore recommend the administration of AED for a period of six months following diagnosis of cSDH [127]. Grobelny et al. found that preoperative AED prophylaxis might reduce the incidence of postoperative seizures in patients treated with BHC [123]. Yet neither pre- or postoperative seizures, nor the decision to initiate AED prophylaxis had a significant effect on outcome [123]. Chen et al. reported a higher postoperative seizure rate in patients with mixed-density cSDH on preoperative CT, and in those with left unilateral cSDH. They therefore propose the use of prophylactic AED in these patients [128].

Further investigation of this topic is necessary prior to establishing definitive recommendations. The indication of AEDs in cSDH patients such as duration of AED therapy, adverse effects of administered AEDs, and efficacy of various AEDs should all be studied through prospective randomized studies.

Until more data is available, AED prophylaxis should be considered for patients with cSDH at high risk for seizures, such as those presenting with seizures, alcoholics and patients with significant underlying brain injury [1] (type C recommendation). In general, the surgeon must balance the potential benefit against possible secondary morbidity of AED administration (Table 4).

13. Postoperative complications, recurrence rates and outcome

Complications

Complications after surgical treatment of cSDH include [5]:

http://dx.doi.org/10.5772/57336

Subject	Recommendation	Type	Studies
Surgical method	• BHC – "first choice" in most patients • TDC – in surgical high-risk patients • Craniotomy – "last choice", with acute shares, significant membranes, multiple recurrences, calcification	B and C	Weigel et al. [11] Santarius et al. [7] Lega et al. [52] Ducruet et al[1]
Amount of burr holes	One or two (both equivalent)	C	Taussky et al.[10] Han et al.[70] Kansal et al.[71] Belkhair et al.[72]
Irrigation	Safe Might lead to less recurrence especially in TDC	B and C	Ram et al.[76] Hennig et al.[77] Aoki et al.[78]
Drainage insertion	Yes	A	Santarius et al.[6]
Drainage location	Subperiosteal (recurrence rate same as subdural, fewer complications)	B	Kaliaperumal et al.[80] Bellut et al.[79] Zumofen et al.[81] Gazzeri et al.[3]
Postoperative mobilization	None Studies in favor of early and late mobilization	A	Abouzari et al.[82] Kurabe et al.[84]
Reversal of anticoagulation	Rapid reversal using PCC and Vitamin K	C	Woo et al.[107]
Resumption of anticoagulation	After 72 hours in high-risk patients In atrial fibrillation compare HAS-BLED score and CHA_2DS_2-VASc score	B C	Kawamata et al.[113] Yeon et al.[114] Chari et al.[111]
Reversal of antiplatelets	Discontinuation 7 days prior to surgery In urgent cases platelet transfusion or desmopressin	C	Mascarenhas et al.[119] Ranucci et al.[120]
Resumption of antiplatelets	No recommendations; case-by-case management	C	Forster et al.[112] Rust et al.[101] Torihashi et al.[121]
Antiepileptic drugs	None, except in high-risk patients (e.g. alcoholics) or presenting with seizure	C	Ducruet et al.[1] Hirakawa et al.[125] Grobelny et al.[123]

Table 4. Summary of recommendations based on the available literature

- Failure of brain to re-expand and/or reaccumulation of blood in the subdural space (leading to recurrent cSDH) [0-30%) [129]
- Seizures (including status epilepticus) (1-23%) [123, 125]
- Intracerebral hemorrhage (0.7-5%, especially in patients over 75 years of age, in rapid decompression of hematoma; 1/3 of the patients die and 1/3 are severely disabled) [5, 129-131]
- Postoperative infections (e.g. wound infection, subdural empyema) (2%) [129]
- Tension pneumocephalus (0-10%) [129, 132, 133]

All complications are more common in elderly, polymorbid, and enervated patients [5]. However, a recent retrospective study evaluating the outcome in patients over 65 years of age who were treated surgically for cSDH, concluded that despite higher rates of anticoagulation and multimorbidity, surgical treatment in elderly patients is safe [134]. Patients above the age of 85 show lower recurrence rates, yet carry a greater risk for complications (e.g. acute SDH) and should be monitored meticulously [134].

Intraoperative imaging, such as intraoperative MRI or CT, might be a useful tool for early detection of surgical complications and drainage malposition and allows for their treatment within the same procedure.

Recurrence rates

Recurrence represents the most common complication following surgical treatment of cSDH, with a median incidence of 15% reported in the literature (range 0%-30%) [8, 129, 135, 136]. Recurrence ensues mainly in hematomas containing a solid or organized clot which cannot be sufficiently cleared by BHC or TDC; or due to the lack of brain expansion after removal of the hematoma resulting in a renewed hemorrhage [129]. Many risk factors for recurrence in surgically treated cSDH have been investigated over the last decades. Although results have been inconsistent, many factors including age, alcoholism, cerebral atrophy, anticoagulation or antiplatelet use, poor performance status on admission, poor GCS and GOS, bilateral cSDH, hematoma width, midline displacement postoperatively, air collection in hematoma cavity, septum formation or multiple membranes in the hematoma cavity, higher hematoma density on CT, hematoma with laminar and separated architecture, and conclusion of BHC or TDC without drainage placement are thought to be associated with higher recurrence rates [12, 46, 82, 121, 129, 135, 137, 138]. The only proven factor leading to lower recurrence rates is the intraoperative placement of a closed system drainage (class I evidence) [6].

Outcome

Morbidity and mortality rates in surgically treated patients with cSDH depends mostly on the surgical technique, the patients' age and co-morbidities [1, 6, 11, 129, 134]. The overall favorable outcome after surgical treatment of cSDH in the literature is reported to be 72-89%, with younger patients generally achieving better outcomes compared to older ones [6, 11, 129, 134, 139]. Worsening neurologic status following drainage of cSDH is estimated at 4% and overall mortality after surgical treatment of cSDH is 0-8% [5]. In their meta-analysis, Weigel et al reported a morbidity of 3%, 3.8%, and 12.3% in patients treated with TDC, BHC and craniotomy respectively. In a further meta-analysis, Ducruet et al. identified morbidity rates of 2.5% in patients treated with TDC, 9.3% in patients treated with BHC and 3.9% in patients treated with craniotomy, while the mortality rates in patients treated with TDC, BHC, and craniotomy were 5.1%, 3.7% and 12.2% respectively. Ramachandran et al. reported a favorable outcome in 66% of patients >60 years vs. 79% in patients between the age of 40 and 60, and 74% in patients <40 years. Borger et al. compared the outcome of 322 surgically treated patients over the age of 65; an overall positive outcome based on Karnofsky performance status (KPS) was seen in 83% of the patients between the age of 65 and 74 years, in 68% of the patients between the age of 75 and 84 years, and in 55% of patients between the age of 85 and 94 years. In their study, mortality rates in patients between the age of 65 and 74 years, 75 and 84 years, and 85 and 94 years were 1.7%, 3.6%, and 3.8% respectively.

14. Conclusion

Chronic SDH represents one of the most frequent entities in neurosurgical patients and a common cause of traumatic brain injury. Since the population will continue to age over the next decades and cSDH primarily affects elderly patients, an increase in incidence rate is expected. cSDH is therefore one of the most significant neurosurgical issues confronting us today. While it is clear that cSDH is mainly caused by minor head trauma, the pathophysiological mechanisms of its maintenance and enlargement over time remain debatable. Neomembranes with fragile neocapillaries forming around the hematoma, inflammation leading to production of VEGF and profibrinolytic and anticoagulation factors produced within the hematoma fluid are elements hypothesized to promote re-bleeding and SDH growth. Clinical presentation varies from general and mild symptoms (e.g. headache, fatigue) to severe symptoms (e.g. hemiparesis, coma). Head CT plays a major role in the initial evaluation of cSDH, because it confirms the diagnosis accurately and hematoma age can be estimated.

The management of cSDH remains controversial. Amazingly, only a few studies with class I evidence evaluating management protocols and surgical techniques exist. Currently, there are several ongoing randomized controlled trials which might provide more clarity for the management and treatment of cSDH. These studies are summarized in Table 5. It is generally accepted that in the presence of neurologic symptoms and radiologic findings, patients should undergo surgical evacuation. Yet the role of conservative treatment (e.g. "wait and scan", corticosteroids, ACE-inhibitors) in asymptomatic patients or in patients presenting with mild symptoms remains unclear. The preferred surgical method seems to be BHC since it produces the best cure to complication ratio in most patients. Yet many questions such as the correct surgical method (BHC vs. TCD and craniotomy), the superiority of two burr holes over one, placement of subdural or subperiosteal drainage, the efficacy of hematoma irrigation, and timing of postoperative mobilization are still insufficiently clarified. The intraoperative placement of closed system drainage for the prevention of recurrence is the only evidence-based recommendation that can be made. Prospective multicenter studies providing type A recommendations for these questions are much needed.

Rationale	Study design	Status	Anticipated number of patients	ClinicalTrail.gov ID
Evaluation of the recurrence rate of cSDH after placing a subperiosteal drainage compared to a subdural drainage	PR	R	400	NCT01869855
Evaluation of the recurrence rate of cSDH after placing a subdural drainage compared to no drainage placement	PR	NYR	260	NCT01785797
Evaluation of the role of CT scanning in the postoperative follow-up after surgical treatment of cSDH	PR, SB	R	400	NCT01624545
To assess whether continued aspirin treatment increases the risk of cSDH in mild head trauma patients 50 years and older who present with negative head CT	PR, DB	NYR	100	NCT01470040
To assess whether treatment with an ACE-inhibitor for 3 months after surgical evacuation of cSDH will decrease the risk of recurrence compared to placebo	PR, DB, Placebo	R	120	NTC00915928
Evaluation of the recurrence rate of cSDH in patients treated postoperatively for 2 months orally with corticosteroids compared to placebo	PR, DB, Placebo	NYR	400	NCT01380028
Evaluation of the recurrence rate of cSDH after placing an active subperiosteal drainage compared to a passive subdural drainage and continuous irrigation	RNR	NYR	950	NCT01930617

PR: prospective randomized, *RNR*: retrospective non-randomized, *SB*: single blinded, *DB*: double blinded, *R*: recruiting, *NYR*: not yet recruiting

Table 5. Summary of ongoing clinical trials evaluating management and treatment of cSDH

In patients with minor symptoms, antiplatelet therapy should be discontinued for 7 days, and anticoagulation converted solely with vitamin K, accompanied by meticulous clinical and radiologic follow-up. For those needing emergency surgery, antiplatelet therapy must be discontinued and platelets could be administered during the procedure. In patients receiving anticoagulants, rapid conversion should be carried out using PCC or FFP, adjuvant to vitamin K. Little evidence exists to determine the optimal timing of postoperative resumption of antiplatelet or anticoagulation therapy. Therefore, case-by-case decision making is necessary. In patients with atrial fibrillation, a comparison of the HAS-BLED score and the CHA_2DS_2-VASc score might be helpful when deciding whether to restart anticoagulation. AED prophylaxis should be considered only in patients at high risk for seizures (e.g. patients presenting with seizure, alcoholics and patients with significant underlying brain injury).

Overall favorable outcome after surgical treatment is described at 72-89%. Mortality rate is estimated at 0-8%. The most frequent surgical complications are: recurrence (15%), seizure (11%), tension pneumocephalus (5%), intracerebral hematomas (2.5%), and infections (2%).

Author details

Jehuda Soleman[1], Philipp Taussky[1,2], Javier Fandino[1] and Carl Muroi[1*]

*Address all correspondence to: carl.muroi@ksa.ch

1 Department of Neurosurgery, Kantonsspital Aarau, Aarau, Switzerland

2 Department of Neurosurgery, University of Utah, Salt Lake City, Utah, USA

References

[1] Ducruet AF, Grobelny BT, Zacharia BE, Hickman ZL, DeRosa PL, Anderson K, et al. The surgical management of chronic subdural hematoma. Neurosurg Rev. 2012 Apr; 35(2):155-69; discussion 69.

[2] Chen JC, Levy ML. Causes, epidemiology, and risk factors of chronic subdural hematoma. Neurosurg Clin N Am. 2000 Jul;11(3):399-406.

[3] Gazzeri R, Galarza M, Neroni M, Canova A, Refice GM, Esposito S. Continuous subgaleal suction drainage for the treatment of chronic subdural haematoma. Acta Neurochir (Wien). 2007;149(5):487-93; discussion 93.

[4] Kudo H, Kuwamura K, Izawa I, Sawa H, Tamaki N. Chronic subdural hematoma in elderly people: present status on Awaji Island and epidemiological prospect. Neurol Med Chir (Tokyo). 1992 Apr;32(4):207-9.

[5] Greenberg MS. Chronic subdural hematoma. Handbook of Neurosurgery. 7th ed. New York, New York: Thieme; 2010. p. 899-902.

[6] Santarius T, Kirkpatrick PJ, Ganesan D, Chia HL, Jalloh I, Smielewski P, et al. Use of drains versus no drains after burr-hole evacuation of chronic subdural haematoma: a randomised controlled trial. Lancet. 2009 Sep 26;374(9695):1067-73.

[7] Santarius T, Kirkpatrick PJ, Kolias AG, Hutchinson PJ. Working toward rational and evidence-based treatment of chronic subdural hematoma. Clin Neurosurg. 2010;57:112-22.

[8] Baechli H, Nordmann A, Bucher HC, Gratzl O. Demographics and prevalent risk factors of chronic subdural haematoma: results of a large single-center cohort study. Neurosurg Rev. 2004 Oct;27(4):263-6.

[9] Nakaguchi H, Tanishima T, Yoshimasu N. Factors in the natural history of chronic subdural hematomas that influence their postoperative recurrence. J Neurosurg. 2001 Aug;95(2):256-62.

[10] Taussky P, Fandino J, Landolt H. Number of burr holes as independent predictor of postoperative recurrence in chronic subdural haematoma. Br J Neurosurg. 2008 Apr; 22(2):279-82.

[11] Weigel R, Schmiedek P, Krauss JK. Outcome of contemporary surgery for chronic subdural haematoma: evidence based review. J Neurol Neurosurg Psychiatry. 2003 Jul;74(7):937-43.

[12] Virchow R. Das Hämatom der Dura Mater. Verh Phys Med Ges Würzburg. 1857(7): 134-42.

[13] Frederickson RG. The subdural space interpreted as a cellular layer of meninges. Anat Rec. 1991 May;230(1):38-51.

[14] Haines DE, Harkey HL, al-Mefty O. The "subdural" space: a new look at an outdated concept. Neurosurgery. 1993 Jan;32(1):111-20.

[15] Drapkin AJ. Chronic subdural hematoma: pathophysiological basis for treatment. Br J Neurosurg. 1991;5(5):467-73.

[16] Sajanti J, Majamaa K. High concentrations of procollagen propeptides in chronic subdural haematoma and effusion. J Neurol Neurosurg Psychiatry. 2003 Apr;74(4):522-4.

[17] Gennarelli TA, Thibault LE. Biomechanics of acute subdural hematoma. J Trauma. 1982 Aug;22(8):680-6.

[18] Maxeiner H, Wolff M. Pure subdural hematomas: a postmortem analysis of their form and bleeding points. Neurosurgery. 2002 Mar;50(3):503-8; discussion 8-9.

[19] Gardner W. Traumatic subdural hematoma with particular reference to the latent interval. Arch Neurol Psychiatr 1932(27):847-58.

[20] Zollinger R GR. Traumatic subdural hematoma, an explanation of the late onset of pressure symptoms. JAMA. 1934(103):245-9.

[21] Yamashima T, Yamamoto S, Friede RL. The role of endothelial gap junctions in the enlargement of chronic subdural hematomas. J Neurosurg. 1983 Aug;59(2):298-303.

[22] Labadie EL, Glover D. Local alterations of hemostatic-fibrinolytic mechanisms in reforming subdural hematomas. Neurology. 1975 Jul;25(7):669-75.

[23] Katano H, Kamiya K, Mase M, Tanikawa M, Yamada K. Tissue plasminogen activator in chronic subdural hematomas as a predictor of recurrence. J Neurosurg. 2006 Jan;104(1):79-84.

[24] Fujisawa H, Ito H, Saito K, Ikeda K, Nitta H, Yamashita J. Immunohistochemical localization of tissue-type plasminogen activator in the lining wall of chronic subdural hematoma. Surg Neurol. 1991 Jun;35(6):441-5.

[25] Hohenstein A, Erber R, Schilling L, Weigel R. Increased mRNA expression of VEGF within the hematoma and imbalance of angiopoietin-1 and -2 mRNA within the neomembranes of chronic subdural hematoma. J Neurotrauma. 2005 May;22(5):518-28.

[26] Suzuki K, Takano S, Nose T, Doi M, Ohashi N. Increased concentration of vascular endothelial growth factor (VEGF) in chronic subdural hematoma. J Trauma. 1999 Mar;46(3):532-3.

[27] Weigel R, Schilling L, Schmiedek P. Specific pattern of growth factor distribution in chronic subdural hematoma (CSH): evidence for an angiogenic disease. Acta Neurochir (Wien). 2001 Aug;143(8):811-8; discussion 9.

[28] Weir B. The osmolality of subdural hematoma fluid. J Neurosurg. 1971 Apr;34(4): 528-33.

[29] Weir B. Oncotic pressure of subdural fluids. J Neurosurg. 1980 Oct;53(4):512-5.

[30] Sambavian M. An overview of chronic subdural hematoma: Experience with 2300 urgical cases. Surg Neurol. 1997(47):418-22.

[31] French BN DA. Infantile chronic subdural hematoma of possterior fossa diagnosed by computerized tomography. Case report. J Neurosurg. 1977(47):949-52.

[32] Kudo H KK, Izawa I. Chronic subdural hematoma in elderly people : Present status on Awaji Island and epidemiological prospect. Neurol Med Chir (Tokyo). 1992(32): 207-9.

[33] Weiner HL CS, Cohen H. Current treatment of normal-pressure hydrocephalus: Comparsion of flow-regulated and differential-pressure shunt valves. Neruosurgery. 1995(37):877-84.

[34] Dietrich U LC, Sprick C. Subdural hematoma in a case of hydrocephalus and macrocrania. Experience with a pressure-adjustable valve. Childs Nerv Syst. 1987(3):242-4.

[35] Kamano S NY, Imanishi T. Management with a programmable pressure valve of subdural hematomas caused by vetriculoperitoneal shunt: Case report. Surg Neurol. 1991(35):381-3.

[36] Markwalder TM. Chronic subdural hematomas: a review. J Neurosurg. 1981 May; 54(5):637-45.

[37] Kotwica Z PL. The association of arteriovenous malformation, aneurysm and chronic subdural hematoma. Case report. Zentralbl Neurochit. 1986(47):158-60.

[38] Pozzatti E TF, Gaist G. Chronic subdural haematoma from cerebral arteriovenous malformation. Neurochirurgia. 1986(29):61-2.

[39] Cinalli G ZM, Carteret M. Subdural sarcoma associated with chronic subdural hematoma. Report of two cases and review of the literature. J Neurosurg. 1997(86):553-7.

[40] Popovic EA LM, Scheithauer BW. Mast cell-rich convexity meningioma: Case report and review of the literature. Surg Neurol. 1994(42):8-13.

[41] Tanaka N YM, Jimbo M. Meningioma associated with chronic subdural hematoma ande meningothelial call cluster within the hematoma capsule- case report. Neurol Med Chir (Tokyo). 1994(34):176-9.

[42] Markwalder TM, Steinsiepe KF, Rohner M, Reichenbach W, Markwalder H. The course of chronic subdural hematomas after burr-hole craniostomy and closed-system drainage. J Neurosurg. 1981 Sep;55(3):390-6.

[43] Teasdale G, Jennett B. Assessment of coma and impaired consciousness. A practical scale. Lancet. 1974 Jul 13;2(7872):81-4.

[44] Jennett B, Bond M. Assessment of outcome after severe brain damage. Lancet. 1975 Mar 1;1(7905):480-4.

[45] Rankin J. Cerebral vascular accidents in patients over the age of 60. II. Prognosis. Scott Med J. 1957 May;2(5):200-15.

[46] Chon KH, Lee JM, Koh EJ, Choi HY. Independent predictors for recurrence of chronic subdural hematoma. Acta Neurochir (Wien). 2012 Sep;154(9):1541-8.

[47] Senturk S, Guzel A, Bilici A, Takmaz I, Guzel E, Aluclu MU, et al. CT and MR imaging of chronic subdural hematomas: a comparative study. Swiss Med Wkly. 2010 Jun 12;140(23-24):335-40.

[48] Osborn AG BS, Salzman KL, Katzman GL, Provenzale J, Castillo N, Heldlund GL, Illner A, Harnsberger HR, Cooper JA, Jones BV, Hamilton BE editor. Diagnostic Imaging Brain. 1st ed. Salt Lake City, Utha, USA: Amirsys; 2004.

[49] Goksu E, Akyuz M, Ucar T, Kazan S. Spontaneous resolution of a large chronic subdural hematoma: a case report and review of the literature. Ulus Travma Acil Cerrahi Derg. 2009 Jan;15(1):95-8.

[50] Parlato C, Guarracino A, Moraci A. Spontaneous resolution of chronic subdural hematoma. Surg Neurol. 2000 Apr;53(4):312-5; discussion 5-7.

[51] Svien HJ, Gelety JE. On the Surgical Management of Encapsulated Subdural Hematoma. A Comparison of the Results of Membranectomy and Simple Evacuation. J Neurosurg. 1964 Mar;21:172-7.

[52] Lega BC, Danish SF, Malhotra NR, Sonnad SS, Stein SC. Choosing the best operation for chronic subdural hematoma: a decision analysis. J Neurosurg. 2010 Sep;113(3): 615-21.

[53] Horn EM, Feiz-Erfan I, Bristol RE, Spetzler RF, Harrington TR. Bedside twist drill craniostomy for chronic subdural hematoma: a comparative study. Surg Neurol. 2006 Feb;65(2):150-3; discussion 3-4.

[54] Santarius T, Lawton R, Kirkpatrick PJ, Hutchinson PJ. The management of primary chronic subdural haematoma: a questionnaire survey of practice in the United Kingdom and the Republic of Ireland. Br J Neurosurg. 2008 Aug;22(4):529-34.

[55] Cenic A, Bhandari M, Reddy K. Management of chronic subdural hematoma: a national survey and literature review. Can J Neurol Sci. 2005 Nov;32(4):501-6.

[56] Neils DM, Singanallur PS, Wang H, Tracy P, Klopfenstein J, Dinh D, et al. Recurrence-free chronic subdural hematomas: a retrospective analysis of the instillation of tissue plasminogen activator in addition to twist drill or burr hole drainage in the treatment of chronic subdural hematomas. World Neurosurg. 2012 Jul;78(1-2):145-9.

[57] Krieg SM, Aldinger F, Stoffel M, Meyer B, Kreutzer J. Minimally invasive decompression of chronic subdural haematomas using hollow screws: efficacy and safety in a consecutive series of 320 cases. Acta Neurochir (Wien). 2012 Apr;154(4):699-705; discussion

[58] Aoki N. Chronic subdural hematoma in infancy. Clinical analysis of 30 cases in the CT era. J Neurosurg. 1990 Aug;73(2):201-5.

[59] Probst C. Peritoneal drainage of chronic subdural hematomas in older patients. J Neurosurg. 1988 Jun;68(6):908-11.

[60] Rodziewicz GS, Chuang WC. Endoscopic removal of organized chronic subdural hematoma. Surg Neurol. 1995 Jun;43(6):569-72; discussion 72-3.

[61] Takeda N, Sasaki K, Oikawa A, Aoki N, Hori T. A new simple therapeutic method for chronic subdural hematoma without irrigation and drainage. Acta Neurochir (Wien). 2006 May;148(5):541-6.

[62] Kubo S, Takimoto H, Nakata H, Yoshimine T. Carbon dioxide insufflation for chronic subdural haematoma: a simple addition to burr-hole irrigation and closed-system drainage. Br J Neurosurg. 2003 Dec;17(6):547-50.

[63] Ishihara H, Ishihara S, Kohyama S, Yamane F, Ogawa M, Sato A, et al. Experience in endovascular treatment of recurrent chronic subdural hematoma. Interv Neuroradiol. 2007 Mar 15;13 Suppl 1:141-4.

[64] Mandai S, Sakurai M, Matsumoto Y. Middle meningeal artery embolization for refractory chronic subdural hematoma. Case report. J Neurosurg. 2000 Oct;93(4):686-8.

[65] Mino M, Nishimura S, Hori E, Kohama M, Yonezawa S, Midorikawa H, et al. Efficacy of middle meningeal artery embolization in the treatment of refractory chronic subdural hematoma. Surg Neurol Int. 2010;1:78.

[66] Takahashi K, Muraoka K, Sugiura T, Maeda Y, Mandai S, Gohda Y, et al. (Middle meningeal artery embolization for refractory chronic subdural hematoma: 3 case reports). No Shinkei Geka. 2002 May;30(5):535-9.

[67] Laumer R, Schramm J, Leykauf K. Implantation of a reservoir for recurrent subdural hematoma drainage. Neurosurgery. 1989 Dec;25(6):991-6.

[68] Sato M, Iwatsuki K, Akiyama C, Kumura E, Yoshimine T. Implantation of a reservoir for refractory chronic subdural hematoma. Neurosurgery. 2001 Jun;48(6):1297-301.

[69] Hallett M, Litvan I. Evaluation of surgery for Parkinson's disease: a report of the Therapeutics and Technology Assessment Subcommittee of the American Academy of Neurology. The Task Force on Surgery for Parkinson's Disease. Neurology. 1999 Dec 10;53(9):1910-21.

[70] Han HJ, Park CW, Kim EY, Yoo CJ, Kim YB, Kim WK. One vs. Two Burr Hole Craniostomy in Surgical Treatment of Chronic Subdural Hematoma. J Korean Neurosurg Soc. 2009 Aug;46(2):87-92.

[71] Kansal R, Nadkarni T, Goel A. Single versus double burr hole drainage of chronic subdural hematomas. A study of 267 cases. J Clin Neurosci. 2010 Apr;17(4):428-9.

[72] Belkhair S, Pickett G. One versus double burr holes for treating chronic subdural hematoma meta-analysis. Can J Neurol Sci. 2013 Jan;40(1):56-60.

[73] Kuroki T, Matsumoto M, Kushida T, Ohtsuka T, Uchino M, Nishikawa H. Nontraumatic subdural hematoma secondary to dural metastasis of lung cancer: case report and review of the literature. No Shinkei Geka. 1994 Sep;22(9):857-62.

[74] Matsumoto K, Akagi K, Abekura M, Ryujin H, Ohkawa M, Iwasa N, et al. Recurrence factors for chronic subdural hematomas after burr-hole craniostomy and closed system drainage. Neurol Res. 1999 Apr;21(3):277-80.

[75] Suzuki K, Sugita K, Akai T, Takahata T, Sonobe M, Takahashi S. Treatment of chronic subdural hematoma by closed-system drainage without irrigation. Surg Neurol. 1998 Sep;50(3):231-4.

[76] Ram Z, Hadani M, Sahar A, Spiegelmann R. Continuous irrigation-drainage of the subdural space for the treatment of chronic subdural haematoma. A prospective clinical trial. Acta Neurochir (Wien). 1993;120(1-2):40-3.

[77] Hennig R, Kloster R. Burr hole evacuation of chronic subdural haematomas followed by continuous inflow and outflow irrigation. Acta Neurochir (Wien). 1999;141(2):171-6.

[78] Aoki N. Subdural tapping and irrigation for the treatment of chronic subdural hematoma in adults. Neurosurgery. 1984 May;14(5):545-8.

[79] Bellut D, Woernle CM, Burkhardt JK, Kockro RA, Bertalanffy H, Krayenbuhl N. Subdural drainage versus subperiosteal drainage in burr-hole trepanation for symptomatic chronic subdural hematomas. World Neurosurg. 2012 Jan;77(1):111-8.

[80] Kaliaperumal C, Khalil A, Fenton E, Okafo U, Kaar G, O'Sullivan M, et al. A prospective randomised study to compare the utility and outcomes of subdural and subperiosteal drains for the treatment of chronic subdural haematoma. Acta Neurochir (Wien). 2012 Nov;154(11):2083-8; discussion 8-9.

[81] Zumofen D, Regli L, Levivier M, Krayenbuhl N. Chronic subdural hematomas treated by burr hole trepanation and a subperiostal drainage system. Neurosurgery. 2009 Jun;64(6):1116-21; discussion 21-2.

[82] Abouzari M, Rashidi A, Rezaii J, Esfandiari K, Asadollahi M, Aleali H, et al. The role of postoperative patient posture in the recurrence of traumatic chronic subdural hematoma after burr-hole surgery. Neurosurgery. 2007 Oct;61(4):794-7; discussion 7.

[83] Choudhury AR. Avoidable factors that contribute to complications in the surgical treatment of chronic subdural haematoma. Acta Neurochir (Wien). 1994;129(1-2): 15-9.

[84] Kurabe S, Ozawa T, Watanabe T, Aiba T. Efficacy and safety of postoperative early mobilization for chronic subdural hematoma in elderly patients. Acta Neurochir (Wien). 2010 Jul;152(7):1171-4.

[85] Nakajima H, Yasui T, Nishikawa M, Kishi H, Kan M. The role of postoperative patient posture in the recurrence of chronic subdural hematoma: a prospective randomized trial. Surg Neurol. 2002 Dec;58(6):385-7; discussion 7.

[86] Voelker JL. Nonoperative treatment of chronic subdural hematoma. Neurosurg Clin N Am. 2000 Jul;11(3):507-13.

[87] Bender MB, Christoff N. Nonsurgical treatment of subdural hematomas. Arch Neurol. 1974 Aug;31(2):73-9.

[88] Decaux O, Cador B, Dufour T, Jego P, Cazalets C, Laurat E, et al. (Nonsurgical treatment of chronic subdural hematoma with steroids: two case reports). Rev Med Interne. 2002 Sep;23(9):788-91.

[89] Delgado-Lopez PD, Martin-Velasco V, Castilla-Diez JM, Rodriguez-Salazar A, Galacho-Harriero AM, Fernandez-Arconada O. Dexamethasone treatment in chronic subdural haematoma. Neurocirugia (Astur). 2009 Aug;20(4):346-59.

[90] Weigel R, Hohenstein A, Schlickum L, Weiss C, Schilling L. Angiotensin converting enzyme inhibition for arterial hypertension reduces the risk of recurrence in patients

with chronic subdural hematoma possibly by an antiangiogenic mechanism. Neurosurgery. 2007 Oct;61(4):788-92; discussion 92-3.

[91] Kurti X, Xhumari A, Petrela M. Bilateral chronic subdural haematomas; surgical or non-surgical treatment. Acta Neurochir (Wien). 1982;62(1-2):87-90.

[92] Suzuki J, Takaku A. Nonsurgical treatment of chronic subdural hematoma. J Neurosurg. 1970 Nov;33(5):548-53.

[93] Coleman PL, Patel PD, Cwikel BJ, Rafferty UM, Sznycer-Laszuk R, Gelehrter TD. Characterization of the dexamethasone-induced inhibitor of plasminogen activator in HTC hepatoma cells. J Biol Chem. 1986 Mar 25;261(9):4352-7.

[94] Gao T, Lin Z, Jin X. Hydrocortisone suppression of the expression of VEGF may relate to toll-like receptor (TLR) 2 and 4. Curr Eye Res. 2009 Sep;34(9):777-84.

[95] Liu Z, Yuan X, Luo Y, He Y, Jiang Y, Chen ZK, et al. Evaluating the effects of immunosuppressants on human immunity using cytokine profiles of whole blood. Cytokine. 2009 Feb;45(2):141-7.

[96] Frati A, Salvati M, Mainiero F, Ippoliti F, Rocchi G, Raco A, et al. Inflammation markers and risk factors for recurrence in 35 patients with a posttraumatic chronic subdural hematoma: a prospective study. J Neurosurg. 2004 Jan;100(1):24-32.

[97] Glover D, Labadie EL. Physiopathogenesis of subdural hematomas. Part 2: Inhibition of growth of experimental hematomas with dexamethasone. J Neurosurg. 1976 Oct; 45(4):393-7.

[98] Berghauser Pont LM, Dirven CM, Dippel DW, Verweij BH, Dammers R. The role of corticosteroids in the management of chronic subdural hematoma: a systematic review. Eur J Neurol. 2012 Nov;19(11):1397-403.

[99] Kageyama H, Toyooka T, Tsuzuki N, Oka K. Nonsurgical treatment of chronic subdural hematoma with tranexamic acid. J Neurosurg. 2013 Aug;119(2):332-7.

[100] Cartmill M, Dolan G, Byrne JL, Byrne PO. Prothrombin complex concentrate for oral anticoagulant reversal in neurosurgical emergencies. Br J Neurosurg. 2000 Oct;14(5): 458-61.

[101] Rust T, Kiemer N, Erasmus A. Chronic subdural haematomas and anticoagulation or anti-thrombotic therapy. J Clin Neurosci. 2006 Oct;13(8):823-7.

[102] Hanley JP. Warfarin reversal. J Clin Pathol. 2004 Nov;57(11):1132-9.

[103] Lankiewicz MW, Hays J, Friedman KD, Tinkoff G, Blatt PM. Urgent reversal of warfarin with prothrombin complex concentrate. J Thromb Haemost. 2006 May;4(5): 967-70.

[104] Lin J, Hanigan WC, Tarantino M, Wang J. The use of recombinant activated factor VII to reverse warfarin-induced anticoagulation in patients with hemorrhages in the central nervous system: preliminary findings. J Neurosurg. 2003 Apr;98(4):737-40.

[105] Mayer SA, Brun NC, Begtrup K, Broderick J, Davis S, Diringer MN, et al. Recombinant activated factor VII for acute intracerebral hemorrhage. N Engl J Med. 2005 Feb 24;352(8):777-85.

[106] Vigue B, Ract C, Tremey B, Engrand N, Leblanc PE, Decaux A, et al. Ultra-rapid management of oral anticoagulant therapy-related surgical intracranial hemorrhage. Intensive Care Med. 2007 Apr;33(4):721-5.

[107] Woo CH, Patel N, Conell C, Rao VA, Faigeles BS, Patel MC, et al. Rapid Warfarin Reversal in the Setting of Intracranial Hemorrhage: A Comparison of Plasma, Recombinant Activated Factor VII, and Prothrombin Complex Concentrate. World Neurosurg. 2012 Dec 5.

[108] Bux J. Transfusion-related acute lung injury (TRALI): a serious adverse event of blood transfusion. Vox Sang. 2005 Jul;89(1):1-10.

[109] Makris M, Van Veen JJ. Three or four factor prothrombin complex concentrate for emergency anticoagulation reversal? Blood Transfus. 2011 Apr;9(2):117-9.

[110] Mayer SA, Brun NC, Begtrup K, Broderick J, Davis S, Diringer MN, et al. Efficacy and safety of recombinant activated factor VII for acute intracerebral hemorrhage. N Engl J Med. 2008 May 15;358(20):2127-37.

[111] Chari A, Clemente Morgado T, Rigamonti D. Recommencement of anticoagulation in chronic subdural haematoma: a systematic review and meta-analysis. Br J Neurosurg. 2013 Jul 8.

[112] Forster MT, Mathe AK, Senft C, Scharrer I, Seifert V, Gerlach R. The influence of preoperative anticoagulation on outcome and quality of life after surgical treatment of chronic subdural hematoma. J Clin Neurosci. 2010 Aug;17(8):975-9.

[113] Kawamata T, Takeshita M, Kubo O, Izawa M, Kagawa M, Takakura K. Management of intracranial hemorrhage associated with anticoagulant therapy. Surg Neurol. 1995 Nov;44(5):438-42; discussion 43.

[114] Yeon JY, Kong DS, Hong SC. Safety of early warfarin resumption following burr hole drainage for warfarin-associated subacute or chronic subdural hemorrhage. J Neurotrauma. 2012 May 1;29(7):1334-41.

[115] Zingale A, Chibbaro S, Florio A, Distefano G, Porcaro S. Management of chronic subdural hematoma in patients treated with anticoagulation. J Neurosurg Sci. 1999 Dec; 43(4):277-84.

[116] Olesen JB, Lip GY, Lindhardsen J, Lane DA, Ahlehoff O, Hansen ML, et al. Risks of thromboembolism and bleeding with thromboprophylaxis in patients with atrial fi-

brillation: A net clinical benefit analysis using a 'real world' nationwide cohort study. Thromb Haemost. 2011 Oct;106(4):739-49.

[117] Pisters R, Lane DA, Nieuwlaat R, de Vos CB, Crijns HJ, Lip GY. A novel user-friendly score (HAS-BLED) to assess 1-year risk of major bleeding in patients with atrial fibrillation: the Euro Heart Survey. Chest. 2010 Nov;138(5):1093-100.

[118] Lindvall P, Koskinen LO. Anticoagulants and antiplatelet agents and the risk of development and recurrence of chronic subdural haematomas. J Clin Neurosci. 2009 Oct;16(10):1287-90.

[119] Mascarenhas L. Illustration of the impact of antiplatelet drugs on the genesis and management of chronic subdural hematoma. Neurochirurgie. 2012 Feb;58(1):47-51.

[120] Ranucci M, Nano G, Pazzaglia A, Bianchi P, Casana R, Tealdi DG. Platelet mapping and desmopressin reversal of platelet inhibition during emergency carotid endarterectomy. J Cardiothorac Vasc Anesth. 2007 Dec;21(6):851-4.

[121] Torihashi K, Sadamasa N, Yoshida K, Narumi O, Chin M, Yamagata S. Independent predictors for recurrence of chronic subdural hematoma: a review of 343 consecutive surgical cases. Neurosurgery. 2008 Dec;63(6):1125-9; discussion 9.

[122] Ratilal BO, Pappamikail L, Costa J, Sampaio C. Anticonvulsants for preventing seizures in patients with chronic subdural haematoma. Cochrane Database Syst Rev. 2013;6:CD004893.

[123] Grobelny BT, Ducruet AF, Zacharia BE, Hickman ZL, Andersen KN, Sussman E, et al. Preoperative antiepileptic drug administration and the incidence of postoperative seizures following bur hole-treated chronic subdural hematoma. J Neurosurg. 2009 Dec;111(6):1257-62.

[124] Ohno K, Maehara T, Ichimura K, Suzuki R, Hirakawa K, Monma S. Low incidence of seizures in patients with chronic subdural haematoma. J Neurol Neurosurg Psychiatry. 1993 Nov;56(11):1231-3.

[125] Hirakawa K, Hashizume K, Fuchinoue T, Takahashi H, Nomura K. Statistical analysis of chronic subdural hematoma in 309 adult cases. Neurol Med Chir (Tokyo). 1972;12(0):71-83.

[126] Rubin G, Rappaport ZH. Epilepsy in chronic subdural haematoma. Acta Neurochir (Wien). 1993;123(1-2):39-42.

[127] Sabo RA, Hanigan WC, Aldag JC. Chronic subdural hematomas and seizures: the role of prophylactic anticonvulsive medication. Surg Neurol. 1995 Jun;43(6):579-82.

[128] Chen CW, Kuo JR, Lin HJ, Yeh CH, Wong BS, Kao CH, et al. Early post-operative seizures after burr-hole drainage for chronic subdural hematoma: correlation with brain CT findings. J Clin Neurosci. 2004 Sep;11(7):706-9.

[129] Gelabert-Gonzalez M, Iglesias-Pais M, Garcia-Allut A, Martinez-Rumbo R. Chronic subdural haematoma: surgical treatment and outcome in 1000 cases. Clin Neurol Neurosurg. 2005 Apr;107(3):223-9.

[130] Ogasawara K, Koshu K, Yoshimoto T, Ogawa A. Transient hyperemia immediately after rapid decompression of chronic subdural hematoma. Neurosurgery. 1999 Sep; 45(3):484-8; discussion 8-9.

[131] Diaz P, Maillo A. (Intracerebral hemorrhage following chronic subdural hematoma evacuation: report of two cases and review of the literature). Neurocirugia (Astur). 2003 Sep;14(4):333-6; discussion 7.

[132] Caron JL, Worthington C, Bertrand G. Tension pneumocephalus after evacuation of chronic subdural hematoma and subsequent treatment with continuous lumbar subarachnoid infusion and craniostomy drainage. Neurosurgery. 1985 Jan;16(1):107-10.

[133] Lavano A, Benvenuti D, Volpentesta G, Donato G, Marotta R, Zappia M, et al. Symptomatic tension pneumocephalus after evacuation of chronic subdural haematoma: report of seven cases. Clin Neurol Neurosurg. 1990;92(1):35-41.

[134] Borger V, Vatter H, Oszvald A, Marquardt G, Seifert V, Guresir E. Chronic subdural haematoma in elderly patients: a retrospective analysis of 322 patients between the ages of 65-94 years. Acta Neurochir (Wien). 2012 Sep;154(9):1549-54.

[135] Ohba S, Kinoshita Y, Nakagawa T, Murakami H. The risk factors for recurrence of chronic subdural hematoma. Neurosurg Rev. 2013 Jan;36(1):145-9; discussion 9-50.

[136] Rohde V, Graf G, Hassler W. Complications of burr-hole craniostomy and closed-system drainage for chronic subdural hematomas: a retrospective analysis of 376 patients. Neurosurg Rev. 2002 Mar;25(1-2):89-94.

[137] Delgado PD, Cogolludo FJ, Mateo O, Cancela P, Garcia R, Carrillo R. (Early prognosis in chronic subdural hematomas. Multivariate analysis of 137 cases). Rev Neurol. 2000 May 1-15;30(9):811-7.

[138] Yamamoto H, Hirashima Y, Hamada H, Hayashi N, Origasa H, Endo S. Independent predictors of recurrence of chronic subdural hematoma: results of multivariate analysis performed using a logistic regression model. J Neurosurg. 2003 Jun;98(6):1217-21.

[139] Ramachandran R, Hegde T. Chronic subdural hematomas--causes of morbidity and mortality. Surg Neurol. 2007 Apr;67(4):367-72; discussion 72-3.

Section 3

Rehabilitation in Traumatic Brain Injury

Chapter 13

Traumatic Brain Injury Rehabilitation: An Overview

Wafa Al Yazeedi, Loganathan Venkatachalam,
Somaya Al Molawi and Fatma Al Kuwari

Additional information is available at the end of the chapter

http://dx.doi.org/10.5772/57527

1. Introduction

Traumatic brain injury (TBI) is a leading cause of seizure, disorders, disability and death worldwide; regrettably, facilities for rehabilitation remain insufficient. One million Americans are treated and released from hospital emergency departments because of TBI [1]. The range of severity of TBI is broad, from concussion through to persistent vegetative states and categories are mild, moderate and severe (Box 1). In the European Union, brain injury accounts for one million hospital admissions per year. Many published studies support the notion that males are far more likely to incur a TBI than females. The highest rate of injury occurs between the ages of 15-24 years. Persons under the age of five or over the age of 75 are also at higher risk. Brain injury can cause many physical, cognitive and behavioural/emotional impairments, as well as common lifestyle consequences that may be either temporary or permanent (Box No 2). Brain injury may also result in seizure disorders. Brain injury is a public health concern that demands ongoing epidemiological study, increased efforts in the prevention of injuries occurring and research to advance medical options and therapeutic interventions.

Box No 1

Injury severity category	Initial Glasgow Coma Scale	Duration of post-traumatic amnesia
Mild	12-15	Less than 24 hours
Moderate	9-11	1-7 days
Severe	3-8	1-4 weeks

Box No 2
Consequences of traumatic brain injury

Neurological impairment (motor, sensory and autonomic)

- Motor function impairment – coordination, balance, walking, hand function, speech
- Sensory loss – taste, touch, hearing, vision, smell
- Sleep disturbance – insomnia, fatigue
- Medical complications – spasticity, post-traumatic epilepsy, hydrocephalus, heterotopic ossification
- Sexual dysfunction

Cognitive impairment

- Memory impairment, difficulty with new learning, attention and concentration; reduced speed and flexibility of thought processing; impaired problem-solving skills
- Problems in planning, organizing, and making decisions
- Language problems – dysphasia, problems finding words and impaired reading and writing skills
- Impaired judgment and safety awareness

Personality and behavioural changes

- Impaired social and coping skills, reduced self-esteem
- Altered emotional control; poor frustration tolerance and anger management; denial, and self-centeredness
- Reduced insight, disinhibition, impulsivity
- Psychiatric disorders – anxiety, depression, post-traumatic stress disorder, psychosis
- Apathy, a motivational state

Common lifestyle consequences

- Unemployment and financial hardship
- Inadequate academic achievement
- Lack of transportation alternatives
- Inadequate recreational opportunities
- Difficulties in maintaining interpersonal relationships, marital breakdown
- Loss of pre-injury roles; loss of independence

2. Body

The recovery process following TBI is a slow process that might take months or years, rather than weeks. Six months following an injury yields a clearer picture of what effect the injury has had, but waiting roughly a year after the accident before making any important decisions regarding the future is considered advisable. Physical recovery will take approximately a year, but psychological recovery might take considerably longer. Factors that may affect the rate of a person’s recovery include injury type, severity, location of the individual's brain injury, age (younger patients tend to have better outcomes), pre-injury health, pre-injury personality, intelligence and lifestyle, social support from family and friends and other factors like alcohol or drug abuse. Depending on the individual's physical abilities and potential, a rehabilitation programme will be initiated accordingly.

http://dx.doi.org/10.5772/57527

The effectiveness of a comprehensive multidisciplinary rehabilitation team (Box No.3) is increased when compared to natural recovery following brain injury. The importance of recognizing the vegetative state and its management is emphasized. Finally, the important, but often neglected area of employment rehabilitation is covered. We plan to explain the rehabilitation process starting from the day of the accident with the aim of recovering from the posttraumatic brain injury sequlae and achieving reintegration into the community.

Box No 3 Rehabilitation team

- Patient and patient's family
- Rehabilitation medicine physician
- Rehabilitation nurse
- Allied health professionals: physiotherapist, occupational therapist, speech pathologist, social worker and orthotics
- Neuropsychologist, clinical psychologist
- Vocational rehabilitation services and counsellors
- Other medical specialties: neurosurgery, orthopaedic surgery and urologist

The iterative process of consultation yielded consensus definitions for each component of the rehabilitation service pathway for head-injured adults (Table 1). This chapter describes not only brain injury due to road traffic accident or fall from a great height but also brain injury in war victims. Modern rehabilitation practices are based around the concepts of impairment, disability and handicap, as outlined by the World Health Organization (WHO) in 1980. Recently, however, the WHO has redefined these concepts.

The modern terminology that will be introduced in the near future still encompasses the term 'impairment', but now replaces 'disability' with 'activity' and 'handicap' with 'participation'. This is not simply a sign of political correctness, but serves to emphasize the positive aspects of disability rather than its negative connotations. The rehabilitation approaches consist of three basic types; 1) reducing the disability; 2) acquiring new skills and their application for reducing the impact of disability; 3) alteration of the environment in physical and social contexts, so that existing disability will carry as little participation as possible. For example, a young male individual had a traumatic brain injury and recovery with residual hemiplegia with spasticity and urinary incontinence. A specific programme tailored to this individual's needs was offered during brain injury rehabilitation management process. (Box No. 4).

Title and description	Sites	Description of rehabilitation input
Minor HI: education – medically stable, requiring 24-48 hrs. observation prior to community rehabilitation, with low probability of acute neurological deterioration	Acute A&E observation ward	Assessment and observation – education, emotional and social support. Planned discharge home or moves to code 30 at 48 h
Supportive rehab – medically unstable, requiring neurosurgical or clinical care	Acute hospital	Identifying and addressing early rehab goals before medically stable and transfer of care to rehab team.
Supportive rehab – medically unstable, not requiring neurosurgical or critical care	Acute hospital	Identifying and addressing early rehab goals before medically stable and transfer of care to rehab team.
Rapid access rehab – medically stable, not (necessarily) able to actively participate due to amnesia, confusion, rejection, low response or awareness	Acute hospital	Needs inpatient care due to physical dependency and requires continuous clinical assessment to facilitate optimal timing for rehab input and detect clinical deterioration. Immediate early rehab delivered and judgment made on timing/ appropriateness of referral to next rehab sector.
Active participation inpatient rehab – medically stable, able to actively participate with and benefit from therapy	Acute or community hospital	Needs inpatient care due to physical dependency, or need for specialist therapy equipment, safe environment, supervision or intensity of therapy, which cannot be provided in community
Behavioural rehab – medically stable, but with prolonged confusion, amnesia or behavioural difficulties, requiring specialist behavioural management, intensive supervision and secure environment	Specialist inpatient unit	Specialist behavioural management, including high staff/patient ratio to ensure intensive supervision and secure environment; access to neuropsychology and neuropsychiatry.
Slow-stream rehab – medically stable, but with low awareness or response persisting beyond, e.g., three weeks after sedation; withdrawn and medically stable. Able to benefit from medical and physical therapy to prevent complications and support recovery	Community hospital or specialist inpatient unit	Assessment/active rehabilitation phase, which needs to be distinguished from long-term care, although planning of care an increasingly important aim after, e.g., six months. Patients may go to active participation unit if sufficient improvement occurs.

http://dx.doi.org/10.5772/57527

Title and description	Sites	Description of rehabilitation input
Community rehab – medically stable, able to actively participate with and benefit from, therapy. Will include spectrum of initial severity of injury with a small minority derived from code 05 category.	Domiciliary or day hospital	Interdisciplinary coordinated management therapy aimed at community reintegration/inclusion by enhancing independence and assisting return to work/education. In collaboration with social services, voluntary and statutory services. Includes treatment of patients in residential care or with live-in carers.
Intensive cognitive rehab – medically stable, independently mobile, primarily cognitive impairments likely to benefit from intensive neuropsychological therapy.	Domiciliary or day hospital	Aiming to return to work, studies or independent community life.
Specialist vocational rehab – medically stable, living in community, aiming to enter/ return to employment.	Domiciliary or residential	Aiming for return to work, where this is influenced by physical or cognitive problems, or needs residential placement.
Maintenance – medically stable but permanently disabled.	Domiciliary, residential or nursing home, respite unit	Prevent deterioration of physical, emotional and behavioural conditions, and long-term management of seating, pressure, spasticity, etc.
Social, patient and carer support – carer support for initial injury, patient support when able to communicate.	All sites	Developing social skills, stamina, confidence, attention and leisure pursuits, sorting out benefits, day supervision and respite care. Specific attention paid to: community involvement and integration (further education, etc.); personal social development; structured daytime activity; family support and outreach.
HI = head injury; rehab = rehabilitation; A&E = accident and emergency		

Table 1. Classification of head injury recovery and actual/potential rehabilitation services (adult).

Box No 4 Specific programme offered during brain injury rehabilitation

Post-traumatic amnesia assessment
Management of post-traumatic agitation
Neuropharmacological management
Cognitive rehabilitation therapy
Coma emergence and rehabilitation of minimally responsive states
Spasticity management including motor point/nerve blocks and Botulinum toxin therapy
Casting, splinting, orthotics, contracture management
Pain management services for cervicogenic headaches, trigger point injections
Balance and vestibular assessment and rehabilitation
Cognitive and behavioural assessment and remediation
Comprehensive dysphagia and speech therapy services
Assistive technology using augmentative and alternative communication
Rehabilitation nursing education
Intrathecal Baclofen pump therapy
Brain injury discharge advice and family education
Social support: discharge planning, caregiver training

Brain injury rehabilitation occurs in the following settings:

- **Inpatient rehabilitation:** This involves intensive specialist rehabilitation for people who are not yet ready to return home after discharge from hospital. Neurological rehabilitation centres provide an ideal setting for further treatment, where a structured rehabilitation programme is in place throughout the day.
- **Outpatient rehabilitation:** Some people may be well enough to return home and receive further treatment as an outpatient, either at a local hospital or at a separate rehabilitation centre.
- **Community rehabilitation:** Following an inpatient rehabilitation stay, some people may be transferred to a residential transitional living unit. Here people can develop their independent living skills so that they may be able to live in a place of their own. Others will go straight back to their homes, with a community rehabilitation team or outreach team helping them to make further progress; this may involve therapists working with the person in their home or community environment [2].

The main basic approach to treatment in the above example was to reduce disability through appropriate medication with which to control spasticity and for applying new skills of physiotherapy and occupational therapy to improve the individual's functional activity of gait training. This was done through means of ambulation in an indoor environment, with physical aid and wheelchair aid for long distances and, by using an adaptive device to facilitate feeding and other self-care activities. Additionally, adapted fittings to bathrooms and kitchens can be installed. Liaising with the injured individual's employer and having a dialogue for initiating his return to work in a part-time capacity or having downtime between working hours to reduce fatigue can also be instigated. The involvement of his family is also important to make them more accepting of his condition and for effecting the necessary adjustments to their own lifestyles.

The basic nature of rehabilitation is to work with the disabled person and family in partnership. The interdisciplinary rehabilitation team should provide accurate information and advice, explain the prognosis and natural history and work with the individual to establish realistic goals within an appropriate social environment. Whatever approaches implemented, setting a realistic goal is key to a good quality rehabilitation programme. For example, the long-term goal of independent walking requires a number of short stages to be implemented, such as sitting, balance training without support, standing balance without support, walking in parallel bar, walking with a person's assistance, walking with aids and lastly independent walking.

The rehabilitation team and disabled person should know what the goal is and when this goal has been achieved. Thus, valid and reliable outcome measures are very important for supporting the rehabilitation process. The most common measures used in the UK is the Barthel Index, but functional independence measures are also very common worldwide. Physiotherapy and occupational therapy using a 10-meter walking test for improvement of walking and a nine hole pick peg test for improvement of hand function, respectively, are also used. There are many other tools for measuring variable functional independence measure (FIM) cognitive and motor functions, the Glasgow coma scale (GCS), Rancho Los Amigos levels of cognitive functioning scale (RLA), disability rating scale and the Coma Recovery Scale-(Revised). It is important to highlight that the use of valid and reliable outcome measures is important in order to observe goals, assess progress and adjust the rehabilitation programme. The author has published articles related to the above-mentioned variables influencing and predicting functional outcomes after TBI [3].

For example, in her study [3], she pointed out that there was positive correlation of functional independence measure on discharge (FIMd) and RLA (Figure1). The researcher postulated GCS as a predictor of functional outcome and showed in her result positive correlation of GCS and FIM cognitive. Rehabilitation should begin as early as possible (even at the level of acute care) (Figure 3) [4]. It is common for individuals to ultimately be transferred to a rehabilitation unit when avoidable complications are already present. Unfortunately, muscle contractures, pressure sores and unnecessary aggressive behaviour are atypical during stay in rehabilitation unit . If the rehabilitation team can be involved in the early stage of acute care setting, perhaps even on the intensive care unit, it is more likely that such complications can be avoided. Studies provide evidence that undertaking rehabilitation within the first days of evolution improved cognition, perception and motor recovery of brain-damaged patients, and led to shorter lengths of stay (LOS) in rehabilitation units [5,6,8].

An early intervention rehabilitation team (EIRT) has been initiated by the author at the level of trauma or surgical intensive care units in Hamad General Hospital (HGH), Qatar. A rehabilitation programme focusing on cognitive stimulation and prevention of musculoskeletal complications of contractures has been offered. Pressure ulcers and aggressive behaviour has been limited by providing appropriate orthoses, frequent positioning, psychotropic

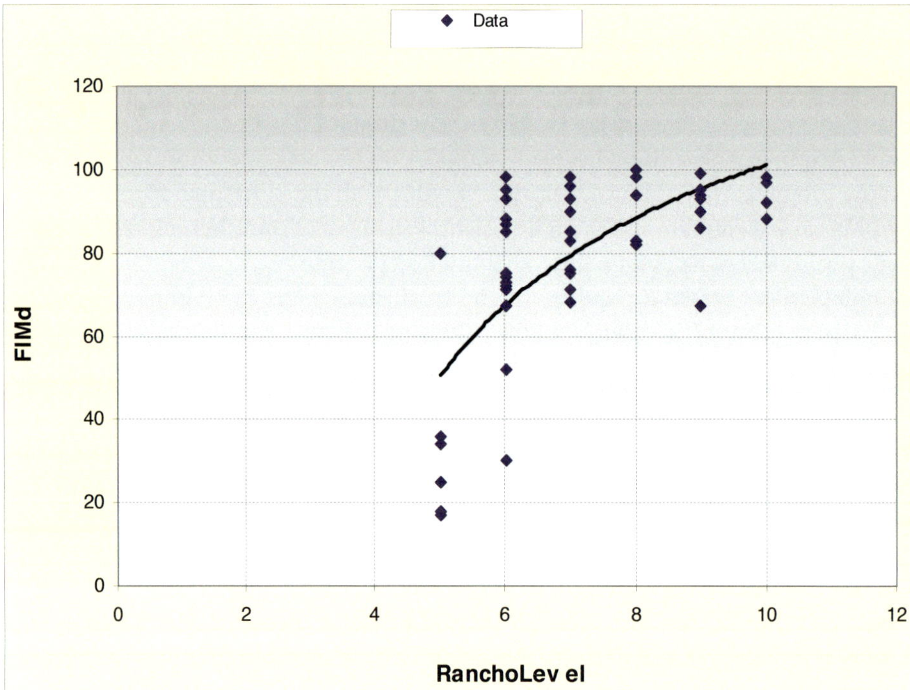

Figure 1. Correlation of FIM d and Rancho

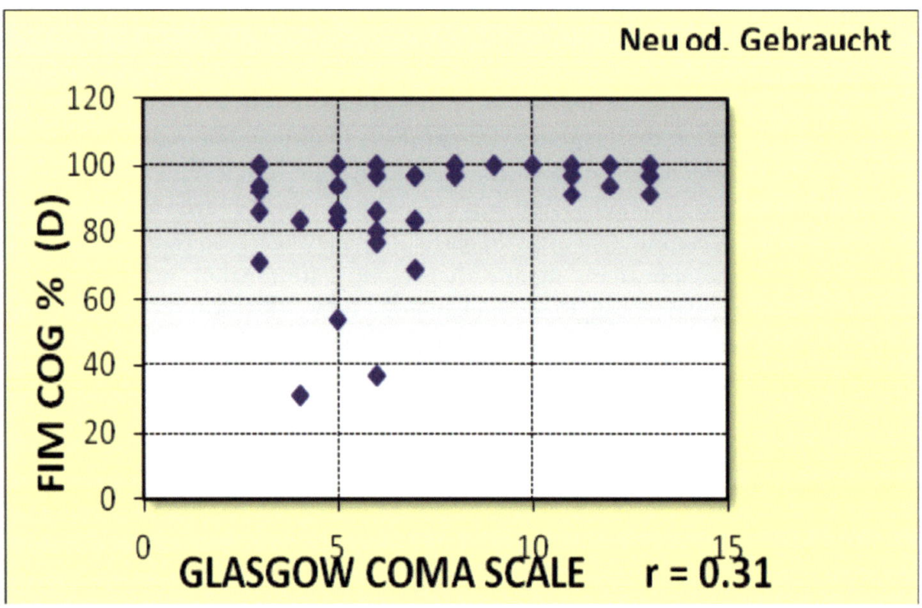

Figure 2. Correlation of FIIM cog D and GCS

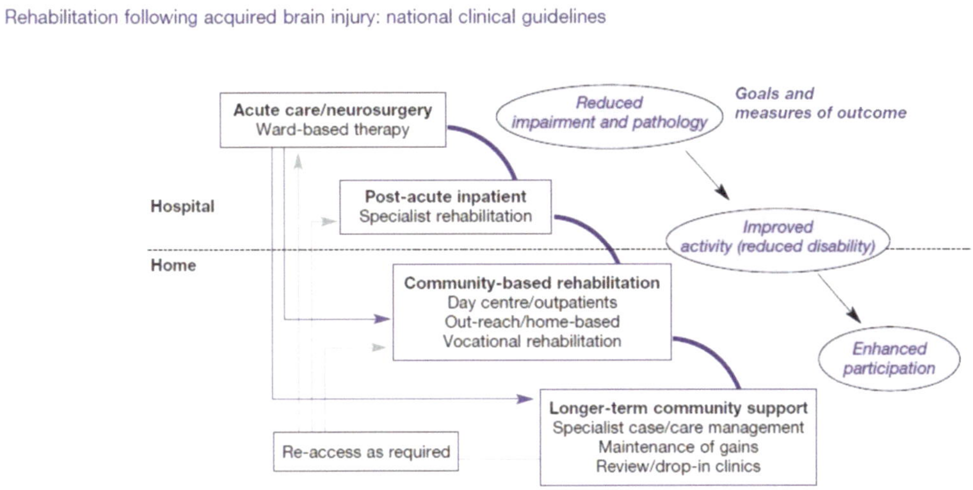

Figure 3. Rehabilitation following brain injury

medication or psychiatric consultation, while transfer to rehabilitation units as early as possible have resulted in shorter LOS in acute care units in HGH and have shown improvement in physical and mental impediments. This chapter will classify and promptly summarize the rehabilitation process of the slow stream rehabilitation programme, the active participation programme and long-term rehabilitation programme, the community based rehabilitation programme and returning to the community or work.

Most rehabilitation units will admit individuals a week or so after injury once they are medically stable. Generally, individuals have an average stay of about three months. However, longer-term rehabilitation is important if short-term gains are not to be lost. Outpatient rehabilitation should continue at least until physical recovery has plateaued. Recovery of cognitive and intellectual problems can take longer than physical problems and it is often such psychological difficulties that cause most handicaps and distress to the injured individual's family. This problem must be compensated by establishing a long-term facility that must be allowed to give the injured individual rehabilitation and medical service for at least one year. Following on, another important step in the rehabilitation setting is that the rehabilitation team will need to clearly establish links with social services as well as other relevant professionals, such as employment rehabilitation experts.

The rehabilitation team recommends developing close links with the established community rehabilitation centre (CRC) in the state or province, with CRC in turn having links with the regional rehabilitation unit or rehabilitation hospital. Most post-acute rehabilitation is conducted in the hospital setting or at the regional rehabilitation unit, before individuals are discharged back into the community. At this point, the community team becomes involved and is able to deliver ongoing physical and psychological rehabilitation through a multidisciplinary team, based in a peripheral hospital, as well as being able to deliver services within the home. The author is working on establishing CRC in Qatar under the national strategy of rehabilitation.

When addressing the efficacy of head injury rehabilitation, there are many problems to be overcome. Randomized and blind study is almost impossible to achieve, as there are very few people who have not received some form of rehabilitation after acute injury. Rehabilitation is obviously a multi-faceted and multi-professional process without clear-cut definitions. M.P. Barnes reported rehabilitation after head injury to be a long-term process with much impairment taking two years or more to achieve full recovery [5]. Thus, long-term follow-up and data collection is important for ideal study. More studies revealed that the standard outcome measures help to improve cognitive and motor functional level. However, no studies have been found that that tracks the continuum of rehabilitation from intensive care to final community re-integration. Though the literature in this area is confusing and hard to come by, some studies are nonetheless worth considering and can begin to provide good evidence for the value of head injury rehabilitation.

McKay and colleagues [6] compared matched groups of severe head injury individuals who did or did not receive formal rehabilitation during their acute trauma centre admission. All TBI rehabilitation individuals received physical, occupational or speech therapy, whereas in the non-rehabilitation group, only a very small minority did. In the rehabilitation group, therapy was also initiated quickly, whereas in the other group, therapy started about three weeks after the acute episode. Overall coma length, rehabilitation stay and cognitive functioning showed a significant benefit in the rehabilitation group. An interesting study by Blackerby [7] demonstrated that an increase in intensity of rehabilitation for five to eight hours a day produced a reduction in the average length of stay.

An important study by Cope and Hall [8] compared 34 head injured people who had either been referred 'early' or 'late' to a comprehensive inpatient rehabilitation programme. The early group had significant reduction in length of stay both in the acute care unit and in the rehabilitation unit. Many studies indicate better functional outcomes following a formal rehabilitation programme. For example, Aronow [9] produced one of the very few case controlled studies, where patients from an inpatient head injury programme were matched with similar patients in a neuro-trauma programme who received no formal rehabilitation. On the outcome scale used (not widely published in terms of validity and reliability), the rehabilitation group had a significantly better outcome than the non-rehabilitation group. Some published studies are supportive of the view that comprehensive TBI rehabilitation programmes produce benefits over and above standard care and spontaneous recovery.

Barnes [5] indicated in his study that early intervention by a specialist head injury service significantly reduced social morbidity and severity of post-concussion symptoms at six months. On referral to this study and her own experience, the author introduced an Early Intervention Rehabilitation Team at the level of intensive care and acute care units in Hamad General Hospital, Qatar, which resulted in reducing the length of stay in acute settings and preventing complications. It is clear that much work still needs to be done in this field. Nonetheless, valuable studies strongly support a comprehensive TBI rehabilitation programme, which produced worthwhile benefits over and above standard care and spontaneous recovery. The rehabilitation team of rehabilitation unit's focus is on a comprehensive assessment for TBI-related neurological and functional impairments and the development of an individualized specific programme based on functional goals and serial monitoring of outcomes (Box No. 4).

What follows will describe the process of a specialized rehabilitation programme for specific rehabilitation problems after traumatic brain injury. Physical disability is divided into three types – mild, moderate and severe. The severe type of disability consists of the following factors of muscle contracture: spasticity, heterotopic ossification and communication problems; these will be discussed in detail.

Severe physical disability. This occurred after severe head injury; relatively few people have had severe physical disability in the long-term. The longer-term problems of traumatic brain injury tend to include cognitive, intellectual, behavioural and emotional difficulties rather than physical problems. This has been confirmed by a number of studies [10, 11]. Spasticity can be particularly problematic after TBI and if not treated persistently, can often lead to muscle contracture and a functionally useless limb. Passive stretching in the acute phase is important, as may be the use of orthoses or even serial splinting and casts in order to prevent such contracture. Whilst there are a number of modern oral anti-spastic agents (e.g., Baclofen, Dantrium and Tizandine), troublesome spasticity tends to be focal and thus better treated by a local modality. Botulinum toxin has recently been introduced as a potent muscle relaxant and a number of studies have now demonstrated efficacy in the management of spasticity [12, 13]. Fortunately, the effects of Botolinium toxin wore off after three months and did not impair long-term recovery, which used to be the case in other focal techniques such as phenol and alcohol nerve blocks.

Heterotopic ossification. Neurogenic heterotopic ossification (HO) is characterized by the formation of bone in soft tissues following traumatic injury to the central nervous system, especially around large joints (CNS). The hip is the most commonly involved joint, but elbow and knee is also common in both spinal cord injury and traumatic brain injury (Figure 4]. HO was first described in 1883 by Reidel and in 1918, Dejerne and Ceillier reported that HO frequently occurred among soldiers who had experienced spinal cord trauma as combatants in World War I [14]. The development of HO is extra-articular and occurs outside the joint capsule. Bone forms in the connective tissue between the muscle planes and not within the muscle itself. The new bone can be contiguous with the skeleton but generally does not involve the periosteum. Mature HO shows cancellous bone and mature lamellar bone blood vessels and bone marrow with a minor amount of haematopoiesis.

Alkaline phosphatase has been recommended as a useful screening tool for HO [15]. Alkaline phosphatase levels become abnormal approximately two weeks after injury. In the typical case of HO, the alkaline phosphatase levels reached approximately 3.5 times the normal value 10 weeks after the inciting trauma, before returning to normal at approximately 18 weeks. The literature suggests reasonably good results for primary and secondary prevention with Non-Steroidal Anti Inflammatory Drugs, bisphosphonates and radiation. Surgical excision and joint release can provide improvements for many patients, but have varied results and often depend

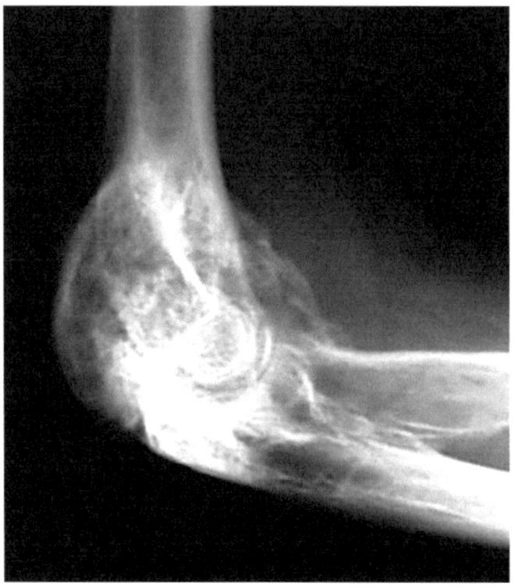

Lateral radiograph demonstrate the anterior and posterior ossification.

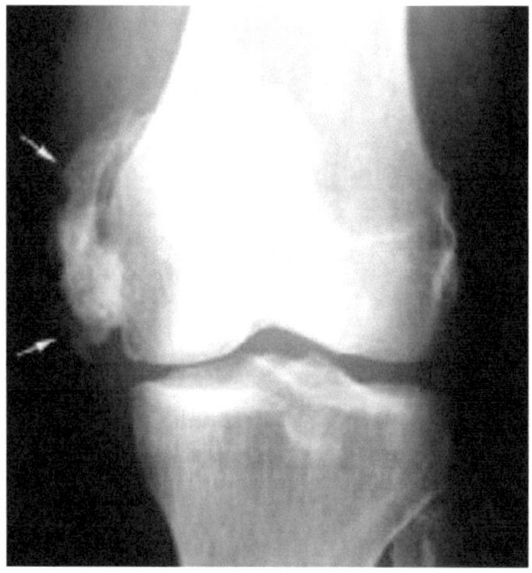

Anteroposterior radiograph of the left knee in a traumatic brain injury. Mature heterotopic ossification surrounds the medial femoral condyle, with a solid Peripheral cortex (arrows).

Figure 4. X Ray of heterotrophic ossification

on the degree of CNS injury. The role of physical therapy in patients with HO is somewhat controversial. There are those who believe that aggressive ROM may lead to increase bone formation. Most, however, agree that physical therapy preserves movement, leading to better function and prevention of ankylosis. Physiotherapy typically involves active and passive ROM, gentle terminal stretching and resisted ROM exercise [16].

Nutrition is a particular problem. The reasons are two-fold – an increased catabolic rate immediately after brain trauma compounded by the common occurrence of swallowing difficulties. If maintenance of good nutritional status is difficult, a judgment that should preferably be made only after adequate dietary advice, then nasogastric feeding can be used in the very short-term. However, if adequate nutrition cannot be maintained within a few days following iTBI, then a fine bore percutaneous endoscopic gastrostomy (PEG) tube should be inserted. This is a relatively simple and straightforward procedure with few complications. If nutrition is not maintained, it can have serious consequences for wound healing and an increased risk of pressure sores.

Pressure sores remained, unfortunately; they are rather common and are nearly always avoidable. Rigid adherence to regular turning regimens, as well as the use of appropriate pressure relieving mattresses and appropriate lifting and handling techniques should help to limit the occurrence of pressure sores. However, risks could be increased by poor nutrition as well as by other factors such as urinary or faecal incontinence. Regrettably, once sores are present, they can be extremely time consuming in terms of healing and often require surgical intervention to excise the ulcer, bony prominence or affected bone and to resurface the defect by skin grafting or other techniques such as a myocutaneous flap [17].

Urinary continence can also be problematic after head injury. Whilst in the short-term, indwelling catheterization can be used, in the long-term this is a most undesirable solution. Urodynamic study regarding the exact nature of the underlying detrusor and/or sphincter problem, combined with appropriate pharmacology, often relieves the situation. However, if impairment of bladder emptying remains, the technique of clean and intermittent self-catheterization can be invaluable - performed either by the patient or sometimes by an appropriate carer [18].

Communication problems can be troublesome after brain injury. Thus, an assessment by a speech therapist is important and various speech and language interventions can obviously be of benefit, particularly for dysarthria and dysphagia. However, those with very severe disablement are often unable to communicate orally and need to revert to an appropriate communication aid. These can vary from simple pointing boards to more complex pre-programmed artificial voice communicators. There are a number of Communication Aid centres who have particular expertise in this field and in the forms of assistive technology, which is becoming increasingly important in reducing disability and participation in those with severe physical problems.

Medical rehabilitation Problems. Urinary tract infections, pulmonary complications and derangement in electrolytes and liver function are common in 60% to 70% of acute TBI cases and may prolong acute hospital stay. Seizures may be reported in 20% of those with a severe TBI. Common neuroendocrine disorders include growth hormone deficiency, syndrome of inappropriate secretion of antidiuretic hormone, diabetes insipidus, secondary amenorrhoea, galactorrhoea and gynaecomastia. The common hormonal insufficiencies were found to be from the pituitary (hypothalamus) region. (Figure 5).In hypo pituitarism, the anterior pituitary insufficiency had higher incidence [19] rather than the posterior. There is an increasing amount of evidence that suggest that post TBI hormonal deficiency syndrome affects many people who have sustained TBI and mild TBI, and research is now beginning to show that replacement of deficient hormones can lead to significant improvements.

To evaluate patients, a full hormonal assessment via a spot or 24-hour urine should be performed to test for hormones and their metabolites (serum testing). It is also important to test for neuroendocrine markers. This can be done with either spot or 24-hour urine testing. It is important to use a laboratory with neurotransmitter capability. To treat hormonal insufficiencies after TBI, early hormonal supplementation should be considered to minimize the physical and psychological sequelae. Hormonal assessments can be done at three-month intervals from the date of injury, or more frequently, based on treatment [20].

Cranial nerve injury in TBI. TBI-complicated cranial nerve injury is subject to a high incidence rate, a high mortality rate and a high disability rate. One study explained that the extent of

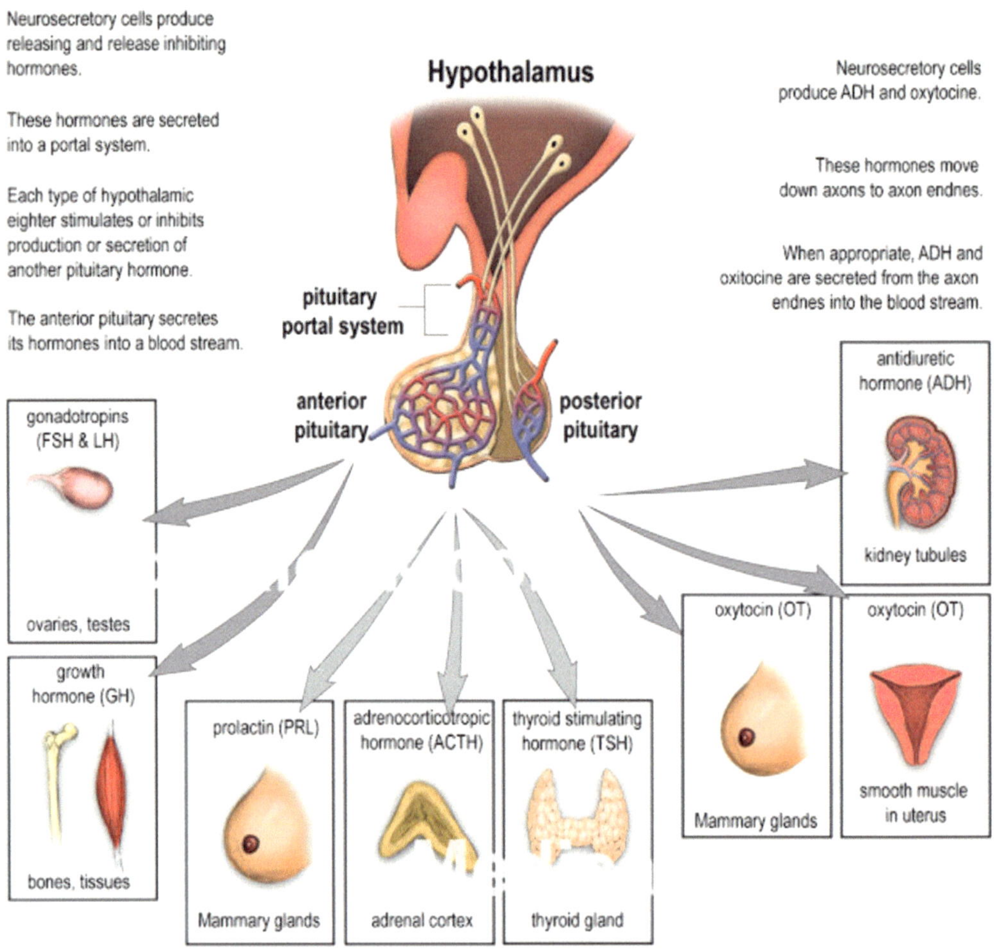

Figure 5. Pituitary hormone15

nerve injury varied and involved the olfactory nerve (66 cases), optic nerve (78 cases), oculomotor nerve (56 cases), trochlear nerve (eight cases), trigeminal nerve (four cases), abducent nerve (12 cases), facial nerve (48 cases), acoustic nerve [10 cases), glossopharyngeal nerve (eight cases), vagus nerve (six cases), accessory nerve (10 cases) and hypoglossal nerve (six cases) [21].

Environmental control equipment provides a means of controlling simple electrical equipment around the house such as the ability to turn the television, lights and other equipment on and off, and the ability to answer the telephone, open the door and adjust the bed. Such independence can be very important to an individual who is otherwise entirely dependent on a third party. The application of relatively simple technology can sometimes make a dramatic difference to level of independence. For example, a number of devices are available that enable a severely disabled person to drive a motor vehicle. These can vary from simple hand controls (switches) to more complex joystick steering with voice controlled accessory equipment (Figure 6]. Overall, those with severe physical disabilities need the support of a rehabilitation expert, which in turn will have access to the necessary wide range of multidisciplinary expertise, facilities and equipment.

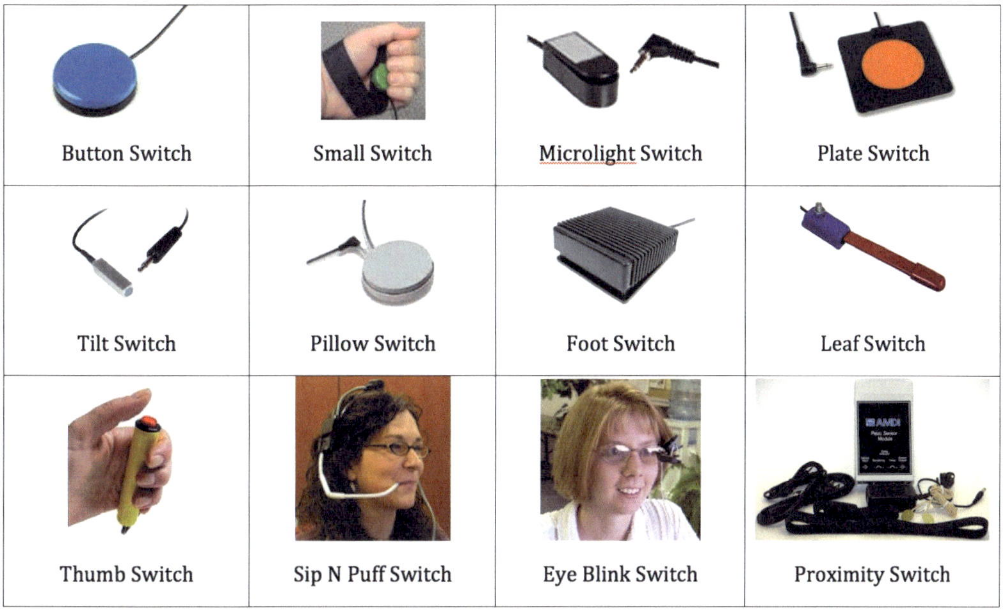

Figure 6. Adaptive devises

Following head injury, a variety of important cognitive impairments can be observed. The commonest are those associated with attention deficits, problems with concentration, memory and perception, information processing speed and problem solving. Normal recovery of neuropsychological difficulties can take place over a prolonged period of time and certainly up to two years post-injury. Obviously, this can be a complex set of impairments and consequent disabilities need to be assessed by a clinical neuropsychologist. It is somewhat controversial whether neuropsychological intervention can actually promote recovery, but there is little doubt that coping strategies can be designed that will effectively reduce disability. There have been very few randomized trials in this field and such trials are likely inappropriate when one is dealing with so many variables. The use of well-designed single case studies is a methodology that is probably best pursued in this area. Most work has been conducted in the field of memory disorders [21].

Rehabilitation can be divided into those techniques involving internal strategies and those dependent on external resources. Internal strategies, for example, can involve the use of various mnemonic techniques such as the use of imagery, methods to organize information in particular sequences (e.g., the PQRST technique), as well as other techniques that involve the use of acronyms, rhymes and systematic queuing. An alternative or even coexistent strategy is to devise interventions for reducing the handicapping effects of amnesic problems. Some may appear simple and obvious, such as planned use of a personal organizer with electronic alarm systems, colour codes around the house or a rigid use of lists, memos and diaries.

However, there is no doubt that such techniques, whilst not influencing memory impairment, can certainly reduce the effects of such impairment and have positive benefits in terms of disability and handicap [22]. Similar approaches have been taken to the remediation of problem solving deficits, attention deficits and perceptual problems [23]

Fortunately, very few individuals remain in prolonged coma or prolonged vegetative state following brain injury. One study [24] found that 0.6% of all brain-injured individuals admitted to a neurosurgical unit remained in prolonged coma (of more than two weeks' duration). Care certainly needs to be taken in the early diagnosis of coma and/or vegetative state, as later recovery has been well documented. A comparison of clinical features associated with vegetative state and minimally conscious state is shown in Table 2. Andrews [25] recently found a very high incidence of misdiagnosis in the so called persistent vegetative state. In view of the level of misdiagnosis, referral to a specialist centre is desirable. Quality of life should be maximized and unnecessary complications avoided, particularly contractures, pressure sores and malnutrition. Prolonged reassessment is necessary in order to detect when some form of cognitive recovery is taking place.

Condition	Consciousness	Sleep /wake	Motor Function	Auditory Function	Visual Fixation	Communication	Emotion
Coma	None	Absent	Reflex and postural responses only	None	None	None	None
Vegetative State	None	Present	Postures or withdraws from noxious stimuli Occasional non purposeful movement	Startle Brief orienting to sound	Startle Brief visual Fixation	None	None Reflexive crying or smiling
Minimally Conscious State	Partial	Present	Localizes noxious stimuli Reaches for objects Holds or touches objects in a manner that accommodate size and shape Automatic movements (e.g., scratching)	Localizes sound location Inconsistent command following	Sustained visual fixation Sustained inconsistent but intelligible verbalization or gesture	Contingent vocalization Inconsistent but intelligible verbalization or gesture	Contingent smiling or crying

Table 2. Comparison of clinical features associated with vegetative state and minimally conscious state

Regrettably, there are a number of case studies that illustrate instances of cognitive recovery when attendants, staff and relatives believed the individual to have still been in a vegetative state. This stage is critical and requires patience and more recovery time. The rehabilitation team should dedicate their specialized skills and techniques to get away from their conditions. The rehabilitation team can apply the appropriate evidence-based pharmacological and non-pharmacological methods. For example, to recover from awareness medication like amantadine, sensory modality assessment and rehabilitation technique (SMART) and coma recovery programme–revised (CRS-R) can be utilized. The rehabilitation team might consider that the involvement of a psychologist or specialized neuropsychology is preferable to the involvement of a psychiatrist.

Many TBI patients are required to undergo appropriate counselling and psychotherapy. Many people with TBI develop behavioural problems in the short-term, particularly whilst emerging from a coma or during the phase of posttraumatic amnesia. A few individuals develop persistent and severe behavioural problems and can be a source of extreme disruption on the acute or rehabilitation ward, and certainly a source of major difficulty for the family. Eames and colleagues [26] stated that the application of behavioural management techniques could be effective in ameliorating difficult behaviour and in improving functional independence levels, as well as improving compliance with physical therapy even years after injury. Staff at a rehabilitation unit should have a degree of expertise in the management of behavioural problems; nevertheless, those with severe and persistent difficulties should be referred to appropriate psychiatric or specialized behavioural units.

The use of drug therapy in the management of such behaviours is best avoided, but in practise, this is unavoidable. Certainly, the use of sedative anxiolytics or psychotropic medication is generally unhelpful and may even worsen behaviour. Occasionally such intervention is essential, because of the proximity of vulnerable people of TBI or because of extreme pressure on staff time and resources. There is very little good quality literature on this subject; however, some studies have indicated an improvement in aggression and episodic dyscontrol by use of serotoninergic anti-depressant Trazodone [27] or the anti-convulsant Carbamazepine [28]. Other authors advocate the use of lithium or beta-blockade with Metoprolol [29]. If TBI patients became severely agitated, some advocate the use of Buspirone, which is chemically distinct from other anxiolytics.

For negative behaviours, some improvement is occasionally noticed following the use of dopamine agonists. A few authors continue to use stimulants such as Dexamethetamine or methyl phenidate, but such medication should be used with caution and only by those with some experience in the field. Other behavioural problems can be less troublesome but nonetheless give rise to marital stress, social isolation and often unemployment. Such problems can include egocentricity, poor judgment, lack of initiation, reduced achievement drive, lethargy, disinterest, lack of depth of feeling, irritability, aggressiveness, reduced tact and increase or decrease in sexual interest. Alongside these problems, both in the patient and the carer, can be the presence of associated mood disorders, particularly depressive illness and anxiety [30]. It is essential for the multidisciplinary team to recognize such problems and treat them appropriately.

There is no evidence that depressive illness responds less well in the context of acquired brain injury than in the context of endogenous depression. Thus, standard approaches, either psychological or pharmacological, should be used as aggressively as needed. Even if some of these problems are not remediable, they should be recognized and explained to the injured individual's family and colleagues. The immediate family will often benefit from counselling and supportive psychotherapy. Nowadays, before initiating pharmacotherapy or psychotherapy, patients should be screened to rule out any history of constipation, retention urine, infection, fracture and musculoskeletal illness. Levy et al. [31] stated that numerous studies have examined the use of medications in the treatment of agitation of post-TBI patients, but that there is limited evidence to help guide the clinician. Thus, the prescription of pharmacotherapy must be closely monitored and a multidisciplinary approach combining both pharmacological and non-pharmacological interventions may be necessary.

In the US, a number of studies have shown the efficacy of an employment support scheme. In such schemes, a trained rehabilitator accompanies the individual back to work and further rehabilitation will take place in the workplace, allowing an opportunity for specific goal orientated re-entry as well as an opportunity for education of the employer and work colleagues. It is regrettable that many rehabilitation facilities feel their job to be completed after the patient has been discharged back home, perhaps having received a follow up after the passing of a few months.

The best long-term outcome in those of working age is to return to their pre-accident employment situation. In many countries, little attention is paid to employment rehabilitation. Such rehabilitation rarely take place at all, or if it does, is the responsibility of a completely separate government department that likely lacks the necessary expertise for the management of those with brain injury problems. Following TBI recovery, rushing back to work too quickly is not advisable. If this happens, symptoms that were on the mend might flare up once more. A period of retraining may be required and adaptations to the workplace may be needed.

If an individual can return to work it will serve as a boost to their self-esteem and independence, particularly from a financial point of view. Such re-employment will clearly be of overall benefit to the state in terms of reduced benefits and may even enable the carer to return to employment as well. Wehman and colleagues [32] have clearly demonstrated the effectiveness of the supported employment programme. In his study forty-one [41] head injured people were included in the initial study;Only 36% of referred clients had achieved any competitive postinjury employment, compared with 91% of the same group who were competitively employed before injury. A job retention rate of 71% was reported, with most jobs in warehouse, clerical, and service-related occupations. This group presented a mean period of seven years from injury and thus the chances of spontaneous recovery were minimal.

The improved integration of employment professionals with health and social service professionals must be a priority in the future. At present, simulator-based education programmes have a strong scientific basis. Virtual reality driving simulation rehabilitation training (Figures 7 and 8) has shown promising results with respect to retraining driving performance and behaviour among military personnel recovering from TBI [33]. This is a new development in advanced driving solutions for disabled individuals after TBI or stroke and spinal cord injury, employing a fully immersive car driving simulator ideally suited for driver assessment and rehabilitation application.

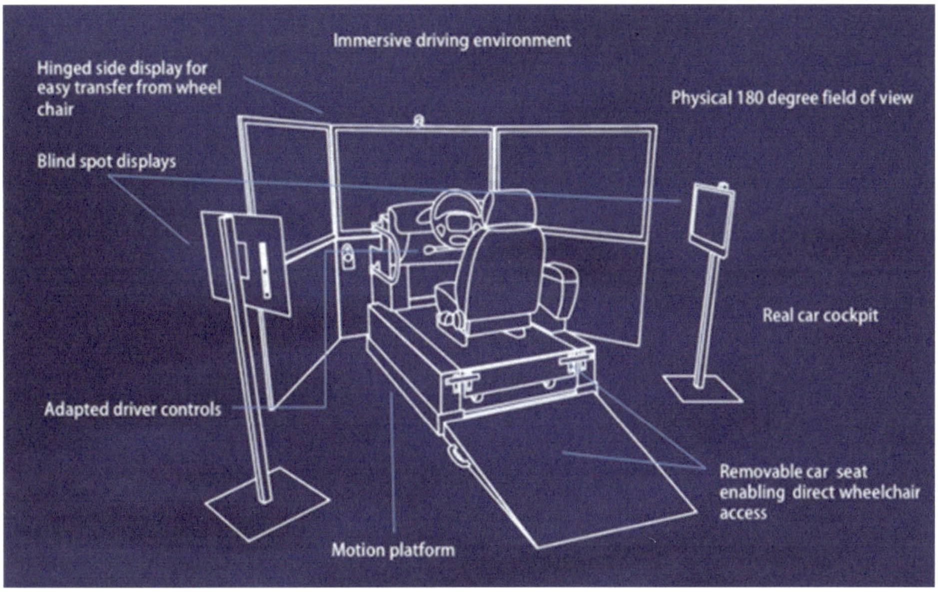

Figure 7. Virtual reality driving simulation rehabilitation training

Figure 8. Virtual reality driving simulation rehabilitation training

3. Conclusion

TBI is a diverse disorder of major public health significance. Rehabilitation services, matched to the needs of people with TBI, as well as community-based nonmedical services are required to optimize outcomes over the course of recovery. Both the person with TBI and their social support networks should have access to rehabilitation services through the entire course of recovery, which will continue for many years after the injury. The services required will change as the person's needs change over time. Survivors of severe TBI face the challenge of resuming a meaningful life for themselves and their families. However, severe TBI is not curable and medical and rehabilitation management may not ultimately be able to provide the improvement desired by the patient and his/her family.

In summary, TBI rehabilitation is a recognized subspecialty of neurorehabilitation and there is increasing awareness of its important role in early management for all severities of injury. While evidence for its effectiveness and specific interventions is limited, emerging therapies need to be subjected to rigorous research. The families of TBI survivors, particularly the severely injured, young TBI patients and those in a vegetative state accept most of the social and societal burdens of long-term care. It is crucial to bear in mind that prevention of TBI is vital, as there remains no cure for the sequelae of either moderate or severe TBI.

Author details

Wafa Al Yazeedi, Loganathan Venkatachalam, Somaya Al Molawi and Fatma Al Kuwari

Department of Physical Medicine & Rehabilitation, Hamad Medical Corporation, Doha, Qatar

References

[1] Data from the National Hospital Ambulatory Medical Care Survey, 1995-1996. National Center for Health Statistics.

[2] https://www.headway.org.uk/rehabilitation-after-brain-injury.aspx

[3] Al Yazeedi W, Venkatachalm L, Georgievski AB. Factors influencing rehabilitation outcome after adult traumatic brain Injury in Qatar. Qatar Medical Journal, 2010, 19(1): 11-16.

[4] urner-Strokes, Lynne. Rehabilitation Following Acquired Brain Injury: National Clinical Guidelines. British Society of Rehabilitation. Clin Med. 2004 Jan-Feb;4(1):61-5.

[5] Barnes MP. Rehabilitation after traumatic brain injury. Br. Med. Bull.1999; 55(4): 927-43.

[6] Blackerby WF. Intensity of rehabilitation and length of stay. Brain In., 1990, 4:167-73.

[7] Cope DN, Hall K. Head injury: benefit of early intervention. Arch Phys Med Rehabil., 1982, 63:433-7.

[8] Aronow HU. Rehabilitation effectiveness with severe brain injury: translating research into policy. Head Trauma Rehabil., 1987, 2:24-36.

[9] McKinlay WW, Brooks DN, Bond LR et al. The short term outcome of severe blunt head injury as reported by relatives of the injured persons. NeurolNeurosurg Psychiatry, 1981, 44:527-33.

[10] Oddy M, Humphrey M, Uttley D. Stresses upon the relatives of head injured patients. Br Psychiatry, 1978, 133: 507-13.

[11] Simpson DM, Alexander DN, O'Brien CF et al. Botuhnum toxin type A in the treatment of upper extremity spasticity: a randomised, double-blind, placebo controlled trial. Neurology, 1996, 46:1306-10.

[12] Yablon SA, Agana BT, Ivanhoe CB et al. Botuhnum toxin in severe upper extremity spasticity among patients with traumatic brain injury: an open labelled trial. Neurology, 1996, 47:939-44.

[13] Dejerne A, Ceillier A. Para-osteo-arthropathies des paraplegiques par lesion medullaire; etude clinique et radiographique. Ann Med., 1918, 5:497.

[14] Orzel JA, Rudd TG. Heterotopic bone formation: clinical, laboratory and imaging correlation. J Nucl Med., 1985, 26: 125-132.

[15] Rogers RC. Heterotopic ossification in severe head injury: a preventive programme. Brain Inj., 1988, 2:169-73.

[16] Bader DL (Ed) Pressure Sores – Clinical Practice and Scientific Approach. London: MacMillan, 1990.

[17] Lipides J, Diokno AC, Lowe BS et al. Follow-up on sterile, intermittent self-catheterization.J.urol /1974, 111:184-7.

[18] Urban RJ, Harris P, Masel B. Anterior hypopituitarism following traumatic brain injury. Brain Inj., 2005, 19(5):349-58.

[19] http://www.worldhealth.net/pdf/Gordon_Thera10.pdf

[20] Jin H, Wang S, Hou L, Pan C Li, B Wang, H Yu. Clinical treatment of traumatic brain injury complicated by cranial nerve injury. Injury, 2010, 41(9):918-23.

[21] Wilson B, Moffat N. The Clinical Management of Memory Problems London: Croom Helm, 1984.

[22] Wilson BA, Baddeley A, Evans J, Shiel A. Errorless learning in the rehabilitation of memory impaired people. Neuropsychol Rehabil., 1994, 4:307-26.

[23] Gianutsos R. Cognitive rehabilitation a neuropsychological speciality comes of age. Bram Inj., 1991, 5:353-68.

[24] Briccolo A, Turazzi S, Fenot HG. Prolonged post-traumatic unconsciousness. Neurosurg., 1980, 52:625-34.

[25] Andrews K, Murphy L, Munday R, Littlewood C. Misdiagnosis of the vegetative state, a retrospective study in a rehabilitation unit. BMJ, 1996, 313:13-6.

[26] Eames P, Cotterill G, Kneale TA et al. Outcome of intensive rehabilitation after severe brain injury: a long term follow-up study. Bram Inj., 1996, 10:631-50.

[27] Simpson DM, Foster D. Improvement in organically disturbed behavior with Trazodone treatment. Clin Psychol. 1986, 47:191-3.

[28] Foster HG, Hillbrand M, Chi CC. Efficacy of Carbamazepine in assaultive patients with frontal lobe dysfunction. Prog Neuropsychopharmacol Biol Psychiatry, 1989, 13:865-74.

[29] Mattes JA. Metoprolol for intermittent explosive disorders. Am J Psychol., 1985, 142:1108-9.

[30] Gualtien CT. Pharmaco therapy and the neurobehavioral sequelae of traumatic brain injury. Bram In., 1988, 2:101-29.

[31] Levy M, Berson A, Cook T, Bollegala N, Seto E, Tursanski S, Kim. Treatment of agitation following traumatic brain injury: a review of the literature. NeuroRehabilitation, 2005, 20(4):279-306.

[32] Wehman P, Kreutzer JS, West MD et al. Return to work for persons with traumatic brain injury: a supported employment approach. Arch Phys Med Rehabil., 1990, 71:1047-52.

[33] Cox DJ, Davis M, Singh H, Barbour B, Nidiffer FD. Driving rehabilitation for military personnel recovering from traumatic brain injury using virtual reality driving simulation: a feasibility study. Mil Med., 2010, 175(6):411-6.

Chapter 14

Physiotherapeutic Procedures for the Treatment of Contractures in Subjects with Traumatic Brain Injury (TBI)

Fernando Salierno, María Elisa Rivas,
Pablo Etchandy, Verónica Jarmoluk, Diego Cozzo,
Martín Mattei, Eliana Buffetti,
Leonardo Corrotea and Mercedes Tamashiro

Additional information is available at the end of the chapter

http://dx.doi.org/10.5772/57310

1. Introduction

Contractures limit free joint movement and are common a consequence of traumatic brain injury. They interfere with activities of daily living and can cause pain, pressure areas, and result in unsightly deformities [1 - 4], affecting patient quality of life and increasing institutionalization rates. Contractures also cause significant secondary impairment which ultimately interferes with the rehabilitation process. Their treatment is therefore an integral part of physical recovery. Before effective intervention can take place, therapists must first determine both the primary cause as well as the specific structures involved. Several different therapeutic modalities exist to treat them, and choice of which to apply will depend on each individual case [5].

The aim of this chapter therefore is to review currently available physical therapy techniques for the treatment of contractures and for prevention of deformity development, in subjects suffering traumatic brain injury (TBI).

2. Generalities

Contractures are a common complication of traumatic brain injury and may occur in up to 84% of cases [4, 6]. The most commonly affected joints are: the hip, shoulder, ankle, elbow and knee, with a significant percentage of patients developing contractures in five or more joints [4].

For purposes of this chapter, we define contracture as any degree of loss in joint range of motion restricting activities of daily living[4]. Movement restriction is not limited only to joints, but will also affect many other body structures including skin, subcutaneous tissue, muscles, tendons, ligaments, joint capsules, vessels and nerves [7].

Contractures are characterized by reduced range of motion (ROM) and increased stiffness. The increased resistance to stretch caused by changes in the mechanical properties of tissues is due to both neurally and non-neurally mediated factors [8, 9]. Non-neural factors include changes in mechanical properties of tissue resulting from stress deprivation, and may be secondary to orthopedic injury, heterotopic ossification, use of a splint or plaster, pain, paralysis, severe spasticity or any disorder that restricts movement. [10]. Contractures also produce structural changes within muscles; myofibril shortening and loss of sarcomeres are often observed, as well as relative increase in connective tissue causing loss of elasticity.

Neural factors are of central origin and cause muscle overactivity. They generate spasticity, increasing interdigitation between actin and myosin, thus reducing muscular concentric contraction range and producing rigidity, as a result of the absence of monosynaptic reflex inhibition [9, 11].

3. Evaluation and diagnosis

Evaluation of patients with head trauma includes examination of joint range of motion, looking for possible contractures. Prior to physical exam, it is important to record general patient posture (decortication, decerebration, etc.) and clinical condition, and especially to register any hyperexcitability such as spasms or clonus triggered by intense stimulation in patients with sensory impairment.

Joints should be observed at rest in order to establish whether muscle overactivity or flaccidity is present; next, voluntary movement should be examined, and presence of dystonia when changing positions recorded, if elicited [12]. These findings are important because contracture development in patients with TBI is related to presence of spasticity and/or dystonia[13]. Joint examination is performed systematically, mobilizing each joint manually and passively. In patients able to follow verbal commands and perform active movements, it is important to encourage them to move each segment to the widest possible range.

Once manual examination has been completed, findings in individual joints should be recorded including quality of resistance and end feel, which refers to examiner perception of a barrier to further motion when exerting passive ROM. For better diagnosis or evaluation, passive range of motion may be measured using an instrument such as a goniometer or applying other objective methods. The universal goniometer (UG) is simple and easy-to-use. Result reliability is acceptable, but only after appropriate user training and always applying standardized measuring methods [14, 15].

http://dx.doi.org/10.5772/57310

When loss of range of motion is detected, differential diagnosis needs to be established between lack of range due to motor deficit (muscle overactivity) or resulting from structural alterations. Clinical assessment of hypertonia can be conducted using a well-established rating scale, the Modified Ashworth Scale. However therapists should be aware that this scale does not distinguish between soft tissue and neural contributions to hypertonia [16]. To make this differential diagnosis, the contracture segments need to be re-mobilized at different speeds in order to unmask muscle overactivity, which if present, can be rated using the Modified Tardieu Scale. This scale is widely used because it renders both quantitative and qualitative measurement of spasticity. Its validity has been studied and has demonstrated high intra and inter-rater reliability when using a standardized protocol. It is highly reliable for assessing spasticity in hamstrings, rectus femoris, gastrocnemius, soleus and tibialis anterior muscles in adults with neurological injury. It has also shown very good intra rater reliability for elbow flexors [17 - 18].

Other important factors that need to be considered in the individual patient and factored into the therapeutic decision-making process include: periods of prolonged bedrest, presence of subcortical lesions, and decreased levels of awareness or lack of voluntary activity in the presence of muscle overactivity, as they are all indicators of greater lesional complexity. Yarkony et al. postulate contracture as clinically evident when coma extends beyond three weeks [4]. Severity of contracture is also generally more pronounced in patients with brainstem lesions and treatment usually begins later and will require more time to achieve gains in ROM [19].

Therefore, one can expect coma patients to suffer greater structural changes in muscle tissues both because of worse neurological status as well as absence of voluntary movement countering effects of immobility.

Motor control and presence of spasticity are important. For example, if no underlying motor control is present and spasticity is not reduced after contracture treatment, probability of sustaining improvement is poor [19]. In contrast, presence of voluntary motor control in muscles antagonist to the contracture, increase the likelihood of maintaining any ROM gain [20].

Cognitive impairment will also influence rehabilitation therapy choice. Sometimes, cognitive impairment may interfere with patient ability to cooperate and interpret pain[19].

Similarly, patients presenting preserved cognitive status, voluntary motor control and minimal or no muscle overactivity, will attain better treatment outcomes.

Once a thorough evaluation has been completed, a treatment plan can be individually designed and the most appropriate technique chosen to treat or prevent contracture. Treatment should be specifically adapted to each patient and will require constant reassessment in light of changes occurring in the course of recovering from head injuries. See table 1.

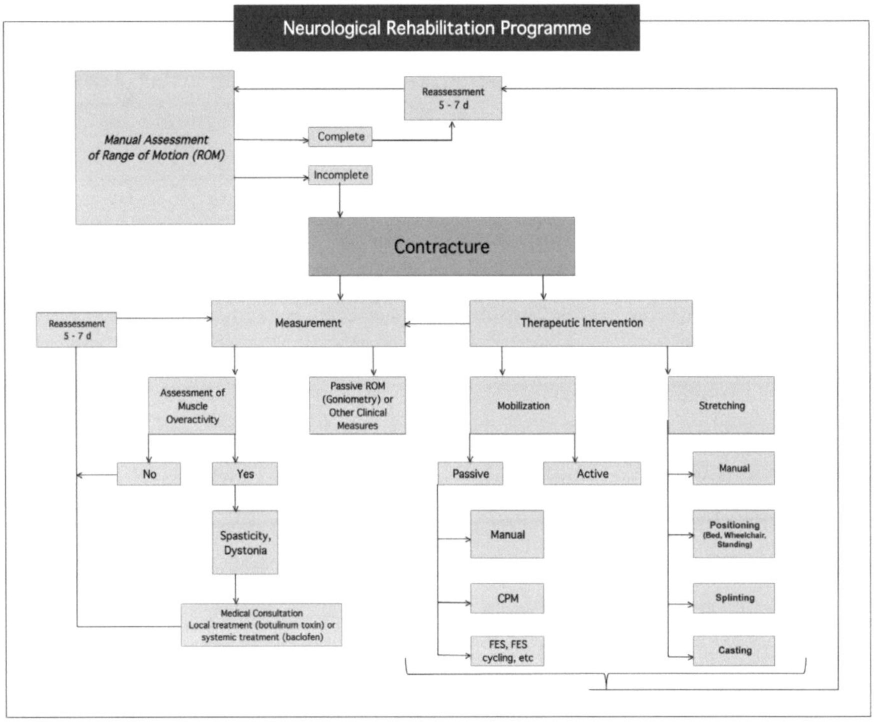

Table 1. Clinical reasoning and therapeutic choice for the contracture. CPM, continuous passive motion. FES, functional electrical stimulation. 5 – 7d, 5 - 7days.

4. Therapeutic modalities

Stretch is one of the most widely used techniques for treatment and prevention of contractures. Its aim is to increase joint mobility and it can be self-administered or applied manually by therapists. Splints, positioning programs or casts changed at regular intervals (serial casting) can also be used. All methods involve mechanical elongation of soft tissues during varying lengths of time. Some can only be applied for short periods, such as manually applied elongations, performed for only a few minutes at a time. Others, such as splints and plasters are used to stretch muscles for longer periods, and sometimes to provide uninterrupted elongation for days or even weeks.

5. Stretching

Research in physiotherapy and neurophysiology has led to the development of some alternatives to passive stretching. Cherry[10] described four approaches to reduce contractures, namely: activating or strengthening the weak agonist, local inhibition, general inhibition and passive lengthening, some of which may be used simultaneously.

The first approach involves activating or strengthening the weak agonist opposing the tight muscle. If tissue innervation is intact and the agonist has the ability to function at all, a double benefit will be gained by improving its ability to contract. If the agonist becomes stronger, it will be able to counter the contracture of the antagonist and pull the joint through more complete range. Also, the contracted muscle (antagonist) will be reciprocally inhibited, allowing itself to be stretched because the stretch reflex is also inhibited.

If the weak agonist can be activated and strengthened, better muscle balance around the joint may result, reducing the potential for myostatic contracture recurrence.

Selection of a technique to strengthen the weak muscle will depend on the nature of the problem and the ability of the patient to cooperate. Effective strengthening methods that employ resistance or load on muscle include maximal resistance in diagonal spiral patterns and progressive resistance exercises. Methods that use unconscious automatic responses to activate the weak agonist require eliciting righting reactions and equilibrium responses.

The second approach may be used when an agonist is unable to contract at all, or the antagonist muscle may be so tight that attempts to strengthen the agonist fail. Also, even if the agonist can be activated, it will move through more complete range of motion if the stretch reflex of the tight antagonist can be inhibited. In this approach, inhibition of the tight muscle is considered. The propioceptive neuromuscular facilitation techniques, such as contraction-relaxation or hold-relax may be used to inhibit the tight muscle selectively, so it will tolerate being stretched without immediate activation of the stretch reflex.

Local inhibition is useful for localized tightness, especially within one muscle group at a single joint, such as after plaster immobilization following injury or surgery. Techniques generating inhibition to the contracted muscle include vibration of the opposite muscle group, prolonged icing and hold-relax, used in traditional rehabilitation.

A third approach to reduce hypertonus, in a single limb or throughout the whole body, consists in allowing tight spastic muscle groups to relax and be lengthened. Some concepts as neuro-developmental treatment (NDT) consider that movement control is achieved in an integrated way when the nervous system works cooperatively. For example the organization of postural systems require interaction between external forces (gravity), body mechanics and kinetics, multisensory inputs, and adaptative responses to voluntary movements. Therefore, inhibition of hypertonus is best when the patient participates actively.

The fourth approach is passive lengthening. If passive lengthening is selected as the appropriate alternative, there are two techniques that may be applied. One involves manual passive stretching and the other prolonged holding of the desired position at the point of maximum tolerated length of the contracted muscle.

Passive stretching is likely to be most effective in individuals whose stretch reflex is inhibited by cortical effect or by peripheral nerve injury.

Adaptative equipment may enable an individual to function more readily in certain positions, for example use of a prone standing board for hip and knee extension, to align lower limb joints in weight bearing patients with lower limb flexor problems. Also, positions for sitting, sleeping, and other daily activities, useful in correcting contracture, can be adopted and held for prolonged periods.

Other techniques of passive lengthening include standing, orthosis, splints and casts with prolonged holding. All require close monitoring (see below). Generally, patients with head trauma present severe injuries. It is very common to find extremely weak agonist muscles to work with, or to attempt to generate antagonist tone inhibition on. When these techniques are employed it is important to consider that any increase in range obtained by forced motion will be lost unless it is maintained by active motion or by use of supportive devices.

6. Standing

TBI patients can be supported while standing using a tilt table, standing frame and\or knee extension splints. Regular standing in the correct position will maintain sufficient range of dorsal flexion of the ankle joint for walking when the patient reaches that stage but, if he has not been made to get out of bed from the beginning; his Achilles tendon may have already shortened [21]. In our experience, the first option for patients with more severe motor deficits or disorders of consciousness is the tilt table. Special attention must be paid to the knee and ankle. If the knee has full range, it should be positioned in slight flexion or in a neutral stance. The ankle must be able to bear sufficient weight on the rearfoot without causing tissue damage or increasing muscle overactivity.

The second option is the standing frame. Patient can be supported in upright position by a solid frame with padded struts in front of his knees to keep them extended and a broad strap behind his hips to prevent them from flexing. Using the standing frame may be useful when the patient has regained consciousness and is able to extend his trunk actively, because it is then possible for him to stand for longer periods.

The third option is standing assisted by the therapist, recommended for patients with better functional status and trunk control. If necessary, extension knee splints can be used to prevent lower limbs from giving way due to lack of motor activity. Knee extension splints should reach from about 8 cm below the ischial tuberosity to 4cm above the malleoli. Made of a firm posterior shell of plaster of Paris or some other hard material, knee-extension splints will need to be bandaged in place with two 10cm wide crepe bandages to give adequate support and maintain optimal leg positioning. Patient should be lying in supine position while therapist bandages on the splints, to be able to correct any inward or outward rotation of the limb [21].

When the patient has marked spasticity in the plantar flexors of the foot, standing upright is often the only way in which therapists can maintain range of ankle dorsiflexion and prevent shortening of the Achilles tendon [21].

In case of equinus deformity that does not allow the heel to contact the ground, a posting on the foot, AFO (anke-foot orthosis) or shoe is needed. Posting entails construction of an interface between the aligned foot and the ground to optimize heel loading and hip and knee alignment while standing or walking (filling in the space under the heel when the ankle is positioned in plantarflexion) [22].

When standing the patient up, one must consider alignment of the pelvis in the frontal plane, since use of a post will be shorten contra-lateral limb length. This must be compensated with an additional pad under the contra-lateral heel, high enough to correct pelvic alignment.

There is evidence to support placing patients in the standing position to prevent loss of calf muscle length. In severe brain trauma cases where little functional recovery is expected, maintaining patient physical condition is challenging. In this context, long term effects of supported standing to maintain muscle length appear more important. Although often prescribed, effectiveness of standing programmes carried out over months or even years remains unknown. Standing may prevent small losses of ankle dorsiflexion, but clinical importance of these effects is uncertain. Future studies should investigate standing in a wider range of settings. Evaluating potential multidimensional effects of standing using standardized measures would provide greater insight and be more effective than studies focused on a single outcome [23].

7. Splinting – Orthosis

Use of splinting for both prevention and reduction of contractures is recommended following traumatic brain injury [24]. Alternatively, orthosis also reduces contractures through prolonged low-load stretch, maintaining joints lengthened [25]. Therapists who apply a biomechanical treatment rationale recommend splinting both to prevent as well as manage length-associated changes in muscles and connective tissue.

7.1. Types of splints and outcome

Splints of various forms, often combined with other passive and/or active tissue 'stretching' procedures (e.g. passive movements, positioning, weight-bearing), are the treatment of choice for physiotherapists. General goals of therapy seek to inhibit/reduce increased muscle tone and/or elongate shortened soft tissues. In the case of some splints, an additional goal may be improving/maintaining appropriate biomechanical limb positioning during (later) functional activity retraining, such as walking [24].

Elongation effects obtained can be sustained by using splints overnight. Advantages to wearing a night splint are: overnight intervention which allows therapy time to be spent on active retraining of everyday tasks, ease of application, and perhaps most importantly, continued long term use after discharge from hospital. The main disadvantage is the risk of pressure sores, particularly when patients have poor vascular supply and/or sensation in the affected areas [26]. It is important for the therapist to monitor splint use with the interdisciplinary team. When indicating a night orthosis, patient tolerance and adherence to use should be closely controlled. In the presence of adverse effects the splint should be removed immediately and its indication reassessed.

A published case report illustrating lower limb orthosis implementation described use of a static adjustable ankle orthosis, placed after administration of a phenol nerve block in conjunction with stretching, strengthening and functional mobility training. The adjustable orthosis was applied to provide low-load prolonged stretch of the ankle and address apparent soft tissue shortening, and the phenol nerve block was administered to address ROM limitations secondary to muscle spasm [27]. Positive results reported in this case would support orthosis as an addition to standard physical therapy stretching regimens in patients with brain injury.

Other published evidence indicates that overnight splinting of an affected ankle in subjects with stroke appears to be as effective as standing on a tilt table in preventing contracture at the ankle [26].

In our experience an alternative intervention to prevent ankle contractures is use of an overnight splint. Although the ankle is not usually positioned in maximum dorsiflexion, the extra time spent in this position helps avoid contracture development.

Indication for upper limb splint placement is common, mostly at rest to keep the joint in extended position, thus halting further shortening.

According to Lannin [27], night splinting the hand in the functional resting position does not produce clinically useful effects in adults with acquired brain impairment who receive a daily stretching programme. In his discussion, the author postulates absence of effect after splint use may have been because routine motor training and upper limb stretch were maintaining muscle length. Therefore, additional stretching provided by the night splint may have been redundant.

Sometimes, therapists indicate additional electric stimulation and splinting for contracture management, but it is not clear if it is more effective than splinting alone after acquired brain injury [28].

7.2. Evidence and conclusion

There is insufficient evidence to either support or refute effectiveness of orthotic devices for contracture treatment or prevention. Further research is needed for better understanding of the influence of elongation on soft tissue, as well as with respect to timing, length of application and efficacy of splinting programmes. In conclusion, splinting options should be carefully analyzed in TBI patients as they may represent a clinically effective strategy when used appropriately.

8. Casting

Casts are a viable option for treating contractures after upper motor neuron injury in adults [19, 30]. These are non-removable external devices, made of plaster or casting tape, applied with intention to change structural or functional characteristics of the neuromuscular system [31]. Casting to control hypertonus (understood as increased resistance to passive movement as a result of spasticity and/or changes in muscular and connective tissue characteristics) was first described in the 1960's in the treatment of children with cerebral palsy [20, 32]. Later, casts were applied to adults with acquired brain injury with different therapeutic objectives. These goals include preventing loss of joint range, helping to cure pressure ulcers associated with severe spasticity, restoration of articular range (ROM) and muscle length, and inhibition of hyperexcitability [20].

Different theories have been proposed to explain underlying neurophysiological and mechanical mechanisms behind the positive effects of plasters, on hyperexcitability reflexes and mechanical changes [32].

The first hypothesis relates to neurophysiology. Plasters prevent changes in muscle length by eliminating excitatory input to muscle receptors, which in turn reduces spasticity. Prolonged elongation stimulates Golgi tendon organs and subsequently afferent Ib fibers, creating an inhibitory response in alpha motor neurons. Type II muscle afferents have also been postulated as inhibitors of alpha motor neurons following prolonged elongation [32]. Casts may also reduce spastic muscle tone through alleviation or reduction of tactile input, proprioceptive input, and temperature receptors. Uninterrupted contact provided by the cast, including the neutral pressure and heat generated, reduces excitability of alpha and gamma motor neurons at the spinal cord level [30 - 32].

A second hypothesis is based on biomechanics. It postulates that casts achieve low load elongation during prolonged periods, able to prevent and correct contractures. Casts are usually applied at the end of joint range of motion because prolonged stretch with low load, generating permanent changes in soft tissues. It has been well established that with prolonged immobilization, specific physiologic changes occur involving connective tissue and muscle remodeling. This remodeling is mediated by fibroblasts in response to physical forces. When a muscle is immobilized in a stretched position, the number and length of sarcomeres in series increases, thus changing the muscle tension / length ratio. Connective tissue is also extended through a process of disorganization of the fibrous matrix [30].

The third hypothesis refers to motor learning. It proposes that casts provide adequate support to proximal joints, until patient gains enough distal control [31]. The external stability provided by the cast to the limb is presumed to allow the patient to receive normal sensory input with appropriate weight bearing and normal reflex patterns, helping to develop normal movement and accommodation in the central nervous system [32].

8.1. Types of casts

Different casts used in TBI patient rehabilitation and described in the literature include: serial casts, inhibitive cast, drop out casts and bivalved casts [30]. Specific types of cast are important in treating contractures and indications will depend on patient characteristics and level of functional recovery.

Serial casting involves application and removal of a series of casts resulting in progressive range of motion increase with the introduction of each cast. Serial casting allows 24 hour a day elongation, with casts changed regularly to maintain gain as the joint becomes more mobile [33].

Inhibition cast. The purpose of an inhibitory cast application is to maintain a position, to reduce spasticity and facilitate improvements in motor function. Inhibitory casts seek to provide stability and inhibition to the treated joint. The inhibition may be achieved by normalizing proprioceptive input, joint alignment, and weight load. Effects of these casts do not seem to last long after they are removed; for this reason, inhibitory casts resulting in a positive impact on functional performance need to be followed by use of an inhibitory splint during a prolonged period [30].

Functional drop out casts are the combination of serial casting and inhibitory casts. When making a functional cast, a portion of a cylindrical cast is removed to allow the involved joint to move beyond the desired range, preventing the joint from pushing back toward the contracted position. This allows passive or active movement in the desired direction and allows the user to gain additional active range while using a cast. These casts also allow application of electrical stimulation or other facilitation techniques [30].

Bivalved casts are casts cut in two halves (front and back), which are then filled and padded at the edges to allow reapplication. Usually this is done when the patient has achieved the desired range and needs to maintain the new position. This is required when tone in the involved limb remains high and there is doubt whether a traditional splint will resist and keep the muscle in the desired position. Bivalved casts provide full contact and have the advantage that they can be removed for hygiene, inspection, active movement, and other dynamic activities [30].

The main indications for use of any type of plaster in general, and of serial casting in particular, is permanent limitation of range of motion [19], or immobility during muscle function only. If other disorders such as muscle weakness or overactivity of some kind are present, these will influence both plastering technique as well as length of treatment.

8.2. Application procedure

The cast-making process follows a basic preparation which adapts to any plastering technique, and consists first in the placement of soft materials such stockinette, foam and cast padding. to protect the skin and bony prominences and avoid friction and pressure points, followed by application of a plaster bandage (fiberglass and/or plaster cast) to achieve the necessary hardness to fix the limb in the desired position.

There two basic objectives sought after by the procedure: to place tension on the muscle and maintain the pressure during a pre established time period.

Tension levels will vary depending on which theory the therapy is based on. If we consider the biomechanics theory, plasters are applied on final range in order to gain maximum length. Neurophysiology-based theory considers tension levels should stretch the joint to within 5 degrees of final existing range. Variability in studies on angle used to make the cast range from 5° and 10° under full range, to neutral, to end of available range. This last option is the one with higher levels of evidence to support it [31]. In our experience, we prefer to use end of range for plaster application. The force applied to elongate the muscle is usually limited by patient pain threshold [31].

With respect to duration of application, it is highly variable, and will depend on the treatment objective and on how much time is needed to achieve the desired effect. There is significant discrepancy in the literature over total length of treatment between published studies [30]. Some publications suggest shorter implementation generates fewer complications and similar results [31].

Significant variation also exists in relation to length of time between cast changes. Duration of casting generally should reflect the pace at which the individual is making progress and the goals of the casting. Tardieu and Tardieu recommend lengthening via casting should be very gradual because careless application may result in muscle fiber break down, but they do not give specific time frames or guidelines. If the patient is experiencing slower progress or is at lower risk for breakdown or complications, casts can be left on longer between changes; this is particularly true if a cast is difficult to apply because of time constraints, cost, patient agitation, or patient need for medication during the switch. If the individual seems to be progressing rapidly or is at increased risk for skin breakdown, if the cast is damaged or wet, or if there is any other concern, it should be changed more frequently [30].

In a serial casting program, we start by placing an initial plaster at rest. This is applied with the extremity positioned at the end of the movement range, but easily reached without applying additional tension. Generally, we use 3 or 4 progressive casts [19]. Some authors such as Davies recommend 6 changes on average [21].

Authors suggest mobilizing passively through the entire range of motion available to maintain full mobility of all immobilized joints and then apply a new cast at a greater angle [23]. In contrast, Davies [21] does not mobilize in between casts, to avoid any chance on losing the range gained. Currently, there is no agreement on whether the extremity should or should not be mobilized between each change of plaster.

The last cast in the series is the supporting or positioning cast. It is cut in two halves making it bivalved, like an anterior-posterior splint [19]. This splint is used in the resting position for as long as deemed necessary until the patient improves and gradually stops using the equipment[19, 20]. If the patient´s muscles are still weak and present overactivity, it is important to continue using the device overnight [19].

Cast progression should be discontinued when there is no further gain of range. Although there is no consensus between authors on this point, some suggest that progression can be interrupted if no quantifiable gain is achieved after two consecutive applications. Others suggest stopping if there is no measurable gain (<5 ° measured with goniometer) in maximum range after three consecutive plaster changes. [20] If the team decides to end the application, it is important the patient be examined by a physician to consider the possibility of orthopedic surgery [20].

Therefore, application, duration and frequency of change will need to be individualized, defining for each protocol: timing, whether it will be a single plaster or several, how often casts will be changed, when joint range will be progressively increased and finally, total duration of single or serial casting [31].

8.3. Precautions, complications and contraindications

There are a number of precautions to consider before placing a cast. With respect to the patient it is important to take into account skin integrity, presence of fluctuating edema, decreased sensitivity, cognitive impairment and agitation [19, 30]. Therapist experience will contribute to establish level of existing complications and whether these can be managed and still continue with the cast. For some authors contracture treatment has priority over any skin condition, because they argue that it can help heal skin lesions [21]. Fluctuating edema and sensitivity disorders can be managed with frequent controls after application and if necessary the plaster may be removed at any time.

In cases of cognitive disorders and agitation, sedation or use of restraints have been described in the literature, so that the patient does not hurt himself or others during the placing of the plaster and is careful while wearing it [21].

Casting contraindications include: uncontrolled hypertension and/or elevated intracranial pressure, open wounds, external fixation or unresolved fractures, ligament injuries, need for access to check vital signs, recent episodes of autonomic dysreflexia, circulatory disorders such as deep vein thrombosis (DVT), acute inflammation, heterotopic ossification [19, 30, 33, 34], tone fluctuation or any unstable medical condition [31]. Some authors also exclude pregnant patients [34].

Adverse effects and/or complications should be considered. Close monitoring, assessing sensitivity, motion, blood flow, skin indemnity and the presence of inflammation is always recommended, checking for presence of vasoconstriction or discoloration of fingers or toes [33]. Temporary discoloration is quite common, but if it persists for more than 20 or 30 seconds, plaster should be removed and a new one applied [19]. It should be noted that patients often report less adverse effects than therapists [33].

In a study by Moseley, adverse effects reported included skin irritation, skin breakdown, pain, inflammation and dysautonomic events [31]. Inflammation and pain can be relieved by limb elevation, applying plaster with less tension and use of analgesia. Irritation and breakdown of skin are serious complications, and often require discontinuation. To avoid complications, patient selection is important together with careful monitor and regular plaster change (once to twice weekly)[33]. If patient develops pain cast may also need to be changed.

8.4. Complementary treatment

Casts are never applied as isolated intervention. Patients who are treated with casts usually also participate in multidisciplinary rehabilitation programs [33].

Techniques complementing treatment with casts include: mobilization, stretching, electrical stimulation, neurodinamics[21], mobility and strengthening exercises and any neuro-rehabilitation techniques such as NDT or PNF (propioceptive neuromuscular facilitation)

It is important to actively or passively mobilize joints adjacent to the cast as a routine practice, as many times a day as the therapist deems necessary.

Electrostimulation is mostly used with dropout techniques, and once the plaster is cut and bivalved can also be used to prevent atrophy or enhance voluntary motor activity.

Exercise plans are designed for each patient, to stimulate activity in paralyzed muscles and encourage voluntary motor control improvement [33].

For upper limb casts we recommend simultaneous exercises such as weight bearing activities, bilateral functional skill development, motor control training on muscles on the periphery of the cast, and voluntary isometric contraction of muscles included in the cast which can be contracted.

For lower limb casts, repetitive exercises are indicated to activate weak muscles in task-related training and improve strength and coordination. It is important that patients be helped to stand with supporting equipment, which will vary depending on motor reserve (tilt table, standing frame, or assisted by the therapist). In this manner patient is forced to use the affected limb. If the patient retains a good level of standing balance, exercises involving weight bearing on the affected limb can be considered, while other limbs perform swings or reaching for objects in the case of the upper extremity.

For subjects who respond minimally, treatment may focus on following simple commands to induce muscle activity in addition to the passive modalities mentioned before [33].

Physical therapists must consider a program of stretching, positioning and indication of night splints, or even use of splints during the day after removal of final cast [31].

Maintaining range gained with cast progression is difficult when muscle overactivity persists as a result of CNS injury and lack of motor control, or of voluntary activity in muscles antagonist to the contracture.

Another common combination is the application of progressive casts with botulinum toxin for muscle overactivity management. Toxin is often used as first line drug therapy for focal spasticity [34]. However, authors disagree on whether toxin use promotes or enhances cast effect [35]. One study attempted to determine whether serial casting combined with botulinum toxin reduced contracture development in calf muscles after severe head injury, and concluded cast alone was sufficient [34].

In the presence of severe or persistent dystonia pharmacological intervention may be required. Neurosurgery is another option to maintain gains obtained with serial casting [20].

The effectiveness of plasters has been compared to that of other therapeutic modalities in the TBI population. According to Moseley, casts are more effective in the short term than positioning for one hour a day to reduce elbow flexion contracture in patients with TBI, observing that treatment difference was not maintained over time [33]. Therefore, although serial casting is effective in reducing contracture deformity after acquired brain injury, and serial casting induces transient increase in range of motion, these effects are not maintained longterm [33].

8.5. Evidence and conclusion

There is insufficient evidence to support or refute the effectiveness of plasters in the upper limbs after an acquired brain injury. Although more evidence has been published for the lower limbs, systematic reviews on the use of casts also conclude there is lack of strong and consistent evidence to support their use in the lower limb [31].

Casts to improve passive joint range have a grade B level of recommendation, and for treatment of spasticity, a grade C. There is also lack of firm evidence therefore to recommend their use to improve function [32]. The development of clinical practice guidelines is limited because most studies are case series. One of the major limitations of research is the method selected for measuring results, very few studies use ROM measurement with controlled torque. For all the reasons mentioned above, authors disagree on when to prescribe casts, and on which approach to use and what underlying theory supports the indication. Currently, decisions on whether to use casts in patients with motor disorders due to CNS injury are based more on clinical judgment than on scientific evidence [31].

Finally, casts should be placed by experienced therapists in patients at early stages of their rehabilitation program. Associated complication rates will decrease as therapists get better at using the technique [19, 30, 34].

This treatment modality should always be part of a global and comprehensive rehabilitation program with a multifactorial vision of patient´s deficits, in particular motor control, in order to maximize functional recovery [19, 30].

9. Case studies

Case 1: Patient with Traumatic Brain Injury and Bilateral Equinus

MC, a 47 year old male diagnosed with multiple trauma and severe TBI after a motor vehicle accident was admitted to the rehabilitation center 137 days post event.

Patient was conscious, oriented in time and space and able to communicate effectively. Physical examination showed mild quadriparesis and loss of trunk control. Superficial and deep sensitivity was preserved with generalized hyperreflexia and normal muscle tone.

Passive mobilization of the soleous and calf showed tension, shortening was observed in both lower limbs, and plantar flexion and contracture of both ankles detected. No signs of muscle overactivity (under the modified Tardieu scale) were found, ruling out presence of spasticity. Voluntary movement was also present in both ankles and feet.

http://dx.doi.org/10.5772/57310

Goniometric measurement with knees extended indicated a right ankle limitation of -24 ° to reach neutral right-angle position and - 34 ° on the left ankle. Ankles remained in plantar flexion even during weight-bearing while standing.

Given the absence of spasticity and the limited range of motion in both ankles, as a result of prolonged immobilization during the acute stage of treatment for multiple complications, the therapeutic team interpreted patient contracture responded to non-neural factors (muscle shortening, structural changes in connective tissue) and implemented the following therapeutic strategies

- Daily stretching in the standing position for 40-50 minutes, first on the tilt table and then using a standing frame.
- Electrical muscle stimulation (EMS) on gastronemius-soleus and pretibial muscles as adjunctive treatment, in order to generate elongation by reciprocal inhibition of both muscle groups, this was implemented while patient was standing and for 20 minutes in each leg.
- Manual passive stretching.

Patient quickly recovered skills such as sitting and standing independently, so the team decided to apply four serial casts (Fig. 1). These were changed every 7 to 14 days for a total program duration of two months.

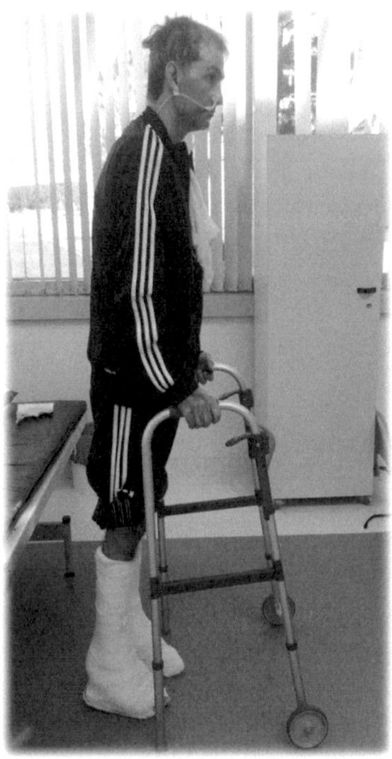

Figure 1. Independent standing with the first of the series of four casts.

Patient was allowed to rest for 2 days between casts, during which time he continued to stand using the standing frame with therapist assistance and received electro-therapy (Fig. 2).

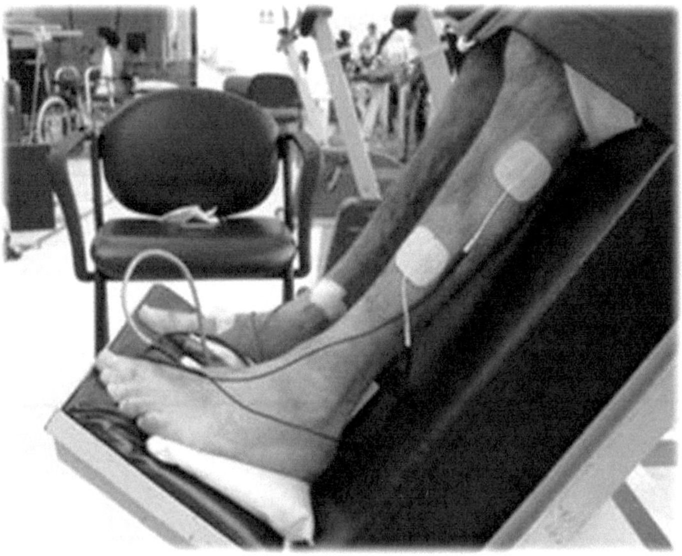

Figure 2. Stretching with electro-therapy on tilt table between casts.

While wearing the casts, patient performed lower limb weight-bearing exercises, balance and gait rehabilitation, with and without walking aids.

Serial cast result was positive with a ROM increase of -4 ° in the right ankle dorsi-flexion limitation, and of -10 ° in left ankle. Functional improvement was also achieved and patient was able to stand and walk unassisted, although use of a posting on both heels was required to compensate limited ankle range of motion (Fig. 3).

Case 2: Young adult with Traumatic Brain Injury and Multiple Contractures

CT, a 27 year-old man who suffered a traumatic brain injury as a result of a motor vehicle accident was referred to our rehabilitation center 53 days later. Imaging studies showed frontal and occipital hemorrhagic contusions had been treated with biparietal cranioplasty.

At time of admission patient was tracheostomized, evaluation of cognitive functioning showed level 3 function according to the Rancho Los Amigos Scale (localized response) and a score of 9/23 on the revised Coma Recovery Scale with visual pursuit, indicating a minimally conscious state. During physical examination, flexed global body posture and spastic flexor pattern was observed in all four limbs, with marked limitation in passive ROM and multiple contractures.

Goniometric measurement of right elbow movement indicated limitation in free range of motion from 10° to full flexion and in left elbow from 70° to full flexion. ROM could not been measured on wrists or hands because patient experienced pain during passive examination.

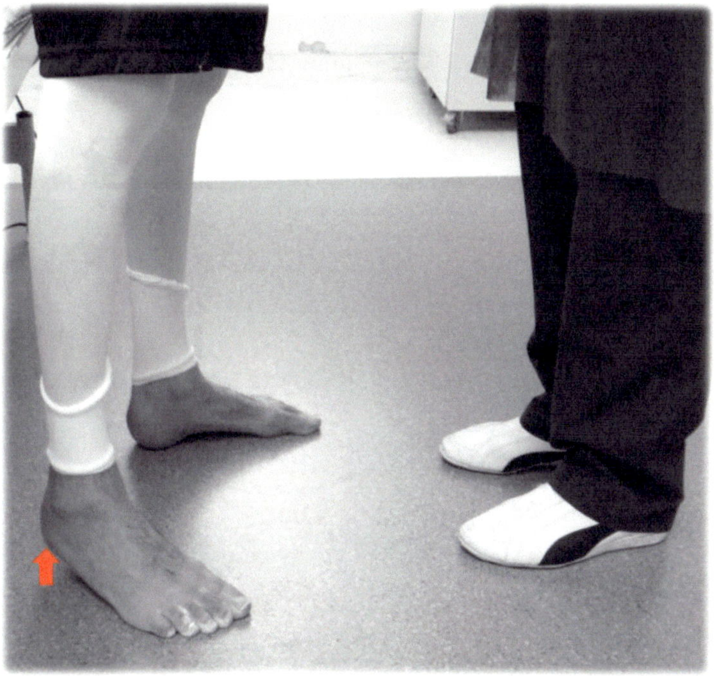

Figure 3. Standing unassisted post final cast without right heel contact on the floor.

Right wrist was extended and fingers flexed. Left wrist was in extreme flexion with metacarpophalangeal (MCP) joints in extension.

In the right lower limb severe hip limitation was present with free range of motion of 30° (from 60° to 90° of flexion), knee had short range of motion, only from 120 ° to full flexion and ankle presented plantar flexion with a 30° angle to reach right-angle neutral position.

In the left lower limb, hip presented severe contracture with range of motion limited to 45° (from 45° to 90° of flexion), knee had reduced range from 130° to full flexion and ankle did not reach the neutral position of (plantar flexion of 40°).

Muscle tone evaluated with the modified Ashworth scale showed a score of 3/4 for elbow flexors and hamstrings, although examination was difficult because of great lost of ROM and presence of heterotopic ossification in the right hip. Voluntary movement was present only in the right shoulder. Patient was admitted to the coma program to increase his level of awareness and receive multi-sensorial treatment (auditory, visual, etc.) to stimulate voluntary motor responses.

To manage contractures we developed the following treatment plan:

- Patient positioning, both in wheelchair and in bed. In the supine position a triangular foam cushion was used to hold legs in maximum possible knee extension and to prevent hip rotation (Fig. 4).

Figure 4. Positioning on bed.

- Manual passive stretching.
- Daily stretching postures achieved through use of the tilt-table, and prone positioning with adaptative equipment because of lack of range in both cases (Fig. 5).
- Knee extension splints were applied for night positioning and standing during therapy.

After two months, patient emerged from minimal consciousness and started a rehabilitation programme. Functional status showed trunk control with independent sitting and voluntary movement in lower limbs performed by imitation and procedural activities.

At this stage, patient muscle contractures were reassessed and a Baclofen pump test performed because of generalized muscle overactivity, with negative results. No changes were observed before or after the test. Given the favorable motor recovery a serial casting program was started for both knees, particularly because absence of spasticity and presence of voluntary movement favored a good outcome.

- Three progressive plasters were applied (changed every 4 to 6 days) on his left knee. The last cast was then bivalved. Initial goniometric measurement for the joint was 65º of flexion, and after final cast reached 40º of flexion. For right knee, initial measurement was 95º of flexion and reached 40º of flexion after six serial casts (lasting on average 6-8 days each). The fourth cast was a dynamic drop out cast. Three months after applying the last plaster, new ROM measurements showed further improvement, attaining 35º of flexion.
- During cast application, patient continued standing and carrying out prone activities. During the drop-out, he also worked on passive and active knee extension (Fig. 6 y 7).

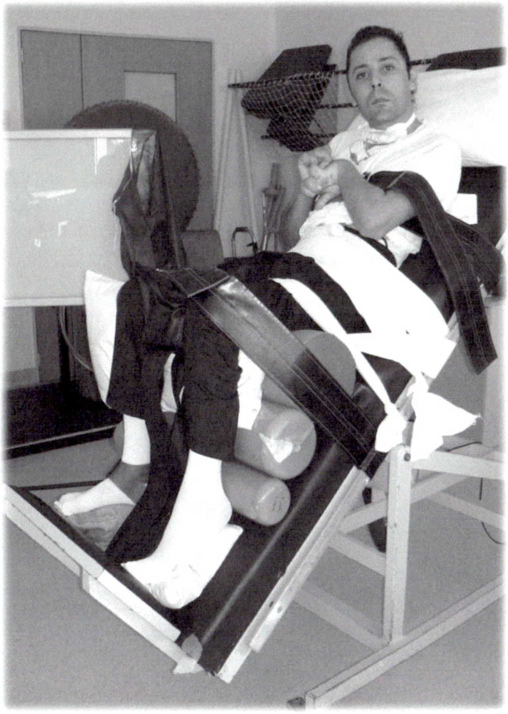

Figure 5. Standing on tilt table with adaptative equipment.

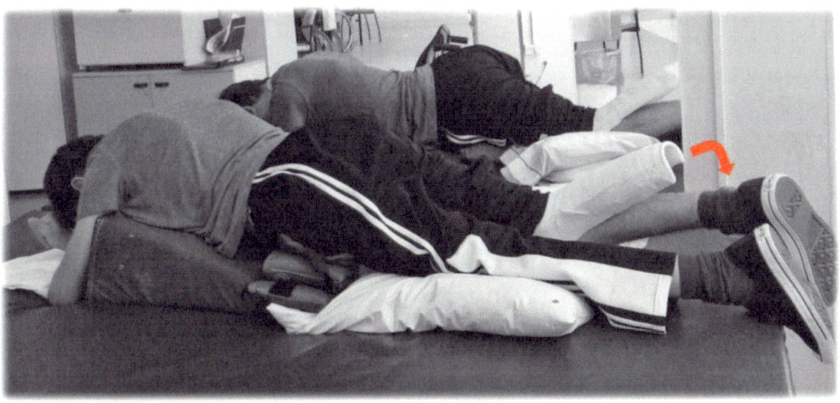

Figure 6. Prone position and exercise with Drop out cast in left knee.

- Electric stimulation was applied to the right quadriceps.

The cast allowed sufficient range to the knees to permit activity performance while standing, training of balance and for assisted walking with armrest support walker. Patient continued rehabilitation with favorable outcome (Fig. 8).

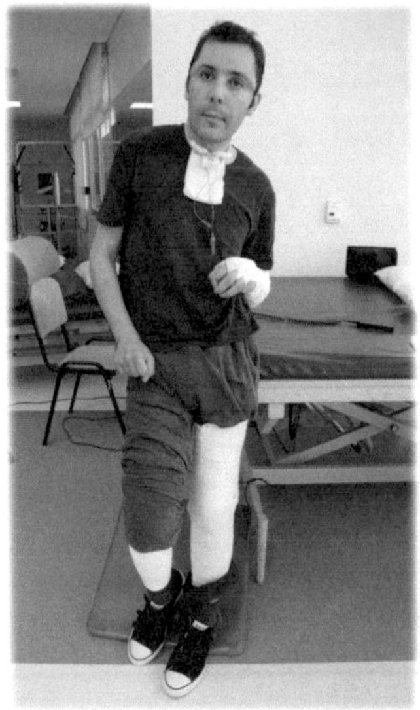

Figure 7. Supported standing with a knee extensor splint in the left leg and knee plaster in the right leg.

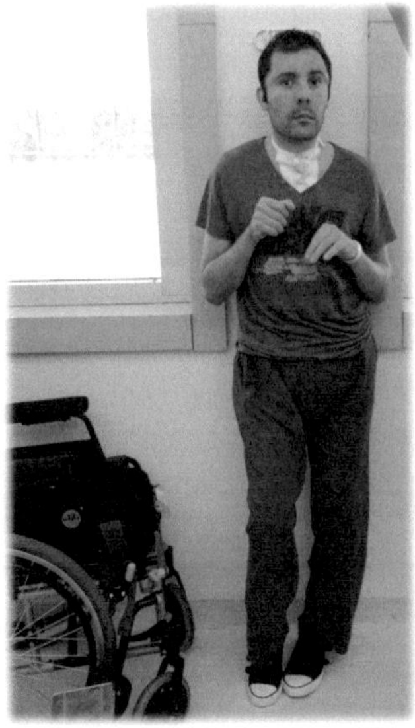

Figure 8. Standing as the final result of the progression of serial casting.

The right hip range limitation was solved surgically several months later. Currently, patient walks independently, requiring supervision for cognitive deficit. As for the upper limbs, right elbow regained normal mobility with conventional therapy and left elbow and wrist flexors required a tendon lengthening surgery.

http://dx.doi.org/10.5772/57310

10. Conclusion

The high prevalence of contractures in patients with head trauma limits functional therapy and presence of multiple complications often increases difficulty of rehabilitation implementation.

Contractures originate as a result of both neurologic disorders specific to central nervous system injuries and of non-neural disorders. These factors alter many anatomical structures that contribute to joint range loss.

While neurologic patients are always included in a comprehensive rehabilitation program, treatment of contractures requires special attention. The goals of the latter are to improve patient quality of life, reduce hospital stay and secondary complications, and promote better mobility and alignment, in order to recover maximum functional potential.

Several physiotherapy procedures exist for specific treatment of contractures, including stretching, positioning, splints and casts. All of them, to greater or lesser degree, involve placing anatomical structures in a position that respects or maintains normal joint range with varying duration of application. The techniques are usually combined with other therapeutic interventions to improve neurological dysfunction in its different presentations (weakness, paresis or paralysis and muscle overactivity). The variety of techniques available for treatment of contractures challenges physical therapists to gain better understanding of the principles on which the techniques are based and to develop skills in their application.

No single method renders final or full solution to contractures and all need further research to establish which procedures are most effective. All existing possibilities are commonly used in any rehabilitation program and represent the current tools until new procedures are developed. Research may also lead to the development of entirely new treatment modalities.

It is very important to address contracture management through a multidisciplinary team approach, and to train therapists for early detection as well as to assess degree of contracture. Treatment of contractures remains an active process conducted under constant monitoring, where team members can make use of different therapeutic procedures at appropriate times and adapt them to changes in clinical responses of each particular patient.

Author details

Fernando Salierno*, María Elisa Rivas, Pablo Etchandy, Verónica Jarmoluk, Diego Cozzo, Martín Mattei, Eliana Buffetti, Leonardo Corrotea and Mercedes Tamashiro

*Address all correspondence to: fsalierno@fleni.org.ar

Physical Therapy Department, FLENI Rehabilitation Institute, Escobar, Prov. de Buenos Aires, Argentina

References

[1] Harvey LA, Herbert RD. Muscle stretching for treatment and prevention of contracture in people with spinal cord injury. Spinal Cord. 2002 Jan; 40(1):1-9.

[2] Mollinger LA, Steffen TM. Knee flexion contractures in institutionalized elderly: prevalence, severity, stability, and related variables. Phys Ther. 1993 Jul; 73(7):437-44; discussion 444-6.

[3] Scott OM, Hyde SA, Goddard C, Dubowitz V. Prevention of deformity in Duchenne muscular dystrophy. A prospective study of passive stretching and splintage. Physiotherapy. 1981 Jun; 67(6):177-80.

[4] Yarkony GM, Sahgal V. Contractures. A major complication of craniocerebral trauma. Clin Orthop Relat Res. 1987 Jun; (219):93-6.

[5] Hellweg S, Johannes S. Physiotherapy after traumatic brain injury: a systematic review of the literature. Brain Inj. 2008 May; 22(5):365-73.

[6] Fergusson D, Hutton B, Drodge A. The epidemiology of major joint contractures: a systematic review of the literature. Clin Orthop Relat Res. 2007 Mar; 456:22-9.

[7] Botte MJ, Nickel VL, Akeson WH. Spasticity and contracture. Physiologic aspects of formation. Clin Orthop Relat Res. 1988 Aug; (233):7-18.

[8] Horsley AS, Herbert DR, Ada L. Four weeks of daily stretch has little or no effect on wrist contracture after stroke: a randomized controlled trial. Australian Journal of Physiotherapy. 2007; 53(4):239-245.

[9] Lieber RL, Steinman S, Barash IA, Chambers H. Structural and functional changes in spastic skeletal muscle. Muscle Nerve. 2004 May; 29(5):615-27.

[10] Cherry DB. Review of physical therapy alternatives for reducing muscle contracture. Phys Ther. 1980 Jul; 60(7):877-81.

[11] Farmer SE, James M. Contractures in orthopaedic and neurological conditions: a review of causes and treatment. Disabil Rehabil. 2001 Sep 10; 23(13):549-58.

[12] Yelnik AP, Simon O, Parratte B, Gracies JM. How to clinically assess and treat muscle overactivity in spastic paresis. J Rehabil Med. 2010 Oct; 42(9):801-7.

[13] Singer BJ, Dunne JW, Singer KP, Jegasothy GM, Allison GT. Non-surgical management of ankle contracture following acquired brain injury. Disabil Rehabil. 2004 Mar 18; 26(6):335-45.

[14] Van Trijffel E, van de Pol RJ, Oostendorp RA, Lucas C. Inter-rater reliability for measurement of passive physiological movements in lower extremity joints is generally low: a systematic review. J Physiother. 2010; 56(4):223-35. Review.

[15] Van de Pol RJ, van Trijffel E, Lucas C. Inter-rater reliability for measurement of passive physiological range of motion of upper extremity joints is better if instruments are used: a systematic review. J Physiother. 2010;56(1):7-17. Review.

[16] Pandyan AD, Johnson GR, Price CI, Curless RH, Barnes MP, Rodgers H. A review of the properties and limitations of the Ashworth and modified Ashworth Scales as measures of spasticity. Clin Rehabil. 1999 Oct; 13(5):373-83.

[17] Shabat E, Palit M, Fini NA, Brooks C, Winter A, Holland AE. Intra and inter-rater reliability of the Modified Tardieu Scale for the assessment of lower limb spasticity in adults with neurological injuries. Arch Phys Med Rehabil. 2013 Jul 11. pii: S0003-9993(13)00522-4. doi: 10.1016/j.apmr.2013.06.026. [Epub ahead of print]

[18] Singh P, Joshua AM, Ganeshan S, Suresh S. Intra-rater reliability of the modified Tardieu scale to quantify spasticity in elbow flexors and ankle plantar flexors in adult stroke subjects. Ann Indian Acad Neurol 2011;14(1):23-6

[19] Booth BJ, Doyle M, Montgomery J. Serial casting for the management of spasticity in the head-injured adult. Phys Ther. 1983 Dec;63(12):1960-6.

[20] Singer BJ, Gnanaketchumy M. Jegasothy, Kevin P. Singer, Garry T. Allison. Evaluation of Serial Casting to Correct Equinovarus Deformity of the Ankle After Acquired Brain Injury in Adults. Arch Phys Med Rehabil 2003;84:483-91

[21] Davies Patricia M. Starting Again: early rehabilitation after traumatic brain injury or other severe brain lesion. Springer-Verlag Berlin Heidelberg 1994.

[22] Cusick Beverly D. Serial Casting and Other Equinus Deformity Management Strategies for Children and Adults with Central Nervous System Dysfunction. Progressive GaitWays, LLC, 2010.

[23] Newman M, Barker K. The effect of supported standing in adults with upper motor neurone disorders: a systematic review. Clin Rehabil. 2012 Dec; 26(12):1059-77.

[24] Watson MJ. Do patients with severe traumatic brain injury benefit from physiotherapy? A review of the evidence. Physical Therapy Reviews 2001; 6: 233–249.

[25] Nuismer BA, Ekes AM, Holm MB. The use of low-load prolonged stretch devices in rehabilitation programs in the Pacific Northwest. Am J Occup Ther. 1997 Jul-Aug; 51(7):538-43.

[26] Robinson W, Smith R, Aung O, Ada L. Aust J Physiother. 2008; 54(1):33-8. No difference between wearing a night splint and standing on a tilt table in preventing ankle contracture early after stroke: a randomised trial.

[27] Blanton S, Grissom SP, Riolo L. Use of a static adjustable ankle-foot orthosis following tibial nerve block to reduce plantar-flexion contracture in an individual with brain injury. Phys Ther. 2002 Nov; 82(11):1087-97.

[28] Lannin NA, Horsley SA, Herbert R, McCluskey A, Cusick A. Splinting the hand in the functional position after brain impairment: a randomized, controlled trial. Arch Phys Med Rehabil. 2003 Feb;84(2):297-302.

[29] Leung J, Harvey LA, Moseley AM, Tse C, Bryant J, Wyndham S, Barry S. Electrical stimulation and splinting were not clearly more effective than splinting alone for contracture management after acquired brain injury: a randomised trial. J Physiother. 2012; 58(4):231-40.

[30] Stoeckman Tina. Casting for the Person with Spasticity. Top Stroke Rehabil 2001; 8(1):27-35.

[31] Lannin NA, Novak I, Cusick A. A systematic review of upper extremity casting for children and adults with central nervous system motor disorders. Clinical Rehabilitation 2007; 21: 963-976.

[32] Mortenson PA, Eng JJ. The Use of Casts in the Management of Joint Mobility and Hypertonia Following Brain Injury in Adults: A Systematic Review. Phys. Ther. 2003; 83:648-658.

[33] Moseley Anne M, Leanne M Hassett, Joan Leung, Jennifer S Clare, Robert D Herbert, Lisa A Harvey. Serial casting versus positioning for the treatment of elbow contractures in adults with traumatic brain injury: a randomized controlled trial. Clinical Rehabilitation 2008; 22: 406-417.

[34] Verplancke, S Snape, CF Salisbury, PW Jones, AB Ward. A ramdomized controlled trial of botulinum toxin on lower limb spasticity following acute acquired severe brain injury. Clinical Rehabilitation 2005; 19: 117 – 125

[35] Yaşar E, Tok F, Safaz I, Balaban B, Yilmaz B, Alaca R. The efficacy of serial casting after botulinum toxin type A injection in improving equinovarus deformity in patients with chronic stroke. Brain Inj. 2010;24(5):736-9.

Chapter 15

Heterotopic Ossification after Traumatic Brain Injury

Jesús Moreta and José Luis Martínez-de los Mozos

Additional information is available at the end of the chapter

http://dx.doi.org/10.5772/57343

1. Introduction

Heterotopic ossification is defined as the formation of bone within soft tissues, most frequently muscle tissue. The heterotopic ossification of muscles, ligaments and tendons is a potential complication following trauma, elective orthopaedic surgery, severe burns and neurological injury. It should not be confused with metastatic calcification associated to hypercalcaemia or dystrophic calcification related to tumours. Although there is an uncommon hereditary disease, called fibrodysplasia ossificans progressiva, most cases result from a local trauma (major local surgery, muscular trauma, fractures) or a neurological injury. The first description of heterotopic ossification after neurological injury was by Dejerne and Ceillier [1] in 1918, in soldiers with spinal cord injuries during World War I. Fifty years later, Roberts [2] described heterotopic ossifications in the elbow after brain injury. Since these articles, several studies have been published to provide information about causes, diagnosis and treatment of this complex entity.

This chapter is meant to serve as a review and an update on the literature regarding aetiology, diagnosis and management of this severe complication after a traumatic brain injury.

2. Epidemiology, aetiology and risk factors

The true incidence rate following neurological injury is difficult to establish clearly. After spinal cord injury, the percentage varies between 16% and 53% [3-6], and after traumatic brain injury has been reported to vary between 11% and 22% [7, 8]. A genetic predisposition has been suggested due to the relationship between neurogenic heterotopic ossifications (NHO) and fibrodysplasia (or myositis) ossificans progressiva [8]. The preferred locations in both cases are the proximal limbs and axial musculature. Moreover, a local trauma could predispose an individual to developing a bony formation. The influence of an association between human leukocyte antigens (HLA) and NHO is not clear. Some authors suggest a high prevalence of HLA-B18 and HLA-B27 in patients with heterotopic ossifications, but these findings should be confirmed [9].

It has been suggested that there is a higher prevalence in younger patients (20-30 years old), which may be because neurological injuries are most frequent in this patients [10, 11]. The incidence rate in relation to gender varies in the literature and the relationship has not been shown clearly.

The development of bone in soft tissues in orthopaedic surgery is a well-known complication. This complication has been associated with total hip arthroplasty, fractures or joint dislocations (mainly in the acetabulum or the elbow) and muscular trauma. For this reason, a relationship seems to exist between local trauma and the development of heterotopic ossifications.

Although the presence of traumatic brain injury alone could be associated with the development of heterotopic bone, it is not an established predictor. Brain trauma is a constant cause but there are specific risk factors implicated, such as spasticity, diffuse axonal injury, decerebrate posture, respiratory ventilation or prolonged immobilization. In a study of patients with NHO after traumatic brain injury, Garland [11] showed that 89% of joints were in spastic limbs. Moreover, diffuse axonal injury seems to be more related than focal brain injury [12]. Therefore, the extent of neurological damage must be taken into account, due to the fact that the risk is correlated with the severity of the brain injury. Flin et al. [13] evaluated patients who suffered from comas after traumatic brain injury in an acute rehabilitation unit. They founded that decerabrate posture could predispose an individual to NHO.

Ectopic bone after a neurological injury could develop spontaneously, but there is a higher risk if a local trauma is involved [14]. In general, prolonged immobility during patients' illnesses, such as multiple fractures, burn injuries or a coma, may contribute to the formation of heterotopic ossifications. In patients with brain injuries, a relatively long period of immobilization is frequent and there is therefore an association with the formation of ectopic bone [15].

3. Pathophysiology

Early in the formation of HO, oedema with exudative infiltrate is present, followed by fibroblastic proliferation and immature connective tissue formation. Posteriorly, osteoid formation is seen with the subsequent deposition of bone matrix. Primitive osteoid is deposited as small masses in the periphery early (within the first two weeks) and osteoblasts are noted, located irregularly. Osteoblasts produce tropocollagen, which polymerizes to form collagen and secrete alkaline phosphatase, which allows calcium to precipitate and the mineralization of bone matrix. As mineralization progresses, amorphous calcium phosphate is progressively replaced by hydroxyapatite crystals. During the following weeks, the lesion matures with a centripetal pattern, and after 6-12 months an appearance of true bone is noted. The new bone is always extra-articular and can be contiguous with the skeleton but generally does not involve the periosteum [16].

Many hypotheses about the physiopathology of neurogenic myositis ossificans have been proposed. Different authors have suggested vasomotor and metabolic disorders induced by immobility, the venous stasis or the induction of enchondral ossification from muscular microtrauma. The cellular mechanism of heterotopic ossification remains unclear. It has been postulated that three conditions must be met to achieve heterotopic bone formation [10]: a stimulating event, oestrogenic precursor cells and a proper environment.

Mesenchymal stem cells are pluripotent progenitor cells with the ability to generate cartilage, muscle, tendon, fat and bone. These cells are ubiquitous in soft tissues and may be induced to differentiate into oestrogenic stem cells capable of producing osteoid. Although they play a key role, these cells alone cannot produce ossification and stimulating factors are necessary. Growth factors, such as bone morphogenetic protein, could cause the differentiation of mesenchymal stem cells in a correct environment. Overexpression of bone morphogenetic protein-4 (BMP-4) has been shown to occur in fibrous dysplasia ossificans progressiva, a genetic form of heterotopic ossification mentioned before [17]. Related to this, inhibition of BMP-4 has shown to prevent ossifications in three separate experimental models [18]. Other stimulating factors such as inteleukin-1, growth hormone, prolactin, prostaglandin E-2, basic fibroblast growth factor and insulin-like growth factor type-1 have been recognized in the formation of ectopic bone after neurological injury [19-21]. After a brain injury, it has been suggested that circulating factors in the blood stimulate heterotopic ossification, but the specific mechanism remains unclear [22-23].

As has been mentioned before, the local environment has been shown to be important. Angiogenesis plays a role through the release of angiogenesis factors and differentiation of pericytes into osteoprogenitor cells [24]. Urist et al. [19] discovered that demineralized bone matrix could cause bone formation in vascular tissue. Otherwise, when they cultured this substance in an avascular system they found the development of cartilage. Other contributing factors in the local environment include hypercalcaemia, tissue hypoxia, pH changes (alkalosis) or prolonged immobilization.

It has also been suggested that there is a relationship between the nervous system and bone metabolism. Several neuropeptides and neurotransmitters have been found in the bone tissue. Glutamate, substance P, leptin, vasoactive intestinal peptide (VIP) and catecholamines have been shown to modify osteoblastic and osteoclastic activity [25-27].

In summary, there is an interaction of systemic and local factors on mesenchymal stem cells resulting in their differentiation into osteoblasts and the formation of heterotopic ossification.

4. Location

In patients who have suffered from a spinal injury, NHO develops in sites distal to the level of the spinal cord damage. After a traumatic brain injury, a considerable percentage of patients (between 10% and 20%) could develop ossifications around different joints. In all joints, the development of NHO is typically extra-articular and occurs in the connective tissue between the muscle planes and not within the muscle itself [28]. Clinically, an NHO is defined by a limited decreased range of movement, usually accompanied with pain.

Although upper and lower extremities could be affected, the most commonly affected joint is the hip. In these patients, the most common location is inferomedial and it seems to be a role of adductor spasticity. The ossified tissues extend from the pubic symphysis to the medial or anteromedial femoral shaft and usually the femoral neurovascular bundle lies anteriorly. Other locations around the hip are the posterior part and the anterior aspect of the joint, below the anterior superior iliac spine [11]. When the location is posterior, the sciatic nerve could be involved and if the ossification develops anteriorly it may involve the femoral neurovascular structure. Gartland et al. have suggested that the anteromedial location is most frequent after spinal cord injury, whereas inferomedial, lateral and posterior planes are usually affected after brain injuries [11, 29].

The next most common sites of involvement in patients with traumatic brain injuries are the shoulders and elbows, with the knees rarely being affected. Nevertheless, knees are frequently involved in patients with spinal cord injuries. After a head injury, the ectopic bone about the elbow usually develops posteriorly from the humerus to the olecranon, and could cause ankylosis of this joint. The ulnar nerve could be affected, due to its close location [30]. Although the knee is a common location after a spinal cord injury, it is uncommon in patients with brain damage. In this case, the ectopic bone may occur in any plane, but the inferomedial aspect of the distal femur seems to be the most frequent [29, 30].

5. Diagnosis

5.1. Clinical presentation

The mean time from injury to development of HO is two or three months, but some studies have reported numbers from two weeks up to one year [31]. Thorough physical examination should be performed because this is a critical point for obtaining an early diagnosis. The onset of a heterotopic ossification may simulate an infection with erythema, warmth and swelling. Lower limb swelling may mimic deep venous thrombosis (DVT), so this entity should be ruled out. Differentiating early heterotopic ossification from lower extremity DVT is sometimes difficult, because the two entities have similar clinical presentation. Furthermore, they also can occur at the same time due to the mass effect of the ectopic bone, inciting venous compression and finally phlebitis and thrombus development.

When the passive mobility is examined there is a decreased range of motion. The examiner may identify a painful joint if the patient is conscious, but the findings described before should be checked, especially if the patient is in a coma.

Although there is no total agreement, accelerated bone healing has been reported [32]. In patients with brain injuries, it could be more callus formation but there is a different histological pattern. For this reason, some authors have suggested that this new bone may be a heterotopic ossification, because its maturation is performed from the periphery. Callus maturation after a fracture occurs from the central area [33].

5.2. Laboratory tests

The serum alkaline phosphatase level reflects osteoblastic activity and is useful for obtaining an early diagnosis, especially in patients who cannot report pain or for whom a physical examination is difficult. This marker begins to rise two or three weeks after the brain injury, and values return to normal at approximately four months [34]. A clinical judgement of this test should be done, because liver disease or associated fractures also could increase the levels. Furthermore, alkaline phosphatase titres do not correlate with peak activity or maturation of HO, so they cannot be used to get information about the evolution of the ectopic bone [35]. Nevertheless, this is a cheap and easy screening tool to detect early HO and begin with treatment. Garland (ref de libro) recommends initiating medical prophylaxis on the basis of serum alkaline phosphatase elevation if there are no fractures associated.

In a study of patients after acute spinal cord injury, the 24-hour urinary excretion of prostaglandin E_2 (PGE_2) showed an increase until the maturation of the HO [37]. In patients with brain injuries, it could also be a valuable indicator before the ectopic bone is established. Indomethacin is a PGE_2- blocking agent, which is a useful medication to prevent this [38].

5.3. Radiography

Plain radiographs are an easy and cheap method to recognize a neurogenic heterotopic ossification but it may take up to six weeks for ossification to be evident. In some cases, the HO could develop in several months (Figure 1). Anteroposterior (AP) and lateral radiographs provide enough information to establish the relationship of the HO to the joint, but further radiological techniques (CT-Scan, MTI) should be performed to get more information on the anatomy and the condition of the joint. The typical radiographic appearance of the ectopic bone is a circumferential ossification with a lucent centre [39].

In the hip, the Brooker classification could be useful, which is commonly used to describe the pattern and extent of ossification around a total hip arthroplasty. The extent of the ectopic bone correlates with the degree of functional impairment. According to this classification, there are four types:

- Type I: Islands of bone within the soft tissues.
- Type II: Bone spurs arising from the pelvis or proximal femur with 1cm or less between adjacent bone surfaces.
- Type III: Bone spurs arising from the pelvis or proximal femur with more than 1cm between adjacent bone surfaces.
- Type IV: Confluent bone bridging the pelvis and proximal femur with apparent ankylosis of the hip.

A recent classification has been introduced by Mavrogenis et al. [40] according to the anatomical location of neurogenic HO in the hip: type 1, anterior; type 2, posterior; type 3, anteromedial; type 4, circumferential. These authors suggested that this classification provides better evaluation of the prognosis after surgical excision than the Brooker classification and they found higher blood loss for patients with anteromedial (type 3) or circumferential (type 4) neurogenic ossifications. They also showed a higher risk of recurrence in patients with brain injuries.

5.4. Bone scans

Bone scans or scintigraphy are considered the most sensitive method of providing early detection of HO, with detection as early as two weeks after the onset of symptoms [39].

Bone scintigraphy using 99mTc-methylene disphosphonate provides imaging in three phases: Phase I (dynamic blood flow study), Phase II (Static study for blood pool) and Phase III (amount of radionuclide in bone). The first two phases allow early detection of incipient ossification at two or three weeks. In Phase III, this test occurs approximately one week later. The level of activity on delayed bone scans usually peaks at two months and then it decreases progressively. Scintigraphy activity returns to normal after approximately 12 months, but it may remain positive even when the ossification has become mature.

5.5. Ultrasonography (US)

This technique has been used in the early diagnosis of HO about the hip joints, showing a chaotic disruption of the normal structure of the muscle in the first stages [41]. Later, a peripherally echogenic and centrally hypoechoic finding can be identified. US has the advantage of the possibility of bedside application and it requires no radiation. Ultrasonography is also a fast and useful tool for diagnosis of DVT, which is associated with patients with spinal cord injury as we mentioned before. Nevertheless, ultrasonography is a difficult technique, which depends on the skills of the examiner.

5.6. Magnetic resonance imaging (MRI) and volumetric computed tomography (CT-Scan)

Although bone scans are considered the most sensitive method for early detection of neurogenic HO, magnetic resonance imaging (MRI) suggests changes compatible with ectopic bone formation, as soon as two days after clinical presentation, especially in the knee [42]. In this joint, MRI findings are joint effusion, thickening of the intermuscular septa and a "lacy-pattern" of the muscles vastus lateralis and medialis in fast spin-echo short time inversion-recovery (STIR) images.

Evaluation with computed tomography (CT) combined with two and three-dimensional surface reconstructions provides good differentiation and the extent of the ectopic bone. Moreover, this technique shows the relationship with the joint, determines the mineral density of the bone and can predict several intra-articular lesions. In a study performed by Carlier et al. [43], they compared the density of the HO around the hip with the iliac wing. Bone mineralisation was classified in four categories: normal (M1), mild demineralisation (M2), significant demineralisation with risk of fracture (M3) and evanescent bone with a high risk of fracture (M4). They correlated these features with surgical findings, so this is an important issue because osteoporosis underlying the ankylosis can lead to fractures, which are a difficult challenge in these patients.

If surgical treatment is required, a thorough vascular assessment should be done, because there is a relationship between the HO and surrounding neurovascular structures (femoral bundle in the hip). Therefore, helical-CT with intravenous contrast or MRI with contrast (Figure 1) should be one of the requirements of the preoperative assessment in patients undergoing surgical excision of HO.

6. Treatment and prevention

The prevention of heterotopic ossification is focused on three basic principles: disrupting the inductive signalling pathways, altering the osteoprogenitor cells or modifying the environment. Although some targets are defined in some diseases such as Progressive Osseous Heteroplasia, the target cell or target tissue in NHO remains unknown [44]. Maintaining joint motion by treating spasticity and gentle physical therapy is the goal of early treatment. When a neurogenic heterotopic ossification is suspected, the treatment should start with medical treatment and when the ectopic bone is established and symptomatic surgical resection should be considered carefully. In Figure 2, there is an algorithm, which could be useful to diagnose and treat this entity.

6.1. Physiotherapy and rehabilitation

Prevention should start with early joint mobilisation. Once the diagnosis of early ossifications is suspected, physical therapy with the help of passive exercises should be done [45]. If the patient is conscious, physiotherapy should involve an assisted range of movement exercises with gentle stretch and terminal resistance training. It is crucial to achieve a good range of motion without generating pain. Although forceful manipulation, even under anaesthesia, could be beneficial, it is advisable to avoid aggressive range-of-motion exercises because they carry the risk of causing more bone formation. This may be due to the effect of repetitive microtrauma to the joint [46].

The treatment of HO by extracorporeal shock wave has been reported. Brissot et al. [47] showed less pain and better range of motion and walking distance with this treatment. If the patient is not a suitable candidate for surgical treatment because of medical comorbidities, shock wave therapy could also be a good alternative [48].

In conclusion, physical therapy aims to preserve movement and good function, avoiding or decreasing the risk of ankylosis of the joint.

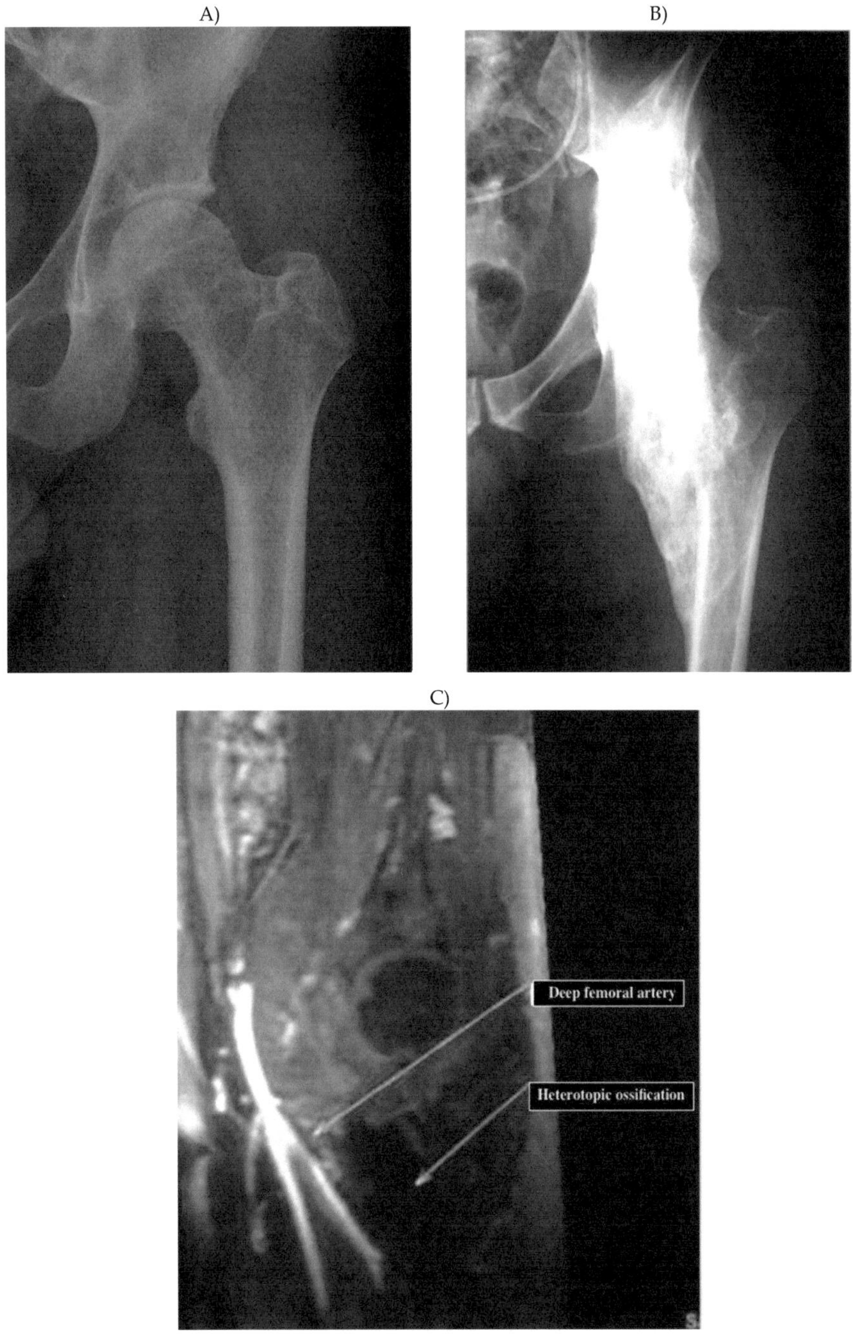

Figure 1. Neurogenic ossification in the hip in a patient after a traumatic brain injury. A) Radiograph of the left hip six months after the brain injury. B) Radiograph showing an anteromedial ossification around the hip causing severe ankylosis, 14 months after the initial injury. C) MRI enhanced with contrast showing a close relationship between the ossification and the deep femoral artery.

http://dx.doi.org/10.5772/57343

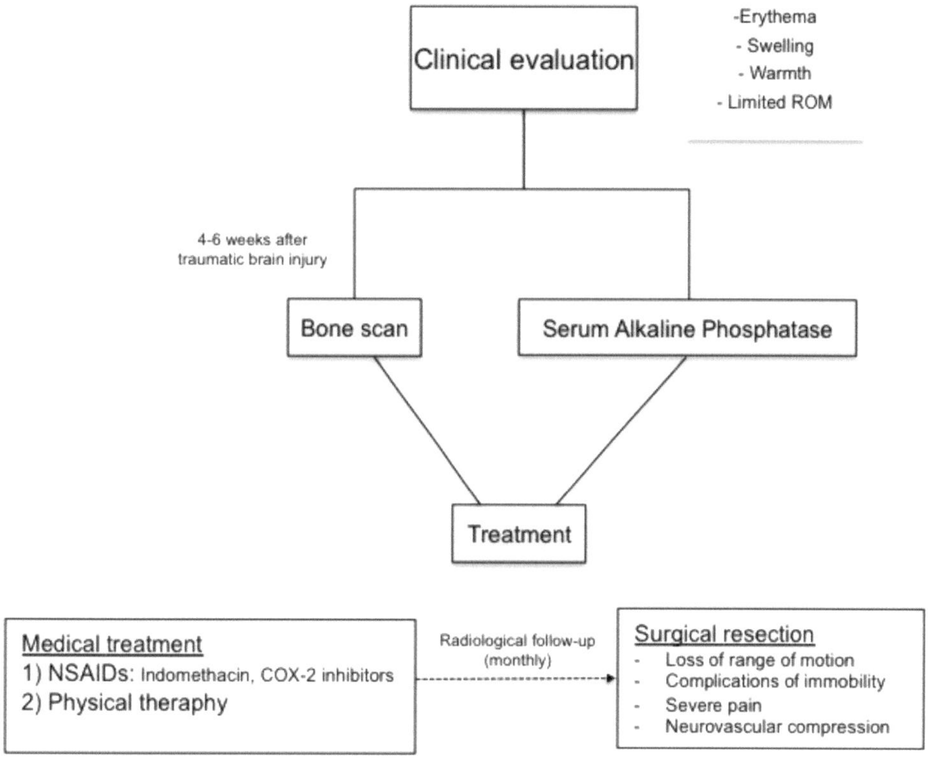

Figure 2. Diagnosis and treatment of neurogenic heterotopic ossifications

6.2. Medical treatment

Medical treatment aims to prevent the formation of HO following brain injury and to avoid recurrence if surgical excision is needed. Due to the fact that the development of HO can occur within one to two months of the initial injury, the prophylaxis should begin relatively early. Although there is no consensus on which medication should be used and when treatment should begin, several drugs have successfully been used.

Nonsteroidal anti-inflammatory drugs (NSAIDs)

These drugs have demonstrated good results in preventing heterotopic ossification after total hip arthroplasty and several studies support their use following traumatic neurological injury [49-51]. A Cochrane Review of HO formation after hip replacements performed in 2004, showed the effectiveness of prevention using NSAIDs perioperatively [52]. A direct effect of NSAIDs on the formation of heterotopic ossification has been described, due to the inhibition of the differentiation of mesenchymal cells into oestrogenic cells. There is also an indirect effect, which refers to the inhibition of bone remodelling by suppression of the prostagland unmediated inflammatory response (specifically PGE-2) [38].

Indomethacin has been considered a useful medication for heterotopic ossification prophylaxis following total hip replacement. In a randomized controlled trial, Banovac et al. [50] showed a lower incidence of early and late HO in the indomethacin group. This drug prescribed for three to six weeks in a dose of 75mg/day, within two months of the injury, may reduce the incidence of heterotopic ossification by two to three times. These same authors performed a trial to determine the effect of COX-2 selective inhibitor rofecoxib on the prevention of HO after spinal cord injury [51]. They concluded that there was a 2.5 times lower risk of developing ectopic bone. COX-2 selective inhibitors could be an attractive option because of less gastrointestinal side effects; however, rofecoxib was taken off the USA market due to elevation of cardiovascular risk. In addition, celecoxib, another selective COX-2 inhibitor, showed the same efficacy as indomethacin in the prevention of HO, but in patients with cardiovascular problems should be used cautiously. Another COX-2 inhibitor, meloxicam, did not reduce the incidence of ossification after total hip replacement compared to indomethacin [53, 54].

In conclusion, indomethacin remains the gold standard for pharmacological prevention of HO, because it is a simple and low cost option. Celecoxib could be a reasonable option in patients without cardiovascular diseases because it has shown good results with fewer gastrointestinal effects.

Bisphosphonates

Bisphosphonates have also been used in the prevention of HO. Etidronate disodium is a well-studied bisphosphonate and it seems to be effective for early ossifications [55, 56] and later phases of this disease. This medication inhibits precipitation of calcium phosphate and blocks the aggregation and mineralisation of hydroxyapatite crystals. In primary prevention, Banovac et al. [56] have suggested that the treatment should begin as soon as elevated alkaline phosphatase is diagnosed or positive findings in ultrasonography or bone scans are shown. In addition, disphosphonates could have long-term effects on prevention when the treatment is finished.

In spite of these results, some authors do not recommend bisphosphonates routinely, because additional fractures are often associated with neurological injuries and the use of bisphosphonates could impede fracture healing [57].

6.3. Radiotherapy

Radiotherapy seems to work by preventing the differentiation of mesenchymal cells into osteoblasts, which could begin the bone formation. This method has been used successfully to prevent heterotopic ossification after hip replacements [58, 59]. In a prospective, randomized study of HO after hip arthroplasty, Knelles et al. found that the best results were obtained with indomethacin and a single irradiation of 7 Gy or four sessions of 3 Gy. These and other authors have suggested that a single irradiation of 7 Gy should be useful in patients at risk, especially if administration of indomethacin is contraindicated [60].

In patients with a traumatic brain injury, radiation therapy may be given as primary (patients with high risk of developing HO) or secondary prophylaxis (those who have been diagnosed with HO and need surgical resection). Sautter-Bihl et al. [62] studied prophylaxis with radiotherapy in 36 patients with spinal cord injuries. These authors used radiotherapy as primary prophylaxis in 27 patients and as secondary prevention (after surgical resection) in 11 patients. With an average follow-up of 23.6 months, 30 of the 36 patients showed no progression of HO and improvements in rehabilitation. It has also been suggested that radiotherapy provides pain relief and decreased serum alkaline phosphatase level in a case report of a patient with HO in both hips and thighs after a brain injury [63]. In this case, the symptoms were refractory to indomethacin, so a dose of 20 Gy in ten fractions was performed.

Like other pharmacological methods of treatment, radiation therapy could affect bone healing. A study focused on HO after elbow trauma has shown that single-fraction therapy leads to an increase risk of non-union [64]. In addition, the location of HO after a total hip arthroplasty is predictable, whereas that after brain injury is difficult to predict accurately where the ectopic bone will develop [65]. For this reason, we do not recommend the routine use of radiation therapy for primary prevention of neurogenic HO, whereas it could be a good method for secondary prophylaxis.

6.4. Surgery

Surgical resection of established ossifications is the treatment of choice to facilitate rehabilitation and is the only effective procedure when the ossification is mature.

The surgery is necessary in patients with symptomatic HO and unsuccessful medical treatment, such as:

- Loss of range of motion and ankylosis with associated functional disability, such as sitting difficulties
- Complications of immobility such as pressure ulcers
- Facilitate rehabilitation and recovery of muscular atrophy secondary to prolonged immobilisation
- Difficulties of appropriate hygiene because access to the perineum or bladder care is needed
- Severe pain refractory to analgesia
- Vascular and/or nerve compression

A careful preoperative assessment is mandatory in order to avoid complications. A CT-Scan is really useful to show intra-articular pathology and the degree of osteoporosis. In addition, a CT-Scan or MRI with contrast describes the relationship with important neurovascular structures.

The timing of operative excision is controversial. Some authors recommend timetables for the best moment to remove these ossifications, but these indications have changed in recent years. Gartland recommended surgery after six months following traumatic heterotopic ossification, one year following spinal cord injury and 18 months after head injury. Another factor to take into account is the neurological condition of the patient. Severe cognitive and physical impairment have poor results from surgery with a high rate of recurrence. However, patients with good neuromuscular control before the surgical procedure have a better functional outcome [65, 66]. In the past, waiting until maturity in order to minimize the rate of recurrence after the surgical treatment was recommended. Nevertheless, recent studies do not confirm a higher rate of recurrence when the HO is excised in an earlier phase [49, 67]. In addition, a long delay before surgery leads to ankylosis, a high degree of osteoporosis and more extensive intra-articular injuries. These findings are associated with bad functional results and potential complications.

Stover et al. [5] showed that the risk of fracture was secondary to osteopenia and this complication increased with the delay in surgical treatment. Perioperative fractures are associated with grade 3 and 4 of osteopenia, according to the classification described by Carlier et al. [43], so therefore this complication could take into account with a preoperative CT-Scan. In a study of HO in 183 hips (143 patients), performed by Genet et al. [49], found 25 perioperative fractures of the femoral neck, all of which occurred in patients with ankylosed hips.

Intra-articular lesions are associated with a smaller range of motion and ankylosis. Some authors considered it to be neurotrophic arthropathy [68], but more recently Genet et al. [49] suggested that it is only ankylosis, which induces articular degradation. In addition, these authors showed that the presence of intra-articular lesions is a cause of worsening of the final results.

In general, the surgical goal is resection of a large enough amount of bone to allow improvement of the range of motion and trying to preserve the joint [49, 69]. A good knowledge of the anatomy is paramount and adequate exposure should be performed with identification of the relationship with neurovascular structures. Intraoperative neurologic monitoring with use of electromyography or somatosensory evoked potentials may be useful, especially in the hip. We use osteotomes and high-speed burrs to remove the ossifications and we frequently use fluoroscopy to check the bone removal (Figure 3). Intraoperative blood loss can be substantial so careful bleeding control should be performed and we recommend that local haemostatic agents should be available, such as bone wax and gelatine-based products. Much of the bleeding is from the osteotomized bone surfaces and is difficult to control with electrocautery. Furthermore, meticulous haemostasis and elimination of dead space decrease the risk of infection.

Wide exposure and identification of major vessels and nerves is extremely important. In many cases, the ossifications have a close relationship with important neurovascular structures and there is distortion of the normal anatomy. Therefore, there is a relatively high risk of iatrogenic injury. Especially in the hip, this complication is extremely important and sometimes life threatening, so therefore we recommend performing this surgical procedure with the assistance of a vascular surgeon [69].

There is a significant risk of recurrence and we recommend prophylaxis methods, such as physical therapy and the administration of indomethacin for two or four weeks. Although no

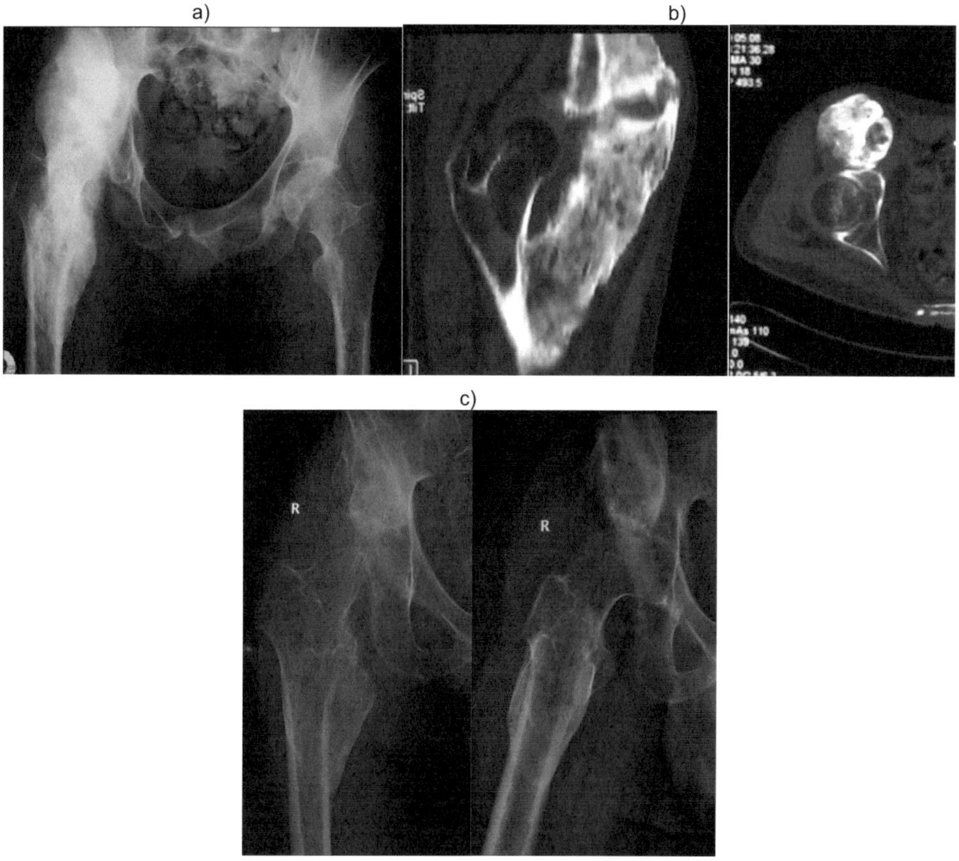

Figure 3. Heterotopic ossification in the right hip causing severe ankylosis. a) Preoperative imaging showing a huge HO from the lesser trochanter to the ilium. b) Preoperative CT-Scan showing involvement of the muscles iliopsoas, vastus medialis and vastus intermedius. b) Radiograph 12 months after surgery.

data have shown the benefit of radiotherapy or medical treatment when ossifications are mature, it is reasonable to use prophylactic treatment postoperatively aiming to prevent recurrence. [66].

7. Conclusion

Heterotopic ossifications after a brain injury present multiple challenges because of the difficult management of this disease. Although it is difficult to predict the development of NHO, it is correlated with the severity of brain damage and several risk factors have been reported. The hip is the most commonly involved joint and the clinical presentation could mimic a septic arthritis or phlebitis in the early stage (3-12 weeks after the neurological injury). Later, although the NHO could be asymptomatic, it can lead to limited range of motion and ankylosis of the affected joint. Early detection and prevention with physical therapy, NSAIDs or bisphosphonates have shown good results. Surgery is recommended when the functional status is affected and thorough preoperative planning should be done to avoid further complications. Although, there is controversy in deciding the timing of surgery, recent articles have suggested that early surgical excision decreases the likelihood of developing osteopenia and intra-articular lesions.

This entity remains a poorly understood condition, so further research in neurogenic heterotopic ossifications should help in continuing to understand pathophysiology, identify pathways and target cells and tissues to find better methods of prevention and treatment.

Acknowledgements

We gratefully acknowledge Dr. O. Sáez de Ugarte and Dr. I. Jauregui for their assistance in the preparation of the manuscript.

Author details

Jesús Moreta* and José Luis Martínez-de los Mozos

*Address all correspondence to: chusmoreta2@hotmail.com

Department of Orthopaedic Surgery and Trauma, Hospital de Galdakao-Usansolo, Galdakao, Spain

References

[1] Dejerne A, Ceillier A. Para-osteo-arthropathies des paraplegiques par lesion medullaire; etude clinique et radiographique. Ann Med 1918;5:497.

[2] Roberts P. Heterotopic ossification complicating paralysis of intracranial origin. J Bone Joint Surg Br 1968;50-B:70-7.

[3] Dai L. Heterotopic ossification of the hip after spinal cord injury. Chin Med J (Engl) 1998;111:1099-101.

[4] Lal S, Hamilton BB, Heinemann A, Betts HB. Risk factors for heterotopic ossification in spinal cord injury. Arch Phys Med Rehabil 1989;70:387-90.

[5] Stover SL, Niemann KM, Tulloss JR. Experience with surgical resection of heterotopic bone in spinal cord injury patients. Clin Orthop. 1991;263:71-7.

[6] Wittenberg RH, Peschke U, Bötel U. Heterotopic ossification after spinal cord injury: epidemiology and risk factors. J Bone Joint Surg (Br). 1992;74-B:215-8.

[7] Citta-Pietrolungo TJ, Alexander MA, Seg NL. Early detection of heterotopic ossification in young patients with traumatic brain injury. Arch Phys Med Rehabil 1992;73:258-62.

[8] Connor JM, Evan DA. Fibrodysplasia ossificans progressiva: the clinical features and natural history of 34 patients. J Bone Joint Surg Br 1982;64:76–83.

[9] Minare P. Neurologic injuries, paraosteoarthropathies and human leukocyte antigens. Arch Phys Med Rehabil. 1984;65:5531.

[10] Genet F, Jourdan C, Lautridou C, et al. The impact of preoperative hip heterotopic ossification extent on recurrence in patients with head and spinal cord injury: a case control study. PloS One 2011;6:23129.

[11] Garland DE. Clinical observations on fractures and heterotopic ossification in the spinal cord and traumatic brain injured populations. Clin Orthop 1988;233:86-101.

[12] Gennarelli TA. Heterotopic ossification. Brain Inj 1988;2:175-8.

[13] Flin C, Curalucci H, Duvovelle A, Viton JM. Heterotopic ossification and brain injury. Ann Readapt Med Phys 2002;45:517-20.

[14] Ebinger T, Roesch M, Kiefer H, Kinzl L, Schulte M. Influence of etiology in heterotopic bone formation of the hip. J Trauma 2000;48:1058-62.

[15] Mielants H, Vanhove E, de Neels J, Veys E. Clinical survey of and pathogenic approach to para-articular ossifications in long-term coma. Acta Orthop Scand 1975;46:190-8.

[16] van Kuijk AA, Geurts AC, van Kuppevelt HJ. Neurogenic heterotopic ossification in spinal cord injury. Spinal Cord 2002;40:313-26.

[17] Hannallah D, Peng H, Young B, Usas A, Gearhart B, Huard J. Retroviral delivery of Noggin inhibits the formation of heterotopic ossification induced by BMP-4, demineralized bone matrix, and trauma in an animal model. J Bone Joint Surg Am 2004;86-A:80-91.

[18] Ahrengart L, Lindgren U, Reinholt FP. Comparative study of the effects of radiation, indomethacin, prednisilone and ethane-1-hydroxy-1, 1-disphonate (EHDP) in the prevention of ectopic bone formation. Clin Orthop 1988;229:265-73.

[19] Urist MR, Nakagawa M, Nakata N, Nogami H. Experimental myositis ossificans: cartilage and bone formation in muscle in response to diffusible bone matrix derived morphogen. Arch Pathol Lab Med. 1978;102:312-6.

[20] Mahy PR, Urist MR. Experimental heterotopic bone formation induced by bone morphogenetic protein and recombinant human interleukin-1B. Clin Orthop 1988;237:236-44.

[21] Wildburger R, Zarkovic N, Egger G, et al. Comparison of the values of basic fibroblast growth factor determined by an immunoassay in the sera of patients with traumatic brain injury and enhanced osteogenesis and the effects of the same sera on the fibroblast growth in vitro. Eur J Clin Chem Biochem 1995;33:693-8.

[22] Cadosch D, Toffoli AM, Gautschi OP, et al. Serum after traumatic brain injury increases proliferation and supports expression of osteoblast markers in muscle cells. J Bone Joint Surg Am 2010;92-A:645-53.

[23] Bidner SM, Rubins IM, Desjardins JV, Zukor DJ, Goltzman D. Evidence for a humorol mechanism for enhanced osteogenesis after head injury. J Bone Joint Surg Am 1990;72-A:1144-9.

[24] Collett GDM, Canfield AE. Angiogenesis and pericytes in the initiation of ectopic calcification. Circ Res 2005;96:930-8.

[25] Wang L, Tang X, Zhang H, Yuan J, Ding H, Wei Y. Elevated leptin expression in rat model of traumatic spinal cord injury and femoral fracture. J Spinal Cord Med 2011;34:501-9.

[26] Hohmann EL, Elde RP, Rysavy JA, Einzig S, Gebhard RL. Innervation of periosteum and bone by sympathetic vasoactive intestinal peptide-containing nerve fibers. Science 1986;232:868-71.

[27] Jones KB, Mollano AV, Morcuende JA, Cooper RR, Saltzman CL. Bone and brain: a review of neural, hormonal, and musculoskeletal connections. Iowa Orthop J 2004;24:123-32.

[28] Jensen LL, Halar E, Little J, Brooke MM. Neurogenic heterotopic ossification. Am J Phys Med Rehabil 1987;66:351-63.

[29] Garland DE. A clinical perspective on common forms of acquired heterotopic ossification. Clin Orthop 1991;263:13-29.

[30] Garland DE, Blum CE, Waters RL. Periarticular heterotopic ossification in head-injured adults: incidence and location. J Bone Joint Surg Am 1980;62-A:1143-6.

[31] Cipriano CA, Pill SG, Keenan MA. Heterotopic ossification following traumatic brain injury and spinal cord injury. J Am Acad Orthop Surg 2009;17:689–97.

[32] Pape HC, Lehmann U, van Griensven M, Gänsslen A, von Glinski S, Krettek C. Heterotopic ossifications in patients after severe blunt trauma with and without head trauma: incidence and patterns of distribution. J Orthop Trauma 2001;15:229-37.

[33] Spencer RF. The effect of head injury on fracture healing: a quantitative assessment. J Bone Joint Surg Br 1987;69-B:525-8.

[34] Orzel JA, Rudd TG. Heterotopic bone formation: clinical, laboratory and imaging correlation. J Nucl Med 1985;26:125-32.

[35] van Kuijk AA, Geurts AC, van Kuppevelt HJ. Neurogenic heterotopic ossification in spinal cord injury. Spinal Cord 2002;40:313-26.

[36] Garland DE, Varpetian A. Heterotopic Ossification in Traumatic Brain Injury. In: Ashley MJ (ed.) Traumatic Brain Injury. Rehabilitation, Treatment, and Case Management. CRC Press; 2004:119-34

[37] Schurch B, Capaul M, Vallotton MB, Rossier AB. Prostaglandin E2 measurements: their value in the early diagnosis of heterotopic ossification in spinal cord injury patients. Arch Phys Med Rehabil. 1997;78:687-91.

[38] Ho SSW, Stern PJ, Bruno LP, et al. Pharmacological inhibition of prostaglandin E-2 in bone and its effect on pathological new bone formation in a rat brain model. Trans Orthop Res Soc. 1988;13:536.

[39] Shehab D, Elgazzar AH, Collier BD. Heterotopic ossification. J Nucl Med 2002;43:346-53.

[40] Mavrogenis AF, Guerra G, Staals EL, Bianchi G, Ruggieri P. A classification method for neurogenic heterotopic ossification of the hip. J Orthopaed Traumatol 2012 13:69-78.

[41] Falsetti P, Acciai C, Palilla R, Carpinteri F, Patrizio C, Lenzi L. Bedside ultrasound in early diagnosis of neurogenic heterotopic ossification in patients with acquired brain injury. Clin Neurol Neurosurg 2011; 113(1):22-7.

[42] Argyropoulou MI, Kostandi E, Kosta P, Zikou AK, Kastani D, Galiatsou E, Kitsakos A, Nakos G. Heterotopic ossification of the knee joint in intensive care unit patients: early diagnosis with magnetic resonance imaging. Crit Care 2006;10:R152.

[43] Carlier RY, Safa DM, Parva P, Mompoint D, Judet T, Denormandie P, Vallée CA. Ankylosing neurogenis myositis ossificans of the hip: an enhanced volumetric CT study. J Bone Joint Surg Br 2005;87-B:301-5.

[44] Kaplan FS, Glaser DL, Hebela N, Shore EM. Heterotopic ossification. J Am Acad Orthop Surg. 2004;12:116-25.

[45] Stover SL, Hataway CJ, Zeiger HE. Heterotopic ossification in spinal cord injured patients. Arch Phys Med Rehabil. 1975;56:199–204.

[46] Garland DE, Razza BE, Waters RL. Forceful joint manipulation in head-injured adults with heterotopic ossification. Clin Orthop 1982;169:133-8.

[47] Brissot R, Lassalle A, Vincendeau S, et al. Treatment of heterotopic ossification by extracorporeal shock wave: 26 patients. Ann Readapt Med Phys 2005;48:581-9.

[48] Reznik JE, Gordon SJ, Barker RN, Keren O, Arama Y, Galea MP. Extracorporeal shock wave therapy (ESWT) as a treatment for recurrent neurogenic heterotopic ossification (NHO). Brain Inj. 2013;27(2):242-7.

[49] Genet F, Marmorat JL, Lautridou C, Schnitzler A, Mailhan L, Denormandie P. Impact of late surgical intervention on heterotopic ossification of the hip after traumatic neurological injury. J Bone Joint Surg Br 2009;91-B:1493-8.

[50] Banovac K, Williams JM, Patrick LD, Haniff YM. Prevention of heterotopic ossification after spinal cord injury with indomethacin. Spinal Cord 2001;39:370-4.

[51] Banovac K, Williams JM, Patrick LD, Levi A. Prevention of heterotopic ossification after spinal cord injury with COX-2 selective inhibitor (rofecoxib). Spinal Cord 2004;42:707-10.

[52] Fransen M, Neal B. Non-steroidal anti-inflammatory drugs for preventing heterotopic bone formation after hip arthroplasty. Cochrane Database Syst Rev 2004, CD001160.

[53] Legenstein R, Bosch P, Ungersbock A. Indomethacin versus meloxicam for prevention of heterotopic ossification after total hip arthroplasty. Arch Orthop Trauma Surg 2003;123:91-4.

[54] Barthel T, Baumann B, Noth U, Eulert J. Prophylaxis of heterotopic ossification after total hip arthroplasty: a prospective randomized study comparing indemethacin and meloxicam. Acta Orthop Scand 2002;73:611-14.

[55] Banovac K, Gonzalez F, Wade N, Bowker JJ. Intravenous disodium etidronate therapy in spinal cord injury patients with heterotopic ossification. Paraplegia 1993;31:660–666.

[56] Banovac K. The effect of etidronate on late development of heterotopic ossification after spinal cord injury. J Spinal Cord Med 2000;23:40–44.

[57] Sullivan MP, Torres SJ, Mehta S, Ahn J. Heterotopic ossification after central nervous system trauma. A current review. Bone Joint Res 2013;2:51–7.

[58] Coventry MB, Scanlon PW. The use of radiation to discourage ectopic bone: a nine-year study in surgery about the hip. J Bone Joint Surg Am 1981;63-A:201-8.

[59] Knelles D, Barthel T, Karrer A, Eulert J, Kölbl O. Prevention of heterotopic ossification after total hip replacement: a prospective, randomised study using acetylsalicylic acid, indomethacin and fractional or single-dose irradiation. J Bone Joint Surg Br 1997;79-B:596-602.

[60] Lo TC. Radiation therapy for heterotopic ossification. Semin Radiat Oncol. 1999;9(2): 163-70.

[61] Sautter-Bihl ML, Hu◎ltenschmidt B, Liebermeister E, Nanassy A. Fractionated and single-dose radiotherapy for heterotopic bone formation in patients with spinal cord injury: a phase-I/II study. Strahlenther Onkol 2001;177:200-5.

[62] Jang SH, Shin SW, Ahn SH, Cho IH, Kim SH. Radiation therapy for heterotopic ossification in a patient with traumatic brain injury. Yonsei Med J 2000;41:536-9.

[63] Hamid N, Ashraf N, Bosse MJ, Connor PM, Kellam JF, Sims SH, Stull DE, Jeray KJ, Hymes RA, Lowe TJ. Radiation therapy for heterotopic ossification prophylaxis acutely after elbow trauma: a prospective randomized study. J Bone Joint Surg Am 2010;92-A:2032-8.

[64] Pape HC, Marsch S, Morley JR, Krettek, Giannoudis PV. Current concepts in the development of heterotopic ossification. J Bone Joint Surg Br 2004;86-B:783-7.

[65] Garland DE, Hanscom DE, Keenan MA, Smith C, Moore T. Resection of heterotopic ossification in the adult with head trauma. J Bone Joint Surg Am 1985;67(8):1261-9.

[66] Ippolito E, Formisano R, Farsetti P, Caterini R, Penta F. Excision for the treatment of periarticular ossification of the knee in patients who have a traumatic brain injury. J Bone Joint Surg Am 1999;81-A:783-9.

[67] Freebourn TM, Barber DB, Able AC. The treatment of immature heterotopic ossification in spinal cord injury with combination surgery, radiation therapy and NSAID. Spinal Cord 1999;37:50–3.

[68] Gacon G, Deider Ch, Rhenter J-L, Minaire P. Ectopic bone formation in neurological lesions. Rev Chir Orthop Reparatrice Appar Mot 1978;64:375-90.

[69] Moreta-Suárez J, Sáez de Ugarte-Sobrón O, Martínez-de los Mozos JL. Osificaciones heterotópicas neurogénicas de cadera. A propósito de 2 casos. Rev esp cir ortop traumatol 2011;55:292-7.

Chapter 16

Auditory/Visual Integration Assessment and Treatment in Brain Injury Rehabilitation

Deborah Zelinsky

Additional information is available at the end of the chapter

http://dx.doi.org/10.5772/57426

1. Introduction

Retinal stimulation should be considered as a possible treatment for brain injury. Recent research has demonstrated that activating the retina – in particular, the *peripheral* retina – can influence inputs to the brain in ways that may be useful in rehabilitation. The specific treatment involves designing non-traditional eyeglasses that intentionally alter signaling pathways, rather than simply sharpening central eyesight. These often overlooked peripheral retinal pathways are activated or inhibited by feedback mechanisms linked with internal body systems, such as metabolism and attention. Changes in peripheral retinal stimulation can affect internal perception and can occur at conscious, or beneath conscious, processing levels.

Traumatic brain injury often disrupts retinal circuitry, because the retina is an extension of brain tissue. This disruption can trigger innate protective mechanisms, some of which can include alteration of peripheral retinal sensitivity. Patients' ability to integrate various sensory, motor, cognitive and emotional pathways is often impaired after both focal and diffuse brain injuries, causing a wide range of symptoms. For instance, after a stroke or concussion, patients may have difficulty making accurate judgments in space and time, misjudging object location and/or speed. Current rehabilitative methods could easily be augmented by the use of selective retinal stimulation to take into account the interactions between external sensory inputs and internal processing.

The mechanism underlying this approach lies in the retinal activity that occurs *before* signals are transmitted to the midbrain, brainstem, limbic system and cortex. The eye's linkage with sounds is concurrently involved in feedback and feedforward signaling pathways from both internal and external sources. Studies that demonstrated correlations between retinal function and auditory stimuli [1] paved the way for research on retinal plasticity. Although it was thought that the visual cortex changed in deaf people to meet increased visual demands of watching sign language, recent research discovered changes at the actual peripheral retinal level [2]. The concept of the retinal cells being plastic caters toward unique internal processing in each patient. Interaction between a person and their external environment can become confusing if incoming sensory information is mismatched, e.g., if the ears are receiving a signal that is not properly linked with visual input. Treatment of brain injury can be more individualized by implementing visual/auditory integration activities during visual rehabilitation.

The organization of this chapter is as follows. Relationships between the eye and internal systems are analyzed in Section 2, retinal processing is simplified in Section 3, visual/auditory connections in are discussed in Section 4 and examples of the usage of visual/auditory interactions in diagnosis and treatment are considered in Section 5. Conclusion, Acknowledgements and References follow.

2. The eye as a modulator between micro and macro environments

Visual pathways are important in studying brain function, yet connectivity between retinal signals and other sensory signals remains minimally explored. The mounting scientific evidence regarding retinal circuitry's effect on body functions suggests that retinal stress tolerance might be a clinical indicator of a person's nervous system's resilience to environmental changes. Acceptable ranges of visual stress vary from person to person, so measuring tolerance to retinal load, using optometric techniques, might be a next step in helping stabilize imbalances after brain injury.

After a traumatic brain injury, eye movements and pupil reactions are often used as diagnostic tools to assess brain function. For instance, sudden onset of unequal pupil size is a known symptom of brain injury. A relationship between the eye and internal systems can be further shown by the significantly higher prevalence of depression, anxiety and sleep disturbances in people with visual field deficits than those without field deficits [3]. Research has even shown that retinal imaging of the eyes can show quantifiable differences between intentionally inflicted injury and accidental injury [4], indicating that the limbic system also plays a role during structural injury.

The eye acts as a modulator between external information and internal systems. Just as cellular metabolism is affected by the surrounding *micro*-environment, mood, attention and behavior are affected by the surrounding sensory *macro*-environment. Interferences in the external sensory environment affect internal processing circuitry, as shown in Figure 1.

The study of the connections in neural networks is termed "connectomics". The Human Connectome Project (HCP) was designed with a goal to map structural and functional connections of the human brain [5,6]. Many of those structural and functional connections are involved with visual processing circuitry.

In the most general sense, the relationship between the eye and internal systems can be conceptualized as indicated in Figure 2 – the somatosensory aspect of retinal stimulation. Note that the term "MIND" does not mean "BRAIN".

http://dx.doi.org/10.5772/57426

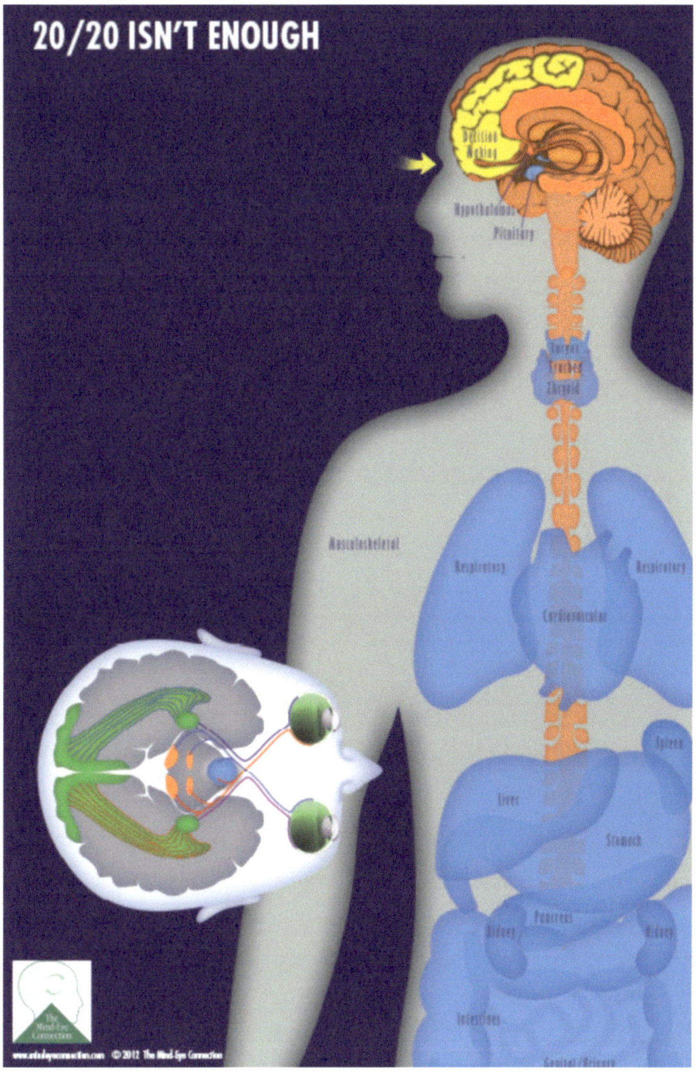

Figure 1. Retinal stimulation affects brain function, which in turn affects body systems. Medical evaluations include review of body systems as a starting point for diagnosis of problems in structural and functional integrity. Internal systems typically run beneath conscious awareness and many have connections with the eye. Multiple neural networks generate feedback and feedforward connections in order to maintain overall homeodynamic balance. When signals shifts out of balance, past habitual comfort ranges, sensory signals eventually seep into the conscious awareness level.

The eye has direct and indirect connections to many structures, allowing internal and external signals to combine and interact. Structures such as the basal ganglia have sensory-motor loops, oculo-motor loops, association loops and limbic loops to assist in sequencing and execution of movements [7]. Structures such as the cerebellum do not have direct connections with the eye, but are indirectly affected by changes in retinal stimulation via such structures as the vestibular nuclei and the inferior olivary nucleus.

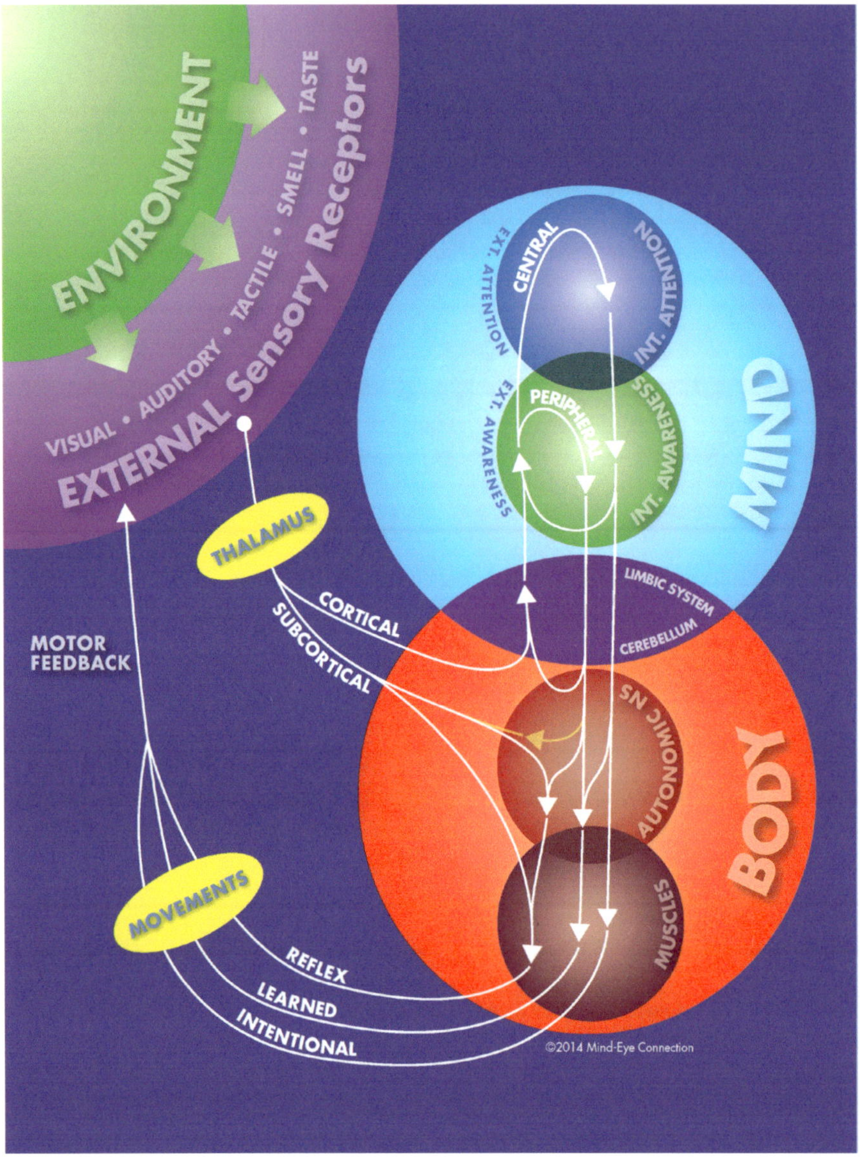

Figure 2. Simplification of Four Main Mechanisms involved in Visual and Auditory Circuitry. Constant signals arise from both internal and external sources. This diagram is a simplification of signaling pathways arising from external sources, showing how eyeglasses can be used in four distinct, yet interconnected, mechanisms. The actual complex bottom-up and top-down processing of simultaneous sensory signals is too extensive for this limited chapter. Many of the pathways have two-way signaling, with fibers returning to the sensory receptors (represented in gold) as well as signals leaving the sensory receptors (shown in white). When external sensory receptors are stimulated by changes in the outside environment, signals are sent to the central nervous system via two main routes, a fast subcortical route and a slower cortical route. The subcortical (reflex) route further separates signals into muscular reflexes and biochemical reactions. The cortical route separates into peripheral and central – Central being a slower route than peripheral. Internal thoughts, cerebellar pathways and the limbic system influence final movements. © 2014 The Mind-Eye Connection. Reprinted with permission.

Smooth interactions between systems are often disrupted after a brain injury, creating frustration. For instance, the interaction of peripheral backgrounds and central targets aids in quick and accurate depth perception and reaction time. Similarly, the motor system relies on the flawless interaction of direct and indirect excitation and inhibition of movement. The eyes and ears have important roles in those navigation systems which require internal perceptions of external space.

Other examples of how internal and external systems affect eye responses can also be cited. Research studies on patients with amblyopia, diabetes and glaucoma demonstrate how the post illumination pupillary response (PIPR) is an indicator of the integrity of intrinsic photoreceptor cells in the retinal ganglion layer, and how pupil measurements have been found to differ in those patients, as compared to the non-affected population [8,9].

Disrupted sensory circuitry can be a warning sign of early stages of certain disease processes. For instance, disturbances in rapid eye movement during sleep can be a warning sign of dopamine levels changing in Parkinson's disease years before obvious signs are noted [10-12]. New research shows that intelligence is linked with the ability to detect motion, because the ability to ferret out salient details (glimpsed by peripheral eyesight) leads to better and faster decisions [13].

3. Retinal processing

More research is needed to determine the exact role that the peripheral retinal circuitry plays in the body. Consider what happens if a person who is afraid of spiders becomes aware of a large spider. The slow movement glimpsed by peripheral eyesight triggers internal reactions such as changes in adrenaline, heart rate, muscle contraction and a shift in mental attention. The body reaction would be completely different if the person were either unaware of the spider's presence or unafraid of spiders. Another example of how peripheral retinal signals interact with the body is the internal discomfort experienced when a person puts on someone else's eyeglasses.

The above scenarios occur because the retina is an extension of brain tissue and part of the central nervous system [14]. Sensory information enters through over 150,000,000 photosensitive retinal sensors and is funneled into approximately 1,000,000 exiting ganglion axon signals through ten layers of retinal processing. Retinal circuitry is so complex that with or without light stimulation its metabolic activity is continuous – occurring even during sleep, when the layperson assumes the eye isn't functioning. The retina contains many cell types, receptors, and significant feedback and feed-forward loops. Retinal processing involves various types of biochemical reactions that are beyond the scope of this limited chapter. However, the key point is that the eye is affected by both internal chemicals that alter retinal sensitivity to the external environment, and by external stimuli that alter internal retinal chemistry.

3.1. Retinal pathways

Traditionally, the eye is viewed as a structure that sends signals into the brain through two main circuits – central eyesight for paying attention to targets, and peripheral eyesight for maintaining awareness of background. Recently, technology has been able to analyze minute changes in electrical and biochemical activity in retinal receptors, demonstrating that the retina itself is much more than an input channel.

Signals leaving the retina through the optic nerve have eight key (and many smaller) destinations:

Image forming (Eyesight):

1. **Visual cortex via thalamus:** The vast majority of signals leaving the optic nerve travel through the lateral geniculate nucleus (LGN) of the thalamus to the visual cortex, as part of the central and peripheral eyesight pathways. There is also a feedback pathway from the visual cortex back to the LGN [15,16].

Non-image forming:

2. **Thalamus:** The contralateral intergeniculate leaflet (a small section of the lateral geniculate nucleus) contributes feedforward information to the hypothalamus regarding body metabolism and sensory conditions. The intergeniculate leaflet (IGL) is also involved in receiving inputs from the vestibulo-visuomotor system, implying that circadian rhythmicity might be influenced by head motion [17,18].
3. **Hypothalamus:** The pacemaker in the suprachiasmatic nucleus (SCN) of the hypothalamus for circadian rhythms [19].
4. **Habenula** (part of the limbic system) has direct connections from a small percentage of retinal ganglion cells and multiple connections in the brainstem. It registers changes in light and is involved in modulation of both dopamine and serotonin systems, playing a role in sleep, depression and schizophrenia, and also is involved in suppression of motor control [20,21].
5. **Nucleus of the optic tract** for smooth pursuit movement, and other pretectal nuclei for reflexive eye movements and visual stability, such as optokinetic nystagmus reflexes and visual-vestibular interactions.
6. **Olivary pretectal nucleus** for pupil constriction. Recent studies have verified that this structure has greater importance previously assumed, in that it has connections with the SCN and IGL mentioned above. The olivary pretectal nucleus links external information regarding luminance with internal metabolism.
7. **Edinger-westphal nucleus,** accessory nuclei of the oculomotor nerve, for pupillary constriction from external light. These nuclei also receive internal information from the olivary pretectal nuclei [22].
8. **Superior colliculus** for posture reflexes [23] and also rudimentary eyesight beneath conscious awareness [24].

3.2. Chemical circuitry

After brain injury, the integration among the various sensory and motor pathways is often disrupted, and sometimes patients are unable to quickly and comfortably readapt to changes. This signal integration occurs in sensory receptors, subcortical structures and cortical structures. Direct structural damage to the eye, such as optic nerve injury, triggers the production of brain-derived neurotrophic factor (BDNF) to promote ganglion cell survival and help preserve the structural integrity of the surviving neurons. This innate immune network in the retina [25] is set up for survival, as are the separate yet interconnected daily visual cycles to replenish the rods and cones with nourishment [26]. The transduction system from chemical to electrical signals in the retina includes image forming pathways (signals being sorted by speed, size, shape, location, color and detail) and non-image forming pathways (posture, pupil function, circadian rhythms, etc).

3.3. Image forming systems

Eyeglasses can be used to selectively alter central and peripheral eyesight. Those two image forming retinal pathways are very different, as the peripheral retinal tissue develops neurologically from a different pathway than the central retinal tissue [27], and the peripheral retina has sections that arise from several different transcription factors [28]. Classically, the central retina is thought of as seeing colors, however, the peripheral retina is also responsive to chromatic shifts [29]. Structurally, there are differences between the distribution of cones in the nasal and temporal peripheral retina [30]. Those structural differences have a functional effect [31]. Patients who have suffered a brain injury very often do not synchronize the inputs from the central and peripheral eyesight pathways. This imbalance in external eyesight has an effect on internal systems, yet often remains undetected.

3.4. Non-image forming systems

It is well established that light activates the image forming photosensitive cells, (i.e., rods and cones). More recently, a small percentage of retinal ganglion cells (ipRGC) have been documented to be intrinsically photosensitive. They are non-image forming and contain melanopsin [32]. Less than 2% of the retinal ganglion cells are these special melanopsin containing ipRGC types. The remaining ganglion cells don't contain melanopsin. The ipRGC information travels to the hypothalamus and activates chemicals involved in circadian rhythms and general health, and also to many other non-visual structures in the brain for non-image forming purposes [33,34].

At this time, five distinct subtypes of ipRGCs have been identified, labeled M1 to M5. Each type has a high sensitivity to blue light (i.e., short wavelengths). IpRGCs have been shown to affect circadian rhythms and melatonin levels, while contributing to a simple awareness of movement via a separate pathway from the rods and cones [35-37]. A function for subtype M3 has yet to be discovered.

While the ipRGCs in the peripheral retina are important, little is known about how eyeglasses affect their purpose. However, it is known that those cells are affected by both intrinsic and extrinsic signals (from rods and cones as well as from circadian functions) [38]. It has been hypothesized that the ipRGCs may be involved in heart rate regulation [39], and that they support spatial visual perception and luminance [40]. It is also believed that the ipRGCs influence an intrinsic retinal circadian clock to regulate retinal melatonin levels by both exogenous light stimulation and endogenous circadian stimulation) [41]. The effects that the ipRGC have on degenerative processes are studied in diseases such as glaucoma and diabetes [42].

3.5. Beyond the retina

In addition to peripheral and central retinal sensory inputs, proprioceptive input is incorporated into spatial judgments [43]. In fact, ocular proprioception is a major player in spatial localization and maintaining clear central eyesight [44,45], dovetailing with other research demonstrating that problems after brain injury can affect signaling in the peripheral retina. A recent study found evidence that retinal ganglion cells in the peripheral retina showed more activity in people with attention deficit disorder, as compared to a normal control group [46].

Both the visual and circadian systems in the retina have well documented connections between vestibular and proprioceptive systems, with a feedback mechanism in addition to the current feed-forwarded information. The visual cortex, rather than simply receiving sensory information from external eyesight, also receives information from thought processes and transfers some signals back (through retinopetal fibers) to either inhibit or excite retinal photoreceptors. These thought processes, include information on space and time from other sensory inputs as well as anticipatory functions, thus, the eye aiming system is guided by both internal and external information involved in spatial navigation.

It suffices to say that a significant amount of chemical signaling occurs at the retinal level before signals exit the optic nerve to be processed further. That signaling combines inputs from the external environment with signals from internal systems. For instance, if a sudden sound is heard, there are reflexes that rotate the neck and eyes to point at the sound source. However, people can become habituated to a sound so that they inhibit those reflex movements. There are also sounds that conjure up thoughts and images in the mind, allowing for individuals to select whether or not they will point their eyes toward or away from the sound. People differ in how they react to the same incoming stimulus. After brain injury, sensory stimuli can be overwhelming, and patients often suppress awareness of external surroundings by inhibitory processing in the peripheral retina. This suppression occurs beneath the conscious level of awareness.

4. Visual and auditory integration

While the eye interacts with multiple internal systems, one of the best-known relationships between seeing and hearing can be demonstrated by the "ventriloquist effect". When an observer's eyes perceive a dummy's mouth rather than the puppeteer's mouth moving, the observer's brain assumes that the voice heard is actually emanating from the dummy's mouth.

http://dx.doi.org/10.5772/57426

Many studies have shown how when faced with a sensory mismatch, the visual system dominates [47]. More recent studies demonstrate that the auditory system can be used to modulate visual attention [48, 49].

A person who is able to pass a vision and hearing screening may not necessarily process visual and auditory signals together in order to effortlessly watch and listen. The blending of those two sensory processing systems is advantageous in such life skills as reading and social interactions. Brain injury often disrupts the developed linkage. Diseases also affect sensory linkages. For instance, auditory and visual circuitries are disrupted in early stages of schizophrenia [50, 51]. Functional differences in connectivity between auditory and visual pathways are quantifiable in early Parkinson's disease – first as different patterns of firing within the same circuitry, later with asymmetries present [52].

Integration between auditory and visual systems has been studied for decades. In 1976, the McGurk Phenomenon [53] showed when auditory and visual signals were mismatched, the brain makes a correction so that the sounds/sights make sense. Later, it was found that children with learning problems weren't using those two sensory systems in the same way as children without a learning disability [54]. It was determined that attentional circuitry modulates audiovisual integration of speech, and synchronization of visual and auditory information is combined to assess spatial awareness in babies [55]. Without the synchronization of visual and auditory signals, attention, behavior and concentration suffer, affecting academic, social and athletic achievement. Stability of brain circuitry between auditory and visual systems is important, as is proprioceptive input [56].

Researchers such as Charles Spence [57] in Oxford, England and David Alais [58] in Australia have laboratories investigating visual and auditory perception, including interactions between the two sensory systems during resolution of sensory mismatches. They also work with internal visualization predicted by auditory cues. For instance, if a person has a roommate, and hears a noise in their home, they will visualize the roommate moving around. However, if they live alone and hear the same noise, they might visualize a thief. The former will not elicit the same stress chemicals that the latter scenario will generate. The incoming auditory signal (the noise) is identical, yet body responses and internal visualization are totally different. New research is finding that the visual cortex is involved in a feedback mechanism for higher cortical functions, such as prediction, attention and imagination. This concept of the visual cortex being not only part of a feedforward system of visual signals from the retina, but also being used during feedback circuitry in higher level processes is relatively new [59,60].

Retinal stimulation affects both chemical and neurological signaling pathways, inducing subsequent behavioral responses. In patients with delicately balanced nervous systems, who are genetically predisposed to certain environmental induced diseases, sensory mismatches can create additional stress. Consider the frustration and additional attention required to watch a badly dubbed movie or to speak with someone on a phone or video chat containing a slowed signal. Retinal stimulation to correct these mismatches may lessen overall stress, potentially making the body more resistant to disease. By implication, an integrated measurement of visual and auditory signals is more effective in diagnosis and treatment than isolated testing of visual and auditory ability alone.

5. Using retinal processing in diagnosis and treatment of brain injury

5.1. Diagnosis of dysfunctions after brain injury

The peripheral retina is activated or inhibited by shifts in internal biochemistry. Research has shown linkages between retinal stimulation and such diagnoses as attention deficit disorder, anxiety, depression, obsessive compulsive disorders, sleep disorders and addictions.

Eyeglasses typically are thought of as bending light to strike the macula, producing clear eyesight. But, ambient light may be selectively harnessed to affect chemical signaling pathways in subcortical non-image forming systems, as well as in classic cortical visual processing systems. Tints can affect visual fatigue [61], and eyeglasses can be designed to disperse light differently on individual areas of the retina affecting the central and autonomic nervous systems, because the eye is connected to both. Measurements of retinal stress tolerated before inducing double vision and blurry eyesight is important in the determination of individualized eyeglasses. On the way to the visual cortex, signals from the retinal ganglion cells travel though specific, mapped areas of the brain. Therefore, light can be intentionally directed toward or away from a damaged cortical area. This concept has been successful in helping impaired spatial navigation skills in patients with brain injury and other conditions.

During standard eye testing with the traditional "which is better, one or two?" method, some patients do not consciously perceive differences between visual targets. In those cases, the eye care professional makes a choice, based on many factors. Most patients simply adapt to whichever lens choice is chosen. However, patients with a fragile interaction between sensory systems will demonstrate more visual/auditory stability with one lens choice than the other. In other words, one of the lens choices might distort auditory localization; the other might enhance it. A minor amount of light striking the retina through a closed eyelid is enough to alter auditory localization. Thus, certain lenses provide integrated sensory information and require less overall energy for processing incoming signals. Those hypersensitive patients would benefit from more in depth testing of brain circuitry via retinal sensitivity. Various tests, such as the Yoked Prism Walk Test, Super Fixation Disparity Test©, the Padula Visual Midline Shift Test, the Van Orden Star Test and Z-Bell© Test are simple ways to assess sensory linkages [62]. By using various eyeglasses and contact lenses to effect auditory localization with retinal circuitry as a way to access brain function, even many non-verbal patients can be accurately measured.

Below are a few of the many ways eyeglasses have been used in ways other than for eyesight to provide improved function in traumatic brain injury patients. The retinal interactions with spatial navigation, posture and limbic system activity have sound neuroscience behind them. For instance, substantial research agrees that light striking the retina stimulates dopamine release, and that darkness stimulates melatonin production. Other research shows that too little dopamine in the mesocortical systems is found in patients with schizophrenia, and too much dopamine is found in mesolimbic systems of patients with hallucinations. It isn't out of the realm of possibility that eyeglasses changing retinal chemistry can alter body chemistry and affect imbalanced systems for the better. Obviously, more research is necessary in order to make any conclusive statements. Yet, patients with brain injury who aren't responding to medication might use this non-invasive method of neuromodulation.

- **Post Concussive Syndrome (PSC) vs. Post Traumatic Stress Disorder (PTSD)**

Post Concussive Syndrome (PCS) and Post Traumatic Stress Disorder (PTSD) are two common occurrences after brain injury. Some patients have both. Their symptoms are similar, but treatment differs. Currently they can be differentiated by fMRI techniques.

Anecdotally, when accounting for shifts in proprioceptive inputs, auditory/visual integration testing has shown that spatial perception in patients with PTSD tends to be symmetrically changed. In PSC, spatial perception has been found to be asymmetric. When a patient has both PSC and PTSD, the perception seems to be both constricted and asymmetric, with proprioceptor involvement showing a different effect than in those patients with PSC alone. Quantifying the interaction between external sensory inputs and internal processing has accurately differentiated between conditions such as post concussive syndrome and post traumatic stress disorder in patients.

- **Vertical Heterophoria**

After traumatic brain injury, patients often see double, or have a slight vertical imbalance between their eyes which creates a discomfort, but not a blur or complete doubling. Sometimes a slight angling of light to balance the two eyes has a far-reaching effect. In 2010, a retrospective study showed that of 83 brain injury patients with post concussive symptoms remaining after standard treatment, 77 of them had vertical heterophoria. When treated, there was a 71% decrease in their symptoms [63].

- **Positional Orthostatic Tachycardia Syndrome (POTS)**

Positional Orthostatic Tachycardia Syndrome (POTS) is a condition where the autonomic nervous system's regulation of cerebral blood flow is dysfunctional, and blood vessels constrict rather than expand when more blood plasma volume is required. This deficiency in cerebrovascular autoregulation is often exacerbated when shifting from a seated to a standing position. A person should be able to maintain normal cerebral blood flow in spite of changing blood pressure, but patients with subtypes of POTS can not autoregulate [64].

Patients who had symptoms of syncope due to autonomic dysregulation were treated by the use of therapeutic eyeglasses. One pair of lenses angled light, affecting expended effort on eye muscle control; the other pair was tinted to filter the incoming light's wavelength. When wearing the lenses, both patients reported cessation of fainting and easier adaptation to sudden posture shifts. The proposed mechanism was the intentional change in dispersion of light on the peripheral retina affecting the exiting signals from the optic nerve. The light was designed to balance the imbalances.

Results suggest the possibility that patients with POTS might have narrow ranges of tolerance to environmental changes. Especially in those patients where medication hasn't been successful, using retinal tolerance tests to design eyeglasses might be beneficial. The ability to expand the patient adaptability to shifts in autonomic nervous system responses could affect syncope and quality of life.

- **Seizures and Post-traumatic Epilepsy**

Many patients experience seizures after brain injury, and a small percentage of brain-injured patients develop epilepsy. The proposed mechanism to help those patients is to divert light onto retinal regions where signals travel through undamaged cortical areas.

Some patients who had frequent seizures were able to match auditory and visual perception of space when a tint or a filter was incorporated into eyeglasses. Several patients have been documented to have lessening of frequency and duration of seizures after being prescribed customized eyeglasses designed for the non-image forming retinal pathways. Further studies need to be developed on a broader scale, but this may be a method to employ when seizure medications are not working sufficiently, or if a patient is pregnant and can't tolerate medications.

- **Auditory and Visual Hallucinations**

After trauma or some medications, patients occasionally experience hallucinations. Research shows that those patients have too much dopamine in the meso-limbic systems. Since light stimulation affects retinal dopamine levels and brain signaling, it is possible that signalling can be modified to address the over activity. Patients with schizophrenia have been documented to have disconnects between auditory and visual signalling systems.

A few patients reported the disappearance of voices in their heads after using eyeglasses customized for balancing the auditory/visual spatial judgments. Perhaps the unknown function of the retinal subtype M3 ipRGC will be discovered to be linked with the auditory localization system.

- **Depression**

Brain injury often results in patients with depression who are less aware of their surroundings than non-depressed patients. Deep brain stimulation to the nucleus accumbens has been noted to help symptoms of depression. The nucleus accumbens interacts with auditory cortices [65] and is indirectly connected to retinal processing through the habenula, which modulates dopamine neurons projecting to it [66].

Many patients who have been clinically diagnosed with depression have been helped by the usage of eyeglasses designed to synchronize their visual and auditory perception of space. Anecdotally, there were two whose EEG findings instantly changed when comparing with and without therapeutic eyeglasses.

- **Anxiety**

Anxiety disorders have been shown to have imbalanced chemical production in the cortex and midbrain. Changing retinal stimulation has an effect on dopamine, serotonin and GABA levels [67]. Many patients who exhibit anxiety have been helped by the use of eyeglasses designed for peripheral retinal calming, rather than for eyesight.

http://dx.doi.org/10.5772/57426

To reiterate, the above usages of eyeglasses appears to have helped many patients in their daily lives after brain injury. The rationale reverts back to the concept of the retina being a modulator between cellular microenvironments and surrounding external sensory environments. Light selectively activates specific retinal regions, sending signals through many feedback and feed-forward circuits in both subcortical and cortical structures. Predictable retinal mapping can shift signaling pathways into an individual's range of tolerance, enabling each patient to have an improved functional outcome. The concept is relatively new, but is non-invasive, and has been successful in many patients with sequellae from brain injuries.

5.2. Treatment – Neuro-optometric rehabilitation

There are several types of eye care providers. Neuro-ophthalmologists identify and treat physical and physiological conditions that manifest in visual impairments, whereas optometrists whose work specifically emphasizes neuro-optometric rehabilitation, use visual pathways to affect changes in physical and physiological functions. Testing can be similar, but the analysis and treatment goals are different. Neuro-optometric rehabilitation uses alterations of light to influence and enhance a person's physical and mental reactions and responses to changes in his environment, enhancing rehabilitative outcomes as an adjunct form of treatment after brain injury. The infinite combinations of internal biochemistry and environmental experiences unique to each person allows for the unconventional usage of eyeglasses on people, even if they have 20/20 eyesight without glasses. Patients with brain injury often fall into a category of not having an eyesight problem, yet still having imbalances in their mind-eye connections.

Neuro-optometric rehabilitation deals with perception and action by measuring a person's awareness of, and responses to, his perceived environment. During an optometric evaluation, this relationship between the "actual" and "perceived" environments is analyzed as well as its effect on the person's internal environment and body posture. By controlling the input of light, and measuring and recording the patient's reaction to new environmental stimuli, optometrists can determine how well the person's visual processing systems are functioning. Measuring the perceived distortion and altering retinal input to change habitual perception, affects comfort, concentration and performance of daily tasks.

Many body functions are directly or indirectly influenced by visual processing at or beneath a conscious level. Signaling is a two-way street – retinal stimulation can influence other systems within the body, and other systems can influence retinal metabolism. Alteration of light on the retina (by varying such factors as the intensity, frequency or direction) changes brain activity. Currently, during standard eye testing, imbalances between right and left and central and peripheral are evaluated, but not necessarily internal vs. external systems or visual vs. auditory systems. Attention can be on internal thoughts or external selected targets. However, if there is a disruption in circuitry, or sensors are stimulated past their range of comfort or tolerance, attention is diverted. People comprehend that glasses might give them a headache or make them nauseous, but the concept of eyeglass prescriptions as a way to alter internal systems beneath conscious awareness has not yet been explored at length. Nevertheless, eye care providers are in a unique position to assess more than simply eyesight and eye health.

SOME COMMON NEURO-OPTOMETRIC INTERVENTIONS

- Angling light toward one edge of the retina. This initially affects the body's positional sense, because reflexive eye movements will point toward the incoming light, triggering internal postural mechanisms in the hips for stability of balance, to counteract the eye movement. Depending on the stability of the person's sense of balance, attention may be shifted off of the body and onto external targets.
- Angling light toward either nasal or temporal retinal sensors. The eyes will also reflexively point toward the light, but this inward and outward movement stimulates different visual and postural mechanisms (shoulders rather than hips), shifting apparent object location.
- Dispersing light toward the edges or the center of the retina tends to make objects appear larger or smaller, by emphasizing or muting the background. This change in light mainly alters the balance between central and peripheral circuitry by having the target and background occupy different percentages of the retinal input.
- Altering either spatial or temporal retinal input with filters affects internal processing and external perception. Tints filter out selected wavelengths of light, stimulating specific retinal cells which alter retinal chemistry (and thus body chemistry), via the autonomic nervous system. Graded occlusion filters alter the spatial components of incoming light.
- Inducing a sensory mismatch between central (target) and peripheral (background) can be accomplished with mirrors. These are used in many aspects of rehabilitation, such as patients with strokes or visual field defects.

Figure 3. Various visual interventions can be used to stimulate the retina through different pathways. ©2010 Mind-Eye Connection Reprinted with permission

6. Conclusion

Traditional eyesight testing is not yet assessing many of the multiple interactions between the peripheral retina and brain function beneath a conscious level of processing. 20/20 eye testing relies on conscious attention to a selected target. As technology advances, by using integrated visual and auditory information, testing eyes and visual pathways might include assessments of such systems as internal visualization, judgments in space and time, balances between dopamine and serotonin pathways, etc. Recent developments supporting this perspective of linking sensory and motor systems as well as internal and external visual systems include three-dimensional movies and Google Glasses, which are designed to verbally access a virtual visual projection. In fact, Google Glasses will soon be in the public domain. Also, the U.S. Food and Drug Administration has recently approved the Vimetrics Central Vision Analyzer (CVA) which tests acuity with contrast and lighting changes rather than the classic black letters on a high contrast white background.

http://dx.doi.org/10.5772/57426

After brain injury, often a person's external environment might remain the same, but the internal perception and interpretation of it differs. By introducing the possibility that the mind-eye connection – how a person's thoughts, movements and behavior are affected by what is in their mind's eye – is unique to each person's processing system and experiences, rehabilitation can utilize this concept of "mind-eye testing" vs. "eye testing" to prescribe eyeglasses on a more individual basis. Eventually, eye testing will be routinely inclusive of other sensory signals, including the linkage between auditory and visual systems. Perhaps the unknown function of the retinal subtype M3 ipRGC will be discovered to be linked with the auditory localization system. Ground-breaking research strongly suggests that the integration of visual and auditory signals in people with an autistic spectrum disorder appears to originate from a timing deficiency, affecting their language and communication [68]. Auditory processing has been shown to be altered by visual deprivation in adult mice, offering validity to new multi-modal rehabilitation methods such as selective retinal stimulation [69].

As always, new concepts require time to travel from bench to bedside, but research continues to demonstrate that the visual system provides much more than simply sending light into the brain to interpret images from external surroundings. New retinal circuitries and feedback pathways are continually being discovered. Retinal stimulation can have significant impacts on physiological systems; however, future research is needed to design specific protocols for this innovative method before these general concepts become more commonplace. Board certification in neuro-optometric rehabilitation has begun, but it will take another generation for the general population to realize the impact eyeglasses have on internal systems in the body, even in patients who measure 20/20 central eyesight.

The brain injured population is a wonderful start for the beginning of this change in diagnostic procedures. Assessment of patient adaptation to environmental changes via retinal stimulation and integration of visual and non-visual pathways will affect future eyeglass prescriptions. These in turn should allow patients to adapt to environmental changes more easily. Neuro-Optometry is complementary with the new field of neurophotonics bridging neuroscience research with optical physics. The impact on processing pathways can be quantified, and measurements can be used to assess and modify internal tolerance to external changes. Diagnostic methods already in use point to a relationship between the eye and internal systems. It follows that testing the relationship between internal and external systems should gain a pivotal role in rehabilitation.

Visual contributions to a vulnerable nervous system should not be ignored. This paper presented findings derived from the author's experience in using retinal stimulation to alleviate patient symptoms, including those caused by brain injuries. While these methods have not yet been rigorously tested in formal clinical studies, the results have been effective enough to allow the author to build a thriving practice that draws patients from all over the United States and many foreign countries. Given this experience and clinical success, which has been replicated by optometrists Benoit Lombaerts in Belgium, Vasillis Kokotos in Greece, Juergen Eichinger in Germany and Stefan Collier in Switzerland, the author believes that retinal stimulation has great potential for becoming a low-cost, low-risk, and effective therapy for a wide range of neurological and biological disorders. Neuro-optometric treatment can provide substantial benefit to patient rehabilitation after brain injury, with minimal additional time expended by the eye care provider.

Acknowledgements

Many people have been involved in retinal research since the discovery of ipRGCs ten years ago, but not many consider how eyeglasses affect those cells. The most significant optometric mentor in my life is Albert A. Sutton, O.D. whose passion for helping others continually guides me to ask about and probe brain function rather than eye function. Thanks to Kelly Richars for helping my many trains of thought reach a final destination, and sincere appreciation to Barbara Apgar, whose mistaken perception of instructions led to the serendipitous discovery of the linkage between retinal stimulation and auditory localization ability. This discovery will eventually provide eye care professionals an accurate method to quickly measure relationships between internal and external systems.

Author details

Deborah Zelinsky*

Address all correspondence to: mindeyeconnection@msn.com

The Mind-Eye Connection, Northbrook, IL, USA

References

[1] Dantsig, I. N. and Diev, A.V. Light sensitivity of the central and peripheral sections of the retina in exposure to noise. *Biull Eksp Biol Med* 1986 101(2): 133-135.

[2] Codina, C., Pascalis, O et al. Visual advantage in deaf adults linked to retinal changes. *PLoS One* 2011 6(6): e20417.

[3] Agorastos, A., Skevas, C., et al. Depression, anxiety, and disturbed sleep in glaucoma. J Neuropsychiatry Clin Neurosci 2013; 25(3): 205-213.

[4] Minns, R.A., Jones, P.A., et al. Prediction of inflicted brain injury in infants and children using retinal imaging. Pediatrics 2012 130(5): e1227-1234.

[5] Van Essen, D.C., Ugurbil, K., et al. The Human Connectome Project: a data acquisition perspective. Neuroimage 2012 62(4): 2222-2231.

[6] Fornito, A., Zalesky, A., et al. Graph analysis of the human connectome: Promise, progress, and pitfalls. Neuroimage 2013 Oct 15;80:426-44.

[7] Miyachi, S. Cortico-basal ganglia circuits--parallel closed loops and convergent/divergent connections. Brain Nerve 2009 61(4): 351-359.

[8] Donahue, S. P., Moore, P., et al. Automated pupil perimetry in amblyopia: generalized depression in the involved eye. Ophthalmology 1997 104(12): 2161-2167.

[9] Kankipati, L., Girkin, C.A., et al. The post-illumination pupil response is reduced in glaucoma patients. Invest Ophthalmol Vis Sci 2011 52(5): 2287-2292.

[10] Kostic, V.S., Susic, V., et al. Reduced rapid eye movement sleep latency in patients with Parkinson's disease. J Neurol 1989 236(7): 421-423.

[11] Gagnon, J.F., Montplaisir, J., et al. Rapid-eye-movement sleep disorders in Parkinson's disease. Rev Neurol (Paris) 2002 158(2): 135-152.

[12] Postuma, R.B. and Montplaisir, J. Potential early markers of Parkinson's disease in idiopathic rapid-eye-movement sleep behaviour disorder. Lancet Neurol 2006 5(7): 552-553.

[13] Melnick, M.D., Harrison, B.R., et al. A strong interactive link between sensory discriminations and intelligence. Curr Biol 2013 23(11): 1013-1017.

[14] Klassen, H., Ziaeian, B., et al. Isolation of retinal progenitor cells from post-mortem human tissue and comparison with autologous brain progenitors. J Neurosci Res 2004 77(3): 334-343.

[15] O'Connor, D. H., M. M. Fukui, et al. (2002). "Attention modulates responses in the human lateral geniculate nucleus." *Nat Neurosci* 5(11): 1203-1209.

[16] Schneider, K. A. and S. Kastner (2009). "Effects of sustained spatial attention in the human lateral geniculate nucleus and superior colliculus." *J Neurosci* 29(6): 1784-1795.

[17] Blasiak, T. and Lewandowski, M.H. Differential firing pattern and response to lighting conditions of rat intergeniculate leaflet neurons projecting to suprachiasmatic nucleus or contralateral intergeniculate leaflet. Neuroscience 2013 228: 315-324.

[18] Saderi, N., Cazarez-Marquez, F., et al. The NPY intergeniculate leaflet projections to the suprachiasmatic nucleus transmit metabolic conditions. Neuroscience 2013 246C: 291-300.

[19] Horowitz, S.S., Blanchard, J.H., et al. Intergeniculate leaflet and ventral lateral geniculate nucleus afferent connections: An anatomical substrate for functional input from the vestibulo-visuomotor system. J Comp Neurol 2004 474(2): 227-245.

[20] Beretta, C. A., Dross, N., et al. Habenula circuit development: past, present, and future. Front Neurosci 2012 6: 51.

[21] Aizawa, H. Habenula and the asymmetric development of the vertebrate brain. Anat Sci Int 2013 88(1): 1-9.

[22] Ishikawa, H. (2013). Pupil and melanopsin photoreception. Nihon Ganka Gakkai Zasshi 2013 Mar; 117(3): 246-268; discussion 269.

[23] Miller, A. M., Obermeyer,W.H., et al. The superior colliculus-pretectum mediates the direct effects of light on sleep. Proc Natl Acad Sci U S A 1998 95(15): 8957-8962.

[24] Leh, S.E., Ptito, A., et al. Blindsight mediated by an S-cone-independent collicular pathway: an fMRI study in hemispherectomized subjects. J Cogn Neurosci 2010 22(4): 670-682.

[25] Templeton, J.P., Freeman, N.E., et al. Innate immune network in the retina activated by optic nerve crush. Invest Ophthalmol Vis Sci 2013 54(4): 2599-2606.

[26] Muniz, A., Villazana-Espinoza, E.T., et al. A novel cone visual cycle in the cone-dominated retina. Exp Eye Res 2007 85(2): 175-184.

[27] Venters, S.J., Mikawa, T., et al. Central and Peripheral Retina Arise through Distinct Developmental Paths. PLoS One 2013 8(4): e61422.

[28] Tombran-Tink, J. and Barnstable, C.J. Visual transduction and non-visual light perception. Totowa, N.J., Humana Press; 2008.

[29] Solomon, S.G., Lee, B.B., et al. Chromatic organization of ganglion cell receptive fields in the peripheral retina. J Neurosci 2005 25(18): 4527-4539.

[30] Visser, E.K., Beersma, D.G., et al. Melatonin suppression by light in humans is maximal when the nasal part of the retina is illuminated. J Biol Rhythms 1999 14(2): 116-121.

[31] Panorgias, A., Parry, N.R., et al. Nasal-temporal differences in cone-opponency in the near peripheral retina. Ophthalmic Physiol Opt 2009 29(3): 375-381.

[32] Hicks, D. Second sight? Graefes Arch Clin Exp Ophthalmol 2011 Mar; 249(3): 313-314. doi: 10.1007/s00417-011-1631-y. Epub 2011 Feb 19.

[33] Van Gelder, R.N. Non-visual photoreception: sensing light without sight. Curr Biol 2008 18(1): R38-39.

[34] Chellappa, S.L., Steiner, R., et al. Non-visual effects of light on melatonin, alertness and cognitive performance: can blue-enriched light keep us alert? PLoS One 2008 6(1): e16429.

[35] Ruan, G.X., Zhang, D.Q., et al. Circadian organization of the mammalian retina. Proc Natl Acad Sci U S A 2006 103(25): 9703-9708.

[36] Zaidi, F.H., Hull, J.T., et al. Short-wavelength light sensitivity of circadian, pupillary, and visual awareness in humans lacking an outer retina. Curr Biol 2007 17(24): 2122-2128.

[37] Markwell, E.L., Feigl, B., et al. Intrinsically photosensitive melanopsin retinal ganglion cell contributions to the pupillary light reflex and circadian rhythm. Clin Exp Optom 2010 93(3): 137-149.

[38] Chen, S.K., Chew, K.S., et al. Apoptosis regulates ipRGC spacing necessary for rods and cones to drive circadian photoentrainment. Neuron 2013 77(3): 503-515.

[39] Hu, K., Meijer, J.H., et al. Fractal patterns of neural activity exist within the suprachiasmatic nucleus and require extrinsic network interactions. PLoS One 2012 7(11): e48927.

[40] Ecker, J.L., Dumitrescu, O.N., et al. Melanopsin-expressing retinal ganglion-cell photoreceptors: cellular diversity and role in pattern vision. Neuron 2010 67(1): 49-60.

[41] Tosini, G., Pozdeyev, N., et al. The circadian clock system in the mammalian retina." Bioessays 2008 30(7): 624-633.

[42] Feigl, B., Zele, A.J., et al. The post-illumination pupil response of melanopsin-expressing intrinsically photosensitive retinal ganglion cells in diabetes. *Acta Ophthalmol* 2012 90(3): e230-234.

[43] Balslev, D., Himmelbach, M., et al. Eye proprioception used for visual localization only if in conflict with the oculomotor plan." J Neurosci 2012 32(25): 8569-8573.

[44] Ventre-Dominey, J., Dominey, P.F., et al. Extraocular proprioception is required for spatial localization in man. Neuroreport 1996 7(9): 1531-1535.

[45] Tong, J., Stevenson, S.B., et al. Signals of eye-muscle proprioception modulate perceived motion smear." J Vis 2008 8(14): 7 1-6.

[46] Bubl, E., Dorr, M., et al. Retinal contrast transfer functions in adults with and without ADHD. PLoS One 2013 8(5): e61728.

[47] Sinnett, S., Spence, C. et al. Visual dominance and attention: the Colavita effect revisited. *Percept Psychophys* 2007 69(5): 673-686.

[48] Recanzone, G. H. Auditory influences on visual temporal rate perception. *J Neurophysiol* 2003 89(2): 1078-1093.

[49] Miller, J. E., Carlson, L.A., et al. When what you hear influences when you see: listening to an auditory rhythm influences the temporal allocation of visual attention. *Psychol Sci* 2013 24(1): 11-18.

[50] Akamatsu, T., Fukuyama, H., et al. The effects of visual, auditory, and mixed cues on choice reaction in Parkinson's disease. J Neurol Sci 2008 269(1-2): 118-125.

[51] Kayser, J., Tenke, C.E., et al. A neurophysiological deficit in early visual processing in schizophrenia patients with auditory hallucinations. Psychophysiology 2012 49(9): 1168-1178.

[52] Husarova, I., Mikl, M., et al. Similar Circuits but Different Connectivity Patterns Between The Cerebellum, Basal Ganglia, and Supplementary Motor Area In Early Parkinson's Disease Patients and Controls During Predictive Motor Timing. J Neuroimaging 2013 May 23. doi: 10.1111/jon.12030.

[53] McGurk, H. and MacDonald, J. Hearing lips and seeing voices. Nature 1976 264(5588): 746-748.

[54] Bastein-Toniazzo, Stroumza, M.A., Cave, C. Audio-visual perception and integration in developmental dyslexia: an exploratory study using the McGurk Effect. Current psychology Letters: Behaviour, Brain and Cognition 2009 Vol. 25, Issue 3.

[55] Rosenblum, L.D., Schmuckler, M.A., et al. The McGurk effect in infants. Percept Psychophys 1997 59(3): 347-357.

[56] Gauthier, G.M., Nommay, D., et al. The role of ocular muscle proprioception in visual localization of targets. Science 1990 249(4964): 58-61.

[57] Spence, C., Ranson, J., et al. Cross-modal selective attention: on the difficulty of ignoring sounds at the locus of visual attention. Percept Psychophys 2000 62(2): 410-424.

[58] Alsius, A., Navarra, J., et al. Audiovisual integration of speech falters under high attention demands. Curr Biol 2005 15(9): 839-843.

[59] Muckli, L. and Petro, L.S. Network interactions: non-geniculate input to V1. Curr Opin Neurobiol 2013 23(2): 195-201.

[60] Muckli, L., Petro, L.S., et al. Backwards is the way forward: feedback in the cortical hierarchy predicts the expected future. Behav Brain Sci 2013 36(3): 221.

[61] Tagami,Y., Ohnuma, T., et al. Visual fatigue phenomenon and prescribing tinted lenses in patients with optic neuritis. Br J Ophthalmol 1984 68(3): 208-211.

[62] Zelinsky, D. Neuro-optometric diagnosis, treatment and rehabilitation following traumatic brain injuries: a brief overview. Phys Med Rehabil Clin N Am 2007 18(1): 87-107, vi-vii.

[63] Doble, J. E., Feinberg, D.L. et al. Identification of binocular vision dysfunction (vertical heterophoria) in traumatic brain injury patients and effects of individualized prismatic spectacle lenses in the treatment of postconcussive symptoms: a retrospective analysis. 2010 *PMR* 2(4): 244-253.

[64] Low, P.A., Novak, V., et al. Cerebrovascular regulation in the postural orthostatic tachycardia syndrome (POTS). Am J Med Sci 1999 317(2): 124-133.

[65] Salimpoor, V. N., Van den Bosch, I., et al. Interactions between the nucleus accumbens and auditory cortices predict music reward value." Science 2013 340(6129): 216-219.

[66] Lecourtier, L., Defrancesco, A., et al. Differential tonic influence of lateral habenula on prefrontal cortex and nucleus accumbens dopamine release." Eur J Neurosci 2008 27(7): 1755-1762.

[67] Nikolaus, S., Antke, C., et al. Cortical GABA, striatal dopamine and midbrain serotonin as the key players in compulsive and anxiety disorders--results from in vivo imaging studies." Rev Neurosci 2010 21(2): 119-139.

[68] Stevenson, R. A., J. K. Siemann, et al. (2014). "Multisensory temporal integration in autism spectrum disorders." J Neurosci 34(3): 691-697.

[69] Petrus, E., A. Isaiah, et al. (2014). "Crossmodal induction of thalamocortical potentiation leads to enhanced information processing in the auditory cortex." Neuron 81(3): 664-673.

Chapter 17

Returning Individuals with Mild to Moderate Brain Injury Back to Work: A Systematic Client Centered Approach

Shaheed Soeker

Additional information is available at the end of the chapter

http://dx.doi.org/10.5772/57309

1. Introduction

Due to an increase in the numbers of individuals who sustained brain injuries due to motor vehicle accidents, trauma induced by violence and substance abuse, has resulted in more disabled individuals becoming non-productive members in society and inactive in the workplace [1]. Research in the field of brain injury rehabilitation internationally is limited, with the majority of research focusing on the medical model of intervention. In the medical model, the disabled or injured individual is regarded as having problems that require medical-biological intervention mainly, with little or no attention given to the difficult process of reintegrating the disabled individual back into society, for example, in resuming their worker roles. The medical approach may result in feelings of disempowerment on behalf of the disabled with regard to the rehabilitation process. The lack of success of current rehabilitation interventions could be seen as a result of an inability to generalize outcomes of rehabilitation in a clinical setting to the skills needed to return to work or re-integrate into the community.

2. Epidemiology

A traumatic brain injury (TBI) is an insult to the brain resulting from external physical forces such as high speed motor vehicle accidents (MVA's) and falling from heights which exceeded the height of the person, as well as sports injuries, gunshot wounds and work related injuries. [1,2] Raskin and Mateer [3] describe TBI as the sudden trauma to the head, which causes injury to the head and the brain. Such injuries can result in impaired physical, cognitive, emotional, and behavioral functioning. According to Gutman [1], Khan et al. [4] and the National Health

Laboratory Service [5], TBI occurs more frequently in men than in women by a ratio of four to one, with 80% of all individuals who sustain TBI being between the ages of 18 and 30 years. Each year, traumatic brain injuries contribute to a substantial number of deaths and cases of permanent disability. Traumatic brain injury (TBI) is a serious public health problem in the United States [6]. Traumatic brain injury is one of the invisible wounds of war, and one of the signature injuries of troops wounded in Afghanistan and Iraq. In 2011, the incidence of traumatic brain injuries of armed forces injured in Afghanistan and Iraq increased to 20 000, a dramatic increase of 5000 incidence since 2007. [7]

3. Return to work

According to Vuadens, Arnold & Bellmann [8] 70% of moderately brain injured individuals do not return to work and that 20% of mildly brain injured individuals are unemployed. Furthermore a total of 10% of brain injury clients get fired and only 2% are employed on a short term basis, one year post trauma. The above evidence indicated that a large number of TBI individuals are unable to return to the vocational roles they established before injury, this has a direct correlation with their volition and self-worth. Cicerone [9] confirms that individuals with brain injury who have failed to return to work have a lowered subjective wellbeing compared to those who have successfully returned to work. Studies have indicated that clients also have harsh unmet needs in association with working at societal level; they no longer fill their roles as breadwinners and lost wages. According to Cicerone [9], individuals with TBI who are unemployed feel as if they are not competent enough to perform at the average work standards, and therefore have a loss of self esteem and an overall decline in volition and occupational engagement not just in vocational activities but in all activities as they no longer see themselves as contributing members of society.

4. Facilitators of return to work for individuals with brain injury

Reduction of travelling expenses to and from the hospitals greatly reduces the financial strain experienced by the individual with the brain injury. Sample and Darragh [10] in their study on perceptions of care access identified that **financial challenges and travelling for services** amongst other challenges are of concern. Easy access to the treatment facilities and open communication between health officials, employers and individuals with brain injury positively contributed to the brain injured individual`s return to work. Clients often indicate that they have to find alternative means of obtaining their medication, some of them often arrange to have their medication collected at their nearest day hospital. **Open communication** is interpreted as communication between the health professional, employer, participant's family and the participant. Effective communication facilitates transparency and allows all stakeholders such as the employer and health professional to know their individual responsibilities. Friesen et al. [11] confirmed that good communication and positive relations between stakeholders (employee, employer and health professional) was important during the return to work process. A **client centered approach** enables clients to take control of their rehabilitation by actively participating in the planning and implementation of their treatment programmes. Townsend and Banks [12] indicates that client centred practice guides the occupational therapist (health provider) to work with clients who are active participants in collaborating their ideas of fulfilling meaningful occupational performance within an environmental context. Schultz-Krohn and Pendleton [13] mentions that client centred practice is guided by the following: the language used to address the client should reflect the person first and then the condition, the client is offered choices and is supported in directing the occupational therapy process, intervention is provided in a flexible and accessible manner, intervention is contextually appropriate and relevant, and there is a clear respect for difference and diversity in the occupational therapy process.

Furthermore clients often mention that **treatment programmes that incorporated home visits** were extremely helpful. Schwartz [14] verified the above in her research where she indicated that home based occupational therapy in addition to traditional clinic approaches can result in meaningful long term improvement in patient performance. Law, Baum and Dunn [15] indicate that work assessment in occupational therapy address specific tasks that contribute to the person's work performance. In the current study the participants indicated that they had to undergo an occupational therapy assessment in order to determine whether they were fit to return to work. These assessments in the occupational therapy department assisted clients in determining whether they had the cognitive, physical and psychological capacity needed to return to their previous job or an alternative job. With regard to vocational rehabilitation programmes the clients are often of the opinion that vocational rehabilitation programmes should not only focus on the **assessment of the client`s work ability rather it should include the use of ergonomics** when designing return to work programmes. Sanders and Wright [16] mentions that a return to work fitness programme offers the worker with a supervised fitness programme that targets specific needs and weaknesses in order to assist transition back to work. These authors further state that partial engagement of actual duties (modified duties) as well as fitness programmes may be beneficial. A gradual return to work process therefore facilitates a better adjustment of the worker when he returns to work after illness or injury. This allows the worker to build up his endurance to the level expected in his job. Sanders and Wright [16] describes ergonomic designs as a health promoting intervention that improves efficiency for all workers performing a job. Ergonomic interventions focus on modifying the work tasks, the work environment and the organization of work in order to minimize risks that may contribute to musculoskeletal pain. Workplace adaptations (ergonomic design of the workplace or work routine) often enables the client to adapt to the workplace demands. A gradual return to work process facilitates a better adjustment of the worker to the workplace when they return to work after illness or injury. It could be argued that this process will allow the client to build up his/her endurance to the level expected in the job. Improving the client's **entrepreneurship skills** as this would enable the client to create their own employment opportunities if they could not find a job. Furthermore entrepreneurship was viewed as a facilitator. Health professionals should thoroughly explore their clients` alternative work skills in that an individual who has worked in a variety of jobs would have the potential to return to the labour market in an alternative capacity despite his or her functional limitations. Sanders and Wright [16] describes employment interests and pursuits as identifying work interests based on an individual's skills, abilities, interests and opportunities available. After a brain injury an individual`s work skills could become compromised resulting in them being dismissed from their work. As a result they viewed self employment or entrepreneurship activities as a measure of sustaining themselves. Individuals with brain injury often view the **support of their family** during the recovery stages after their injury to be extremely beneficial. Family assistance is often seen as essential for the clients to resume their worker roles without the assistance of their family. Rehabilitation professionals should incorporate the family of the individual with the brain injury as a form of support during the intervention process. Piece and Salter [17] states that the goals of a support group is to provide a safe accepting environment in which to express feelings; to give the participants the opportunity to hear others express similar feelings and conflicts; to ask help and to provide an environment for problem solving. It could therefore be argued that a support group would be extremely helpful when designing a rehabilitation programme.

Friesen et al. [11] confirmed that **good communication and positive relations** between stakeholders (employee, employer and health professional) was important during the return to work process. This good communication was seen as a measure of being transparent. Transparency with their employers about their condition then employers was aware of their functional limitations. The clients felt that they could resume their worker roles when they had the experience of working in different jobs. This means that health professionals such as occupational therapists need to focus on the other skills that clients may have as this could aid, if individuals with brain injury was transparent with the employers their clients in finding alternative employment. Watt and Penn [18] found that individuals with brain injury who had an education of Grade 12 or less and who were unskilled were significantly less likely to return to work than those with tertiary education and who had managerial or professional jobs.

5. Barriers to return to work for individuals with brain injury

Barriers relating to the health system included the **poor administration procedures** relating to the access of individuals with brain injury to the hospital system or when they have to apply for a disability benefit. Poor administration systems could be linked to **poor communication** between the various departments as well as the work pressure that staff works under. Another concern is that these staff members become **emotionally depleted** because of the psychological and physical energy that they utilize in treating patients. When this occurs they become frustrated and treat patients without sympathy. McGee [19] attributes this to a concept called depersonalization which is described as a state in which the helping professional no longer has sympathy, respect or positive feelings for clients. In brain injury rehabilitation research conducted, Abreu, Seale, Podlesak and Hartley [20] mention that the best case scenario in developing quality intervention would involve adequate funding for patient stay in the hospital and ideal clinical treatment which should be provided in terms of personnel and adequate therapy for rehabilitation. Other barriers included the clients **receiving incompetent medical and rehabilitative services** from hospital staff. The high costs of medical or rehabilitative services prevent client`s access to hospital services. Other barriers that prevented individuals with brain injury from returning to work included client`s not being explained the **side effects of medication,** this often caused the brain injured individual to be dizzy, and tired in the workplace therefore negatively affecting the client`s work performance. Individuals with brain injury in a study conducted by Soeker [21] indicated that they were not given thorough explanations regarding the medications that were prescribed to them. As a result they developed side effects such as stomach complications that had a long term effect on their health. Frieg and Hendry [22] identified that health care professionals should educate both the client and caregivers (family) about the use of medication. A **lack of communication about reasonably accommodating the injured worker** in the workplace negatively affected their chances of returning to work. Friesen et al [22] confirmed that good communication between all stakeholders greatly enhanced the potential of the injured worker to return to work. Finally the failure of health professionals to recognise the **patients` rights charter** caused the individual with the brain injury to lose confidence in rehabilitation programmes. The right to have a second opinion about their medical condition closely relate to the patients' rights charter which states a number of patient rights for example the right to be treated with dignity and respect, the right to be counselled about your condition and the right to have a second opinion. [23]

6. Current rehabilitation approaches

Rehabilitation in the context of occupational therapy is currently defined as 'a service that aims to enable and empower people whose occupations are restricted because of disadvantage, illness or disabling physical or social barriers, to adapt to restrictions in function. [24] In the North American health system, rehabilitation, inclusive of cognitive rehabilitation, starts immediately after the patient has been medically stabilised in the intensive care unit. Patients are then transferred to an in-patient rehabilitation facility and thereafter, they enter a transitional living programme as an out-patient. [25]

Two of the major cognitive rehabilitation approaches are the remedial approach and the adaptive approach. The remedial approach is characterised by attempts to improve memory and perceptual skills. [26] The remedial approach is also referred to as the restorative approach focussing on attempting to remediate core areas of cognitive dysfunction by means of systematic training. Assessment procedures entail a sequence of highly structured psychometric tasks. These tasks are chosen in order to exercise the identified area of cognitive impairment and are graded by complexity, quality, speed or presentation and the cuing needed to complete the task. The therapeutic modalities include pencil exercises, computerised soft ware, table top tasks and graded occupations of daily living. [27] A popular critique of this approach is that the amount of transference of the learned skill to functional settings is minimal. [27]

The compensatory approach, which is also known as the adaptive approach, is generally geared toward the facilitation of activities of daily living. [27] This approach capitalises on the intact area of cognitive abilities and attempts to bypass the area of cognitive impairment. The emphasis of this approach is on successful participation in daily occupations rather than specific cognitive skills underlying task performance. This approach is further characterised by internal compensatory strategies such as verbal description, rehearsing and mnemonics. Whereas the external compensatory approach is characterised by memory aids such as diaries, calendars and electronic cuing devises. [27, 28] Critique of these approaches include that they offer therapeutic intervention during the early stages of recovery but they fail to meet the client's needs in the later stages of recovery. Blundon and Smits [27] state that there is no strong evidence supporting the effectiveness of either of these approaches in enhancing occupational performance. Neither approach is client centred, nor do they take the client's personally felt or expressed needs into consideration.

7. Return to work programmes

7.1. Holistic return to work programmes

Another programme that is used to gain employment among the brain injured population is the holistic cognitive rehabilitation programmes. It is normally described by three phases, namely, holistic remedial intervention focussing on the general strategies to aid daily living; guided occupational trials in vocational placement and support for the maintenance of employment. Ben-Yishay, Silver, Piatsetsky and Rattok [29] investigated the return to work success rates of 94 participants who participated in a head trauma programme which utilised a holistic cognitive approach. The study results revealed a 63% return to competitive work at the levels (academic, skilled and unskilled).

Sarajuuri, Kaipio, Koskinen, Niemelä, Servo and Vilkki [30] describe a comprehensive neurorehabilitation programme as an alternative to returning to work. The above programme, which could be classified as a holistic cognitive programme, consisted of a post-acute, intensive interdisciplinary six week rehabilitation programme. In this programme a treatment group was compared to a control group. The treatment group received neuropsychological rehabilitation, psychotherapy, vocational intervention and follow up support. The control group received conventional care and rehabilitation. In the latter study productivity was defined as working, studying or participating in volunteer activities. The results indicated that 89% of treated patients returned to a productive pursuit in comparison to 55% of the control group. One critique of this study is that of the 19 participants, only three participants returned to part time work and one to full time work.

7.2. Supportive employment

Supportive employment is defined as competitive employment in an integrated setting with ongoing support services for people with the most severe disabilities. [31] Jones, Perkins and Born [32] further described supportive employment as programmes that promote self-sufficiency and improve the quality of life of people with disabilities by motivating them to pursue work in the traditional environment at equal pay to non-disabled people. Supported employment programmes provides assistance with job coaches, transportation, assistive technology, specialized job training and tailored supervision. [33] Wehman, West, Kregal, Sharron and Kreutzer [34] conducted a study utilising the supportive employment framework in which 87 participants participated. The results indicated that only 51.3% of the participants were employed after a period of 12 months. The above authors attributed the poor return to work rates to medical/health problems, economic lay off, slow/poor quality work, poor job match and inappropriate behaviour by the participants. In a study conducted by Wehman, Sharron, Kregal, Kreutzer, Tran and Cifu [35] of the 80 participants, the monthly employment ratio increased from 13% before services to 67% after participation in the supportive employment programme. However, only 46% of these participants maintained continual employment. The authors attributed the poor return to work rates to psychiatric/psychological complications, social adjustments and substance abuse. Another study by Preston and Ulicny [36] showed that in a sample of 124 participants, 61% were either placed in a competitive job setting or were considered job ready at the time of program completion with half of those placed in competitive employment found employment with their former employers, even though some job modifications were required. In a more recent study Gamble and Moore [37], followed 1073 participants with TBI of which 78% received supported employment services during the vocational rehabilitation process. Of the participants, 48.6% were competitively employed by the time their cases were closed and 51.4% were not employed. Of the participants who were not employed, 7.3% were provided with additional supported employment services and 92.7% were not provided with supported employment. Of the clients who were provided with supported employment, 67.9% of these clients were placed in competitive employment.

7.3. Summary of return to work programmes

Holzberg [33] states that the most effective treatment approaches in North America are holistic cognitive rehabilitation and supported employment services. These approaches consist of elements of the remedial and adaptive approaches but go further in assisting the client to gain or maintain employment. However, neither the holistic cognitive rehabilitation nor the supported employment services reveal highly successful return to employment rates. [29, 34] According to Sarajuuri et al. [30] employment rates for patients with traumatic brain injury have ranged from 19% to 99%, thus indicating that there is disparity pertaining to return to work rates of this population.

7.4. A need for the exploration of the personal perspectives of brain injured individuals

There is a lack of research that address the personal experience of brain injured individuals when adapting to their worker role after rehabilitation. Johansson and Tham [38] indicate that one area in which there is a lack of knowledge is the meaning of work to people with brain injuries. It could be argued that it is important to take the perspectives of the individual living with the brain injury into account especially when developing rehabilitation models of intervention programmes. Similarly, there are minimal occupational therapy studies that focus on the lived experience of brain injured individuals. [38] The literature suggests that the quality of intervention programmes and services tend to be ineffective when the health professional does not take the brain injured individual's self-identified needs into consideration, hence it needs to be client centred. [39] My exploration of the literature also revealed only one study that focused on the best practice for maintaining employment. [33] The study described the best practice for gaining and employing people with brain injuries from a developed world perspective. However the study lacked information on how brain injured individuals adapt to different working environments and it lacked the incorporation of the personal perspectives of the brain injured individuals themselves.

8. Description of the model of occupational self efficacy

The structure of the model is a spiral which indicates that the stages of the model are not linear (see Figure 1). The individual can fluctuate between the stages due to his or her level of Occupational Self Efficacy. In between each stage is another spiral representing the influence of the environment on the individual's performance. The environment is also presented by a spiral as the environment will affect the person's performance throughout the four stages of the model. The environment may present family members, structural barriers, workplace, work colleagues, health professionals and external organizations. Throughout the process of developing Occupational Self Efficacy, there are critical contacts. These contacts serve as points of activation which set the process into action. These contacts include the contact with the occupational therapist or health therapist that facilitates the first stage of the model. Other forms of contacts may include the person himself, family members, health care team, other brain injured individuals who had completed rehabilitation and work colleagues. It is envisaged that these critical contacts could be present throughout the four stages of the model.

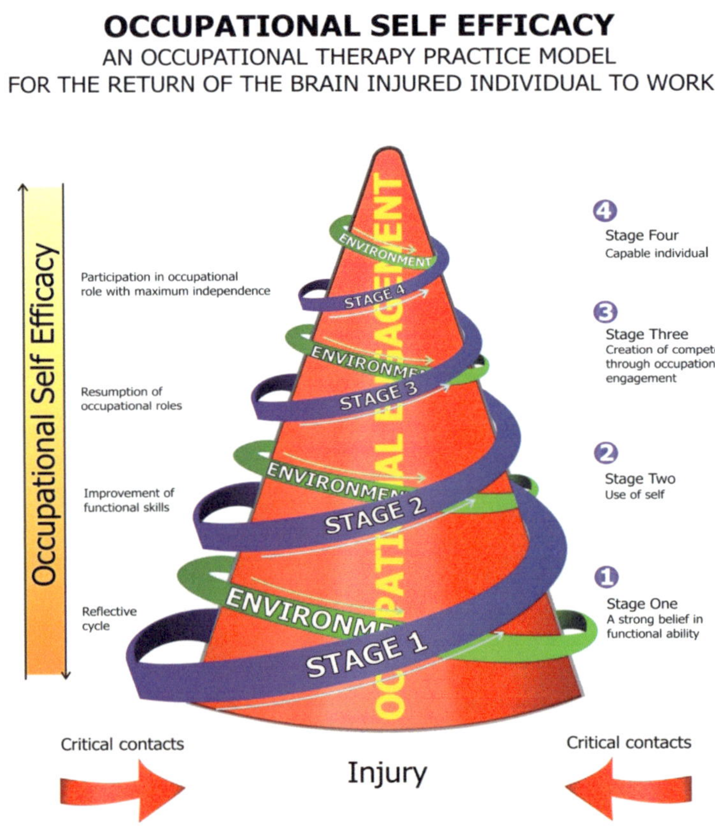

Figure 1. A graphical description of Occupational Self Efficacy: An occupational therapy practice model to facilitate returning to work after a brain injury.

STAGE ONE

During this stage the brain injured individual would be seen as an outpatient in the rehabilitation unit, a client that is receiving home based intervention in the community and or a client that has already resumed employment. Regarding the participant's cognitive status it is envisaged that he or she should be classified on level VIII of the Ranchos Los Amigos cognitive scale. The scale describes an individual that is alert and orientated, is able to recall and integrate past and recent events, and is aware of and responsive to his or her culture. [40] Based on introspection and reflection the client would be able to develop new insights into his or her ability to cope within the environment. This process will enable the client to develop inner strength and a sense of efficacy. Ultimately the client would be able to better plan his or her choices and future actions through this process. The occupational therapist facilitates the process of reflection as advocated by Gibbs [41]. This process of reflection would in turn encourage introspection.

Reflection is described by six steps namely:

Step one: Description of the event or what happened:

During this step the occupational therapist would request that the client give a detailed description of the event or concern that he or she may have. This concern may be related to feelings regarding the acceptance of the brain injury, barriers that he or she may be experiencing relating to occupational roles and community re entry or return to work. During this step the client will be encouraged to reflect on the environment, context of the event or action, other people's roles and his or her role as well as the outcome of the event.

Step two: Feelings

During this step the occupational therapist will enable the client to explore his or her thought processes. The client will explore his or her feelings regarding the actual event or stressors. They may want to know how the event or people made them feel and also how they felt about the outcome of the event.

Step three: Evaluation of the circumstances

During this step the client will be requested to evaluate his or her circumstances or make a judgment about his or her experience regarding the event or phenomena of interest. The occupational therapist will enable the client to consider what was good and bad about the experience. For example if the client was reflecting about a problem in performing tasks at work, the occupational therapist would ask him to think about what is required to do the tasks. Is it the process of doing the task that is difficult, is it the tools or equipment that is difficult to manage or is it the instructions that are difficult to understand? Once the problem is thoroughly evaluated then the individual needs to determine which aspects of the processes he or she did successfully.

Step four: Analysis of the situation or problem

During this step the client will be encouraged to break the problem into its component parts. The client may have to thoroughly analyze the problem as a whole. Here he will ask questions such as what went well and what did not go well. During this process the client may have to determine who would be able to assist him in rectifying the problem and also what he or she needs to do in order to rectify or minimize problems. For example the client may need to seek further training in order to improve his skills, he or she may need to adapt his tools or her or work routine.

Step five: Conclusion

During this step the client reflected on his or her problem situation by exploring it from different perspectives. The client has now accumulated a lot of information which will enable them to develop insight into their problem. The occupational therapist needs to encourage the client to be as honest as possible in his or her reflection about him or herself and others regarding the issue of concern. This honest exploration of the problem should be reflected in all the stages as inaccurate information may decrease the valuable opportunities for learning.

Step six: Action plan

The client will be requested to think about him or herself in the same situation or experiencing the same problem. He or she will then have to determine whether he would manage the problem or situation in the same way as before or would he or she manage the problem or situation differently. The above process could take place with the client alone or in the presence of his or her family. His or her family could assist the client in the reflection process and in goal setting if appropriate. Stage one focused on introspection and reflection, successfully working through this phase would enable the client to move to the next phase of the model.

STAGE TWO

Through the process of introspection and inner strength development, the client would be able to realize his autonomy to participate more in occupational activities of choice (i.e. activities of daily living, leisure and work). During this stage the occupational therapist would continue to act in the role of a facilitator as the self reflection process would have enabled the client to develop a plan of overcoming the barriers that they experienced at that point in time. Specific areas that need remediation according to the needs of the client will be focused upon. During this stage, specific components of function may need to be enhanced. For example, clients, in collaboration with the therapist, may decide that they need continued rehabilitation in order to improve their range of motion, muscle strength, tone, co ordination and balance. They may also require continued cognitive behavioural therapy whereby the client's memory, concentration and frustration tolerance are improved. At this point, the occupational therapist will be acting in a dual role that of a facilitator and of a case manager. In the role of a case manager the occupational therapist would enable the client to contact other role players such as a speech therapist, physiotherapist or physician who may be able to assist the client. Ultimately the goal would be for the client to act as his own case manager however, initially the occupational therapist would be able to facilitate the process. During this step, the client in collaboration with the occupational therapist would utilize a transdisciplinary[1] approach whereby all of the stakeholders may it be health professionals, employer or family should be aware of the client's goals. For example, if the client's goal is to return to work then the physiotherapist, occupational therapist, family and employer should be aware of this goal. This means that even if the physiotherapist focuses on balance and the speech therapist on improving communication skills, the ultimate goal of these health professionals would be to return the client to the work place. The client through participation in meaningful occupation would realise his strengths, weaknesses and potential. Engagement in occupations of choice would also enable the client to revise their self concept and ultimately improve their self esteem. Calhoun and Acocella [42] defines the mental self portrait as comprising three dimensions namely knowledge, expectations and evaluation of self.

Knowledge of self is described by what the person knows about him/herself. It is envisaged that through the participation in occupation the client revises their knowledge themselves. For example, through a simple activity such as dressing, the client would be able to get an idea of their functional limitations.

Expectation of self is described by the person's perceptions of what he or she could be. These expectations in turn propel the client into the future and guide his or her actions. The client, who has an expectation of returning to his or her role as a worker, would be able to visualize what actions will be required to return to work. These expectations become realistic expectations when the client is allowed to actually engage in work tasks. Through engagement in these tasks the client would develop the insight to adjust his expectations of self.

Evaluation of self is described by the person's judgment about him or herself, measuring what he or she is against, his or her expectations of self or his or her standards for self. The client's perception of their satisfaction with themselves facilitates their self esteem. It is important to note that there should be balance between the client's actual self or functional ability in occupational tasks and their expectations of themselves to enable them to develop a realistic self esteem and self concept.

STAGE THREE

This stage is described as the creation of competence through participation in occupation. During this stage the client will focus on a specific occupational performance area (i.e. work). If the client has not resumed his worker role he or she will gradually be reintegrated into this role. In stage two the client had already participated in intervention programmes aimed at improving their functional performance. He or she now has the functional skills to resume his or her occupational roles. The occupational therapist will continue to act as a facilitator and case manager where the client will be encouraged to self reflect and problem solve the manner in which he would like to resume employment. The client will be encouraged to utilize their social relations by initiating contact with stakeholders such as the employer, colleagues, health professionals and family for the purpose of participation in their worker role and in order to improve their support systems. This stage will place emphasis on improving the client's knowledge base. The occupational therapist will encourage clients to improve their problem solving skills. The client will be encouraged to use the reflective process as a method for solving problems. The occupational therapist may want to refer the client to another occupational therapist who specializes in vocational rehabilitation, work assessment or screening or they

1 A transdisciplinary model of functional rehabilitation is described as a model where health team members conduct an integrated evaluation that results in the collaboration of assessment information. [43]

could initiate this process themselves. The client will be requested to demonstrate a problematic workplace scenario. For example, a client who had worked as a sales person may indicate that he or she has a problem in coping with difficult customers. The occupational therapist will request that the client verbalizes the actual workplace problem and will then advise him or her on various coping strategies that may assist him or her. The client will be asked to analyze the reasons why he or she struggled to cope with customers. Possible reasons could be a lack of assertiveness or poor communication skills.

The client will be asked to role play a scenario where they did not cope with a difficult customer. He or she will then be requested to identify the reasons why he or she could not cope. Thereafter they will be asked what they could have done to change their interaction with the difficult customers (if role playing a conflict situation between service provider and a difficult customer). The client in collaboration with the occupational therapist will then physically and practically role play a scenario where they expresses the desired behaviour to improve their interaction with the difficult customer. Feedback will then be given to the client regarding his or her behaviour and approach. The family and or other patients who have suffered from a brain injury could give feedback if this stage is done in a group set up. The client and the occupational therapist would then be responsible for setting up a work test placement with his existing employer. The work test placement will entail that the client perform the actual duties of his occupation under supervision of the occupational therapist who will be acting as a job coach. At this stage the employer or a designated person from the workplace could be present to ensure that the work is performed according to the required standard. The client's work performance would then be monitored with a schedule. The schedule would assist the occupational therapist in observing the client's occupational (work) behaviour, components of function (i.e. physical components of function and psychological components of function), his or her work endurance and productivity. After the work test placement which may be 1-3 days in duration, the results will be discussed with the client. The client and occupational therapist would engage in the reflective process where their opinions about their performance will be explored. Any problems or aspects that did not go well and aspects that did go well will be discussed. After the discussion the client and occupational therapist may have to explore the use of assistive devices in order to make the job easier, or consider workplace accommodation strategies and adaptation to workplace routines. Furthermore the client and occupational therapist may explore his or her legal rights within the workplace, possibly in the form of their right to work in a safe environment, their right to be reasonably accommodated in the workplace and their right to access disability pension benefits, if applicable. Based on the client's perceptions of their performance and realistic expectations they may choose to seek another form of employment to accommodate their current functional capacity.

During this stage the client would develop renewed confidence and knowledge of their ability to resume their occupational role as a worker. This paves the way for stage four which is the development of the client into a capable individual.

STAGE FOUR

During this stage clients would be encouraged to undergo self reflection about the previous stages and about their ability to participate in the occupational role as a worker. The client

http://dx.doi.org/10.5772/57309

ultimately would synthesize and internalize the actions that they undertook and skills that they learnt during the previous stages. He or she would be able to conceptualize his or her ability to overcome various barriers to participation in his or her worker role. It is important to take note of the model's dynamic and spiral nature. This means that a client could revert back to a previous level based on his ability to meet the challenges of the various stages. Ultimately this stage emphasizes an individual that has fully accepted their condition and that has developed a strong occupational efficacy to overcome various barriers to the worker role.

This stage is also described by prolonged participation in the satisfactory worker role whereby the client experiences meaning and fulfilment. There may be a positive interaction between the client and the environment, which may consist of the family system, work system and health system. During this stage the client would view themselves as capable and would be able to engage in the worker role with maximum independence. The occupational therapist`s role is gradually withdrawn. This process will be a unique experience for each client.

9. Guidelines for the operationalization of Occupational Self Efficacy: An occupational therapy practice model to facilitate returning to work after a brain-injury

9.1. To facilitate a strong personal belief

The following guidelines should be implemented in order to achieve the above objective:

- An occupational therapist that specializes in vocational rehabilitation and or the treatment of the brain injured individual should facilitate the process of introspection.
- The occupational therapist should have insight into the brain injured individual's social, community and cultural dynamics.
- The occupational therapist should have in depth knowledge about the treatment of the brain injured individual, his workplace and job description.
- Through collaboration with the individual and his family the occupational therapist should facilitate the process of reflection. The steps as advocated by Gibbs [38] should be used.

9.2. To encourage the client's use of him or her self

The following guidelines should be implemented in order to achieve the above objective:

- The occupational therapist continues to function in the role of a facilitator. He or she continues to facilitate the process of introspection of the client.
- The occupational therapist encourages occupational engagement in tasks such as activities of daily living, using transport and vocational related activities.

- The occupational therapist should have good communication skills, problem solving skills, negotiation skills, empathy and be transparent. He or she should be a role model to the brain injured individual.
- In addition to the role of a facilitator the occupational therapist would act as a case manager in that he or she would have to be able to provide the participant with choices relating to continued rehabilitation or return to work.
- Furthermore the occupational therapist, in their role in the medical team would facilitate a client centred approach to intervention.

9.3. To enhance competency through occupational engagement

The following guidelines should be implemented in order to achieve the above objective:

- The occupational therapist will continue to act in his or her role as a facilitator and the participant will be guided into participation in his or her worker role.
- Client centred practice will enable the client and therapist to identify various needs that will enhance competency in occupational roles. For example, if there is a need to improve the client`s life skills such as coping skills or assertiveness skills then this will be a focus of intervention.
- The occupational therapist will continue to act as a case manager in that he or she would enable the client to identify and utilize resources that will enable him or her to resume their worker roles. The occupational therapist will put the client into contact with stakeholders such as the employer, relevant people in the medical sector and family members as this will form a base for long term support.
- Supportive employment workshops will be held with the client and his employer whereby gradual return to work will be emphasized. During this process the client and occupational therapist will identify further needs that the client may require such as the use of compensatory equipment or techniques.
- Work test placement with the client's previous employer will be initiated and workplace accommodation should be encouraged

9.4. To develop a capable individual

The following guidelines should be implemented in order to achieve the above objective:

- Self reflection pertaining to the client`s performance in work related tasks should be emphasized
- Prolonged participation in the occupational role as a worker should be encouraged and positive interaction/communication between the client, environment, worker and family should continue
- Transformation of the client into a capable person, who would be able to participate in the worker role with maximum independence, is the final goal.

http://dx.doi.org/10.5772/57309

10. Application of the case study

10.1. Case study 1

J.K. Is a 20 year old female who suffered a moderate brain injury at the age of three, a taxi driver lost control of his vehicle and drove into JK who was walking with her mother on a pavement. She has a Grade 11 level of education which she completed at a school for children with physical and mental disabilities. JK lived with her parents and 3 siblings in a low socio economic area, both her parents were unemployed and the family was dependent on a government support grant. As JK was injured while being a pedestrian, she qualified to have compensation from the South African Road Accident Fund. This compensation was +/- R1000 (equivalent to U$100) per month.

Results: Individual is currently employed for 8 months at a Fast Food Restaurant.

- **Stage 1** A strong personal belief

  – Ranchos Los Amigos scale VIII

  –**Introspection**, Gibbs reflection cycle- Initially it was very difficult to get JK to trust the health therapist as she was protected by her parents all her life. The health therapist had to arrange intervention sessions with JK at her home as well as at the restaurant where she was going to work. It was difficult to get JK to attend the sessions on her own initially as she did not feel competent in using transport independently. The type of questions asked included: Describe the incident, 2) Practically how has the incident changed your life/ circumstances? 3) Emotionally, do you think that your ability to do tasks is different when compared to your ability to do tasks before the accident?

  – JK slowly responded to the questions and really had to introspect and accept what had happened to her and focus on improving her work skills. JK decided that she would want to work in the food retain industry. She initially wanted to work in a bakery where she could bake and prepare various delicacies.

- **Stage 2** To encourage the client's therapeutic use of him or her self

  – Introspection continued: During stage to JK and the health therapist focused on improving her work skills such as numeracy and comprehensive ability. Furthermore JK was walking with an "abnormal gait" as a result of the TBI. The health therapist had to assess whether JK`s abnormal gait was going to affect her ability to initiate tasks such as standing for more than 30 minutes and walk over short distances while carrying weights and trays. At this stage it was felt that no further rehabilitation was going to improve her gait and that compensatory measures had to be used by the participant. These compensatory measures included the use of memorization/visualization techniques, whereby JK had to remind herself about the pace at which she was walking, the structural barriers such as steps in the workplace, distances between restaurant tables, danger of certain equipment example stoves and fryers, distance of transportation pick up points for employees of the etc.

– During this stage the health therapist acted in the capacity of a case manager the aim was for the client to take responsibility for his or her own rehabilitation.

- **Stage 3**: To enhance competency through occupational engagement

  – Focus on the occupational area of work. During this phase the health therapist and JK focused on improving JK`s work skills such as coping skills, problem solving skills, use of transport, money management, setting up a CV, basic work abilities (social presentation, communication skills, ability to work an 8 hour work shift) and in-service-training. During this stage the client (JK) was asked to reflect back onto her initial goals i.e. working in a food retail or fast food restaurant environment. The health therapist and JK identified the requirements of working in a super market and a fast food store. A job analysis was performed whereby the physical layout of the workplace, equipment used, hours worked and job tasks were explored in detail. The health therapist communicated with possible employers in the food retail industry in order to provide JK with a work experience opportunity. JK was provided with an opportunity to work for 5 hours initially in order to determine how she will cope in the new work environment. A meeting was set up with the restaurant manager, JK and her mother, and the health therapist. The type of work duties, hours of work as well as what should be done if a problem occurs. During stage two, the health therapist arranged a meeting with staff members of the restaurant in order to prepare them for the arrival of JK.

  – The type of intervention used before the provision of the work experience included role plays of different problems that could occur in the work place. These roles plays helped improve the client`s life skills such as coping skills, problem solving skills and assertiveness. Her work performance was monitored with a work schedule (see Appendix 1), the work schedule could give the client a visual concrete view of how their performance was improving during the work skills training process. During this period of time, the client worked reduced hours initially, she also only worked day shifts, however as her confidence and skills improved, her work hours increased and she was allowed to work night shift duty. The client was provided with weekly supervision initially thereafter, the supervision was decreased to twice monthly.

- **Stage 4**: To develop a capable individual

  – During this phase the client was requested to undergo self reflection about the previous phases (i.e. she internalised the skills that she learned). She was asked to identify how her work skills and confidence improved during the four phases. JK indicated that she was able to use transport independently, she was able to work any work shift provided and that she has mastered most of the work tasks in the fast food restaurant. She also indicated that she would follow a systematic problem solving process should she experience any difficulty at work e.g. write down the difficult problem that she experienced and explore options of solving it. The supervision process was reduced to once every three months with on- going telephonic supervision when required. At the end of one year the supervision was drastically reduced as the client did not require constant supervision.

http://dx.doi.org/10.5772/57309

10.2. Case study 2

Application of the case study

S.S is a 24 year old male who suffered a moderate brain injury at the age of 20, he sustained a stab wound to the head by gangsters. SS has a Grade 12 level of education and was going to study at a local University. Due to the extent of his injury, he decided to rather discontinue schooling in order to recover and find employment. After participating in physiotherapy and occupational therapy intervention, SS tried to find employment, however was unsuccessful. Due to him struggling with comprehension and numeracy skills he struggled to obtain work that required administrative skills. SS was dependent on a disability grant which is equivalent to U$100. SS is single but has 2 dependents, he is currently living in a low socio economic area. SS has had no formal work training prior to the injury, however he reported that he was working at a fast food restaurant as a general assistant.

Results: Individual is currently employed for 8 months at a Fast Food Restaurant.

- **Stage 1** A strong personal belief

  – Ranchos Los Amigos scale VIII

  –**Introspection**, Gibbs reflection cycle- During the stage of reflection and introspection, SS clearly indicated that he wanted to improve his employability skills by improving his level of education. He felt that he would not be able to resume employment in a "well paid" job without a tertiary qualification. However during the work skills assessment it was obvious that he needed intense remediation regarding his reading and writing skills. The type of questions asked included: Describe the incident, 2) Practically how has the incident changed your life/ circumstances? 3) Emotionally, do you think that your ability to do tasks is different when compared to your ability to do tasks before the accident?

  – SS responded openly to the questions and it was clear that he was determined to find employment in the Open Labour Market. As SS was determined to improve his work skills, he was informed of the possibility of completing a short course in the area of entrepreneurship with the South African Department of labour. As an incentive to complete the course he was provided with a monthly stipend of U$200 in addition to his disability grant. The initial intervention sessions took place in a hospital environment.

- **Stage 2** To encourage the client's therapeutic use of him or her self

  – Introspection continued: During stage to SS was provided with the details of the short course especially the duration of the course as well as where the course was going to be offered. He was encouraged to make contact with the service provider and ask questions related to the course. This would therefore enable him to take responsibility for improving his own work skills. After engaging in tasks such as developing a CV, doing simulated administrative tasks (arranging files and post alphabetically) and doing inventory related work such as collecting stock from shelves and placing stock on shelves using a product list; he increased his confidence to engage in administrative tasks. SS successfully completed the entrepreneurship short course with the Department of labour over a period of 12 months.

Some of the questions asked to the participants during this phase included: 1) How did you find the rehabilitation experience so far? 2) What in your opinion was beneficial and not relevant about the experience, 3) How do you think you could use what you were taught practically in a work experience or at home? 4) Do you think you have improved as a worker?

- **Stage 3**: To enhance competency through occupational engagement

  – Focus on the occupational area of work. During this phase the health therapist and SS focused on providing SS with a work experience and integrating him into the Open labour market. A company that manufactures soft drinks indicated that they were willing to provide him with this work experience over a period of 2 months. An initial meeting was arranged with the staff members and Human Resource manager of the company. The purpose of the project was explained as well as there was a discussion that focused on the fears that staff had in working with individuals who was diagnosed with a brain injury. This initial discussion with staff members was fruitful in that in improved the insight of staff members regarding the strengths and limitations of individuals with brain injury.

His work tasks included operating a conveyor belt as well as separating poor quality bottles from good quality bottles. He also had to keep a schedule of the amount of bottles packed into crates and had to collect raw material from the company stores. The health therapist regularly contacted the line manager in order to monitor SS`s work performance. His work performance was measured according to the company work performance chart. As he performed at the same level as his colleagues, this enhanced the confidence of SS as he could measure his own work standards to the standards required in the Open Labour Market. In terms of the work shifts, SS was only required to work night shifts.this period of time, the client worked reduced hours initially, however as his confidence and skills improved, his work hours increased and he was allowed to work night shift duty. The client was provided with weekly supervision initially thereafter, the supervision was decreased to once monthly.

- **Stage 4**: To develop a capable individual

  – During this phase the client was requested to undergo self reflection about the previous phases (i.e. he internalised the skills that she learned). He was asked to identify how his work skills and confidence improved during the four phases. SS indicated that he was able to use transport independently, that he completed the training course successfully and that he can relate to work related tasks. He could use transport independently, he was able to work any work shift provided and that transport was made available at the end of their night shift. She also indicated that she would follow a systematic problem solving process should she experience any difficulty at work e.g. write down the difficult problem that she experienced and explore options of solving it. The supervision process was reduced to once every three months with on- going telephonic supervision when required. At the end of one year the supervision was drastically reduced as the client did not require constant supervision. SS was offered a 12 month contract at the soft drink company.

http://dx.doi.org/10.5772/57309

11. Conclusion

The model of Occupational Self Efficacy could be used in both a private and public hospital or rehabilitation setting. It could also be implemented in the homes or workplaces of the clients. The model could used in conjunction with other treatment modalities such as biomedical and cognitive rehabilitation approaches. Furthermore the model focuses on using reflection and improving self efficacy beliefs through engagement in work related tasks. Inevitable the model seeks the bridge the gap from rehabilitation and returning to work. Finally the model provide a step by step process for developing occupational self efficacy and resuming the worker role of the individual with the brain injury.

Appendix

WORK ABILITY SCREENING TOOL

Key:

1-INCAPABLE WEAK	2
3- BELOW AVERAGE AVERAGE	4
5- ABOVE AVERAGE EXCEPTIONAL	6

WORK HABITS	
PERSONAL PRESENTATION	*Remarks (if any)*
Date:	
Attendance	
Punctuality	
Self – discipline	
Helpfulness	
Attitude	

SOCIAL PRESENTATION	
Date:	*Remarks*
Ability to work with supervisors	
Ability to work with co- workers	
Need for strict limit setting	

Ability to handle criticism	
General social interaction with co- workers and supervisors in the unit	
Co – operation	
Ability to fulfil a supervisor role	
WORK COMPTENCY AND SKILLS	
Date:	Remarks
Ability to follow verbal instructions	
Demonstrated	
Written	
Illustrated	
Ability to follow and execute instructions 1-3 steps	
Ability to follow and execute instructions 1-6 steps	
INSTRUCTION RETENTION	
Date	Remarks
Ability to listen	
Comprehension	
Interpretation	
Accurate execution of instructions with minimal or no prompts of the assessor	
TASK PLANNING AND EXECUTION	
Date:	Remarks
Orientation to work area	
Planning of work area in relation to structuring	
Efficiency	
Ergonomics	
Practical planning of the work task in relation to steps of activity	
TASK COMPLETION	
Date:	Remarks
Ability to start the task	
Tolerance to maintain task	
Ability to complete task	
Ability to retain task	

Ability to recognise errors	
Ability to correct errors	
SAFETY	
Date:	Remarks
Knowledge of equipment and tools needed for task	
Ability to handle tools and materials safely	
Knowledge of where to collect equipment	
Knowledge and implementation of storing equipment safely and clean after use	
Implementation of basic precautions of own safety in the area	
Awareness and implementation of safety precautions of tools	
WORK TOLERANCE	
Date:	Remarks
Physical endurance in extended sitting posture	
Physical endurance in prolonged sitting	
Physical endurance in repetitive mobilisation from station to station	
Psychological endurance (mood)	
Psychological endurance (ability to work under pressure)	
Ability to sustain work effort	
Number of rest periods	
Reasons for rests	
Prompt return to work after rests	
MOTIVATION	
Date:	Remarks
Intrinsic motivation	
Extrinsic motivation	
Derives satisfaction from being productive	
PERFORMANCE	
Date:	Remarks
Quality of repetitious tasks	
Assembly- line operations	

Packing- assembly- inspecting
Physical/ manual labour tasks
Quantity produced
Hours lapsed to reach above quantity
Speed
General comments (if any):

Author details

Shaheed Soeker*

Address all correspondence to: msoeker@uwc.ac.za

Occupational Therapy Department, University of the Western Cape, South Africa

References

[1] Gutman, S.A. Traumatic brain injury. In L.W. Pedretti & M.B. Early (Eds.). Occupational therapy: Practice skills for physical dysfunction (5th edition). St. Louis: Mosby; 2001.

[2] Urban, R.J., Harris, P. & Masel, B. Anterior hypopituitarism following traumatic brain injury.Brain Injury. 2005;19(5):349 – 358.

[3] Raskin,S. & Mateer,C. Neuropsychological Management of Mild traumatic Brain Injury. England: Oxford University press; 2000.

[4] Khan, F. Baguley, I.J. & Cameron, I.D. Rehabilitation after traumatic brain injury. Medical Journal of Australia. 2003;178:290 -295.

[5] National Health Laboratory Services. Traumatic Brain Injury (head Injuries) - World Head Injury Awareness [Online]. Available: http://www.nioh.ac.za//?page=topical&id=13&rid=214 [Accessed 23 April 2013]

[6] Faul M, Xu L, Wald MM, Coronado VG. Traumatic brain injury in the United States: emergency department visits, hospitalizations, and deaths. Atlanta (GA): Centers for Disease Control and Prevention, National Center for Injury Prevention and Control; 2010.

[7] Traumatic Brain Injury: Department of defence special report [Online] http://www.defense.gov/home/features/2012/0312_tbi/

[8] Vuadens,P. & Arnold,P & Bellmann,A. Return to work after a traumatic brain injury-Vocational Rehabilitation. Pari: Springer Paris; 2006.

[9] Cicerone, K.D. Cognitive Rehabilitation for traumatic brain injury and stroke: Updated review of the literature from 1998 through 2002 with recommendations for clinical practice. America: Archives of Physical medicine and rehabilitation, 2000;92,(4), 1596-1615.

[10] Sample, P.L. & Darragh, A.M. Perceptions of care access: the experience of rural and urban women following brain injury. Brain injury, 1998;12(10): 855-874.

[11] Friesen, M.N., Yassi, A. & Cooper, J. Return to work: The importance of human interactions & organisational structures. Work;2001, 17, 11-22.

[12] Townsend, L. & Banks, S. Exploring client centred practise. The national, 1992;9(5):8

[13] Schultz- Krohn, W. & Pendleton, M.H. Application of the occupational therapy practise framework to physical dysfunction. In H.M. Pendleton & W. Schultz- Krohn (Eds.). Pedretti`s occupational therapy: Practise skills for physical dysfunction (6th edition). Philadelphia. Elsevier Mosby; 2005.

[14] Schwartz, S.M. Adults with traumatic brain injury: three case studies of cognitive rehabilitation in the home setting. The American journal of occupational therapy, 1995;49 (7): 655-667.

[15] Law, M., Baum, C.M. & Dunn, W. Occupational performance assessment. In C.H Christiansen, C.M Baum & J. Bass- Haugen (Eds.). Occupational therapy: performance, participation & well being (3rd edition). Thorofare New Jersey: Slack Incorporated; 2005.

[16] Sanders, M. & Wright, R. Work. In C. Meriano & D. Latella (Eds.). Occupational therapy interventions: functions and occupations. New York: Slack; 2008.

[17] Pierce, L.L. & Salter, J.P. Stroke support group: A reality. Rehabilitation nursing, 1988;13(4): 189-190.

[18] Watt, N. & Penn, C. Predictors and indicators of return to work following traumatic brain injury in South Africa: Findings from a preliminary experimental database. South African Journal of Psychology, 2000, 30 (2): 27-37.

[19] McGee, R.A. Burnout and professional decision making: an anolog study. Journal of counselling psychology. 1989;36: 345-351.

[20] Abreu, B.C., Seale, G., Podlesak, J. & Hartley, L. Development of critical paths for post acute brain injury rehabilitation: Lessons learned. The American journal of occupational therapy, 1996, 50(6):417-427.

[21] Soeker, M.S, Van Rensburg, V & Travill, A. Are our rehabilitation programmes enabling our clients to return to work? Return to work perspectives of mild to moderate

brain injured individuals in South Africa. Work, A Journal of Prevention, Assessment and Rehabilitation. 2012;43(02), 171-182

[22] Frieg, A. & Hendry, J.A. Disability grant and recipients and caregiver utilisation. South African journal of occupational therapy, 2002;32(2): 15-18.

[23] Patients` rights charter [Online].Available:www.doh.gov.za/docs/legislation/patientsright/chartere. html [Accessed: 8/10/2009]

[24] Watson, R.M. Occupational therapy defined for district rehabilitation. South African journal of occupational therapy, 2004;34(1): 11-13.

[25] Zoltan, B. & Rykeman, D.M. Head injury in adults. In L.W. Pedretti & B. Zoltan (Eds). Occupational therapy: Practice skills for physical dysfunction. St Louis: Mosby; 1990.

[26] Lee, S.S., Powell, N.J. & Esdaile, S. A functional model of cognitive rehabilitation in occupational therapy. Canadian journal of occupational therapy, 2001;68 (1): 41-50.

[27] Blundon, G. & Smits, E. Cognitive rehabilitation: A pilot survey of the therapeutic modalities used by Canadian occupational therapists with survivors of traumatic brain injury. Canadian journal of occupational therapy, 2000;67(3): 184-196.

[28] Giles, G.M. & Wilson, J.C. Brain injury rehabilitation: a neurofunctional approach. London: Chapman & Hall; 1993.

[29] Ben-Yishay, Y., Silver, S.M., Piatsetsky, E. & Rattok, J. Relationship between employability and vocational outcome after holistic cognitive rehabilitation. Journal of head trauma rehabilitation, 1987;2: 35-48.

[30] Sarajuuri, J.M., Kaipio, M.L., Koskinen, S.K., Niemelä, M.R., Servo, A.R. & Vilkki, J.S. Outcome of a comprehensive neurorehabilitation program for patients with traumatic brain injury. Archives of physical medicine rehabilitation, 2005;86: 2296- 2302.

[31] Cook, J.A. & Burke, J. Public policy and employment of people with disabilities: exploring new paradigms. Behavioural sciences and the law, 2002;20(6): 541-557.

[32] Jones, C.J., Perkins, D.V. & Born, D.L. Predicting work outcomes and service use in supported employment service for persons with psychiatric disabilities. Psychiatric rehabilitation journal, 2001;25(1):53-59.

[33] Holzberg, E. The best practise for gaining and maintaining employment for individuals with traumatic brain injury. Work, 2001;16: 245-258.

[34] Wehman, P., West, M., Kregel, J., Sherron, P. & Kreutzer, J.S. Return to work for persons with severe traumatic brain injury: a data based approach to programme development. Journal of head trauma rehabilitation, 1995;10: 27-39.

http://dx.doi.org/10.5772/57309

[35] Wehman, P., Sharron, P., Kregel, J., Kreutzer, J., Tran, S. & Cifu, D. Return to work for persons following severe traumatic brain injury. The American journal of physical medicine and rehabilitation, 1997;72: 355-363.

[36] Preston, B. & Ulicny, G. (1992).Vocational placement outcomes using a transitional job coaching model with persons with severe acquired brain injury.Rehabilitation Counseling Bulletin. 35 (4): 230.

[37] Gamble and Moore (2003), Supported employment: Disparities in vocational rehabilitation outcomes, expenditures and service time for persons with traumatic brain injury, 2003;19 (1): 47-57

[38] Johansson, U. & Tham, K. The meaning of work after acquired brain injury. The American journal of occupational therapy, 2006;60: 60-69.

[39] Darragh, A.R., Sample, P.L. & Krieger, S.R. Tears in my eyes: Cause somebody finally understood: Client perceptions of practitioners following brain injury. American journal of occupational therapy, 2001;55 (2), 191-199.

[40] Tipton- Burton, M. McLaughlin, R. & Englander, J. Traumatic brain injury. In Pedretti`s occupational therapy: Practise skills for physical dysfunction (6th edition). H.M. Pendleton & W. Schultz- Krohn (Eds.). Philadelphia. Elsevier Mosby; 2005.

[41] Gibbs, G. Learning by Doing: A Guide to Teaching and Learning. London: Further Educational Unit; 1998.

[42] Calhoun, J.F. & Acocella, J.R. Psychology of adjustment and human relationships. (10th edition). New York: McGraw- Hill Publishers; 1990.

[43] Jansen, P. Vocational rehabilitation following brain damage: alternative approaches and functional models of retraining. South African journal of occupational therapy, 1994;24 (2): 20-24.

Section 4

Cognitive Impairment in Traumatic Brain Injury

Chapter 18

Understanding, Assessing and Treating Prospective Memory Dysfunctions in Traumatic Brain Injury Patients

Giovanna Mioni, Shawn M. McClintock and
Franca Stablum

Additional information is available at the end of the chapter

http://dx.doi.org/10.5772/57307

1. Introduction

Our capacity to shape and direct our future behaviour is of fundamental importance in the development, pursuit, and maintenance of independence and autonomy from early childhood to late adulthood. A cognitive ability required for those functions is prospective memory (PM), which is the ability to form and remember to prospectively perform the intended action [1, 2]. Researchers have extensively focused on PM impairment in patients with traumatic brain injury (TBI) [3]. However, there has been limited research into the assessment and treatment of PM impairment in TBI patients. Reliable and valid tests with normative data are necessary for health professionals working with people with PM impairments. This chapter reviews the principal findings on PM impairment in TBI patients, and the main procedures used to assess and rehabilitate PM.

2. Prospective Memory

Prospective Memory (PM) refers to the cognitive ability to form and remember to perform an intended action at a specific moment in the future [1, 2]. That cognitive ability is essential for many daily activities, such as remembering to pick up something at the market after work, send a gift for a birthday, or call a friend at a specific time. Although past researchers tended to characterize PM as a unitary process, it is very complex and comprised of various components [4]. For example, consider the situation in which a principal asks a teacher to relay a message to a student. The teacher has committed him/herself to two memory tasks: one is to remember the content of the message, and the second is to deliver it as soon as he/she sees the student. Remembering the content of a PM task is essential to perform the PM task, which researchers refer to as the *content* of a PM task (*retrospective component*). However, remembering only the content of the PM task will not produce successful performance. The action must also be performed at the appropriate moment in the future. This is referred to as the *prospective component* of PM tasks. Thus, a critical aspect to success on PM tasks is not only recalling the content of the intended action, but also performing the action at the appropriate moment in the future [4].

A four stages model has been proposed to explain the functioning of PM [1, 4-6]:

1. *Intention formation* - the first phase consists on the formation of the delayed intention, which often involves forming a plan. Different degrees of motivation may influence the strength of encoding the delayed intention. In fact, the strength of an intention may reflect not only the personal importance, but also the potential benefits and costs of realizing the delayed intention.

2. *Intention retention* - the retention intervals describe the delay between the formation of and execution of the intended action. The retention interval, most of the time, is filled with an "ongoing task" [7]. Some authors found lower performance after a long relative to a short delay [8], though others failed to find an association between delay length and decline [9, 10] or improvement in performance [11]. In addition to the length of the delay interval, the cognitive load added during the delay interval (filled or un-filled intervals) may influence performance [7, 9].

3. *Intention initiation* - the point in time at which execution of the intention is (or should be) initiated. Successful performance on PM tasks, not only required to accurately encode the intended action and maintain the intended action active during the interval delay, but also to recognize the cue, retrieve the action associated with that intention, and perform the action. People may fail on their performance not only because they fail to recall the intended action when the target occurs, but also because their cognitive resources were captured by the demands of the ongoing task.

4. *Intention execution* - where and when the intended action is executed. This may be considered the final stage with the execution of the delayed intention. However, successfully completing all previous phases, errors can occur and the delayed intention may not be performed. Distractions or failure to complete the task due to external circumstances may compromise performance. For example, the intention to phone a friend may fail because of interruption by the doorbell or the friend was not at home. When the delayed intention is not executed, it is necessary to re-establish the intended action and re-form the plan [1].

A fifth phase, *Evaluation of outcome* was proposed and is concerned with monitoring the output of the execution of the intended action. The evaluation of outcome describes the process by which a person checks if the intended action has been accurately performed [1].

According to Einstein and McDaniel's model [12], there are two types of PM targets, event-based and time-based tasks[1], that trigger the execution of our delayed intention. In conditions that require event-based PM, a person performs an action when a specific event occurs (i.e.

passing a message when a friend calls); while in situations that require time-based PM, the action has to be performed at a specific time in the future (i.e. remembering the appointment with a friend at 4:00 p.m.) [12]. Event-based PM tasks are considered to be less cognitively demanding than time-based PM tasks because they require less self-initiated retrieval and external cue(s) are available to help recall the task [13, 14].

2.1. Prospective memory in traumatic brain injury patients

It has been suggested that the prefrontal cortex predominantly mediates executive control mechanisms [15], which are required to successfully initiate and execute the intended action [5, 13, 14, 16]. Because of this link between executive functions and the prefrontal cortex, and because prefrontal areas are commonly damaged after a TBI, it is not surprising that individuals with TBI are often impaired on PM tasks [3]. For individuals with TBI, frequent PM failures (e.g., forgetting to repay a loan to a friend, maintain an appointment, take medication, turn off the stove) can be frustrating, embarrassing and in some cases, life threatening. These failures have the potential to limit the independence of these individuals, causing them to rely on a carer for prompting and completion of activities and instrumental activities of daily living. Moreover, these failures may affect their opportunity to return to work or start a new vocation [3, 17, 18].

Impairment of PM in TBI patients would differ according to the complexity and requirement of the tasks [7, 19]; moreover, when tested both on event- and time-based tasks, TBI patients are particularly impaired on time-based tasks consistent with the idea that those tasks require more self-initiation. Dysfunctions on time-based tasks may also be due to less strategic monitoring behaviour engaged by TBI patients [1, 4, 16].

The earliest studies that examined the effect of TBI on PM performance used self-rating scales [20] or few PM items [21, 22]. In those studies, the Prospective Memory Questionnaire (PMQ) was tested along with attentional and memory tasks [20]. Adults with TBI and older adults performed more poorly than younger adults on short-term PM tasks, and TBI patients rated themselves more poorly than younger adults. Two PM tasks were used to assess PM performance [22]: the first task involved telling the participants about a self-report memory questionnaire at the beginning of an assessment session and instructing them to ask for the questionnaire at the end of the session (even-based task). The second task involved asking the participants to return (by mail) an evaluation form with the date written in the top corner (time-based task). Because of the small number of items used and the correct/incorrect nature of these items, the PM scores obtained with these tasks were limited in range and thus were unreliable [23].

To assess PM more reliably and accurately, it is necessary to increase either the number of items used or the number of responses required for each item. Recent studies considered these limitations and developed more appropriated tasks to investigate the effects of TBI on event- and time-based PM performance (Table 1).

---

1 A third prospective remembering task, called, activity-based [1]. An activity-based prospective memory task required to carry out an intended action at the end of an activity (i.e., torn off the oven when you have finished cooking).

		TBI characteristics				Type of prospective memory tasks	
Study[1]	Year	Sample	GCS	Time since injury (months)	NPS assessment[2, 3]	Event-based	Time-based
[25]	1999	12 TBI and 12 controls	8 or less	24.58 (15.58)	-	Response to questions and the target word	Press a key every 5 min and monitor the time
[26]	2001	13 TBI and 13 controls	6.6 (3.3)	1.8 (0.7)	WCST, COWAT and DASS	Deliver a message (unfilled and filled conditions)	Stop the timer after 5 min (unfilled and filled conditions)
[28]	2003	14 TBI and 14 controls	3 to 15	9.71 (7.5)	TOL, COWAT and LNST	Response to a cue (low and high cognitive demand)	-
[11]	2004	16 TBI and 16 controls	8 or less	108	-	Perform actions when the target appeared (related-unrelated cues) (10 vs. 45 min delay)	Perform actions after the interval (related-unrelated cues) (10 vs. 45 min delay)
[19]	2004	7 TBI, 21 older and 19 younger controls	severe	41.71 (11.79)	Verbal intelligence and WCST	Complex tasks following the multiprocess model	-
[29]	2004	24 TBI and 24 controls	8 or less	25.75 (21.63)	Digit span forward and backward, TMT, LM Visual recognition, COWAT, Stroop, WCST	Press a designed key on focal cue or peripheral cue	-
[27]	2005	25 TBI and 25 controls	8 or less	7	Visual (Story sub-test, RAVLT) and verbal (Pictures, Faces and Routes sub-tests of RBMT) declarative memory, digit span, TMT, COWAT and WCST	RBMT	Stop a timer and send the envelop
[7]	2007	17 TBI and 16 controls	11.67 (3.50)	41.28 (32.52)	-	Response to a cue (one or four cue targets)	-
[18]	2008	44 TBI but no controls	7.08 (3.89)	-	TMT, COWAT and metacognitive variables	CAMPROMPT	CAMPROMPT

http://dx.doi.org/10.5772/57307

			TBI characteristics			Type of prospective memory tasks	
Study[1]	Year	Sample	GCS	Time since injury (months)	NPS assessment[2, 3]	Event-based	Time-based
[24]	2010	31 TBI and 31 controls	13-15	0.89 (0.22)	-	MIST	MIST
[16]	2012	18 TBI and 18 controls	6.77 (2.13)	24.50 (19.67)	Stroop and N-Back	-	Press a key every 5 min and monitor the time

1 Studies included here have been conducted with behavioral measures.

2 References of neuropsychological tests are not reported because the authors referred to different versions. Please refer to the specific articles for the appropriate references.

3 Appendix A reports the full name of neuropsychological tasks employed in the studies with a brief description.

Table 1. Studies that used behavioural tasks to investigate PM performance on TBI patients. GCS = Glasgow Coma Scale (Scores < 8 = severe; scores 9-12 = moderate; scores 13-15 = mild); Time refers to the time post injury (months); NPS assessment refers to the neuropsychological tasks included in the studies; WCST = Wisconsin Card Sorting Test; COWAT = Controlled Oral Word Association Test; DASS = Depression, Anxiety, Stress Scale; TOL = Tower of London; LNST = Letter-Number Sequencing Test; TMT = Trial Making Test; RBMT = Rivermead Behavioural Memory Test; CAMPROMPT = Cambridge Behavioural Prospective Memory Test; MIST = Memory for Intention Screening Test.

Among the studies identified[2], only one included patients with mild TBI [24]. Most studies included have evaluated event-based PM performance, 6 examined both event- and time-based performance, and one only analyzed time-based PM performance. When both event- and time-based tasks were included in the studies, TBI patients obtained lower performance on time-based PM tasks [11, 25-27], which substantiated that time-based tasks are more difficult to execute probably due to more self-initiated retrieval impairment [12]. Generally, TBI patients showed PM dysfunctions compared to healthy controls, but the degree of impairment varied according to the characteristics of the employed tasks. By varying the cognitive demand on the ongoing task TBI patients performed more poorly than controls on the high, but not low demand condition [28]. Some authors also manipulated the salience of the event cue in PM performance [29]. The cue was either integrated (focal cue) or peripheral (no-focal cue) to the ongoing working memory task. TBI patients and controls showed no differences on the ongoing task, but PM performance was poorer in TBI patients in both focal and peripheral cue conditions. That finding suggested that even with highly salient cues, TBI patients exhibit PM failures. Moreover, both TBI and controls self-reported that greater monitoring effort was required for the peripheral, rather than for the focal-cue condition. Instead, when the cue is maintained in a focal condition, but is varied the number of distractions during the ongoing task, TBI patients performed more poorly both on one and four target conditions [7]. Finally, some authors have varied the delay between encoding and task performance (10 vs 45 min) expecting lower performance in the longer delay condition. Researchers have also manipulated the functional link (semantic inter-item associative link) between intended actions with the expectation that

2 Studies were included if conducted with TBI patients and match controls. Research was conducted on Psychinfo and PubMed with key words: "Prospective Memory", "Traumatic brain injury", "Event-based" and "Time-based".

performance would improve. Patients with TBI performed significantly lower than controls; however, there were no significant effects of delay interval or functional link of the intended actions [11].

A subset of studies included executive functions tasks to determine whether impairment on PM was associated with executive dysfunctions. The available evidences were mixed regarding the relationship between PM and executive functions, as some studies found significant [16, 22, 28] and others no [27] association between those two constructs. For example, we found [16] that working memory significantly correlated with PM accuracy only in TBI patients, while inhibition was involved in controls. Also, working memory significantly correlated with PM performance in the low cognitive demand condition in both TBI and controls [28]. These results suggested that for the employed dual-task event-based paradigm, participants had to actively maintain in memory the requirement while undertaking the ongoing task [16, 28]. Spontaneous flexibility assessed through the Controlled Oral Word Association Test (COWAT) [30] significantly correlated with event-based PM performance in the high-demand condition in TBI patients. TBI event-based PM performance significantly correlated with attention/speed processing tasks, which added further support to the notion of a strong relationship between attention and PM performance [29, 31]. In particular, the authors pointed out that those PM dysfunctions may be due to momentary lapses of attention to task details rather than to complete forgetting of the task instructions. Retrospective memory also accounted for a significant amount of variance in PM tasks of TBI patients independent of executive functions [25]. In particular, it was found that executive functions were involved both on time- and event-based PM performance, but it particularly influenced time-based PM performance, whereas retrospective memory contributed most to event-based performance. Finally, a complex PM paradigm was adopted [19, 32] to investigate which of the four phases of PM were affected in TBI patients. Impairment of PM was found on intention formation, intention re-instantiation, and intention execution. That finding confirmed that executive functions are involved on these three phases rather than only on intention retention. Only one study investigated the involvement of temporal abilities in time-based PM performance and found that adequate temporal abilities were involved in monitoring behavior rather than PM [16].

In addition, two studies have investigated PM performance according to demographic, clinical and metacognitive variables [18, 25]. Depressive symptoms were found to adversely affect successful performance on the timing component of the time-based task [20], while anxiety symptoms adversely affected performance on event-based tasks [25]. In a more recent study, the localization of the damage (mainly frontal) significantly correlated with event-based Cambridge Test of Prospective Memory (CAMPROMPT) scores, whereas duration of post-traumatic amnesia (PTA) and metacognitive variables significantly correlated with both time- and event-based scores [18].

2.2. Virtual tasks to investigate prospective memory in TBI patients

For adequate clinical assessment and rehabilitation programs, neuropsychologists need information about how a patient functions in routine daily activities, and laboratory-based tasks may be unable to always provide such information [33]. Chaytor and Schmitter-Edgecombe (2003) found that the relationship between neuropsychological tests and measures of outcome is often limited. Recently, it was concluded that most PM tasks lack adequate reliability [23]. As a consequence, it is often impossible to translate test scores into either goals for rehabilitation or conclusions about the level of impairment; that is, many conventional tests of memory-related abilities lack ecological validity [23, 33, 34]. To solve this important discrepancy between performance on neuropsychological tests and performance in everyday life, researchers, following the evolution of computer programs, have developed tasks that can bridge conventional neuropsychological tests with behavioral observations. Virtual tasks, in fact, can simulate the activity of everyday life in a controlled setting [33].

Study	Year	Sample	GCS	Time since injury (months)	NPS assessment[1]	PM Tasks
[38]	2005	25 TBI and 20 controls	-	113.76 (75.13)	NART, WCST, LM, COWAT and TMT	Videotape: Performing activities when the cue appeared on the screen by saying to the experimenter the activities required
[37]	2005	2 brain injury patients[2] and 3 controls	Patients SR GCS = 3 and Patients TJ GCS = 4	Patients SR 6 months; Patients TJ 3.5 months	Word list from WMS, Digit span, Stroop, WCST and NART	Virtual Street: Imagine walking on a street and performing 3 activities. The list of activities was either visible or hidden after instruction
[39]	2006	20 TBI and 20 controls	-	160.2 (91.66)	DEX, Logical memory from WMS and Ruff 2&7 selective attention test	Virtual Street: Imagine being a street inspector and performing 3 activities under low and high distraction. Moreover performing 10 errands
[40]	2009	16 TBI and 16 controls	8 or less	at least 3 month	HVLT, TMT and Digit span	The supermarket shopping trip: "buy" 8 items while watching a 8-min DVD of a supermarket trip
[44]	2013	18 TBI and 18 controls	4.54 (1.37)	66.94 (95.22)	Phonemic and Semantic fluency; TMT; WCST	Virtual Week

[1] References of neuropsychological tasks are not reported because the authors referred to different version of the tasks. Please refer to the specific articles for the appropriate references.

[2] The study includes 3 patients with brain injury: SR suffered a severe TBI as a result of assault, TJ suffered a TBI as a result of motor vehicle accident and RW suffered cerebellar and sub-arachnoid haemorrhages. Only information on patients SR and TJ were included because TBI patients.

Table 2. Studies that used ecological tasks to investigate PM performance on TBI patients. GCS = Glasgow Coma Scale; NPS assessment refers to the neuropsychological tasks included in the studies; NART = National Adults Reading Test; WCST = Wisconsin Card Sorting Test; LM = Logic Memory; COWAT = Controlled Oral Word Association Test; TMT = Trial Making Test; WMS = Wechsler Memory Scale; DEX = questionnaire from the Behavioural Assessment of Dyexecutive Syndrome; HVLT = Hopkins Verbal Learning Test.

In Table 2 are revised studies that were conducted with virtual tasks. This literature is presented separately from other studies on PM because it is more representative of everyday PM performance, and focused on filling the gap between performance on laboratory tasks and PM performance in real life.

The Prospective Remembering Video Procedure (PRVP) was designed to test PM abilities in virtual street simulation [35]. The task involves remembering to carry out a set of activities while watching a videotape filmed by a person walking slowly through a shopping complex. Participants virtually visited a shopping centre, with a set of event-based activities, each of which involved an action (`buy a record') and a cue (`from the record stand'). As each cue comes into view, the associated action must be recalled. The view that was presented focused mostly on the shop-front, although the footage also included views of passengers crossing the street. Studies conducted with health students and mix-brain injury patients demonstrated that the PRVP procedure was reliable and easier to complete when the video was set in a familiar location. Moreover, it was found that the video-based task was correlated with performance on an equivalent real life memory task, and provided evidence for criterion validity [35, 36, 37]. Twenty severe TBI patients and 20 matched controls were also tested with PRVP and completed the ongoing and PM tasks while "walking" along the street. Severe TBI patients showed poorer performance and were more affected by distractions on the PM tasks compared to controls [38]. Previous findings were confirmed and extended with a Virtual Street task [37, 38, 39]. Results showed that TBI patients performed more poorly than the control group on the PM tasks, but had similar performance on the Wechsler Memory Scale (3rd Edition) Logical Memory subtest. The authors discussed this result in favor of sensitivity of the Virtual Street task to measure PM performance. These results also suggested that TBI patients may show normal performance on memory tests, particularly in a controlled and quiet setting, but when memory performance is tested in a less controlled setting with distractors and tasks that require more strategic processes (like in real life conditions), real TBI dysfunctions may emerge [39].

Interesting, following the idea to create tasks more representative to real life situations, other authors have investigated the "generation effect", which refers to the effect of people demonstrating better memory for self-generated material than for experimenter-generated materials [40, 41, 42]. Considering the instructions often presented in PM setting to perform the action at a specific time or when a specific cue occurs, these instructions may allow control over the tasks, but they are unable to produce the same level of motivation in participants compared to self-generated activities [42]. Some authors reasoned whether people with TBI were more likely to remember a PM task if it is self-generated as opposed to experimenter-generated [40]. TBI patients were less accurate than controls, but the source of the to-be-remembered item exerted minimal influence on PM performance in both TBI patients and controls. Additional analyses were conducted to further investigate the source of errors, and it was found that both groups were able to recognize at post-test the shopping item they had intended to buy, but TBI patients failed to activate their intention to buy the intended item at the appropriate time. Moreover, significant correlations were found between working memory, attention-set

http://dx.doi.org/10.5772/57307

shifting and PM performance, which provided further support to the notion that higher order cognitive functions are necessary in PM [19, 40, 42].

Although they showed a high level of correspondence with real-life situations, the PRVP and the Virtual Street required participants to perform only event-based tasks. This was a limitation as PM performance should be investigated both with event- and time-based tasks. To resolve these limitations, the Virtual Week task was developed [42]. This task simulates daily life activities in a board game, where participants move around the board with the roll of a dice. Each circuit around the board represents one virtual day. Participants are required to perform 10 PM activities each virtual day (5 virtual days from Monday to Friday; additional information about the task procedures are presented in the section "Behavioral measures"). We found that people with TBI had significant difficulties executing PM tasks, which were more pronounced for time-based than event-based tasks [44]. These data suggested a relatively global PM deficit in people with TBI. Of particular interest was the finding that the magnitude of TBI impairment was consistent across regular and irregular tasks. Because the key distinction between regular and irregular tasks was that they place low and high demands on retrospective memory, respectively, these data suggested that failures of retrospective memory were not the major cause of TBI-related impairment in PM [44].

Taken together, TBI patients showed PM impairment when measured with laboratory based PM tasks. Some studies have reported that TBI patients were particularly impaired when they performed time-based tasks [11, 25-27] consistent with the view that time-based tasks are demanding and require more self-initiated processes [12]. However, other studies found a more generalized PM impairment in TBI patients [3, 44]. General agreement was observed between authors concerning the involvement of executive functions in PM performance; participants with lower executive functions would also present lower PM performance [16, 22, 25, 26, 28, 29, 31, 44]. The degree of the involvement of executive functions in PM performance varies according to the task employed and cognitive load. It is important to note that some authors have observed lower correspondence between PM performance obtained with laboratory-based tasks and real life situations [23, 33, 34, 45, 46]. Studies conducted with virtual reality tasks are trying to fill this gap, but further research is required to provide reliable and valid scores, and to develop tasks that can be used in clinical settings to screen for PM performance.

3. Assessment of prospective memory dysfunctions in clinical settings

Reliable and valid PM tools with normative data are necessary for health professionals working with people with any type of neurological disorder to acquire reliable information regarding PM performance. Such psychometrically sound tools are important not only to evaluate the degree of PM impairment, but also to provide useful information to inform the appropriate rehabilitation program. This section provides the most commonly used tools to investigate PM in clinical settings.

3.1. Self-report questionnaires

Self-report questionnaires are the most commonly employed tools to assess PM in real life situations. They assess PM dysfunctions by asking participants to rate their memory abilities by answering questions regarding their frequency of forgetting or remembering. They are particularly useful in clinical settings and in the rehabilitation process, because they can provide information about the patients' PM dysfunctions. However, it is very important to consider that a common problem after TBI is an inability to recognize impairments or disabilities that resulted from injury. This phenomenon has been termed "lack of insight" [22] or impaired "self-awareness" [47]. Self-report questionnaires have been recognized for their value in contributing to the understanding of patients' everyday PM dysfunctions; however, when used with TBI patients, impaired self-awareness may compromise the validity of self-report measures. Useful information can be acquired from relatives or caregivers. Questionnaires completed by significant others of the patients' general PM function have been found to correlate with objective PM test scores, and reliably address the patients' PM dysfunction in daily life [17, 18, 45].

The first developed questionnaires investigated general memory dysfunction in daily life with no specific distinction between retrospective or PM dysfunctions. These questionnaires included the Memory Functioning Questionnaire [48], Inventory of Everyday Memory Functioning [49], Cognitive Failure Questionnaire [50], Everyday Memory Questionnaire [51] and the Subjective Memory Questionnaire [52].

The Prospective Memory Questionnaire (PMQ) is a 52-item questionnaire that includes four PM subscales: 1) the *long-term episodic subscale* is related to memory for irregularly scheduled tasks, which require completion some hours or days after a cue to perform it (i.e. "I forgot to return books to the library by the due date"), 2) the *short-term habitual subscale* addresses memory for tasks to be completed shortly after the relevant cue that occur on regular basis (i.e. "I forgot to put a stamp on a letter before mailing it"), 3) the *internally cued subscale* addresses memory for tasks that do not have a clear external cue (i.e. "I was driving and temporarily forgot where I was going"), and 4) the *techniques to remember subscale* measures the use of strategies to aid prospective memory (i.e. "I rehearse things in my mind so I will not forget to do them") [20]. The PMQ questionnaire was administrated to 114 younger adults, 27 older adults and 15 patients with TBI [20]. Coefficient alpha was.92 for the PMQ total score, and ranged from.78 to.90 for the subscales. Factor analysis confirmed the 4-subscale structure. Test-retest reliability was found to be high, r =.88for the PMQ total and ranged from.64 to.88 for the subscales. Overall, the results found that the ratings of TBI patients were significantly different from those of controls on one of the short-term habitual subscales of the PMQ, however, very few studies have used the PMQ with TBI patients [53].

The Comprehensive Assessment of Prospective Memory (CAPM) measure was developed to evaluate the frequency of PM failure, and to evaluate the perceived amount of concern about these memory lapses and the reasons why people are successful or unsuccessful in performing PM tasks [54]. The items include both basic and instrumental activities of everyday life and are divided into three sections. Section A analyses frequency of failure (39 items; i.e. "Forgetting an appointment with your doctor or therapist", "not locking the door when leaving home"), and is considered to be the best section to evaluate PM dysfunction and outcome after rehabilitation [55]. Section B analyses the amount of concern about these failures (39 items; the same items used in the first session), and Section C analyses the reasons associated with the success or failures of PM tasks (15 items; "I rely to other people to remind me when I have to remember to do things"). Section A was used with 33 severe TBI patients (GCS = 5.1; PTA = 6.66 days), 33 relatives and 29 healthy controls to investigate frequency of PM failure [54]. Analyses were conducted on self-rating and informant-rating, and the difference between these two indices provided a measure of self-awareness on PM dysfunction. The results showed no significant differences between the TBI and controls' self-ratings of the frequency of PM failure. However, a significant difference was found between the informant-ratings, which suggested that relatives and/or caregivers reported higher levels of PM failure. These findings are likely to be due to the commonly found self-awareness difficulties in patients with TBI [47, 54]. The section C was used to investigate how TBI patients perceived and self-rated the reasons for PM failure [56]. Relatives and caregivers were also asked to rate TBI patients' PM failures. One-hundred thirty-six participants were included in the study and divided into four groups: 38 severe TBI (GCS = 5.1, PTA = 62.6 days), 34 significant others of TBI patients, 34 controls, and 30 significant others of controls. Results showed that patients with TBI were more likely to forget planned activities; moreover, TBI patients forgot to do things when they focused on other tasks or if they considered other activities unimportant. Patients with TBI seemed also to rely more on relatives to remind them to do things. However, no significant differences were found between TBI patients and controls on reporting reason of remembering or forgetting, and indicated that TBI patients tended to overestimate their PM abilities [54, 56]. It is important to note that authors reported that some TBI patients included in the study indicated that they have difficulty understanding some questions included in section C and suggested to further investigate the internal consistency of the section C [56]. The validity of CAMP (section A) was investigated in a study conducted with 45 patients with moderate and severe TBI (age M=30.02 years; GCS = 7.08; PTA = 47.89) and their relatives [45]. Participants also performed two tests of PM, the CAMPROMPT [57] and the Memory Intentions Screening Test (MIST) [58]. Concurrent validity was investigated by the comparison of scores on the CAPM with scores on the CAMPROMPT and MIST. Results showed that self-report CAPM scores did not significantly correlate with scores on the CAMPROMPT or MIST. TBI patients reported very few PM failures on the CAPM, but demonstrated impairments on neuropsychological testing. Interestingly, results showed that the responses provided by the relatives on CAPM correlated with CAMPROMPT and MIST performance. These findings suggested that the relative version of the CAPM has some concurrent validity when compared with performance on neuropsychological assessments, and that the relative version of the CAPM is an objective and valid measure of PM failure in TBI patients [17, 45]. To our knowledge, no studies have been conducted with TBI on section B of CAMP.

Finally, the Prospective and Retrospective Memory Questionnaire (PRMQ) was developed to investigate the rate of frequency with which patients make particular types of memory errors [59]. The PRMQ consists of 16 items that are divided into 8 categories (two items in each category) that tap different aspects of memory failure: prospective short-term self-cued, prospective short-term environmentally-cued, prospective long-term self-cued, prospective long-term environmentally cued, retrospective short-term self-cued, retrospective short-term environmentally-cued, retrospective long-term self-cued, and retrospective long-term environmentally-cued [59]. Different forms of the PRMQ are available for patients and caregivers. In addition to the 16 items included in the patient form, caregivers are asked to respond to four additional questions. Two of the questions concern the caregivers' frustration and two ask the caregivers to rate the patients' frustration. The PRMQ has never been used with TBI patients, but was tested with 242 healthy older adults (age M = 72.74 years) and 155 patients with Alzheimer disease (AD; age M = 73.95 years). One-hundred fifty-five caregivers of AD patients (age M= 56.85) and younger participants (age M = 44.19) were also included in the study. The split half reliability of the PRMQ conducted in younger and older adults (n = 406) compared the two items within each category was found to be.84. Results showed that memory dysfunction was highest for AD patients and lowest for caregivers, with younger and older adults in between. Moreover, caregivers rated PM dysfunctions of AD patients more frustrating than patients did, and PM errors were rated as significantly more frustrating than retrospective memory errors. The authors concluded that the PRMQ might be a potentially useful tool for investigating differences in frequency of prospective and retrospective memory failure in non-patient samples, but that the questionnaire needed to be adapted to increase its sensitivity to allow for more extensive clinical use [59]. These results were extended through the testing of 87 younger adults (age M = 44.11 years) with PRMQ and with a standard laboratory procedure of PM [60]. The reliability of PRMQ was.86 (Cronbach's alpha) for the total scale,.72 for the PM scale, and.72 for the retrospective memory scale. Participants reported higher scores (higher memory failure) for the PM than retrospective memory scale. To examine whether the PRMQ scores may predict actual performance in the laboratory PM tasks, correlation analyses were conducted and showed that only the PM scale was a significant predictor for the standard laboratory PM performance. In sum, the study extended initial findings and added additional support for the validity of the PRMQ. In addition, it provided the first evidence for the utility of the PRMQ subscales in differentiating between prospective and retrospective memory task performance [59, 60].

To our knowledge, no other studies appear to have examined the relationships between PM performance and measures of self-awareness in TBI patients. This may reflect the unreliability of self-rated PM measures [47]. Despite previous studies that have provided good validity and reliability of questionnaires, the results are mixed when self-rated scores were compared with laboratory-based measures of PM performance [3, 56, 60, 61]. There are two important caveats to the questionnaires. First, it is important to consider that some authors have reported lower correspondence between the PM tasks used in the examination condition, and PM activities in the everyday situation. Indeed, in real word situations, people required higher levels of executive functioning to perform multiple tasks than in a controlled experimental standardized setting. Therefore, some authors have found that TBI patients showed unimpaired performance on neuropsychological PM tests, but reported impaired PM performance in real life [45, 46]. Second, it is important to consider the self-awareness dysfunction often observed in TBI patients that might compromise the validity of TBI participants' self-report measures [22, 47].

3.2. Behavioral measures

In 2002, Shum et al. [61] published the first review about PM dysfunctions in TBI patients. The authors identified three neuropsychological tests that provided objective scores of PM, and that were used in clinical settings. These tests included the Rivermead Behavioural Memory Test (RBMT) [62], Cambridge Behavioural Prospective Memory Test (CBPMT) [63], and Memory Intentions Screening Test (MIST) [58]. More recently, a second review about PM dysfunctions in TBI patients was published [3]; despite the increasing number of studies conducted, very few studies were dedicated to the development of valid and reliable measures to investigate PM failure in clinical settings. We briefly present the main tools that were highlighted in the prior reviews [3, 61] and a newly developed task.

The RBMT was the first widely used task in clinical practice that explicitly tests PM performance. It includes three PM tasks: (1) remembering where a belonging is hidden and asking for it to be returned, (2) asking for the next appointment time when an alarm sounds, and (3) delivering a message. Six neurologically impaired adults (age M = 42 years) with a length of stay in the post acute rehabilitation program that ranged from 22 to 101 days (M = 66.2 days) were tested with the RBMT [64]. Participants were required to perform RBMT and some real-life activities every morning with or without verbal or visual cues. Low or no significant correlations were found between the number of activities correctly performed and RBMT score, which indicated that the RBMT was an invalid predictor of PM performance. Different results were obtained in a study with a larger sample size [65]. One-hundred-nineteen participants were divided into four groups: 20 severe TBI, 29 moderate TBI, 39 mild TBI, and 31 controls. Participants performed the RBMT together with other clinical memory measures (Wechsler Memory Scale-Revised and Luria Nebraska Neuropsychological Battery). The authors found the RBMT to be an accurate and valid measure to test PM memory failure in everyday life. However, RMBT had ceiling effects when used with 25 TBI patients (10 moderate and 15 severe; age = 28.6 years) and 25 controls [27]. In addition, the authors found that only one of the three RBMT PM items (RBMT delayed message) successfully differentiated between TBI patients and controls. Although the RBMT was widely used in clinical settings and in several neuropsychological studies, it provided only a limited range of scores and is unlikely to be sensitive to deficits in non-severe patients [27, 61, 64]. Importantly, the test does not assess time-based PM performance [62]. A new version of the RBMT, The Rivermead Behavioural Memory Test Extended Version (RBMT-E) was created by combining two parallel forms of the RBMT [66]. The idea was to develop a more demanding test that eliminated the ceiling effects and thus would be more sensitive to minor PM impairments. The RBMT-E includes 8 subtests: (1) First and second names: remembering the names of three people; (2) Appointment and belongings: retrieving to ask for two belongings previously hidden; (3) Picture recognition: recognizing 20 previously-seen pictures and distinguishing them from distractors; (4) Face recognition: recognizing 15 previously-seen faces and distinguishing them from distractors; (5) Story: recalling a story immediately and after a delay; (6) Route: remembering 7 routes immediately and after a delay; (7) Message: remembering to pick up and deliver to the correct location a message immediately and after a delay and (8) Orientation and date: answering 12 orientation questions and giving the correct date. The RBMT-E was tested with 16 TBI patients (age M = 40.5 years; time since injury M = 47.5 months, range 8-92 months) and 16 controls [67]. Patients included in the study were all previously evaluated with RBMT and all scored in the normal range (screening score between 9-12), but all lamented PM dysfunctions in daily life. The authors intended to compare the performances obtained with RBMT and the RBMT-E. The results, however, were mixed and inconclusive. In fact, when patients were retested with the RBMT, some participants scored outside the normal range, which indicated poor test-retest reliability. Moreover, when the analyses of RBMT-E performance were conducted on those patients who scored in the normal range at test and retest (data collected with RBMT) and compared with controls, no differences were found between groups [16].

The CBPMT includes four time-based and four event-based tasks and requires approximately 40 minutes to be completed [63]. Participants were allowed to use any strategy to remember the tasks. The time-based tasks included (1) remembering the experimenter to not forget the keys after 15 minutes; (2) requesting the tester for a newspaper after 20 minutes; (3) switching task after working for 20 minute and (4) stop working on the booklet after 3 minutes. The event-based tasks included (1) reminding the tester about five hidden objects; (2) when the alarm rings, putting a briefcase under the desk; (3) Changing pens after having completed seven filler assignments and (4) give the envelope with the message. The CBPMT was used with 36 people with mixed brain injury (age M = 35.61 years; time since injury 75.56 months) and 28 control participants [63]. Participants also performed the EMQ along with general intelligence, attention, working memory, executive functions and retrospective memory tasks. Results showed good correlation between the CBPMT and measures of working memory and executive functions. No correlations were found between the CBPMT and EMQ in brain injury patients, but significant correlations were found with controls. Subsequently, the CBPMT was used to create the CAMPROMPT [47], which included three time-based and three-event based tasks to be completed in 25 minutes. PM tasks are executed while performing a filler activity, and participants are allowed to spontaneously use strategies (i.e. take notes). The time-based tasks included: (1) remembering his/her belongings when there were 7 minutes left to the end of the session; (2) when the timer shows 16 minutes, the participants had to remember to stop the task in 7 minutes time and (3) at 10:11 and 5 minute after the end of the session, remembering to call the reception. For the event-based tasks participants were asked (1) to return a book to the examiner when he/she came to a question about the television program 'EastEnders'; (2) to return an envelope with "MESSAGE" written on it and (3) to remind the examiner to pick up five objects that had been hidden when the session is over. The CAMPROMPT was used with 44 moderate and severe TBI patients (age M = 29.64; GCS = 7.08) [18]. Participants also completed executive functions tasks and the CAMP section A. Results showed that PTA was a good predictor of PM performance, and patients with lower executive functioning showed lower PM performance with a higher degree of impairment on the time-based tasks. Overall, the CAMPROMPT showed very high reliability (Pearson r = .99) and high internal consistency (Cronbach's alpha = .75), but moderate test–retest reliability (Kendall's Tau-b = of .64) [57, 68].

http://dx.doi.org/10.5772/57307

The MIST includes four event-based and four time-based tasks to be performed in 30 minute sessions while playing a word-search puzzle that serves as a distractor [58]. The 8 PM activities are balanced in terms of delay interval (2 or 5 minutes delay), cue (time-based or event-based), and response modality (verbal or a physical response). The MIST also includes an 8-item multiple-choice recognition test and a more naturalistic task that has to be performed in a 24-hour delay (calling the examiner the next day and report how many hours of sleep the night before). The MIST test has been widely used with healthy older adults and various clinical populations [58]. No significant correlation was found between the MIST score and CAMP in 45 moderate and severe TBI patients (age M = 30.02 years; GCS = 7.08; PTA = 47.89) and their relatives, but TBI patients showed lower performance on the time-based tasks [45]. The MIST task was used with 38 mild TBI (mTBI) patients (age M = 40.6 years, GCS range 13-15, no PTA, loss of consciousness < 20 minutes) and matched controls were included in the study [24]. Patients were tested within a month of injury (M = 26.6 days) and 3 months post-injury. Mild TBI (mTBI) patients performed more poorly than controls on the MIST task within the first month following injury, which indicated that PM impairment is part of the acute cognitive dysfunction profile of mTBI. PM dysfunctions were also observed after 3 months post-injury, suggesting that PM may be a sensitive indicator of cerebral compromise in mTBI patients. Taken together, these studies conducted with mild and severe TBI patients [24, 45, 58, 69] showed a consistent deficit on the MIST summary score and total number of errors. Both TBI patients and controls demonstrated higher performance when they executed activities after a short time delay. Better performance was obtained with event-based than with time-based tasks in both TBI patients and controls. The controls demonstrated higher performance for action responses compared to verbal responses, though no effect of response type was observed in TBI patients. Overall, the MIST demonstrated good validity and test-retest reliability when performed across a two-week interval [70]. The split half reliability was .70 (Spearman-Brown coefficient), internal reliability (Cronbach's alpha = 0.89) of the six subscales was high, but the internal reliability (Cronbach's alpha = .47) for each trial was poor.

The Royal Prince Alfred Prospective Memory Test (RPA-ProMem) includes three alternative forms, each of which has two time-based (i.e., "in 15 minutes time I would like you to tell me it's time for a coffee break") and two event-based tasks (i.e., "At the end of our session today, I would like you to ask me for an information sheet on note-taking strategies") that have to be performed within the session (short-term: "In 15 minute interval I would like you to remind me to move my cat so I don't get the ticket") or at a later moment (long-term: "When you arrive home today, I want you to phone and leave a message on my voice mail, telling me your mother's name") [69]. External aids are permitted during the testing session. The RPA-ProMem was used with 20 patients with brain lesions (stroke n = 7, epilepsy n = 5, tumour/cyst n = 2, TBI n = 2, arteriovenus malformation removal n = 2, encephalitis n = 1 and systemic lupus erythematosus n = 1) and 20 controls [69]. Participants also performed the MIST, CAMP section A, and EMQ. Patients with brain lesions showed lower performance on the RPA-ProMem compared to controls, even when they rated their own PM performance in daily life as normal. No differences were found between the scores of the three parallel form of RPA-ProMem, and they showed good alternate form reliability (Spearmen-Brown coefficient = .71). These finding are consistent with those obtained from studies with self-report questionnaires [3, 61] and support the use of RPA-ProMem as an objective clinical measure of PM performance.

Finally, Virtual Week was developed as a laboratory PM task that would more closely represent PM activities in everyday life [43]. Two clocks are placed above and below the dice. The clock above the dice is a chronometer that starts at the beginning of each virtual day and the clock below is a virtual clock that moves when participants roll the dice and indicates the virtual time of the day (the clock moves 15 minute every two squares). Participants are required to perform 10 activities every virtual day. The original version [43] included 7 virtual days, but the later versions include 3 or 5 days and maintained task validity [71]. A training day is included at the beginning of the task to familiarize patients with task procedures. Interestingly, Virtual Week also includes regular and irregular tasks. Every virtual day, participants are required to perform 4 regular tasks that simulate regular activities that occur as one undertakes normal health care duties every day (i.e. "take medications every day at breakfast"), and 4 irregular tasks that simulate the kinds of occasional activities that are new (i.e., Monday: pick up the laundry when shopping). Within the regular and irregular tasks, two are time-based (i.e., "call the plumber at 3pm") and two are event-based (i.e., "return the book to Philip when you see him").. We tested 18 TBI patients (age M = 31.72 years; GCS M = 4.54, time since injury M = 66.94 months) and 18 controls with Virtual Week and other neuropsychological tasks [44]. Results showed a generalized PM impairment in TBI patients, and significant correlations were found between PM performance, indices of cognitive recovery (Level of Cognitive Functioning and Functional Independence Measure/Functional Assessment Measure), and semantic fluency. These results indicated that participants with better cognitive recovery obtained better PM performance. Since the first publication [43], Virtual Week has been widely used with healthy and clinical populations [71], with corresponding evidence indicated good psychometric properties [71, 72]. The reliability of Virtual Week was investigated in a study that involved younger and older adults [72]. Across the entire sample, reliability estimates ranged from .84 to .94 for the regular, irregular, and time-check tasks. In another study, the split-half reliability for the overall Virtual Week measure was estimated to be: 74 in a clinical group with schizophrenia [73]. In a recent study, test-retest reliability was investigated in older adults by use of the same (Study 1) or parallel (Study 2) versions across two testing time points separate by 1 month [74]. Preliminary results in Study 1 showed good internal consistency (Cronbach's alpha) of .64 at the initial and .83 at the retest session. In Study 2, for version A, internal consistency was .70 at the initial test and .89 at the retest session, and for Version B was .82 at the initial test and .65 at the retest session.

4. Prospective memory trainings

Traditionally, the rehabilitation of memory impairment in TBI patients has focused on retrospective memory. Only recently has the importance of PM impairments following TBI been recognized and resulted in the development of specific training programs [17, 18]. Considering that PM is not a unitary process, it is difficult to address the optimal rehabilitation paradigm that can be solely focused on treating PM as a unitary system, or can focus on treating specific PM components. In this section, we review the main rehabilitation approaches (Remedial vs. Compensatory) used in the treatment of PM impairment.

4.1. Remedial/Restorative approaches

Cognitive interventions have been developed to restore "the underlying defective cognitive functions usually via repetitive drills or training activities designed to stimulate damaged neural networks or establish new networks" [61, pp 9]. The first studies investigated the effect of repeating PM tasks after different retention intervals on PM performance. The Spaced Retrieval Technique (SRT) required participants to remember a task over an interval of time. If the task was recalled correctly, the retention interval was an increase. The SRT was tested with two TBI patients over 4.5 and 3.5 months of training. Results showed positive effects on PM performance, in fact, TBI patients increased their performance from being unable to complete a PM task following a 60 seconds delay to being able to complete a PM task following 8 minute delay, for patients 1; and an increment from 4 to 8 minutes delay was observed in patients 2 [75]. Based on these encouraging results, the authors further investigated the generalization effects of SRT with one TBI patient [76]. The training paradigm consisted of repetitive administration of PM tasks, with systematic lengthening of the delay period between task administration and task execution. Probes were taken that evaluated generalization to performance on naturalistic PM and retrospective memory tasks. Results confirmed previous findings and showed that patients were able to increment of temporal interval between task administration and task execution an maintain good PM performance. However, the results were less encouraging as tasks showed poor ecological validity [76]. Moreover, two TBI patients were tested with SRT training and retrospective memory drills [77]. The first patient increased the delay interval from 1 to 5 minutes, while the second patient increased the delay interval from 2 to 10 minutes. Results were consistent across the two TBI patients who showed improvement on PM task performance at the delayed target time in both experimental and real-life settings. A more ecologically valid training program called the Prospective Memory Re-training Package (PMRP) was developed [78]. Six TBI patients were first tested with the Prospective Memory Screening test (PROMS) [75] and then were trained on the PMRP. The training required the patients to perform real-life activities (i.e., putting a stamp on an envelope) at specified target times, with a maximum delay of 6 minutes. The delay between task administration and task execution increased depending on correct or incorrect responses. Results showed better PM performance after 2 months of training [78]. In addition, positive effects were found when space retrieval was combined with Error Learning (EL) technique, which was achieved by the discouragement of participants from guessing if they were uncertain of a response [79].

The effects of cognitive strategies on PM performance were also investigated [80]. Some authors showed that participants had better performance on PM tasks when they had previously made a strong and explicit association between PM cue and intended action through imagining themselves performing that PM action. By reinforcing the cue-action association during the encoding phase, future intentions are more likely to be supported by automatic cognitive processes according to the McDaniel & Einstein's Multiprocess Model [4]. The authors [4] showed that less attentional resources and executive processes are required to perform the PM action because the intended action is spontaneously retrieved when the prospective cued is encountered. Following this finding, a training program was developed with visual imagery technique [81]. The training included three successive phases. The first phase evaluated the individual visual imagery capacity, the second focused on acquisition of visual imagery skills using spaced retrieval techniques, and the third focused on visual imagery techniques that are progressively applied to real-life situations in order to promote learning transfer. During the third phase, participants are encouraged to identify challenging daily-life situations and to develop more appropriate visual imagery strategies to successfully complete the real-life activities. Twenty-one patients with various brain injuries were included in the study: 12 participants were included in the control group (age M = 36.6 years; time since injury = 60 months) and 9 were included in the experimental group using visual imaginary technique (age M = 41.9 years; time since injury = 65 months). Participant included in the control group performed different memory trainings respective with the training used in their centre. The trainings were different for each patient, but intensity and frequency of training were the same in both groups. Results showed that visual imagery training significantly improved delayed recall of everyday verbal information (i.e., stories, appointments). Relatives also reported a reduction in PM problems and the effects were stable after a 3-month follow-up. Successive, visual imagery technique was used with 30 severe TBI patients [82]; 10 of who were assigned to the experimental condition using visual imaginary technique (age M = 35.00 years; PTA = 30.60 days; time since injury = 43.40 months) and 20 to the control condition (age M = 30.90 years; PTA = 29.20 days; time since injury = 34.00 months). The program required the patients to learn flexible cognitive strategies by creating gradually more complex and ecological associations between cue and intended action that could be used in real life situations. Participants who took part in the rehabilitation program improved their PM performance (measured by Test écologique de mémoire prospective; TEMP [83]), and also reported the use of imagery techniques in everyday life situations with corresponding improvement in PM function. Relatives also confirmed this improved performance [82].

The positive results obtained with visual imagery techniques seem to rely on the strong cue-action association and engagement required by the training. Repeated imagination of the retrieval situation and increased cue-action association might increase the detection of retrieval cues with the occurrence of the situation. This technique seems to be particularly effective for event-based PM tasks in which the PM cue is available. For time-based actions, maybe additional effort is needed and patients should try to associate a visual cue even if the action is time-based. For example, if the action is to make a call at 3 pm, participants should look at the phone and remember to make the phone call. Thus, imagination may help to translate a time-based into an event-based task. However, such a transference of time to an event based task presumably will only be possible in patients with mild impairment.

Some studies have attempted to resolve PM dysfunction with a different approach. Considering that PM is not a unitary system [4], one line of rehabilitation might be to restore the components of PM. These approaches followed the idea that retrospective memory and/or executive dysfunction may underlie PM impairment [16, 25, 27], and that improvement of these cognitive components should result in better PM performance [84]. Hypothetically, training programs devoted to improve attention, working memory, and executive functions could have positive effects on PM performance [3, 61, 84]. In general, approaches to the rehabilitation of executive deficits have focused on training patients to follow step-by-step problem solving strategies (i.e., define the problem and generate potential solutions) [85]. Positive effects were found when retrospective memory dysfunctions were rehabilitated [86, 87]; however, the value of such an approach for PM improvement remains to be investigated. Other studies have found that structured group experience facilitate the PM rehabilitation. Participants who enrolled in Goal Management Training (GMT) were encouraged to highlight common executive difficulties and discover which were the best strategies (i.e., breaking down goals into sub-goals, using mental imagery) [88]. Attempts to promote generalization included homework exercises and recording of everyday errors [85, 89]. Despite efforts to foster generalization and maintenance, the everyday benefits of such training still largely depend on strategies that spontaneously "come to mind" in everyday life [90]. Remedial strategies seem to work on the mental organization and encoding of prospective actions that should help recall. However, PM not only involves encoding and storing the content of PM action, but also recalling the intended action at the appropriate moment in the future. The remedial strategies previously presented may have more application to the rehabilitation of PM by enhancing recall of the content (the retrospective component) than enhancement of the intention. These techniques have demonstrated good effects on PM performance but need to be further developed in line with PM models and real life situations [1, 4, 5, 91]. For example, training paradigms need to include both event- and time-based tasks, with the knowledge that routine tasks (i.e., daily regular tasks [43]) can be better encoded, remembered, and performed. Novel activities can be difficult to remember, but in such cases researchers and clinicians can manipulate the salience of the PM cue [4, 72]. Preliminary data suggested that PM training based on remedial strategies can improve individual's PM performance, but further research is needed to investigate the generalizability to real-life situations.

4.2. Adaptive/Compensatory approaches

In contrast to the remedial approaches that are aimed to restore the impaired PM function, adaptive/compensatory strategies are aimed to compensate the PM dysfunction. These techniques are widely used in clinical practices and patients often spontaneously report the use of compensatory strategies such as a timer or diary. Memory aids are widely available, can be inexpensive, and have the potential to be highly effective in compensating for PM problems in mild to moderately impaired patients. Although patients can find it difficult to learn and remember to use such aids, it is possible to implement an effective system, even in patients with profound amnesia [61].

There is a substantial body of evidence supporting the use of external memory devices [90, 92], which can be categorized into non-electronic and electronic external memory devices. *Non-electronic memory devices* include a diary, calendar, and to-do lists. These are easily available, but are only beneficial if patients make the effort to use them at the appropriate time [92]. Despite the fact that patients often spontaneously report to take notes on calendar and use a pin-board, very few studies have investigated the effectiveness of non-electronic memory devices on PM rehabilitation. Of the few conducted. one study used a memory book in training with TBI patients with amnesic disorders [93]. Patients were first introduced to the different functions and learned how to use the memory book. Then, patients used the memory book in a controlled laboratory environment followed by use in everyday life. Patients had to demonstrate mastery of criteria before they moved to the next step (use the memory book in real-life condition). For severe patients, the training lasted 6 months. Memory improvement was observed and effects were stable over a long time period. Previous results were extended to evaluate the effectiveness of the memory book in comparison to a more social therapy in which patients were trained with problem-solving strategies [94]. Eight TBI patients were randomly assigned to one of the two experimental conditions: memory book (mean age = 29.9 years; days of coma = 39.7; time since injury = 77.7 months) and supportive therapy (mean age = 26.8 years; days of coma = 37.5; time since injury = 86.8 months). After the training, the authors reported reductions in the number of errors in everyday life in patients who used the memory book. However, 6-months later, there were no longer differences between the two groups. This may raise the problem of generalization and long-term use of learned memory strategies. Some authors have noted that many patients quickly abandon diary use once they are discharged [95]. To overcome such problems, several formal training approaches worked also on self-awareness of PM deficits in order to minimize patients' resistance to use memory aids when back in their home environment. In fact, some studies showed the importance of involving family and friends in the training and the need to have practice sessions in community-based settings [96]. Taken together, these results are encouraging regarding the utility of non-electronic memory aids; also, a later study showed that participants who spontaneously took notes (13 out of 36 people with brain injury and 14 out of 28 controls) performed significantly better on a PM test than those who did not take notes [63]. Results lead to the conclusion that note-taking significantly benefits PM performance. The suggestion is therefore to integrate non-electronic memory aids in other memory training programs such as group or remedial therapies [3, 92, 97].

Considering the increasing use of electronic devices in everyday life, it has been observed a consequent increase of electronic memory devices in clinical training and PM rehabilitation. *Electronic memory devices* have the clear benefit of not merely telling the patients what he/she is intended to do, but also drawing the patient's attention to this information at the appropriate time. A brief summary of electronic memory aids in PM training for TBI patients is provided in Table 3.

Probably the most used and commercialized electronic memory device used to compensate for PM deficits is the NeuroPage system [98, 99, 100], which enables an individual to receive timed reminders or cues that appear on a portable pager carried by the person. The aim is to help individuals maintain independence by reminding them to carry out everyday tasks. Messages appear on a big screen and the user can control the system with a single large button, which makes the NeuroPage suitable for patients with motor difficulties. Messages have to be confirmed by the patients by calling the central system, in the case of no answer, the message is displayed again. All messages are also stored in the central system. The authors also

http://dx.doi.org/10.5772/57307

Memory Aids	References	Advantages	Disadvantages
NeuroPage	[98, 99, 100]	Very small and little learning is required. The system can ring or vibrate and a text message appeared on the screen.	The appointments are changed through a paging company and user cannot change them.
Mobile Extensible Memory and Orientation Sustem (MEMOS)	[102, 103, 104]	Therapists and caregivers can access to the system from any computer with internet and modify to encode data.	Problem with connection between server and devices.
Memojog	[105, 106]	Users, therapists and caregivers can access to the system. Possibility to add others information beside PM tasks (i.e. birthdays, addresses).	Problem with connection between server and devices. Difficulties with touch screen system.
Voice Organizer	[107, 108]	User can autonomously record new messages.	Patients with speech and motor impairments might have difficulties in using the system.
Palmtop	[109]	User can autonomously add new messages.	Too difficult to use for patients with severe impairments

Table 3. Overview of electronic memory devices.

presented the results of a large study conducted with 143 patients aged between 8 and 83 years [98]. All patients had brain damage (mainly TBI or stroke) and memory, attentional, and planning deficits. Results showed that patients generally benefitted from the use of NeuroPage with improvement in everyday goals by an average of 30% when the pager was used relative to baseline performance. Importantly, in the same patients, there was evidence that the benefits persisted when the system was no longer used, except in those with more pronounced executive impairment [84; 98, 99, 101]. Many patients suggested the pager served as a training function (i.e., consolidating intentions into a everyday life).

The Mobile Extensible Memory and Orientation System (MEMOS) [102, 103, 104] is a system specially designed for patients with brain injuries. The system includes a central server to communicate with a personal memory assistant over a wireless cellular phone. Similar to the NeuroPage, MEMOS includes a large screen with few buttons to accommodate patients with cognitive and motor impairments. Of interest, the device allows patients to leave messages for future appointments that are transferred to the central system where the caregiver can input them. The system sends reminders for PM tasks with feedback and step-by-step guidance. Important tasks, as taking medication, can be set as “critical tasks” for which the caregivers can be advised if the patient executed them.. The effectiveness of MEMOS was compared with a Palm organiser in 13 patients with brain injury [104]. Patients reported positive effects from both devices, but during the interview after the training, patients reported that the MEMOS was more helpful in the execution of PM tasks and desired to continue using it. The long-term use of MEMOS was evaluated with 3 TBI patients [103]. The authors reported that the MEMOS alone was unable to compensate for memory deficits in patients with severe TBI. For those patients, additional support for relatives is necessary [103].

Another interesting memory aid is Memojog [105, 106], which is similar to the MEMOS device [102]. Memojog is a remote and interactive communication system that provides prompting for people with memory impairments. A central system is used to deliver text prompts to a mobile device that are announced by an alarm. Interestingly, both the user and the caregivers can add new information into the devices from any internet station. Also, new tasks can be phoned in by others to add them into the system. The reminders are then wirelessly transmitted at the appropriate times and users have to respond to them. As for MEMOS, "crucial tasks" can be monitored with particular attention. Moreover, the system can record other information such as birthdays, addresses, contact information, pictures, and details of any appointments. Limitations of the system are the connectivity problems, thus authors and caregivers are unsure if the patients are doing something wrong or if the system is malfunctioning. Besides that, users (both older adults and patients with brain injury) are motivated by the system and haven shown improvement in memory performance with Memojog [106].

Also commercialised are systems in which it is possible to record messages that are played back at the appropriate time. One of them, Voice Organiser is a phone-type device that requires no input messages trough the central system, but the user can autonomously add new task and reminders [107]. Five TBI patients (mean age = 42.6 years; time since injury = 38.6 months) were trained in an A-B-A experimental design and performance was monitored for 9 weeks. Results showed that all but one patient improved in PM performance. The improvement was evident immediately at the beginning of the training. A limitation of the Voice Organiser device is that patients with significant speech disorders or poor motor ability might have difficulties in using it. In such conditions, relatives can program the system [107]. More recently, 8 patients with brain damage (mean age 46.25 years) and PM impairments were trained with an IC Recorder for a maximum of 3 months [108]. The names of the experimenter and patients were said before the alarm to attract attention and motivate the user. Soon after, the recorded message was said, indicating the task to be performed. Results showed that tasks prompted by the IC Recorder were better remembered and executed for five of the eight patients. The study added further support to the high quality of memory aids and suggested that the IC Recorder has great potential as a voice memory aid to assist patients with PM impairments [108].

Finally, microcomputers can be used to facilitate everyday PM performance. They have the advantage to be hand-held computers that can be easily carried around. Moreover, they provide visual as well as auditory messages to prompt the PM actions. Twelve patients (11 TBI patients and 1 with cerebrovascular accident; mean age = 49.6 years) were trained with a palmtop organiser and evaluated at follow-up between 2 months and 4 years after initial training [109]. Participants were asked about prior experiences with computer devices, then whether the device was useful, and if and how often they still used it. Nine out of 12 patients reported that the palmtop computer was very useful during the training, and they were still using it. Of the remaining three patients, 1 never learned how to use it and 2 found no appropriate situation in their daily life where use it. Later PM performance was investigated in 12 patients (mean age = 4 years; time since injury = 49.5 months) using Palm organiser or mobile phone. The authors were not only interested in training PM performance, but also in training patients to enter appointments into the devices. The study showed that only patients with mild impairment were able to correctly enter the appointments [110].

The electronic market moves very quickly and memory devices need to be updated accordingly. We would like to report two interesting devices tested with older adults, but that can be extremely promising with patients with brain injuries: Autominder [111] and Coach [112]. Autominder is a mobile robot platform with an integrated screen display and portable mobile device designed to assist older adults in their home. As with the other memory systems, Autominder transmits reminders for PM activities, but interestingly has the ability to adapt to the user's schedule depending on the behaviour detected by sensors installed throughout the home. Coach consists of a video camera installed in the home and provides support for both prospective and retrospective memory deficits. In fact, Coach provides cues to initiate future actions and helps users to execute the action by suggesting the procedure step-by step. Interestingly, Coach is requires no learning on its use and thus has limited demands of the patient in terms interfacing with the device. The system verbally communicates the reminders so no reading is required. The advantages from using these systems are clear and users showed improvement in their PM performance. However, for some older adults, technology is still intimidating and work with it may generate apprehension. Future generations would benefit more from high quality technology [3, 84, 92, 113].

5. General conclusion and future directions

This chapter reviewed the literature regarding assessment and treatment of PM impairment in TBI patients. Dysfunctions in PM are often observed in TBI patients, they can be very frustrating and limit the independence of TBI patients, causing them to rely on caregivers for prompting. Moreover, these failures may affect their chance to return to work or start new activities after injury [2, 17, 18]. The PM impairment in TBI patients may be due to underling dysfunctions in working memory, inhibition, and executive functions [16, 22, 25, 28, 29].

Evaluation of PM performance in clinical setting was initially assessed with questionnaires, the most known are the PMQ [20], CAPM [54] and PRMQ [59]. Questionnaires have the advantage to be easly administrated, but they are unreliable and non-objective measures of PM dysfunction [47]. Despite some previous studies that have provided good reliability and validity of questionnaires [3, 56, 60, 61], the results are mixed when self-rated scores are compared with laboratory-based measures on PM performance. The main issue concerning the employment of questionnaires to evaluate PM performance is the impairment in self-awareness that is often observed in TBI patients [22, 47] that might compromise the validity their responses to the questions.

Laboratory based tests to investigate PM performance have also been developed. The most popular are the RBMT [62], CAMPROMPT [45], and MIST [58]. More recently, to overcome the limitation of a few numbers of trials observed in the previous tests, two new testes have been developed: Virtual Week [71] and RPA-ProMem [69]. These tasks have the advantage of assessing PM performance under more controlled conditions. Nevertheless, they often lack ecological validity. More studies should be dedicated to develop reliable and ecologically valid measures to assess PM performance in clinical settings. The evolution of new technologies might help in this direction (i.e., virtual reality) to bridge the gap between performance in experimental controlled settings and real life situations. Researchers also need to develop PM test according to PM models [4, 32, 91] and ensure measurement of both event-based and time-based activities.

In terms of PM rehabilitation, two main approaches can be followed: remedial and compensatory. Remedial strategies are dedicated to restore the underlying cognitive functions with the intent to restore the damage neural networks or to create new networks. Compensatory strategies are implemented to modify the environment to overcome patients' limitations [3, 84, 92]. Most patients report the use of calendars and diaries to keep track of their duties, but these tools are limited as they require the patient to initiate using them. Electronic memory devices have the advantage to attract attention and display messages that indicate the action that has to be conducted. The most popular and commercialized electronic memory devices are the NeuroPage [98] and MEMOS [102], but many more devices are available, and some have the option to record messages as Voice organizer, [108] and add birthdays, addresses, or photos [105, 106, 109]. However, the problem with electronic devices may be that most patients are unable to learn how to properly operate them. This may be of less relevance in upcoming years when there are more individuals who are familiar with the use of such technological devices.

Aside from technology, however, the most relevant aspect seems to be whether or not patients are aware of their deficits, and therefore willing to use any kind of compensatory strategy. In any case, training with memory aid needs to be integrated into a more general neuropsychological treatment. Before administering any kind of treatment, the therapist needs to analyze what the patient is capable of accomplishing in everyday life. Therapy has to reflect the fact that acceptance of any device is highly related to awareness and acceptance of impairment.. In addition, external memory aids may be combined with other compensatory strategies such as changes in work behaviors and management of work breaks. Finally, caregivers need to work together with patients and and the treatment team to identify the more suitable memory aids and training.

Appendix A

Description of neuropsychological tests and questionnaires included in the studies.

	Full name	Characteristics
NEUROPSYCHOLOGICAL TESTS		
	Executive functions	

WCST	Wisconsin Card Sorting Test	The materials included four stimulus cards and two identical decks of 64 fixed sequence response cards. Participants were required to place the card according to a criterion and they were notified only whether the responses were right or wrong. Once 10 consecutive correct matches are made, the sorting criterion is changed, without warning.
COWAT	The Controlled Oral Word Association Test	It is measures rapid strategy formation, organization, cognitive flexibility and generation of words given a phonetic (letter) cue (C, F, L or F, A, S, depending on the study).
VERBAL FLUENCY		The test includes phonemic and semantic sub-tests. In the phonemic sub-test participants are required to produce word starting with a certain letter (as in the COWAT). In the semantic sub-test participants are required to produce word part of a category (animal, cities and color).
TOL	Tower of London	Participants are presented with a standard array of four colored disks (white, yellow, blue and black) on three pegs of varying lengths and asked to achieve a given arrangement in a stated number of moves by shifting the disks one at a time from one peg to another.
Attention		
TMT (Part A and B)	Trail Making Test	Required participants to rapidly draw lines between 25 consecutively numbered circles or between alternating numbers and letters.
Stroop		Participants were instructed to identify the color of the central word. Two possible conditions (congruent and incongruent) are included. In the congruent condition, the color of the central word corresponded with the written word; whereas, in the incongruent condition, the written word appeared in a different font color.
RUFF 2 & 7		Participants are asked to cross out the digits 2 and 7 in two conditions. In one condition, the 2s and 7s are randomly interspersed among other digits, and in the second condition the distractors are made up of alphabetical letters.
Memory		
LNST	Letter-Number Sequencing Test	It involves the auditory presentation of lists of letters and numbers. Participants are then required to repeat the numbers first in ascending order, and then the letters in alphabetical order.
Digit Span	Digit Span (Backward and forward from Wechsler Memory Scale)	Participants were required to immediately recall strings of numbers that increased in length (maximum score 16).

Word List	Word Lists sub-tests from the Wechsler Memory Scale	
LM	Logical Memory sub-test from Wechsler Memory Scale	The task includes immediate (LM-I) and 30-min delayed (LM-II) prose recall.
N-Back		Common bi-syllabic word stimuli are presented sequentially. Participants were instructed to press a designed keys to indicate when they recognized a word that was the same as one of the N words back.
HVLT	*The Hopkins Verbal Learning Test*	The test includes three learning trials of 12 orally presented words. A yes/no recognition task was administered immediately following learning trial.
Verbal declarative memory		
Story sub-test of RBMT		Recall a short passage, both immediately and after a delay of approximately 15 minutes.
RAVLT	Rey Auditory Verbal Learning Test	Learning and recalling lists of words both immediately and after a delay.
Visual declarative memory		
Pictures, faces and routes sub-tests of RBMT		Participants have to remember a series of 10 line drawings of common objects and to identify them from a larger group of 20 drawings; remember five photographs, which they had to identify from a set of 10 photographs, finally, retracing a five-part pathway around the room.
QUESTIONNAIRES		
DASS	The Depression, Anxiety, Stress Scales	Three self-report scales designed to measure the negative emotional states of depression, anxiety and stress.
DEX	The Dysexecutive Questionnaire from Behavioral assessment of the dysexecutive syndrome (BADS)	20-item questionnaire focusing on the symptoms of dysexecutive syndrome.
BDI	Beck Depression Inventory	Self-report questionnaire in which each item consists of four statements indicating different levels of severity of a particular symptom experienced over the past week. Scores for all items are summed to yield a single depression score.
PREMORBID INTELLIGENCE		

http://dx.doi.org/10.5772/57307

NART	National Adult Reading Test	The test requires subjects to read out loud a set of 50 words which are irregular in terms of their grapheme–phoneme correspondences. The responses are individually scored as correct or incorrect, according to their pronunciation. This score can then be used to derive a premorbid IQ estimate.

Author details

Giovanna Mioni[1,2*], Shawn M. McClintock[3] and Franca Stablum[2]

*Address all correspondence to: mioni.giovanna@gmail.com

1 École de Psychologie, Université Laval, QC, Canada

2 Department of General Psychology, Padova, Italy

3 Neurocognitive Research Laboratory, Division of Brain Stimulation and Neurophysiology, Department of Psychiatry and Behavioral Sciences, Duke University School of Medicine, Department of Psychiatry, UT Southwestern Medical Center, USA

References

[1] Ellis JA. Prospective Memory or the Realization of Delayed Intentions: A Conceptual Framework for Research. In: Brandimonte M, Einstein GO, McDaniel MA (eds.) Prospective memory: Theory and Applications. Mahwah, NJ: Lawrence Erlbaum; 1996. p1–22.

[2] Kliegel M, McDaniel MA, Einstein GO. (eds). Prospective Memory: Cognitive, Neuroscience, Developmental, and Applied Perspective. Mahwah, NJ: Lawrence Erlbaum; 2008.

[3] Shum D, Levin H, Chan RCK. Prospective Memory in Patients with Closed Head Injury: A Review. Neuropsychologia 2011; 49 2156-2165.

[4] McDaniel MA, Einstein GO. Strategic and Automatic Processes in Prospective Memory Retrieval: A Multiprocess Framework. Applied Cognitive Psychology 2000; 14 127-144.

[5] Kliegel M, Martin M, McDaniel MA, Einstein GO. Complex Prospective Memory and Executive Control of Working Memory: A Process Model. Psychologische Beiträge 2002; 44 303–318.

[6] Ellis JA, Freeman JE. Ten Years of Realizing Delayed Intentions. In: Kliegel M, McDaniel MA, Einstein GO. (eds.) Prospective Memory: Cognitive, Neuroscience, Developmental, and Applied Perspective. Mahwah, NJ: Lawrence Erlbaum; 2008. p1-22.

[7] Henry JD, Phillips LH, Crawford JR, Kliegel M, Theodorou G, Summers, F. Traumatic Brain Injury and Prospective Memory: Influence of Task Complexity. Journal of Clinical and Experimental Neuropsychology 2007; 1 1-10.

[8] Meacham JA, Leiman B. Remembering to Perform Future Actions. In Neisser U. (ed.) Memory Observed: Remembering in Natural Contexts. San Francisco: W. H. Freeman; 1982. p327-336.

[9] Einstein GO, Holland LJ, McDaniel MA, Guynn MJ. Age-Related Deficits in Prospective Memory: The Influence of Task Complexity. Psychology and Aging 1992; 7 471-478.

[10] Guynn MJ, McDaniel MA, Einstein GO. Prospective Memory: When Memories Fail. Memory and Cognition 1998; 26 287-298.

[11] Carlesimo GA, Formisano R, Bivona U, Barba L, Caltagirone C. Prospective Memory in Patients with Severe Closed-Head Injury: Role of Concurrent Activity and Encoding Instructions. Behavioural Neurology 2004; 22 101-110.

[12] Einstein GO, McDaniel MA. Normal Aging and Prospective Memory. Journal of Experimental Psychology: Learning, Memory and Cognition 1990; 16 717-726.

[13] McDaniel MA, Guynn MJ, Glisky EL, Rubin SR, Routhieaux BC. Prospective Memory: A Neuropsychological Study. Neuropsychology 1999; 13 103-110.

[14] McFarland CP, Glisky EI. Frontal Lobe Involvement in a Task of Time-based Prospective Memory. Neuropsychologia 2009; 47 1660-1669.

[15] Stuss DT. Traumatic Brain Injury: Relation to Executive Dysfunction and the Frontal Lobes. Current Opinion in Neurology 2011; 24, 584–589.

[16] Mioni G, Stablum F, McClintock SM, Cantagallo A. Time-based Prospective Memory in Severe Traumatic Brain Injury Patients: The Involvement of Executive Functions and Time Perception. Journal of the International Neuropsychological Society 2012; 18, 697–705.

[17] Fleming JM, Shum D, Strong J, Lighthouse S. Prospective Memory Rehabilitation for Adults with Traumatic Brain Injury: A Compensatory Training Program. Brain Injury 2005; 19 1-13.

[18] Fleming JM, Riley L, Gill H, Gullo MJ, Strong J, Shum D. Predictors of Prospective Memory in Adults with Traumatic Brain Injury. Journal of the International Neuropsychological Society 2008; 14 823-831.

[19] Kliegel M, Eschen A, Thöne-Otto AIT. Planning and Realization of Complex Intentions in Traumatic Brain Injury and Normal Aging. Brain and Cognition 2004; 56 43–54.

[20] Hannon R, Adams P, Harrington S, Fries-Dias C, Gibson MT. Effects of Brain Injury and Age on Prospective Memory Self-rating and Performance. Rehabilitation Psychology 1995; 40 289-297.

[21] Bisiacchi PS. The Neuropsychological Approach in the Study of Prospective Memory. In: Brandimonte M, Einstein GO, McDaniel MA. (eds.) Prospective Memory: Theory and Applications. Mahwah, NJ: Lawrence Erlbaum; 1996. p297–317.

[22] Kinsella GJ, Murtagh D, Landry A, Homfray K, Hammond M, O'Beirne L, Dwyer L, Lamont M, Ponsford, J. Everyday Memory Following Traumatic Brain Injury. Brain Injury 1996; 10 499-507.

[23] McDaniel MA, Einstein GO. Prospective Memory: An Overview and Synthesis of an Emerging Field. Thousand Oaks, CA: Sage; 2007.

[24] Tay SY, Ang BT, Lau XY, Meyyappan A, Collinson SL. Chronic Impairment of Prospective Memory After Mild Traumatic Brain Injury. Journal of Head Trauma 2010; 27 77–83.

[25] Shum D, Valentine M, Cutmore T. Performance of Individuals with Severe Long-term Traumatic Brain Injury on Time-, Event-, and Activity-based Prospective Memory Tasks. Journal of Clinical and Experimental Neuropsychology 1999; 21 49-58.

[26] Kinch J, McDonald S. Traumatic Brain Injury and Prospective Memory: An Examination of the Influence of Executive Functioning and Retrospective Memory. Brain Impairment 2001; 2 119-130.

[27] Mathias JL, Mansfield KM. Prospective and Declarative Memory Problems Following Moderate and Severe Traumatic Brain Injury. Brain Injury 2005; 4, 271–282.

[28] Maujean A, Shum D, McQueen R. Effect of Cognitive Demand on Prospective Memory in Individuals with Traumatic Brain Injury. Brain Impairment 2003; 4 135-145.

[29] Schmitter-Edgecombe M, Wright MJ. Event-based Prospective Memory Following Severe Closed-head Injury. Neuropsychology 2004; 18 353–361.

[30] Spreen O, Strauss E. A Compendium of Neuropsychological Tests (2nd ed.). New York: Oxford University Press; 1998.

[31] Maater CA, Sohlberg MM, Crinean J. Focus on Clinical Research: Perceptions of Memory Function in Individuals with Closed-head Injury. Journal of Head Trauma Rehabilitation 1987; 2 74-84.

[32] Kliegel M, McDaniel MA, Einstein GO. Plan Formation, Retention, and Execution in Prospective Memory: A New Approach and Age-related Effects. Memory and Cognition 2000; 28 1041-1049.

[33] Knight RG, Titov N. Use of Virtual Reality Tasks to Assess Prospective Memory: Applicability and Evidence. Brain Impairment 2009; 10 3–13.

[34] Burgess PW, Alderman N, Forbes C, Costello A, Coates LM, Dawson DR, Anderson ND, Gilbert SJ, Dumotheil I, Channon, S. The Case for the Development and Use of 'Ecologically Valid' Measures of Executive Function in Experimental and Clinical Neuropsychology. Journal of the International Neuropsychological Society 2006; 12 194–209.

[35] Titov N, Knight RG. A Video-based Procedure for the Assessment of Prospective Memory. Applied Cognitive Psychology 2001; 15 61-83.

[36] Titov N, Knight RG. A Procedure for Testing Prospective Remembering in Persons with Neurological Impairment. Brain Injury 2000; 14 877-886.

[37] Titov N, Knight RG. A Computer-based Procedure for Assessing Functional Cognitive Skills in Patients with Neurological Injury: The Virtual Street. Brain Injury 2005; 19 315-322.

[38] Knight RG, Harnett M, Titov N. The Effects of Traumatic Brain Injury on the Predicted and Actual Performance of a Test of Prospective Remembering. Brain Injury 2005; 19 27–38.

[39] Knight RG, Titov N, Crawford M. The Effects of Distraction on Prospective Memory Remembering Following Traumatic Brain Injury Assessed in a Simulated Naturalistic Environment. Journal of the International Neuropsychology Society 2006; 12 8-16.

[40] Kinsella GJ, Ong B, Tucker J. Traumatic Brain Injury and Prospective Memory in a Virtual Shopping Trip Task: Does it Matter who Generates the Prospective Memory Target? Brain Impairment 2009; 10 45–51.

[41] Ellis JA, & Kvavilashvili, L. Prospective memory in 2000: Past, Present and Future Directions. Applied Cognitive Psychology 2001; 14 1-9.

[42] Bertsch S, Pesta BJ, Wiscott R, McDaniel MA. The Generation Effect: A Meta-analytic Review. Memory & Cognition 2007; 35 201-210.

[43] Rendell PG, Craik FIM. Virtual and Actual week: Age-related Differences in Prospective Memory. Applied Cognitive Psychology. Special Issue: New Perspective in Prospective Memory 2000; 14 S43-S62.

[44] Mioni G, Rendell PG, Henry JD, Cantagallo A, Stablum F. An Investigation of Prospective Memory Functions in People with Traumatic Brain Injury Using Virtual Week. Journal of Clinical and Experimental Neuropsychology (in press).

[45] Fleming JM, Kennedy S, Fisher R, Gill H, Gullo M, Shum D. Validity of the Comprehensive Assessment of Prospective Memory (CAPM) for Use with Adults with Traumatic Brain Injury. Brain Impairment 2009; 1, 34-44.

http://dx.doi.org/10.5772/57307

[46] Fleming JM, Doig E, Katz N. Beyond Dressing and Driving: Using Occupation to Facilitate Community Integration in Neurorehabilitation. Brain Impairment 2000; 1 141-150.

[47] Fleming JM, Strong J, Ashton R. Self-awareness of Deficits in Adults with Traumatic Brain Injury: How to Best Measure? Brain Injury 1996; 10 1-15.

[48] Gilewski MJ, Zelinski EM. Memory Functioning Questionnaire. Psychopharmacology Bulletin 1988; 24 665-670.

[49] Hermann DJ, Neisser U. An Inventory of Everyday Memory Experience. In: Gruneberg MM, Morris PE, Sykes NR (eds.) Practical Aspects of Memory. New York: Academic Press; 1978.

[50] Broadbent DE, Cooper PF, FitzGerald P, Parkes KR. The Cognitive Failures Questionnaire and its Correlates. British Journal of Clinical Psychology 1982; 21 1-16.

[51] Sunderland A, Harris JE, Gleave J. Memory Failures in Everyday Life Following Severe Head Injury. Journal of Clinical Neuropsychology 1984; 6 127-142.

[52] Bennett-Levy J, Powell GE. The Subjective Memory Questionnaire: An Investigation into the Self-reporting of 'Real Life' Memory Skills. British Journal of Social and Clinical Psychology 1980; 19 177-188.

[53] Buchnan T, Ali T, Hefferman TM, Ling J, Parrott AC, Rodgers J, Scoley AB. Nonequivalence of On-line and Paper and-pencil Psychological Tests: The Case of the Prospective Memory Questionnaire. Behavior Research Methods 2005; 37 148-154.

[54] Roche NL, Fleming JM, Shum D. Self-awareness of Prospective Memory Failure in Adults with Traumatic Brain Injury. Brain Injury 2002; 16 931-945.

[55] Chau L, Lee J, Fleming J, Roche N, & Shum D. Reliability and Normative Data for the Comprehensive Assessment of Prospective Memory (CAPM). Neuropsychological Rehabilitation 2007; 17 707-722.

[56] Roche NL, Moody A, Szabo K, Fleming JM, Shum D. Prospective Memory in Adults with Traumatic Brain Injury: An Analysis of Perceived Reasons for Remembering and Forgetting. Neuropsychological rehabilitation 2007; 17 314-334.

[57] Wilson BA, Emslie H, Foley J, Shiel A, Watson P, Hawkins K, Groot Y, Evans J. The Cambridge Prospective Memory Test, London: Harcourt; 2005.

[58] Raskin SA. Memory for Intentions Screening Test: Psychometric Properties and Clinical Evidence. Brain Impairment 2009; 1 23-33.

[59] Smith G, Della Sala S, Logie RH, Maylor EA. Prospective and Retrospective Memory in Normal Ageing and Dementia: A Questionnaire Study. Memory 2000; 8 311–321.

[60] Kliegel M, Jäger T. Can the Prospective and Retrospective Memory Questionnaire (PRMQ) Predict Actual Prospective Memory Performance? Current Psychology: Developmental, Learning, Personality and Social Fall 2006; 25 182-191.

[61] Shum D, Fleming J, Neulinger K. Prospective Memory and Traumatic Brain Injury: A Review. Brain Impairment 2002; 3 1-16.

[62] Wilson BA, Cockburn JM, Baddeley AD. The Rivermead Behavioural Memory Test. Bury St Edmunds: Thames Valley Test Co; 1985.

[63] Groot YCT, Wilson BA, Evans J, Watson P. Prospective Memory Functioning in People with and without Brain Injury. Journal of the International Neuropsychological Society 2002; 8 645–654.

[64] Mills V, Kixmiller JS, Gillespie A, Allard J, Flynn E, Bowman A, Brawn CM. The Correspondence between the Rivermead Behavioural Memory Test and Ecological Prospective Memory. Brain and Cognition 1997; 35 322-325.

[65] Makatura TJ, Lam C, Leahy BJ, Castillo M, Kalpakjian C. Standardized Memory Tests and the Appraisal of Everyday Memory. Brain Injury 1999; 13 355-367.

[66] Wilson BA, Clare L, Baddeley AD, Cockburn J, Watson P, Tate R. The Rivermead Behavioural Memory Test Extended Version (RBMT-E). Bury St Edmunds: Thames Valley Test Company; 1999.

[67] Wills P, Clare L, Shiel A, Wilson BA. Assessing Subtle Memory Impairments in the Everyday Memory Performance of Brain Injured People: Exploring the Potential of the Extended Rivermead Behavioural Memory Test. Brain Injury 2000; 14 693-704.

[68] Delprado J, Kinsella G, Ong B, Pike K, Ames D, Storey E, Saling M, Clare L, Mullaly E, Rand E. Clinical Measures of Prospective Memory in Amnesic Mild Cognitive Impairment. Journal of the International Neuropsychological Society; 2012; 18 295-304.

[69] Radford KA, Lah S, Say M, & Miller LA. Validation of a New Measure of Prospective Memory: The Royal Prince Alfred Prospective Memory Test. The Clinical Neuropsychology 2011; 25 127-140.

[70] Woods SP, Moran LM, Dawson MS, Carey CL, Grant I, The HIV Neurobehavioral Research Center (HNRC) Group. Psychometric characteristics of The Memory for Intentions Screening Test. Clinical Neuropsychology 2008; 22 864-878.

[71] Rendell PG, Henry JD. A Review of Virtual Week for Prospective Memory Assessment: Clinical Implications. Brain Impairment 2009; 10 14-22.

[72] Rose NS, Rendell PG, McDaniel MA, Abele I, Kliegel M. Age and Individual Differences in Prospective Memory During a "Virtual Week": The Role of Working Memory, Task Regularity and Cue Focality. Psychology and Aging 2010; 25 595-605.

[73] Henry JD, Rendell PG, Kliegel M, Altgassen M. Prospective Memory in Schizophrenia: Primary or Secondary Impairment? Schizophrenia Research 2007; 95 179–185.

http://dx.doi.org/10.5772/57307

[74] Mioni G, Stablum F, Rendell PG, Gamberini L, Bisiacchi PS. Test-retest Reliability of Virtual Week: A Task to Investigate Prospective Memory (in preparation).

[75] Sohlberg MM, White O, Evans E, Mateer C. Background and Initial Case Studies into the Effects of Prospective Memory Training. Brain Injury 1992a; 6 129-138.

[76] Sohlberg MM, White O, Evans E, Mateer C. An Investigation of the Effects of Prospective Memory Training. Brain Injury 1992; 6 139-154.

[77] Raskin SA, Sohlberg MM. The Efficacy of Prospective Memory Training in Two Adults with Brain Injury. Journal of Head Trauma Rehabilitation 1996; 11 32-51.

[78] Hannon R, Feliciano L, Messner A. Prospective Memory Retraining in Adults with Traumatic Brain Injury. Archives of Clinical Neuropsychology (abstract).

[79] Kixmiller JS. Evaluation of Prospective Memory Training for Individuals with Mild Alzheimer's Disease. Brain and Cognition 2002; 49 237–241.

[80] Chasteen AL, Park DC, Schwarz N. Implementation Intentions and Facilitation of Prospective Memory. Psychological Science 2001, 12, 457–461.

[81] Kaschel R, Della Salla S, Cantagallo A., Fahlböck A, Laaksonen R, Kazen M. Imagery Mnemonics for the Rehabilitation of Memory: A Randomised Group Controlled Trial. Neuropsychological Rehabilitation 2002; 12 127-153.

[82] Potvin MJ, Rouleau I, Sénéchal G, Giguère JF. Prospective Memory Rehabilitation Based on Visual Imagery Techniques. Neuropsychological Rehabilitation 2011; 21 899-924.

[83] Potvin MJ, Rouleau I, Audy J, Charbonneau S, Giguère, JF. Ecological Prospective Memory Assessment in Patients with Traumatic Brain Injury. Brain Injury 2011; 25 192-205.

[84] Fish J, Wilson BA, Manly T. The Assessment and Rehabilitation of Prospective Memory Problems in People with Neurological Disorders: A review. Neuropsychological rehabilitation 2010; 20 161-179.

[85] von Cramon, DY, Matthes-von Cramon GM, Mai N. Problem-solving Deficits in Brain Injured Patients: A Therapeutic Approach. Neuropsychological Rehabilitation 1991; 1 45-64.

[86] Hutchinson J, Marquardt TP. Functional Treatment Approaches to Memory Impairment Following Brain Injury. Topics in Language Disorders 1997; 18 45-57.

[87] Raymond MJ, Malia KB, Bewick KC, Bennett TL. A Comprehensive Approach to Memory Rehabilitation Following Brain Injury. The Journal of Cognitive Rehabilitation 1996; 14 18-23.

[88] Levine B, Robertson IH, Clare L, Carter G, Hong J, Wilson BA, et al. Rehabilitation of Executive Functioning: An Experimental Clinical Validation of Goal Management Training. Journal of the International Neuropsychological Society 2000; 6 299-312.

[89] Levine B, Schweizer T, O'Connor C, Turner G, Gillingham S, Stuss DT, Manly T, Robertson IH. Rehabilitation of Executive Functioning in Patients with Frontal Lobe Damage with Goal Management Training. Frontiers in Human Neuroscience 2011; 5-9.

[90] Bellezza FS. Mnemonic Devices: Classification, Characteristics, and Criteria. Review of Educational Research Summer 1981; 51 247-275.

[91] Smith RE, Persyn D, Butler P. Prospective Memory, Personality, and Working Memory: A Formal Modeling Approach. Zeischrift fur Psychologie 2011; 219 108-116.

[92] Thöne-Otto AIT, Walther K. Assessment and Treatment of Prospective Memory Disorders in Clinical Practice. In: Kliegel M, McDaniel MA, Einstein GO. (eds.) Prospective Memory: Cognitive, Neuroscience, Developmental, and Applied Perspective. Mahwah, NJ: Lawrence Erlbaum; 2008. p321-345.

[93] Sohlberg MM, Mateer CA. Training Use of Compensatory Memory Books: A Three Stage Behavioural Approach. Journal of Clinical and Experimental Neuropsychology 1989; 11 871–91.

[94] Schmitter-Edgecombe M, Fahy JF, Whelan JP, Long CJ. Memory Remediation After Severe Closed Head Injury: Notebook Training Versus Supportive Therapy. Journal of Consulting Clinical Psychology 1995; 484-489.

[95] Lynch WJ. You must Remember this: Assistive Devices for Memory Impairment. Journal of Head Trauma Rehabilitation 1995; 10 94–97.

[96] Burke JM. Danick JA, Bemis B, Durgin CJ. A Process Approach to Memory Book Training for Neurological Patients. Brain Injury 1994; 8 71-81.

[97] Ownsworth TL. McFarland K. Memory Remediation in Long-term Acquired Brain Injury: Two Approaches in Diary Training. Brain Injury 1999; 13 605–626.

[98] Wilson BA, Evans JJ., Emslie H et al. Evaluation of NeuroPage: A New Memory Aid. Journal of Neurology, Neurosurgery and Psychiatry 1997; 113-115.

[99] Wilson BA, Emslie HC, Quirk K, Evans JJ. Reducing Everyday Memory and Planning Problems by Means of a Paging System: A Randomised Control Crossover Study. Journal of Neurology, Neurosurgery, and Psychiatry 2001; 70 477-482.

[100] Evans JJ, Emslie H, Wilson BA. External Cueing System in the Rehabilitation of Executive Impairments of Action. Journal of the International Neuropsychology Society 1998; 4 399-408.

[101] Fish J, Manly T, Emslie H, Evans JJ, Wilson BA. Compensatory Strategies for Acquired Disorders of Memory and Planning: Differential Effects of a Paging System

for Patients with Brain Injury of Traumatic versus Cerebrovascular Aetiology. Journal of Neurology, Neurosurgery, and Psychiatry 2008; 79 930-935.

[102] Walther K, Schulze H, Thöne-Otto A. An Interactive Memory Aid Designed for Patients with Head Injury: Comparing MEMOS with Two Commercially Available Electronic Memory Aids. Poster at First Congress of the European Neuropsychological Societies in Modena, Italy, April 2004.

[103] Walther K, Thöne-Otto AIT. Long-term Usage of the Interactive Memory Aid System MEMOS. Poster presented at the Second International Conference on Prospective Memory, Zurich, Switzerland. Retrieved August 23, 2005.

[104] Thöne-Otto AIT, Walther K. Evaluation of MEMOS, an Interactive Memory Aid System for Brain Injured Patients: Personal Memory Assistant vs. Palm Organizer. Poster presented at the Second International Conference on Prospective Memory, Zurich, Switzerland. Retrieved August 23, 2005.

[105] Morrison K, Szymkowiak A, Gregor P. "Memojog–An Interactive Memory Aid Incorporating Mobile Based Technologies." Mobile Human-Computer Interaction-MobileHCI 2004. Springer Berlin Heidelberg, 2004. 481-485.

[106] Szymkowiak A, Morrison K, Gregor P, Shah P, Evans JJ, Wilson BA. A Memory Aid with Remote Communication Using Distributed Technology. Personal and Ubiquitous Computing 2005; 9 1-5.

[107] van den Broek MD, Downes J, Jhonson Z, Dayus B, Hilton N. Evaluation of an Electronic Memory Aid in the Neuropsychological Rehabilitation of Prospective Memory Deficits. Brain Injury 2000; 14 455-462.

[108] Yasuda K, Misu T, Beckman B, Watanabe O, Ozawa Y, Nakamura T. Use of an IC Recorder as a Voice Output Memory Aid for Patients with Prospective Memory Impairment. Neuropsychological Rehabilitation 2002; 12 155-166.

[109] Kim HJ, Burke DT, Dowds MM, Boone KA, Park GJ. Electronic Memory Aids for Outpatient Brain Injury. Brain Injury 2000; 14 187-196.

[110] Thöne-Otto AIT, Walther K. How to Design an Electronic Memory Aid for Brain Injured Patients: Considerations on the Basis of a Model of Prospective Memory. International Journal of Psychology 2003; 38 236-244.

[111] Pollack ME, Brown L, Colbry D, McCarthy CE, Orosz C, Peintner B, Ramakrishnan S, Tsamardinos I. Autominder: An Intelligent Cognitive Orthotic System for People with Memory Impairment. Robotics and Autonomous Systems 2003; 44 273-282.

[112] Mihailidis A, Fernie GR, Claghorn WL. The Development of a Computerized Cueing Device to Help People with Dementia to be More Independent. Technology and Disability 2000; 13 23-40.

[113] Caprani N, Greaney J, Porter N. A Review of Memory Aid Devices for an Ageing Population. PsychNology Journal 2006; 4 205-243.

[114] Shum D, Sweeper S, Murray R. Performance on Verbal Implicit and Explicit Memory Tasks Following Traumatic Brain Injury. The Journal of Head Trauma Rehabilitation 1996; 11 43-53.

Chapter 19

Approaches to Managing Executive Cognitive Functioning Impairment Following TBI: A Focus on Facilitating Community Participation

Steven Wheeler

Additional information is available at the end of the chapter

http://dx.doi.org/10.5772/57395

1. Introduction

Successful and satisfying community living represents the ultimate outcome of traumatic brain injury (TBI) rehabilitation. Unfortunately, for the vast majority of affected individuals, full participation in pre-injury roles and responsibilities is negatively impacted by a complex mix of cognitive, physical, emotional, psychological, and behavioral impairments. Each individual living with TBI is unique and a reflection of numerous factors including pre-injury characteristics, severity of injury, home and community supports, and related internal and external factors. The TBI literature often depicts life following TBI as unproductive, including the need for varying levels of caregiver assistance and supervision. Long-term recovery following TBI is characterized by reduced independence in home and community activities including employment, parenting, driving, and participation in leisure activities [1]. Additionally, TBI is frequently associated with profound social isolation and decreased life satisfaction.

Problems of community re-entry following TBI carry enormous personal and societal costs. This chapter will focus on rehabilitation approaches for executive cognitive functions, a set of cognitive abilities that control and regulate goal directed behavior. Often considered *higher level* cognitive abilities, executive functions include the ability to initiate, modify, and stop actions, to monitor behaviors and adjust them to appropriately suit a situation, to plan future behavior when faced with unfamiliar situations, and to anticipate outcomes and adapt to changing situations. They include the self-regulation or self-control functions that enable individuals to understand their strengths and limitations, formulate goals, devise ways to achieve them, and effectively implement the plans. They also influence more basic cognitive abilities such as attention and memory and are essential to adult role participation, involving the ability to engage in independent daily activities. One's ability to initiate, plan, set goals, monitor performance, anticipate consequences and respond flexibly and adaptively are all dependent on executive cognitive functions.

Executive functions are commonly impaired following moderate to severe TBI due, at least in part, to damage of the frontal lobes and subcortical limbic system. As "control" functions, executive cognitive functions enable people to engage in controlled behavior beyond that based on instinct, habit, or impulse alone [2], facilitating successful adaptation to unpredictable environments, stressful situations, and challenging tasks. People with poor executive functions commonly have difficulty interacting with others both in terms of general social competence and the ability to inhibit socially inappropriate behaviors. A mature executive system enables people to a) inhibit socially impulsive behavior, b) guide behavior in social contexts by applying learned rules of appropriateness, c) consider other people's perspectives and interests when making decisions, and d) foregoing immediate gratification in the interest of long-term gain. In many respects, the executive system differentiates adult social behavior from that of normally developing young children [2]. Common terms to describe persons with executive impairments that may also characterize the very young or immature child include being concrete, egocentric, socially inappropriate, disorganized, childlike, impulsive, inflexible, and impatient [2]. Executive function impairments are considered among the most disabling of all cognitive impairments following TBI because they pervade virtually all aspects of a person's ability to function in his or her personal or professional life [3]. They are inherently tied to academic, vocational, and social success, roles associated with independence and life satisfaction. Impaired executive functions are sometimes referred to as *executive dysfunction* or *dysexecutive syndrome.*

2. Understanding executive functions within the continuum of rehabilitation following TBI

Executive functions are thought to be heavily influenced by the frontal lobes and subcortical limbic system, which govern most cognitive functions, especially those involving new and unstructured situations. Commonly accepted components of executive functions are listed and defined in table 1.While definitional problems exist, there is little disagreement regarding the impact of executive dysfunction on emotional, behavioral, and social outcomes following TBI. During normal development, many executive functions do not fully develop until late adolescence and early adulthood and represent higher level adult thought, reasoning, and decision making. When damaged, an individual's behavior and social functioning appears more reflective of earlier stages of development. Examples include difficulty delaying gratification, inability to anticipate consequences, verbal and behavioral disinhibition, impaired awareness of and/or concern for social norms, apathy, and indifference. Adult family, social, and vocational roles are dependent upon higher level cognitive skills and, when compromised, have significant social, financial, and personal consequences.

http://dx.doi.org/10.5772/57395

The stages and settings associated with recovery following TBI are summarized in figure 1. The early stages of recovery following moderate to severe TBI generally involve a period of acute care and rehabilitation which are hospital based and include medications, surgery, and therapeutic exercises/activities directly targeted to managing identified symptoms. In the United States, there is a trend towards decreasing length of inpatient stay following TBI. According to a study by Turkstra [4], the average length of stay in inpatient TBI rehabilitation was 18 days. She noted that, compared to previous decades, patients with brain injury are "sicker when they are admitted to inpatient rehabilitation and sicker when they leave" (p. 333). Rehabilitation during the inpatient recovery period tends to be governed by the medical model, which emphasizes the patient's dependency on health care professionals who act in the interest of the patient in an effort to preserve and minimize the severity of impairments. While this approach is recognized as a necessary component of TBI treatment, it has been argued that by fostering the 'rehabilitation patient' role, the inpatient setting is poorly suited to address executive cognitive functions and facilitating the transition to important life roles.

Term	Definition
Self-awareness	The ability to accurately recognize one's abilities and limitations. It is influenced by many factors including memory, decreased sensory or perceptual abilities, impulsivity, and inability to plan for the future. Deficits include displaying very limited or no apparent appreciation with regard to the nature of difficulties or impairments and the impact of these on everyday functioning. They impact ability to acceptance of feedback from others related to limitations, affecting the ability to benefit from rehabilitation [5]. Impaired self-awareness impacts performance of social interactions, including difficulties with personal boundaries, appreciating their impact on others, managing arguments, anger management, and adjusting to unexpected changes [6].
Goal Setting	The ability to determine what one wants and to foresee the future realization of those needs. Goal attainment involves estimating task difficulty and self-evaluating performance so that adjustments can be made, both of which are impacted by impaired self-awareness of deficits or abilities.
Initiation	The ability to start an activity. Deficits in this area include lacking spontaneity, being slow to respond, or lacking initiative. The individual may verbalize an intent to act but fail to follow through [7]. Deficits may be misinterpreted as lack of motivation or drive and can be a considerable source of frustration for both client and family.
Inhibition	Deficits include problems stopping – demonstrating disinhibition, impulsivity, and quick shifts in behavior and emotional tone [7].
Planning	The ability to organize the steps to complete an action, weigh alternatives, develop a framework for carrying out the plan, prepare for setbacks in carrying out the plan, assemble the needed materials and possess the skills necessary to carry out the plan [6]. Deficits impair the ability to perform adult home, academic, or vocational roles.
Problem solving / Decision making	Requires integration of several cognitive skills, including attention, initiation, impulse control, organization, categorization, mental flexibility, reasoning skills, and the ability to self-evaluate. Deficits impact ability to select among various options to make a decision [8].

Mental flexibility	The ability to initiate, stop, and switch actions depending upon feedback from the environment during goal oriented task performance [8]. Deficits result in rigid, inflexible, or perseverative behavior, seemingly becoming fixed to a behavioral set that is no longer productive or appropriate [7]. Deficits result in behavior that appears perseverative and concrete with limited ability to generalize current information for future problem solving. Tasks such as meal preparation, driving, financial planning and many IADLs are affected.
Attention	Frontal lobe involvement results in distractibility and poor selective and divided attention.
Self-monitoring / Self correction	The ability to evaluate and regulate the quality and quantity of one's behavior to allow identification and correction of incorrect responses. Deficits impact an individual's ability to manage his or her own learning and apply correct strategies to accomplish goals.
Concept formation / abstraction	The ability to make inferences from information. Deficits contribute to limited imagination, problems generalizing from individual events, failing to plan ahead, and difficulty explain ideas [8]. Information is viewed in a concrete manner, contributing to a rigid approach to thinking [10].
Categorization	The ability to find commonalities among large amounts of information and assign objects and events into groups [8]. Deficits impact all cognitive skills and abilities [11].
Generalization	The ability to use a newly learned strategy in novel situations. Deficits impact ability to learn a skill in one setting and apply it elsewhere.

Table 1. Categories of Executive Cognitive Functions [12].

Willer and Corrigan [13] argue that the longer the individual is hospitalized, the longer it can take to return to pre-injury roles. In this respect, decreasing length of inpatient stay could have potential benefits in the treatment of executive dysfunction if community based models were to receive the financial and professional support required to meet the needs of survivors and caregivers. Such a belief is supported by the extensive accounts of poor adjustment to the community following TBI, leading to discussions about TBI rehabilitation models that may better suited to the goal of community re-integration. In this respect, decreasing length of inpatient could have potential benefits in the treatment of executive dysfunction if community based models are adequately were to receive the financial and professional support required to meet the needs of survivors and caregivers. However, the development, implementation, and support for such models must be considered to be early in terms of their contributions to the TBI continuum of recovery. Gordon and colleagues [14] summarized these historic service delivery problems, along with the need to extend services beyond the medical model of symptom management, as follows:

> *... although individuals with TBI have been described as being sometimes "unaware," this term can just as easily be applied to rehabilitation service providers, insurance companies, and government officials in years past in describing their understanding of the impairments and service needs of individuals with TBI (p.321).*

Unfortunately, this sentiment continues to be expressed today as community services remain inaccessible for many who could benefit from them [4]. Willer and Corrigan [13] postulated that the challenging transition to the community following TBI is best achieved through recognizing individual differences and implementing client centered interventions that involve the individual and the environments within which they participate. Since the time the model was first described, it has garnered considerable support and has served as a foundation for various assessment and treatment approaches designed to facilitate community re-entry. Their *Whatever it Takes* approach is based upon the following principles:

1. No two individuals with acquired brain injury are alike;
2. Skills are more likely to generalize when taught in the environment where they can be used;
3. Environments are easier to change than people;
4. Community integration should be holistic;
5. Life is a place-and-train venture;
6. Natural supports last longer than professionals;
7. Interventions must not do more harm than good;
8. The service system presents many barriers to community integration;
9. Respect for the individual is paramount; and
10. Needs of individuals last a lifetime; so should their resources.

The model has supported the emergence of community based rehabilitation approaches, including various coaching models such life skills training and supported employment, both of which have demonstrated positive rehabilitation outcomes. Through its emphasis on natural environments and holistic recovery, it is also well suited to the assessment and treatment of executive functions.

The progression from acute care to community is generally marked by a reduction in environmental structure. Institutional based rehabilitation tends to be marked by pre-established schedules – wake-up times, meal times, therapy times, and visiting times. In addition, patients tend to be taken to therapy sessions, appointments, meals and activities by staff and rarely participate in menus or the nature of service providers. In such settings, the reduced need for independent goal setting, decision making, and problem-solving limit opportunities to utilize higher level cognitive functions. Consequently, the extent of executive dysfunction tends to manifest later in the rehabilitation continuum, often challenging family members and providers to address previously undetected deficits and behaviors.

3. Executive functions as barriers to participation

The International Classification of Functioning Disability and Health (ICF) (Figure 2) represents an important contribution to the process of understanding and documenting the structural, functional, personal, and social manifestations associated with disability. The

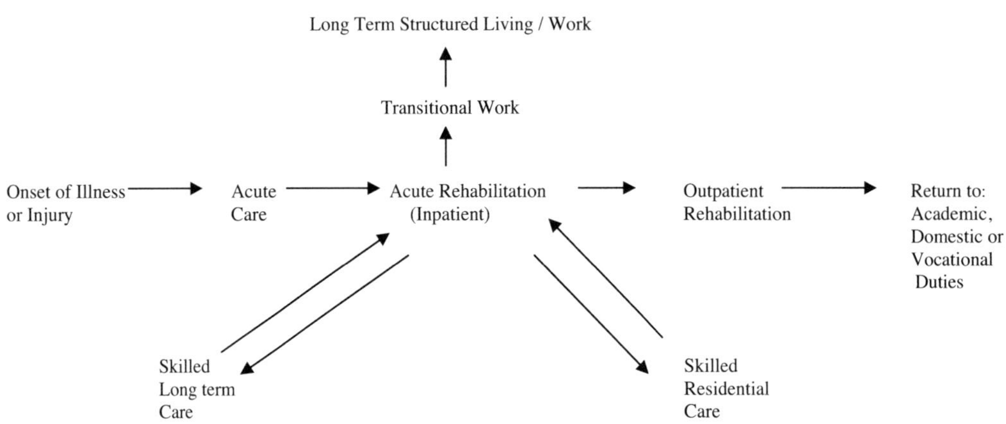

From "Living with Brain Injury: A Guide for Families (2nd Edition)" [15]. Copyright 2001 by Healthsouth Press. Reprinted with permission of the author.

Figure 1. Traumatic Brain Injury Recovery Tree

model is also well suited to conceptualizing the impact of impaired executive functions following TBI and facilitating more effective methods of evaluation and intervention. In the ICF, the construct *body structures and functions (impairments)* addresses the functional and structural integrity of the body systems while considering duration of condition and developmental stage of the individual [16]. *Activities* represent the performance of the person acting within the context of their culture. These two areas are emphasized in the earlier stages of the TBI rehabilitation continuum. The ICF defines an individual's *participation* as:

... the nature and extent of a person's involvement in life situations in relation to Impairment, Activities, Health Conditions and Contextual Factors. Participation may be restricted in nature, duration, and quality, e.g. participation in community activities, obtaining a driver's license ... [17].

As opposed to activities (based upon the ICF), which occur at the level of the individual, participation occurs at the societal level and is assessed by comparing participation in life activity of persons with and without disability within the context of society [18]. As is commonly reported following TBI, a participation reduction involves the social, familial, educational, vocational, or other role disadvantage associated with a disability such as failure in school or loss of a job due to a communication deficit [19]. The ICF serves as a useful guide for research, assessment and intervention of executive cognitive functions, placing the ultimate rehabilitation focus on community participation over impairment remediation. It has also influenced the rapid development of measures looking beyond impairments and activity limitations to include evaluation of the environments by which individuals participate in meaningful roles.

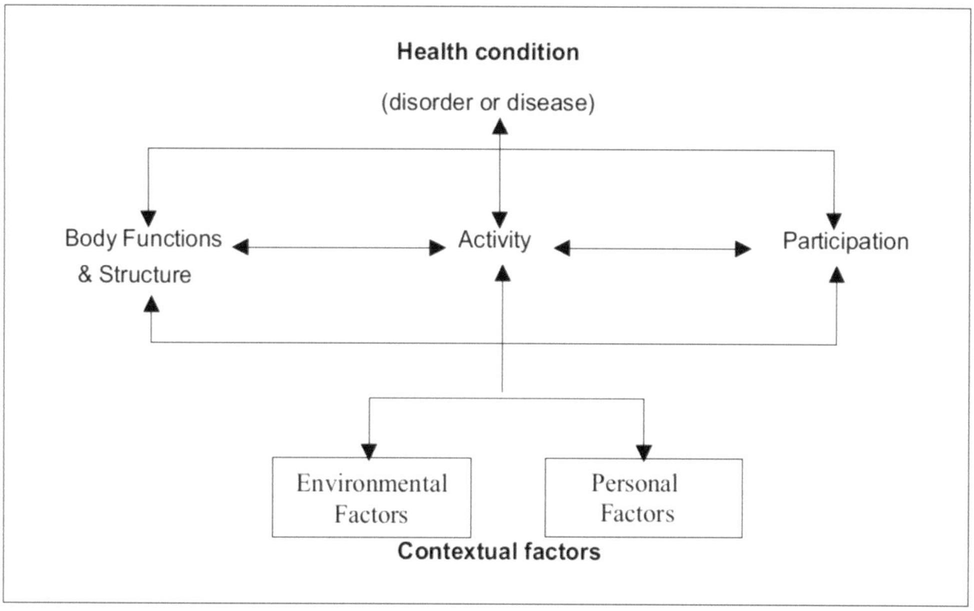

Figure 2. The International Classification of Functioning, Disability,and Health [16]. Copyright 2001 by the World Health Organization. Reprinted with permission of the author.

Executive dysfunction represents a participation level condition involving difficulties performing tasks within the context in which the person lives [16]. Participation focused assessment requires a top-down approach, starting with gathering information on what the individual wants to do, followed by an analysis of the person as he or she performs the desired tasks. In contrast to assessment within the medical model, consideration of the causes of the problem, including client factors is the final stage of the top-down evaluation process [20]. Contextual factors within the environment that contribute to societal participation including features of the physical, social, and attitudinal world together with the attributes of the individual [16]. Evaluation of such factors and considering them within the treatment plan is consistent with the *Whatever it Takes* approach and advocated by many experts in the field of TBI rehabilitation [4,21].

Given the contributions of executive cognitive functions to social competence and adult social participation, it is no surprise that social isolation is among the most reported and profound life changes for persons with TBI. According to Rowlands [22], with the passage of time, the depth of social relationships tends to decrease, taking on a transient quality and contributing to the common complaint of loneliness. Studies looking at marital relationships following TBI have highlighted the coping challenges that spouses face dealing with behaviors such as emotional unpredictability, self-centeredness, and 'child-like' behaviors. Poor communication skills following TBI results in reduced participation in social activities and decreased life satisfaction [23]. Research has also suggested that, two years post-injury, persons with TBI still experience difficulties with social skills, which often lead to social isolation and depression [24].

Attention, memory, learning, and social-emotional impairments coupled with executive function and self-regulation impairments place individuals with TBI at unique risk for failure when attempting to return to work or school [25]. Individuals with TBI often experience difficulty securing and/or returning to competitive employment post-injury and maintaining employment for extended periods of time. Estimates of the employment rate for person with TBI range from 20% to 50% depending on the severity of injury, the prior work experience of the individual, and demographic characteristics (e.g. age, education, and socioeconomic status). The National Longitudinal Transition Study-2 reported that students with TBI whose injuries occurred before the onset of postsecondary education demonstrated significantly lower college graduation rates than their non-disabled peers [26]. If residual cognitive and social impairments remain unidentified by the academic institution, the student may not be offered needed accommodations and supports. These students are at greater risk for frustration, failure, social rejection, and placement in special education settings with students with dissimilar learning and social issues [27]. Students injured late in high school frequently struggle with more problems in college, where the levels of special services and TBI awareness are often even lower [28].

4. A comprehensive model for approaching evaluation and treatment of executive functioning deficits

Executive functions pose both assessment and intervention challenges for the rehabilitation professional. The manifestation of executive dysfunction is frequently context dependent and dynamic, rendering narrow band, static, office based assessment tools poorly suited to accurate measurement of the impact of impairments on community participation [21]. Compounding the issue is the need for rehabilitation professionals, particularly in inpatient settings, to quantify progress to third party payers, contributing to the use of didactic worksheets and activities that reduces the intervention process to a series of checklists of cognitive benchmarks [4]. As discussed previously, executive cognitive functions are more heavily relied upon during unstructured daily social, vocational / educational, and community activities. As such, many have argued for a more real-world approach to evaluation involving observations in the context of everyday activities, away from the acute care and structured hospital environments. To facilitate this means of evaluation, Burgess and colleagues [29] outline the following characteristics of everyday activities:

1. Many tasks
2. Interleaving – Performance on these tasks needs to be dovetailed in order to be time effective.
3. One task at a time – Due to cognitive or physical constraints, only one task can be performed at a time.

http://dx.doi.org/10.5772/57395

4. Interruptions and unexpected outcomes – Unexpected, sometimes high priority, interruptions occur, and things will not always go as planned.

5. Delayed intentions – The time to return to a task that is already in progress is not signaled directly by the situation.

6. Differing task characteristics – Tasks usually differ in terms of priority, difficulty, and the length of time they will occupy.

7. Self-determined targets – People decide for themselves what constitutes adequate performance.

8. No immediate feedback – No minute to minute performance feedback. Failures are not signaled at the time they occur.

Figure 3 illustrates a proposed hierarchical model for the treatment of executive functions. Its organizational structure is influenced by the belief that rehabilitation of executive dysfunction begins with an appreciation, understanding, or *awareness* of deficits by the individual with TBI along with a perceived need for personal change [30]. During recovery from TBI, the convergence of emotional states and memory of abstract mental states facilitate the emergence of awareness. Decreased awareness of one's cognitive, behavioral, physical and emotional impairments is commonly cited after TBI and recognized by researchers and clinicians as one of the greatest obstacles in brain injury rehabilitation [31-32]. Not only does it affect the individual's belief about their impairments, but their belief in their ability to benefit from rehabilitation. Individual's post-TBI who show an accurate perception of themselves, as well as a greater willingness to change, demonstrate a better transition home and to the community than those who inaccurately perceive their limitations [33].

Within the model, rehabilitation to facilitate self-awareness begins with therapeutic relationship building, goal setting, and participation in activities meaningful to the client. Decreased self-awareness should not be viewed as all or none. Various models support the notion of some form of hierarchy or degree of severity that can vary depending on the area of function being assessed and the approaches used to make the assessment [32]. Three levels of awareness have been described – intellectual, emergent, and anticipatory – progressing from having an understanding of the limitations to being able to anticipate that a problem is likely to happen because of some deficit. [33] Through the process of participating in individual, group, and community activities, with formal and informal feedback provided regarding performance, treatment goals are revised to reflect emerging self-awareness. The process continues with instruction in other aspects of executive dysfunction implemented on an ongoing basis. Failures and consequences foster awareness of deficits. Successes build the self-esteem and self-confidence needed to take risks and persevere in the face of challenge. As is the case with all forms and learning, both successes and failures are normal aspects of development and, in the case of TBI rehabilitation, essential to the recovery process. Such a formula may seem simple, but it is not. The following sections expand upon this process, highlighting the challenges facing rehabilitation providers and the effort and commitment required by all members of the treatment team.

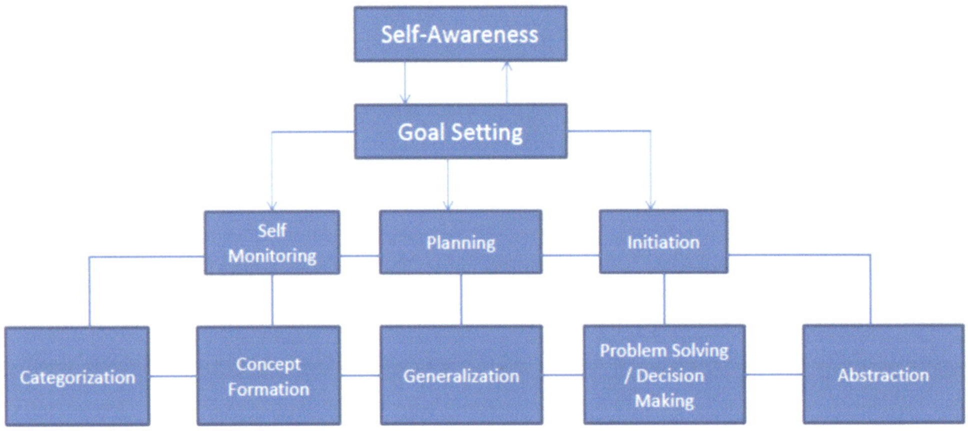

Figure 3. Proposed Model of Rehabilitation for Executive Dysfunction following TBI

4.1. Exploring the relationship between self-awareness, goal setting, and life satisfaction

Goal setting, as it pertains to executive dysfunction following TBI, can be considered both a stage in the rehabilitation process and a specific deficit area associated with impaired executive functions. The two are inter-related in "client centered" goal setting whereby treatment priorities reflect those areas deemed important to the client. Individualized treatment planning, as emphasized by Willer and Corrigan [13], involves active participation by the client in the establishing of treatment goals and priorities. Individuals lacking accurate awareness of deficits, as is common following moderate to severe TBI, are unable to effectively engage in this process, impacting goal identification, motivation, compliance, and outcomes. Trudel, Tryon, and Purdum [34] found a direct relationship between unawareness of impairments and poorer recovery seven years post-injury while Koskinen [35] reported that impaired awareness of deficits was strongly correlated with a decrease in family well-being ten years after TBI.

When impaired self-awareness impedes motivation and interest in rehabilitation, gaining insight into problem areas become a necessary element of recovery. It is generally recognized that life experiences following TBI that are relevant to a person's goals may increase that individual's understanding of injury related deficits and, subsequently, contribute to more realistic expectations about future outcomes. According to Toglia and Kirk [36], self-awareness increases through participation in different types of tasks. During normal development, the frontal lobes and other cortical areas become "wired" by ones learning history. Damage to this cortical area necessitates a need for "re-wiring," a process that occurs through the consistent application of positive and negative consequences. However, adults post TBI are generally ill prepared for needed feedback when lacking awareness of deficits and after being accustomed to pre-injury independence. As such, they are likely to reject feedback from family and to resent authority. Central to the success of rehabilitation of executive cognitive functions is the establishing of a supportive, therapeutic relationship whereby the client has an understanding that they can fail at a task or make mistakes without losing the support of the rehabilitation therapist. Participation in tasks that can expose the severity of an individual's deficits and "restructuring of self-knowledge," [37] can represent a potential threat to self-esteem, personal control, and sense of independence, so coping strategies and emotional reactions must be monitored closely. Clients at this stage of rehabilitation may "test" a therapist's commitment to the therapeutic relationship by sabotaging tasks, or present with defiant, aggressive, or insulting behavior. If the therapist understands such actions and behaviors as a component of the relationship building process, then the client's behaviors can be interpreted as aspects of recovery as opposed to noncompliance or the need for a staffing change. Approaches to fostering the therapeutic relationship include negotiating to achieve a common understanding, validating the client's perspective, fostering self-advocacy, and providing encouragement and feedback [38]. The strength of the relationship is established over time through shared experiences and continued support. The process is rarely a smooth one, often characterized by a *two steps forward, one step backwards* progression involving challenges presented to the client, a response by the client, the provision of feedback, and another response. While thoughts and behaviors of persons with executive dysfunction may at times appear child and adolescent-like, it is critical that the use of a developmental template to understand such behaviors not result in infantilization. Interacting with clients in a disrespectful manner by professionals and family members runs opposite to the needed approach to facilitate positive change, including adult level interactions and social role modeling.

While a necessary element to addressing other elements of executive dysfunction, the emergence of self-awareness marks a vulnerable period of recovery, often characterized by frustration, anger, and emotional distress. It is at this time that the professional skill of the therapist can be a determining factor as to whether the recovery process continues or comes to a halt. In Wheeler's [39] study involving a group of 40 individuals with severe TBI enrolled in a community based residential rehabilitation program, a statistically significant decrease in overall self-reported life satisfaction was found at 90 day follow-up. This decrease occurred despite an increase in community integration as measured by improved independence in home management, social participation, and productive activity. It was hypothesized that participation in intensive individual and group activities, program participants became increasingly dissatisfied with life despite functional improvements that were deemed clinically significant but were well below pre-injury levels of functioning. The development of self-awareness post-TBI is generally considered to be a gradual process that involves comparing performance on functional tasks in a familiar setting with one's pre-morbid functional level [40] and "coming to terms with their new selves" [30]. The early stages of increased self-awareness may involve an appreciation of physical deficits before cognitive, emotional, behavioral, or psychological difficulties. According to study by Trahan, Pepin, and Hopps [41], persons with TBI tend overestimate emotional, cognitive, and behavioral difficulties when compared to clinician ratings. Given the manner in which impaired awareness of deficits can impede treatment goal setting and implementation of specific interventions, efforts to facilitate engage the client in activities to increase self-awareness warrant considerable emphasis during community re-entry following TBI. Unfortunately, this process tends to be complex, unpredictable, and affected by cognitive, emotional, and psychological factors.

It is essential that the clinician appreciate the unique role of decreased life satisfaction to recovery and be prepared to work through this transition period using additional professional supports as needed. It would be easy to view decreased satisfaction as a negative and avoid putting the patient in challenging situations that contribute to frustration, sadness, and ultimately, improved awareness of deficits. However, significant life changes are often fueled by a desire to overcome a dissatisfying situation. In this respect, a patient who lacks awareness of deficits and is generally satisfied with life may pose a significant challenge for the rehabilitation therapist. How do we engage an individual in challenging treatment goals when they don't perceive a problem? The decision to make major life changes and the pursuit of challenging goals, such as career changes, relationship changes, or personal fitness programs are generally sparked by some element of dissatisfaction. In Wheeler's [39] study, the initial 90 day follow-up period characterized by dissatisfaction was followed by a non-significant increase in satisfaction at one year follow-up. Given the importance of life satisfaction as a rehabilitation outcome, it is difficult to consider a decrease in this area to be a positive progression in the treatment progress. However, if viewed as a temporary step in the recovery process, the rehabilitation provider is better suited to managing and supporting the client as well as educating others. Such a finding also helps the clinician educate family, other professionals, or payer sources who may become discouraged by the apparent emotional regression displayed by the client. Recognizing signs of emotional distress is a necessary aspect of this process.

Those with TBI and executive dysfunction are considered at risk for depression, suicidal ideation, and suicidal behavior during the recovery process. For many, the residual deficits associated with TBI are debilitating and persistent. Major depression is one of the most frequently reported behavioral sequelae after TBI. In a multi-center study involving 666 non-acute individuals with moderate to severe TBI, fatigue (29%), distractibility (28%), anger-irritability (28%), and rumination (25%) were identified as the most common depressive symptoms [43]. Theories attempting to explain the nature of the relationship between TBI and depression include pre-injury depression, pre-injury personality type, social integration after injury, family support, neurochemical imbalances, and site of anatomical damage [37, 43-44]. Depression complicates the process of recovery and rehabilitation because it contributes to increased effort in information processing and by creating general apathy towards rehabilitation [43]. Additionally, unemployment, social isolation, and emotional distress, combined with reduced cognitive resources increases the susceptibility of persons with TBI to contemplate suicide and engage in suicidal behavior [45]. The majority of studies on this topic report on the presence of suicidal ideation, estimated to be approximately 18%-23% among persons with TBI [45, 46] compared to approximately 3% in the general population [47]. Simpson and Tate [49] identified four pre-injury risk factors for suicidal ideation for persons with TBI: suicide attempts, alcohol abuse, drug abuse, and emotional / psychiatric disturbance. Of these, only pre-injury alcohol and substance abuse was related to suicidal ideation in a separate study [49]. Identification of specific post-injury risk factors has proven to be illusive to researchers. Research suggests that there is no critical period post-injury during which suicidal ideation is more likely to occur [45] and is hence is considered equally likely at any time after injury.

It is important to make a distinction between organic self-awareness deficits (i.e., anosognosia), a component of executive cognitive functions, and psychological denial, a psychological defense mechanism [36]. Both can be present and whether together or in isolation, complicate recovery. When the nature of a traumatic event exceeds one's capacity to cope, psychological denial represents an involuntary and useful coping strategy during the early stages of recovery. Denying that the event has occurred allows one to consciously maintain a sense of competence instead of overwhelming depression and anxiety. However, continued non-acceptance of the TBI beyond the acute stages of recovery is associated poorer life outcomes [50-51]. Individuals in denial of their disability have partial knowledge of their impairments and struggle to accept and deal with this new information [52]. Psychological interventions to decrease denial should be incorporated into post-acute rehabilitation.

4.2. Individual and coaching approaches

Optimal treatment for executive dysfunction following TBI involves a combination of individual and group interventions that afford participation in home, community, vocational / school, and leisure activities with ongoing feedback provided in regards to performance. One-to-one approaches are well suited to specific functional IADL skills training but have limitations, especially in regards to elements of social behavior. Many intervention models in occupational therapy and psychology involve the therapist "coaching" the client with TBI in both clinic and community settings. Sloan and colleagues [38] described skills development following TBI within a Community Approach to Participation (CAP). The authors detail the use of the approach within an occupational therapy treatment program, emphasizing the following components:

- Skills being addressed are related to the individual's goals: The top-down approach is based on the notion that clients are more motivated when there is an explicit link between the skill and the desired goal.
- Simplify and structure the task in which the skills are to be learned: Tasks should be simplified to increase the likelihood of success. Strategies may be utilized to compensate for impairments. Self-confidence and skill level increase and the individual becomes better able to meet everyday challenges.
- Build on strengths: Observing the client allows the therapist to identify personal strengths and skills that can serve as a positive platform for skill development. Pursuing pre-injury activities may facilitate success, based on the notion that it is easier to relearn an old skill than learn a new one.
- Develop routines: The goal of skill development is automatic performance within the context of the client's daily routine. Routines provide opportunities for repeated, consistent practice and contexts provide natural cues and prompts.

Wheeler [39] described a life skills training program that, when combined with daily group treatment, resulted in significant improvements in areas related to social integration and productive activity. The intensive life skills training utilized a one-on-one Life Skills Trainer (LST) to maximize the client's level of personal accountability, to provide immediate and consistent feedback regarding the social appropriateness of the client's behavior, and to provide ongoing training in the use of compensatory cognitive strategies. The LST provided continual intervention to facilitate and enhance the client's independent living skills via verbal cuing, training in compensatory skills, structuring of daily activities, redirection, assistance with problem-solving, encouragement of targeted behaviors, and cuing for safety awareness. The role of the LST or "Life Coach" is based on the psychotherapeutic relationship; i.e., it is the therapeutic bond between the LST and the client that provides the reinforcement for even the smallest of gains and the anticipated loss of this relationship that serves as a disincentive to unwanted behaviors. It is the essential collaboration or partnership between the client and the LST that provides the client with the confidence (i.e., sense of self-efficacy) to strive to conquer the brain injury and its associated deficits. In the context of this therapeutic relationship, the LST provides immediate, direct contingencies for behavior in a natural environment. The one-on-one nature of the relationship permits the LST to remain "in tune" with the client at all times, thereby maintaining the very delicate balance between praise and extinction/ punishment that is need to provide encouragement while at the same time increasing awareness of deficits.

The curative role of the relationship between the client and the LST is rooted in the principles of social learning. This theory maintains that behaviors are strengthened/increased or weakened/decreased as a function of events which follow them. Positive reinforcers increase the frequency of the behavior they follow. There are two types: a) Primary or unconditioned reinforcers do not require special training to acquire reinforcing value (e.g., food, water); b) Secondary or conditioned: reinforcers acquire reinforcing value through learning (e.g., praise, grades) ; that is, neutral stimuli acquire reinforcing when repeatedly paired with events that are reinforcing. Social reinforcers, such as verbal praise, attention, physical contact, and facial expression, are conditioned reinforcers. For social reinforcers to have reinforcing value, they must be paired with a variety of positive experiences. Thus, in order for the relationship with the LST to matter to the client (i.e., to take on reinforcing value), the LST must first take the time to engage in mutually enjoyable experiences with the client - spend meaningful time listening to the client, engaging in the client's preferred activities. Through this process, the client and the LST form a therapeutic bond or alliance. Once formed, this relationship may then be utilized - through the application of traditional behavioral techniques such as extinction, shaping, fading, differential reinforcement of other behavior (DRO) - the reinforcing value of this relationship can be utilized to increase the frequency of wanted behaviors and decrease the frequency of unwanted behaviors.

The involvement of a one-on-one LST on a 24 hour per day basis in the home and in the community can provide the opportunity for continual intervention to address common neurobehavioral problems via verbal cuing and training in compensatory skills. For example, many individuals with brain injuries exhibits verbal disinhibition resulting in inappropriately sexual, hostile, or irrelevant speech. To address this behavior, the LST establishes a nonverbal cue (e.g., tug of the ear) that signals the client to monitor their speech and engage in a predetermined self-talk strategy such as "Is it my business? Does it pertain to topic? Is it sexual? Will it make people think more or less of me?" This type of self-cuing is very effective when cuing is provided immediately and consistently in a variety of situations. Similarly, the presence of the LST provides effective intervention for explosive outbursts. In this situation, the LST provides ongoing cuing in the home and community to assist the client in identifying the initial signs of sympathetic nervous system arousal (e.g., motor tension, accelerated heart rate). The client is then cued to remove themselves from the situation and initiate relaxation techniques such as controlled breathing, visual imagery, attention-diversion, or muscle relaxation. In-vivo rehearsal of these techniques enhances the client's ability to rapidly identify when they are becoming overwhelmed and agitated, thereby enabling the client to remove themselves from stressful situation before explosive behavior occurs.

Client centered goal setting has been presented in this chapter as a challenging but necessary component of TBI rehabilitation for executive dysfunction. Goal based measures, such as the Canadian Occupational Performance Measure (COPM) [53] have been used effectively to facilitate a sense of goal ownership with clients and measure client progress over time. Studies using these techniques report that individuals, including those with TBI, who generated their own goals, are more likely to want to work on the goals and report that the goals were important to them [54]. The COPM is a semi-structured interview whereby clients identify problem areas then rate their current level of performance and satisfaction with each area on a scale of 1 to 10. On re-assessment, respondents review their goals and again rate their performance and satisfaction on the goals identified in the initial assessment [55]. Studies using the COPM in a variety of brain injury rehabilitation settings have found it to be sensitive to change and that its use provides a sense of satisfaction with progress by both clients with TBI and their significant others [56]. It is important to note that client-centered goal setting is not synonymous with relinquishing total decision making to the client or doing whatever treatment the client believes is worthwhile [57]. Therapists have the responsibility to determine situations that place clients at risk, are fiscally irresponsible, or which have ethical or legal implications and to assist clients in examining and understanding such issues.

Ylvisaker and Feeney [2] utilized a similar approach that they termed a *goal-plan-do-review* routine. The approach involves the following steps:

1. Goal – What do I want to accomplish?
2. Plan – How am I going to accomplish my goal?
3. Identification of materials / equipment and steps / assignments necessary to act on plan.
4. Prediction – How well will I do? How much will I get done?
5. Do
6. Identification of problems that arise.
7. Formulation of solutions to identified problems.
8. Review – How did I do?
9. Self rating / Other rating (teacher, therapist, peer, family member)
10. What worked? What didn't work?

11. What will I try differently the next time?

The authors note that executive system habits are more likely to become internalized when activities begin with the formulation of a goal and a plan, and end with a review that includes both a general rating of success and a listing of effective and ineffective strategies.

Goal attainment scaling (GAS) is another approach that has been effectively used to motivate participants, foster awareness of deficits, and provide a structured means to gauge progress. Within Dahlberg et al's [23] social communication skills training program, goals were developed with input from participants and scaled into five steps with progress evaluated by the client themselves, the group leaders, and a significant other. An example of a cognitive social goal from this study is as follows:

GOAL: I will interrupt less often during a 10 minute conversation.

1. I will interrupt 3 or more times during a 10 minute conversation with one prompt.
2. I will interrupt less than 3 times during a 10 minute conversation.
3. I will interrupt less than 2 times during a 10 minute conversation.
4. I will interrupt only one time during a 10 minute conversation.
5. I will not interrupt during a 10 minute conversation.

Goals focused on numerous aspects of social communication including self-awareness (i.e., "I will be able to identify social skill strengths and weaknesses"), speech (i.e., I will speak clearly enough to be understood 90% of the time"), and interpersonal ("I will be able to name 2 places to meet new people, and will visit 1 of those places). GAS can be used alongside instruments like the COPM to translate the broad client-centered goals from the COPM into specific behavioral actions. In a study combining use of the COPM and GAS with both individuals with TBI and caregivers, Doig and colleagues [54] concluded that the measures were sensitive to change for persons with TBI. They noted however that, in some clients with moderate to severe awareness deficits, client ratings on the COPM did not reflect positive change even though objective assessment and significant other ratings indicated otherwise. By receiving feedback about performance during the therapy and GAS process, the clients' self-awareness improved over time and post-intervention COPM ratings became more realistic.

Turner and colleagues [1] developed a framework for classifying self-identified goals following TBI that can be utilized during rehabilitation to monitor the scope of an individual's treatment plan. The categories, focusing on common community reintegration problems following TBI, are as follows:

1. Relationships (e.g., family, interpersonal relationships, friendships, and social activities);
2. Work and education (i.e., returning to previous employment or school, seeking new employment, considering new educational pursuits or training, and volunteer activities);
3. Injury rehabilitation;
4. Health and leisure;

5. Daily life management; and

6. General life / personal goals.

When engaged in goal setting, clients with TBI at both post-acute and long term phases of recovery tended to focus more on "injury" goals consistent with the physical aspects of recovery. This heightened focus on physical deficits over cognitive, social, and emotional difficulties, was more pronounced in the post-acute group and consistent with research on self-awareness deficit and TBI. Continued life experiences relevant to an individual's goals may increase one's understanding of the implications of the injury, leading to more realistic expectations about future outcomes and subsequently, expansion in the variety of identified goal areas. Goverover and colleagues utilized self-awareness training during instrumental activities of daily living (IADL) tasks [58]. The program involved having participants define the goals of the task, predict task performance, anticipate and preplan for any obstacles that they feel will encounter during task performance, choose a strategy to circumvent such difficulties, and anticipate the amount of assistance they feel they will need to successfully perform the task. Then, after performing the IADL task, participants completed a self-evaluation of the task, engaged in a discussion with the therapist regarding task performance and performance relative to pre-task questions, received feedback from the therapist regarding observed performance, and finally, wrote about the experience in a journal. While improvements were observed in specific task awareness, self-regulation, and functional performance, improvements were not observed in general awareness and community integration.

4.3. Vocational / Educational rehabilitation

Being employed is consistently associated with better quality of life in TBI survivors [59] and represents an important rehabilitation outcome. The capacity to work contributes to an individual's self-esteem and gives a person a sense of control over his or her life from the standpoint of economic independence, and structured, purposeful daily routine. Additionally, work affords social opportunities, the many benefits of which were discussed previously. Unemployed survivors with moderate to severe TBI have been found to be mildly clinically depressed, more fatigued, and experienced a variety of other symptoms compared to their employed counterparts, despite equivalent levels of cognitive functioning or pre-injury job satisfaction [60]. Individuals with executive dysfunction generally experience considerable difficulty securing and/or returning to competitive employment post-injury. Those that do find a job tend to have difficulty maintaining it for extended periods of time. Estimates of employment rates for persons with TBI range from 20% to 50%, depending on factors such as injury severity, prior work experience, and demographic characteristics including age, education, and socioeconomic status [61]. Positive factors influencing one's capacity to return to work after brain injury include a younger age, a higher level of education, better post-injury cognitive abilities, and the absence of a psychiatric disorder or problem with substance abuse. Inconsistent links have been found between injury severity and return to work, but generally speaking, more severe injuries have a more negative impact on return to work, in part because of greater cognitive impairment.

Principles of skills coaching are central to many return to work programs and models of supported employment [62], school re-entry [25], social integration, and general home and community functioning [38-39]. Achieving independence in basic self-care and some advanced living skills sets the ideal foundation for community participation and more challenging cognitive tasks. Essential to this is self-awareness of capacities and limitations and acceptance of one's disabilities. As with other aspects of community integration, poor self-awareness has been suggested to be a major barrier to the successful return to productive activity [63]. In such cases, productive volunteer activity may be utilized to promote self-awareness, acceptance, and psychological well-being as well as promote social roles during the recovery process. Volunteer work allows persons with brain injury to test their limits and abilities, to explore new avenues of work, to develop new interests, and to discover new sets of capacities that may lead to productive activity despite existing limitations [64]. Despite having a larger portion of more severely injured individuals on long-term disability, which one might associate with higher risk for psychological distress, a group of individuals with TBI engaged in volunteer activities was found to be comparable in their psychological adjustment to a group of TBI survivors who were competitively employed or engaged in academic pursuits [64].

Kennedy and Krause [25] presented a dynamic coaching model of supported education, involving self-regulated learning, with two college students with severe TBI. The intervention involved the following:

1. Guided conversations about strengths and weaknesses based on assessment test results;
2. Establishing academic and specific goals around themes of studying and learning, time management, and relating to others;
3. Discussion of skills related to studying and time management – useful approaches versus ineffective ones pertaining to each course;
4. Identification of human supports such as friends, family, disability service counselors, vocational rehabilitation counselors;
5. Identification of, and plan to use, academic accommodations;
6. Direct individualized coaching including direct instruction on agreed upon strategies and evaluation of the effectiveness of such strategies (i.e., what is working and what is not) related to assignment and test performance; and
7. End of semester portfolio preparation where students outline relative strengths and weaknesses, along with descriptions of various study and learning strategies and time management tools that were beneficial throughout the academic year.

These authors reported that the approach yielded positive results for both individuals as measured by improved performance on tests and assignments, good academic standing, completion of the majority of credits attempted, and positive academic decision making.

For people with TBI, one determinant of satisfaction with life in general is the resumption of leisure activities and, more specifically, involvement in activities outside the home and involvement with friends [65]. Studies show a disruption in leisure engagement after TBI, with changes in the frequency of socialization, time spent on leisure type activities, and the types of activities done compared to prior to the traumatic injury. Over time, the successes derived from leisure participation can help build self-esteem, recognized as an important attribute in the challenging process of community reintegration. Involvement in leisure activities helps to create a broader social network while providing respite or additional social opportunities for family and friends. Additionally, leisure represents an occupation through which therapeutic goals can be achieved [66], and self-awareness fostered, and an opportunity for physical and cognitive rehabilitation whereby the individual can find meaning and enjoyment in life while testing identity changes following TBI in a less threatening setting [67-68]. Leisure also fosters the development of skills to deal with environmental influences, increases social interaction, and enhances skills that can be transferred to other activities of daily living.

4.4. Group approaches to facilitate social participation

As discussed throughout this chapter, interventions to address social competence and social participation represent a critical element of any rehabilitation program addressing executive cognitive functions. The effect of impaired social skills manifest in virtually every discussion of important community based TBI outcomes – family, friendships, employment, school, and leisure [69-70]. For example, while most adult survivors remember how to do their pre-injury job and, at some point, may be able to return to work, their success tends to be temporary because of interpersonal difficulties, especially in the presence of deficits in executive cognitive functioning [71]. Social skills include both basic competencies and situational relevant behaviors that enable a person to be accepted and liked in chosen social settings [69,72]. Socially skilled people are capable of influencing others in a positive manner and with the effect that they intended. They are also able to be affected positively by others in the way that others would like to affect them. Ylvisaker and Feeney [2] conceptualized the qualities of socially skilled people as including the following:

1. Cognition – social cognition (i.e., knowledge of relevant social rules, roles, and routines and ability to effectively perceive and interpret the social behavior of others), general cognition, and decision making;

2. Communication (i.e., language and non-language behavior) – Communicating intentions effectively and in ways that are considered situationally appropriate by others; knowing how to enter into an interaction with others, comfortably maintaining that interaction, and negotiating conflicts;

3. Self – Awareness of interactive strengths and weaknesses, being reasonably comfortable with self as a social agent; and

4. Person support for interaction – Dressing and grooming self in ways considered appropriate in their social milieu, and understood and accepted by members of the social groups with which they choose to affiliate.

Similarly, Hawley and Newman [73] described social skills as including communicating needs and thoughts, listening and understanding others, giving and interpreting non-verbal communication, regulating emotions during social interactions, assertiveness, working with others to solve problems, and following social boundaries and rules.

Clients with TBI who demonstrate communication impairments and / or who experience social isolation and limited participation in their communities may benefit from group intervention [74]. As a *social microcosm,* therapy groups provide an ideal setting for interpersonal learning, social skill development, and interpersonal relationships that can exert a powerful influence on the individual [75]. Group interventions provide important social feedback from other group members, both positive and negative, allowing the client to address problematic social behaviors within the supportive structure of the group. While group topics can provide opportunities information to enhance social skills, it is through group dynamics and the "therapeutic community" that behavior changes occur. Therapist group leaders need to ensure that group structure be flexible enough to allow relationship building among participants so that the group serves to reinforce desirable social behavior and extinguish undesirable behaviors.

In their *functional group model,* Schwartzberg, Howe, and Barnes [76] present a task oriented approach that encourages active participation among members to achieve common group goals. Basic assumptions of the model include:

1. The goal of the functional group is not the product of the group, even though the group may have a meaningful product, but rather the learning process that occurs through active participation.

2. Functional groups nurture interpersonal and intrapersonal development through activity choice, climate, and goals.

3. Functional groups make use of both the human and nonhuman environment and object relations. Attention is directed to attachments to people and objects, separations from people and objects, and the symbolic nature of attachment.

4. Functional group leaders are cognizant of the individual's need for self-motivation and desire for mastery and guide the activity of the group accordingly.

Therapy group models are well suited to addressing many aspects of executive dysfunction can be integrated into various community based TBI treatment settings. Simmons and Griswold (2010) implemented a group intervention as part of a community based day program in an effort to empower members and improve their performance in daily life roles. The program involved both casual interactions and skill training groups pertaining to areas such as computer use, social skills, current events, or movie discussions. Structured group sessions focused on areas that were a) creative, b) emotional / spiritual, c) functional, d) vocational, e) recreational / leisure, and f) related to physical fitness. The researchers reported significant improvements over an eight week period. Similarly, Dahlberg et al. [23] conducted a twelve week social communication skills group (1.5 hour weekly sessions) that significantly improved both communication skills and self-reported life satisfaction. The topics of the group sessions followed the program described in *Social*

Skills and Traumatic Brain Injury [73]. The program addresses numerous components of executive cognitive functioning and is arranged as follows:

1. Group overview: Learning the skills of a good communicator;
2. Self-assessment and setting goals;
3. Presenting yourself successfully and starting conversations;
4. Developing conversation strategies and using feedback;
5. Being assertive and solving problems;
6. Practice in the community;
7. Developing self-confidence through positive self-talk;
8. Setting and respecting social boundaries;
9. Video-taping and problem solving;
10. Video review and feedback;
11. Conflict resolution; and
12. Closure and celebration.

The program uses a group co-leader, allowing two clinician perspectives and two role models. Emphasis was placed on facilitating self-awareness, progressing to individual goal setting that utilized goal attainment scaling. The group process was used to foster interaction among group members, creating an environment conducive to feedback, problem solving, social support, and awareness that one is not alone. Finally, the group involved the family and friends and included homework assignments to facilitate generalization of skills to home and community environments. At each session, participants were provided with time to discuss events from the previous week, time in the middle of the session for an unstructured break, and time towards the conclusion of each session to summarize the meeting and plan for the following week.

Wheeler [39] emphasized the importance of weekly goal setting as part of any group curriculum for persons with TBI. This program involves clients and group leaders gathering so that clients may set attainable goals (typically 5-10 specific objectives) for the following week and then subsequently receive positive or negative feedback - depending on their success at meeting their weekly goals - from the remainder of the group when progress toward goals is reviewed each week. The practice of setting and evaluating weekly goals within the group is based in the observation that, after the initial 6-9 months post-injury, an individual's rate of recovery tends to slow down and appear to plateau. Perceiving that their rehabilitation efforts are of no avail, clients are prone to give up at this point. Setting weekly attainable goals provides the individual with a brain injury with the very important feedback that they are continuing to recover - thereby increasing their self-efficacy, i.e., their belief in their ability to do what is necessary to cope with their brain injury. The weekly goal-setting session permits the client to have an active role in determining their own course, in that the client is working on their own personal goals, rather than goals established by a therapist. Furthermore, setting attainable goals each week allows for the client to be praised and reinforced for even very small gains, so that most of the feedback the client receives is positive. For example, in this goal-setting group, a client may set a goal such as "No more than three angry outbursts per day" and then receive praise and applause for meeting this goal. By contrast, it is hard to imagine that family members would praise an individual with a brain injury for having 21 angry outbursts in a week. Thus, the setting of weekly attainable goals ensures that the feedback that the client receives is predominantly positive, thereby increasing their self-confidence and their belief in their ability to overcome their brain injury

5. The role of the family in the recovery process

Optimal community participation for the individual with TBI cannot be achieved without consideration of the role of the family and family dynamics. Understanding the severity of an individual's executive cognitive impairment may only become evident once the individual with TBI has been discharged from the rehabilitation facility, leaving the family and caregivers to deal with the challenges of community re-entry. The emergence of executive functions at this stage can give the appearance that the individual's deficits are getting worse when, in fact, they are a product of the challenges imposed by the demands and expectations of a typical daily routine. Ongoing emotional distress within the family has been reported to be a contributing factor to the breakdown in marital relationships, reduced personal health, and decreased social contacts experienced by family members [77]. Family members also report increased financial strain, altered or reduced job responsibilities, and reduced free time due to requirements associated with caring for the individual with TBI [77-79].

Turkstra [4] proposes increasing the role of the family during cognitive rehabilitation at the inpatient stage as a means of facilitating a more favorable transition to home and community. The family based approach is summarized below based upon the following goals:

1. Establishing a therapeutic alliance with patients and families, so that they identify occupational and speech and language pathologists as an ongoing resource as needs arise post-discharge.
2. Help patients and families observe and understand the natural history of cognitive recovery after traumatic brain injury to help them interpret behaviors they are seeing each day and appreciate their family members progress.
3. Provide patients and families with some tools to help manage everyday consequences of the patient's cognitive impairments at home.
4. Minimize bad habits that can develop during the early days post-injury when patients have normal implicit learning of habits but impaired declarative memory and reasoning.

5. Help patients and families be advocates for their own needs post-discharge and educated consumers of cognition- related resources.

The significant strain placed on family members caring for injured relatives is well documented in the literature, negatively impacting physical health and contributing to emotional distress and the likelihood of depression [80-82]. Unfortunately, many family members and friends report that they still manage all of the activities of daily living and instrumental activities of daily living of the individual with a TBI 10 years after the trauma and that they are exhausted by having to do so [83]. Research suggests that the effects of TBI within the family structure are reciprocal. The well-being of the person with TBI affects the well-being of the family, and as one would expect, the well-being of the family affects the well-being of the injured person [84]. Hence the family system is more than a recipient of stress and strain following TBI, it most likely influences outcome. As such, involvement of family and caregivers by the rehabilitation team throughout the recovery process is critical. That may involve assisting families and caregivers in their own coping so that they are in a psychological state appropriate to the challenges that community re-entry may present.

6. Conclusions

Despite a growing interest in executive cognitive functions and their effects on community participation and life satisfaction following TBI, precise definitions of executive functions continue to elude clinicians and researchers. Disorders of executive functions, as is commonly seen following moderate to severe TBI, tend to be context dependent and present unique challenges in terms of both assessment and rehabilitation. This chapter presented an overview and hierarchical approach to the rehabilitation of executive cognitive functions. Beginning with therapeutic relationship building, the clinician engages the client in a collaborative program of activity designed to facilitate self-awareness and meaningful goal setting. The process tests the commitment and professionalism of the interdisciplinary team and may contribute to struggle, frustration, and periods of dissatisfaction by both client and their support system. By gaining an appreciation of deficits, clients are better suited setting realistic goals and making the social, cognitive, behavioral, and physical adjustments necessary for participation in interventions to specifically geared to the performance of important adult roles.

Acknowledgements

The author would like to thank Dr. James Phifer and the staff of Radical Rehab Solutions, LLC for their expertise and helpful insights pertaining to the content of this chapter. I would also like to thank Breanna Hart, OTS for her work in reviewing and editing the chapter.

Author details

Steven Wheeler

Division of Occupational Therapy, West Virginia University School of Medicine, West Virginia University Injury Control Research Center, Morgantown, West Virginia, USA

References

[1] Turner, B., Fleming, J., Cornwell, P., Haines, T., & Ownsworth, T. (2009).Profiling early outcomes during the transition from hospital to home after brain injury. *Brain Injury, 23,* 51-60.

[2] Ylvisaker, M. & Feeney, T. (1998). Collaborative Brain Injury Intervention: Positive Everyday Routines. San Diego, CA: Singular Publishing Group.

[3] Tsaousides, T., & Gordon, W. (2009). Cognitive rehabilitation following traumatic brain injury: assessment and treatment. *Mount Sinai Journal of Medicine, 76,* 173-181.

[4] Turkstra, L. (2013). Inpatient cognitive rehabilitation: Is it time for a change? *Journal of Head Trauma Rehabilitation, 28*(4), 332-336.

[5] Fischer, S., Gauggel, S.,&Trexler, L. (2004).Awareness of activity limitations and prediction of performance in patients with brain injuries and orthopedic disorders. *Journal of the International Neuropsychological Society, 10,* 190-199.

[6] McDonald, B., Flashman, L.,& Saykin,A. (2000). Executive dysfunction following traumatic brain injury: Neural substrates and treatment strategies. *Neurorehabilitation, 17,* 333-344.

[7] Mateer, C. (1999). The rehabilitation of executive disorders. In Stuss, D., Winocur, G., & Robertson, I. (Eds.), *Cognitive NeuroRehabilitation* (pp. 314-332). Cambridge, UK: Cambridge University Press.

[8] Parente, R.,& Hermann, D. (2003).*Retraining cognition: Techniques and applications* (2nd. edition). Austin, TX: Pro-Ed.

[9] Goverover, Y.,& Hinjosa, J. (2002). Categorization and deductive reasoning: Predictors of instrumental activities of daily living performance in adults with brain injury. *American Journal of Occupational Therapy, 53,* 509-515.

[10] Golisz, K.,& Toglia, J. (2003).Perception and cognition. In: E. Crepeau, E. Cohn, & B. Schell (Eds.), *Willard and Spackman's Occupational Therapy.* Philadelphia, PA: Lippincott, Williams, and Wilkins.

[11] Ashley, M. (2004).Evaluation of traumatic brain injury following acute rehabilitation. In M. Ashley (Ed.), *Traumatic brain injury rehabilitative treatment and case management.* Boca Raton, FL: CRC Press.

[12] Zoltan, B. (2007). *Vision, perception, and cognition* (4th Ed.).Thorofare, NJ: Slack Incorporated.

[13] Willer, B., & Corrigan, J. (1994). A model for community based services: The Whatever It Takes Model. *Brain Injury, 8,* 647-659.

[14] Gordon, W., Hibbard, M., Brown, M., Flanagan, S., & Korves, M. (1999). Community integration and quality of life of individuals with traumatic brain injury. In Rosenthal, M., Griffith, E., Kreutzer, J. & Pentland, B. (Eds.) *Rehabilitation of the Adult and Child with Traumatic Brain Injury* (pp.312-325). Philadelphia, PA: FA Davis.

[15] Senelick, R., & Dougherty, K. (2001).*Living with brain injury: A guide for families*. Birmingham, AL: Healthsouth Press.

[16] World Health Organization [WHO].(2001). *International classification of functioning, disability, and health.* Geneva: author.

[17] World Health Organization. (1999) International Classification of Functioning and Disability: ICIDH-2, Beta Draft Short Version. Geneva, Switzerland: WHO

[18] Gray, D. & Hendershot, G. (2000). The ICIDH-2: Developments for a new era of outcomes research. *Archives of Physical Medicine and Rehabilitation, 81,* S10-S14.

[19] Ylvisaker, M., Hanks, R., & Johnson-Greene, D. (2002). Perspectives on rehabilitation of individuals with cognitive impairment after brain injury: Rationale for reconsideration of theoretical paradigms. *Journal of Head Trauma Rehabilitation, 17*(3), 191-209.

[20] Simmons, C., & Griswold, L. (2010).Using the evaluation of social interaction in a community-based program for persons with traumatic brain injury. *Scandinavian Journal of Occupational Therapy, 17,* 49-56.

[21] Gioia, G., Kenworthy, L., Isquith, P. (2010). Executive function in the real world: Brief lessons from Mark Ylvisaker. *Journal of Head Trauma Rehabilitation, 25*(6), 433-439.

[22] Rowlands, A. (2001). Ability or disability? Strength-based practice in the area of traumatic brain injury. *Families in Society: The Journal of Contemporary Human Services,82,* 273-286.

[23] Dahlberg, C., Cusick, C., Hawley, L., Newman, J., Morey, C., Harrison-Felix, C., & Whiteneck, G. (2007). Treatment efficacy of social communication skills training after traumatic brain injury: A randomized treatment and deferred treatment controlled trial. *Archives of Physical Medicine and Rehabilitation,88,* 1561-1573. doi:10.1016/j.apmr.2007.07.033.

[24] Engberg, A.,& Teasdale, T. (2004). Psychosocial outcome following traumatic brain injury in adults: A long-term population based follow-up. *Brain Injury, 18,* 533-545.

[25] Kennedy, M.,& Krause, M. (2011). Self-regulated learning in a dynamic coaching model for supported college students with traumatic brain injury: two case reports. *Journal of Head Trauma Rehabilitation, 26*, 212-223.

[26] Wagner, M., Newman, L., Cameto, R., Garza, N., & Levine, P. (2005). After high school: A first look at the postschool experiences of youth with disabilities. A report from the National Longitudinal Transition Study-2 (NLTS2). Menlo Park, CA: SRI International. Retrieved August 18, 2012 from http://www.nlts2.org/pdfs/afterhighschool_report.pdf

[27] Deaton, A. (1990). Behavior change strategies for children and adolescents with traumatic brain injury. In E.D. Bigler (Ed.), *Traumatic brain injury* (pp.231-249). Austin, TX: Pro-Ed.

[28] Kennedy, M., Krause, M., & Turkstra, L. (2008). An electronic survey about college experiences after traumatic brain injury. *NeuroRehabilitation, 23*, 511-512.

[29] Burgess, P., Veitch. E., deLacy Costello, A., & Shallice, T. (2000). The cognitive and neuroanatomical correlates of multitasking. *Neuropsychologia, 38*(6), 848-863.

[30] O'Callaghan, A., McAllister, L., & Wilson, L. (2012). Insight vs readiness: Factors affecting engagement in therapy from the perspectives of adults with TBI and their significant others. *Brain Injury, 12*, 1-12. doi:10.3109/02699052.2012.698788

[31] Noe, E., Ferri, J., Caballero, M., Villodre, R., Sanchez., A., &Chirivella, J. (2005). Self-awareness after acquired brain injury: Predictors and rehabilitation. *Journal of Neurology, 252*, 168-175.

[32] Sherer, M., Bergloff, P., Boake, C., High, W., & Levin, E. (1998). The Awareness Questionnaire: Factor structure and internal consistency. *Brain Injury, 12*, 63-68.

[33] Flashman, L., & McAllister, T. (2002).Lack of awareness and its impact in traumatic brain injury. *NeuroRehabilitation, 17*, 285-296.

[34] Barco, P., Crosson, B., Bolesta, M., Wets, D., & Stout, R. (1991).Training awareness and compensation in postacute head injury rehabilitation. In J. Kreutzer & P. Wehman (Eds.), *Cognitive rehabilitation for persons with traumatic brain injury: A functional approach* (pp. 129-146). Baltimore, MD: Paul H. Brookes Publishing Co.

[35] Trudel, T., Tryon, W.,& Purdum, C. (1998).Awareness of disability and long-term outcome after traumatic brain injury. *Rehabilitation Psychology, 43*, 267-281.

[36] Koskinen, S. (1998). Quality of life 10 years after a very severe traumatic brain injury (TBI): The perspective of the injured and the closest relative. *Brain Injury, 12*, 631-648.

[37] Toglia, J., & Kirk, U. (2000).Understanding awareness deficits following brain injury. *Neurorehabilitation, 15*, 57-70.

[38] Ownsworth, T., & Clare, L. (2006).The association between awareness deficits and rehabilitation outcome following acquired brain injury. *Clinical Psychology Review, 26*, 783-795. doi:10.1016/j.cpr.2006.05.003

[39] Sloan, S., Winkler, D., & Callaway, L. (2004). Community integration following severe traumatic brain injury: Outcomes and best practice. *Brain Impairment, 5*, 12-29.

[40] Wheeler, S. (2012). The impact of intensive community based rehabilitation on community participation and life satisfaction following severe traumatic brain injury. In A. Agriwal (Ed.), *Brain Injury: Functional Aspects, Rehabilitation, and Prevention* (pp. 95-120). Rijeka, Croatia: Intech Publishers.

[41] Dirette, D., & Plaisier, B. (2007). The development of self-awareness of deficits from 1 week to 1 year after traumatic brain injury: preliminary findings. *Brain Injury, 21*, 1131-1136

[42] Trahan, E., Pepin, M., &Hopps, S. (2006). Impaired awareness of deficits and treatment adherence among people with traumatic brain injury or spinal cord injury. *Journal of Head Trauma Rehabilitation, 21*, 226-235.

[43] Seel, R., Kreutzer, J., Rosenthal, M., Hammond, F., Corrigan, J., & Black, K. (2003). Depression after traumatic brain injury: A NIDRR model systems multi-center investigation. *Archives of Physical Medicine and Rehabilitation, 84*, 177-184.

[44] Jean-Bay, E. (2000). The biobehavioral correlates of post-traumatic brain injury depression. *Journal of Neuroscience Nursing, 32*, 169-176.

[45] Rosenethal, M., Christensen, B., & Ross, T. (1998).Depression following traumatic brain injury. *Archives of Physical Medicine and Rehabilitation, 79*, 90-103.

[46] Simpson, G. & Tate, R. (2007).Suicidality in people surviving traumatic brain injury: Prevalence, risk factors and implications for clinical management. *Brain Injury, 21*, 1335-1351.

[47] Kishi, Y., Robinson, R., & Kosier, J. (2001). Suicidal ideation among patients during the rehabilitation period after life threatening physical illness. *Journal of Nervous and Mental Disorders, 189*, 623-628.

[48] Kessler, R., Berglund, P., Borges, G., Nock, M.,& Wang, P. (2005).Trends in suicidal ideation, plans, gestures, and attempts in the United States, 1990-1992 to 2001-2003.*JAMA, 293*, 2487-2495.

[49] Simpson, G. & Tate, R. (2002).Suicidality after traumatic brain injury: Demographic, injury and clinical correlates. *Psychology and Medicine, 32*, 687-697.

[50] Tsaousides, T., Cantor, J., & Gordon, W. (2011). Suicidal ideation following traumatic brain injury: Prevalence rates and correlates in adults living in the community. *Journal of Head Trauma Rehabilitation, 26*.265-275.

[51] Curran, C., Ponsford, J., & Crowe, S. (2000). Coping strategies and emotional outcome following traumatic brain injury: A comparison with orthopedic patients. *Journal of Head Trauma Rehabilitation, 15*, 1256-1274.

[52] Moore, A., & Stambrook, M. (1995). Cognitive moderators of outcome following traumatic brain injury: a conceptual model and implications for rehabilitation. *Brain Injury, 9*, 109-130.

[53] Prigatano, G., & Klonoff, P. (1998). A clinician's rating scale for evaluating impaired self-awareness and denial of disability after brain injury. *Clinical Neuropsychology, 12*, 56-67.

[54] Law, M., Baptiste, S., Carswell, A., McColl, M. A., Polatajko, H., &Pollock, N. (2005).*The Canadian Occupational Performance Measure* (5th ed.). Toronto, Canada: Canadian Association of Occupational Therapists

[55] Doig, E., Fleming, J., Kuipers, P., & Cornwell, P. (2010).Clinical utility of the combined use of the Canadian Occupational Performance Measure and goal attainment scaling. *American Journal of Occupational Therapy, 64*, 904-914.

[56] Law, M., Baptiste, S., Carswell, A., McColl, M.A., Polatajko, H., & Pollock, N. (1998). *Canadian Occupational Performance Measure* (2nd ed. Rev.) Ottawa, ON: CAOT Publications ACE.

[57] Doig, E., Fleming, J., Cornwell, P., & Kuipers, P. (2009). Qualitative exploration of a client centered, goal directed approach to community-based occupational therapy for adults with traumatic brain injury. *American Journal of Occupational Therapy, 63*, 559-568.

[58] Law, M., & Baptiste, S. (2002).Working in partnerships with our clients. In M. Law, C. Baum,& S. Baptiste (Eds.).*Occupation based practice: Fostering performance and participation*. Thorofare, NJ: Slack Incorporated.

[59] Goverover, Y., Johnston, M., Toglia, J., & Deluca, J. (2007). Treatment to improve self-awareness in persons with acquired brain injury. *Brain Injury, 21*, 913-923.

[60] Nalder, E., Fleming, J., Foster, M., Cornwell, P., Shields, C., & Khan, A. (2012). Identifying factors associated with perceived success in the transition from hospital to home after brain injury. *Journal of Head Trauma Rehabilitation, 27*, 143-153.

[61] McCrimmon, S.,&Oddy, M. (2006). Return to work following moderate to severe traumatic brain injury. *Brain Injury, 20*, 1037-1046.

[62] Keyser-Marcus, L., Bricout, J., Wehman, P., Campbell, L., Cifu, D., Englander, J., High, J., & Zafonte, R. (2002). Acute predictors of return to employment after traumatic brain injury: A longitudinal follow-up. *Archives of Physical Medicine and Rehabilitation, 83*, 635-641.

[63] Wehman, P., Kregel, J., Keyser-Marcus, L., Sherron-Targett, P., Campbell, L., West, M., & Cifu, D. (2003).Supported employment for persons with traumatic brain injury:

A preliminary investigation of long term follow-up costs and program efficiency. *Archives of Physical Medicine and Rehabilitation, 84,* 192-196.

[64] Wehman, P., Targett, P., West, M.,& Kregel, J. (2005). Productive work and employment for persons with traumatic brain injury: What have we learned after 20 years? *Journal of Head Trauma Rehabilitation, 20,* 115-127.

[65] Ouellet, M., Morin, C.,& Lavoie, A. (2009).Volunteer work and psychological health following traumatic brain injury, *Journal of Head Trauma Rehabilitation, 24,* 262-271.

[66] Anke, A.,& Fugl-Meyer, A. (2003). Life satisfaction several years after severe multiple trauma: A retrospective investigation. *Clinical Rehabilitation, 17,* 431-442.

[67] Blacker, D., Broadhurst, L., & Teixeira, L. (2008). The role of occupational therapy in leisure adaptation with complex neurological disability: A discussion using two case study examples. *Neurorehabilitation, 23,* 313-319.

[68] Lyons, R. (1993). Meaningful activity and disability: Capitalizing upon the potential of outreach recreation networks in Canada. *Canadian Journal of Rehabilitation, 6,* 256-265.

[69] Malley, D., Cooper, J., & Cope, J. (2008).Adapting leisure activity for adults with neuropsychological deficits following acquired brain injury. *NeuroRehabilitation, 23,* 329-334.

[70] Dahlberg, C., Hawley, L., Morey, C., Newman, J., Cusick, C., & Harrison-Felix, C. (2006). Social communication skills in persons with post-acute traumatic brain injury: three perspectives. *Brain Injury, 20,* 425-435

[71] Doble, S. E., Bonnell, J. E., & Magill-Evans, J. (1991). Evaluation of social skills: A survey of current practice. *Canadian Journal of Occupational Therapy, 58,* 241–249.

[72] Arlinghaus, K., Shoaib, A., & Price, T. (2005).Neuropsychiatric assessment. In J.M. Silver, T.W. McAllister, & S.C. Yudofsky (Eds.), *Textbook of traumatic brain injury* (pp. 59-78). Washington, DC: American Psychiatric Publishing.

[73] Feeney, T.,Ylvisiker, M., Rosen, B., & Greene, P. (2001). Community supports for individuals with challenging behavior after brain injury: An analysis of the New York State Behavioral Resource Project. *Journal of Head Trauma Rehabilitation, 16,* 61-75

[74] Hawley, L., & Newman, J. (2006).*Social skills and traumatic brain injury: a workbook for group treatment.* Denver, CO: authors.

[75] Golisz, K. (2009). *Occupational therapy practice guidelines for traumatic brain injury.* Bethesda, MD: American Occupational Therapy Association.

[76] Yalom, I. (1995). *The theory and practice of group psychotherapy.* New York, NY: Basic Books.

[77] Schwartzberg, S., Howe, M.,& Barnes, M.A. (2008).*Groups: Applying the functional group model.* Philadelphia, PA: F. A. Davis Company.

[78] Hall, K., Karzmark, P., Stevens, M., Englander, J., O'Hare, P. & Wright, J. (1994). Family stressors in traumatic brain injury: A two year follow-up. *Archives of Physical Medicine and Rehabilitation, 75*(8), 876-884.

[79] Tyerman, A. & Booth, J. (2001). Family interventions following traumatic brain injury: A service example. *Neurorehabilitation, 16*(1), 59-66.

[80] Perlesz, A., & O'Loughlan, M. (1998). Changes in stress and burden in families seeking therapy following traumatic brain injury: a follow-up study. International Journal of Rehabilitation Research, 21(4), 339-54.

[81] Knight, R., Devereaux, R., & Godfrey, H. (1998).Caring for a family member with traumatic brain injury.*Brain Injury, 12*, 467-481.

[82] Leathem, J., Heath, E., & Woolley, C. (1996).Relatives' perceptions of role change, social support, and stress after traumatic brain injury. *Brain Injury, 10*, 27-38.

[83] Connolly, D., & O'Dowd, T. (2001).The impact of different disabilities arising from head injury on the primary caregiver.*British Journal of Occupational Therapy, 64*, 41-46.

[84] Lefebvre, H., Cloutier, G., & Levert, M. (2008).Perspectives of survivors of traumatic brain injury and their caregivers on long-term social integration. *Brain Injury, 22*, 535-543.

[85] Vangel, S., Rapport, L., & Hanks, R. (2011).Effects of family and caregiver psychosocial functioning on outcomes in persons with traumatic brain injury. *Journal of Head Trauma Rehabilitation, 26*, 20-29.

Chapter 20

Communication Disorders Following Traumatic Brain Injury

Edilene Curvelo Hora, Liane Viana Santana,
Lyvia de Jesus Santos, Gizelle de Oliveira Souza,
Analys Vasconcelos Pimentel,
Natalia Tenório Cavalcante Bezerra,
Sylvia Rodrigues de Freitas Doria,
Tiago Pinheiro Vaz de Carvalho,
Afonso Abreu Mendes Júnior,
Jessica Almeida Rodrigues,
Renata Julie Porto Leite Lopes and Ricardo Fakhouri

Additional information is available at the end of the chapter

http://dx.doi.org/10.5772/57321

1. Introduction

Traumatic brain injury (TBI) constitutes a public health problem of great significance with importance in both morbidity and mortality, accounting for approximately 15 to 20% of deaths in people between five and 35 years of age and responsible for 1% of all adult deaths.[1-2]

In the United States, about 59 million people every year die following TBI. This represents one third of all injury-related deaths. Furthermore, 90,000 individuals suffer from permanent incapacity related to TBI.[3]

In Brazil, data from the Ministry of Health (2011)[4] shows that 145,842 deaths occurred due to external causes, and amongst all injuries associated with these external causes, TBI stood out in terms of magnitude, as it was one of the most common injuries.[5]

Trauma caused by traffic accidents is one of the most frequent causes of death in individuals aged 10 to 24. There is an imbalance in the prevalence of the risk of traffic-related trauma in developed countries and in developing ones, with the higher risk being found in the latter.

Factors include the infrastructure of a country and the accelerated motorization of its population.[6]

The main causes of TBI include violent aggression, transportation accidents, and falls, the former two being the most prevalent. TBI victims are mostly young males.[7-10] TBI-related consequences extend beyond recorded fatalities and can also be observed in victims who survive trauma.

These individuals may present physical, cognitive, communication, and behavioural disabilities and incapacities at several levels, in addition to suffering problems on social and occupational levels.[11-12] The consequences of trauma consequences also touch the victims' families, who can be considered hidden victims. A crisis in the family system often arises, as well as the emergence of diseases that compromise the family's ability to function and recover.[13]

The participation of the speech therapist in the multidisciplinary team providing care to TBI victims is of great importance because this professional will be able to assess the specific needs of the victims at an early stage, regarding their communicative skills and other related problems (eating and swallowing difficulties), in order to prevent, minimize, or eliminate possible trauma sequelae.[14]

According to the American Speech-Language-Hearing Association (ASHA), patients with traumatic brain injury may experience difficulties in finding words to express themselves or in understanding an idea through speech, writing, and/or reading. Additionally, the muscles of the mouth, face, and the respiratory system can present changes in tone and coordination such that a speech motor disorder called dysarthria may result.[15] These language and/or speech and cognitive alterations compromise an individual's communication to varying degrees, ranging from minimal to extensive.

Speech and language production processes include distinct activities in the cerebral cortex. Therefore, different types of alterations in the Central Nervous System (CNS) may result in various kinds of language and/or speech disorders.[15]

For better understanding of communication disorders this chapter will address the importance of communication and language, as well as interdisciplinary approach victims of traumatic brain injury.

2. Communication: An essential instrument in human relations

Communication is a method by which the sharing of thoughts, feelings, ideas, and messages occurs, and it can influence the behaviour of those who respond according to their own beliefs, values, cultures, and life stories. Communication can be defined as "the capacity to exchange or discuss ideas, to dialogue, to converse with the aim of an understanding between the parties".[16] Communication, then, is an instrument of great importance in the realization and development of work, leisure, education, relationships, conversation, and negotiation.[17]

Communication, an instrument that is indispensable to interpersonal interaction is not only restricted to verbal language and the utilization of vocabulary. Rather, it also comprises other

http://dx.doi.org/10.5772/57321

methods, such as gestures and body language, facial expressions, signs, figures, objects, colours, which is to say, it also includes visual signs, which are responsible for ensuring efficacy in the conversation process.[17] Visual signs are often the only components of certain conversations, a fact which highlights their great importance in the realization of communication.

The linguistic component is characterized by biological and social aspects and it characteristically favours adapting to the surrounding environment. These components include phonological, syntactic, semantic, and pragmatic aspects of language, described below. [18]

The capacity to surpass auditory perception and develop metaphonological abilities constitutes, in effect, a reflection upon the phonological aspects of a language, which is referred to as "phonological conscience", and which is an ability that is an important prerequisite in the development of written language.[19]

Syntax is the part of linguistics that is dedicated to the study of the rules and principles that govern the organization of a sentence's constituents. Semantics, on the other hand, examine the meaning of a word and of the relation between signs and its referents.

Pragmatic theories basically focus on two factors: communicative functions and conversational skills. Communicative functions are broad and abstract, and they reflect the communicative intention of the speaker; they involve motivation and the goals the speaker aims to achieve in communicating with the other individual. Conversational skills refers to a subject's capacity to participate in an interactive sequence of spoken acts, with the objective of communicative exchange.[19]

The ability to converse involves an interactive sequence of spoken acts and is the result of communicative exchange between two or more interlocutors inserted into a social context.[20] Besides this sequence, efficient conversation requires the interlocutors' compliance with the rule of taking turns, their commitment to the theme being discussed, and the capacity to adapt to participants and situations.[21]

Language is thus conceptualized as a "privileged instrument of inter-human communication and the privileged vehicle of thought".[22] Its organized use as a social rule and communicative mode of interaction are expressed and perceived in social culture. Social rules are organized by signs that express ideas that are manifested in speech, which makes communication with others possible[18]

The symbol comes from imitation, and representation is the use of the symbolic function of language. During the sensorimotor period, children's mental structures improve, and they form new constructions and start to associate and represent new signs. In this stage, the roots of thoughts are found in actions and in the mechanisms overriding linguistic fact, and so the role of language is an accessory in the construction of knowledge. In the transitory period between the sensorimotor and pre-operatory stages, the use of gestures with communicative interaction is accompanied by speech (babbling). Oral expression and the use of gestures develop in parallel with the development of communication. Communication is thereby characterized by the capacity to exchange information, express oneself, and interact with others, developing knowledge bases and expanding one's concepts in order to develop language.[23]

In any sphere, communication is necessary, and such communication only occurs with the participation of two or more elements (transmitter and receiver) that transmit information using the means of language both verbal and nonverbal, resulting in an exchange of knowledge, which is to say that in order for the message to be transmitted and understood, it is necessary for the transmitter and receiver to promote mutual interaction.[24]

Among other forms of communication, nonverbal communication is a means of transmitting information. Vivacious gestures elicit responses, and one could almost argue that happens in accordance with an elaborate secret code that remains unwritten, acknowledged by none, but understood by all.[25]

Scholars have sought to demonstrate that children, through symbolic activity, develop instruments of language during their interactions with each other. As the children appropriate this language, they become capable of controlling their environment using gestures, facial expressions, and the emission of sounds, which function as a diffuse communication that constitutes the pre-verbal phase. After that, practical intelligence and speech are integrated, making verbal thought and language more rational.[26]

B. F. Skinner's book *Verbal Behavior* (1957), which deals with verbal behaviour and its function in children's acquisition of language, presents the idea of language reduced to a set of verbal responses associated with situations, in accordance with the Stimulus-Response model of conditioned reflexes. According to this theory, children learn language through imitation and reinforcement; in other words, the acquisition of language is considered a process of storing reinforcements.[27]

Interpersonal relations are revealed in thoughts, ideas, and feelings, and they transmit information that permeates the human condition. Scientific evidence suggests that the ability to develop language is innate and that the universal aspects of language acquisition are established in the structure of the human brain and are susceptible to sociocultural influences.[28]

Evidence collected from clinical observation shows how behaviour relates to mental processes and has long aroused interest. Different philosophical trends were instrumental in developing contemporary neurology. In the early 19th century, Franz Joseph Gall (1757-1828) spread the idea of phrenology that claimed an analysis of the skull's surface made it possible to determine whether or not a mental function had been developed. Followers of this philosophy attributed 35 different types of "organs" in the brain, each of which corresponded to a certain function. According to the theory, the "sense of language and of the word" was located in the frontal lobe of the brain.[22, 29-31] The scientific community eventually abandoned such ideas due to the lack of substantiated hypotheses.

In the mid-19th century, anthropologist Paul Broca's (1824-1880) important discoveries favoured the theory of cerebral localization, as suggested by Gall in his studies. Broca described patients who presented with impairment in the production of speech and in the preservation of language comprehension, and he identified lesions in the frontal lobes of the left cerebral hemisphere, an area which became known as "Broca's area" and which is responsible for "the functional center of language" and the syndrome became known as "Broca's aphasia".[22,29-30]

http://dx.doi.org/10.5772/57321

The neurologist Carl Wernicke (1848-1904) awoke an interest in types of injuries that were different from those described by Broca, but that also presented impairments in language comprehension. These lesions corresponded to the temporal cortex of the left cerebral hemisphere (referred to as "Wernecke's area"), and the syndrome became known as "Wernecke's aphasia".[22,29-30] The two areas, Broca's and Wernicke's, are integrated by bundles of nerve fibres known as the arcuate fasciculus (Figure 1). In the majority of individuals, the cortical language areas are located in the left hemisphere of the brain.

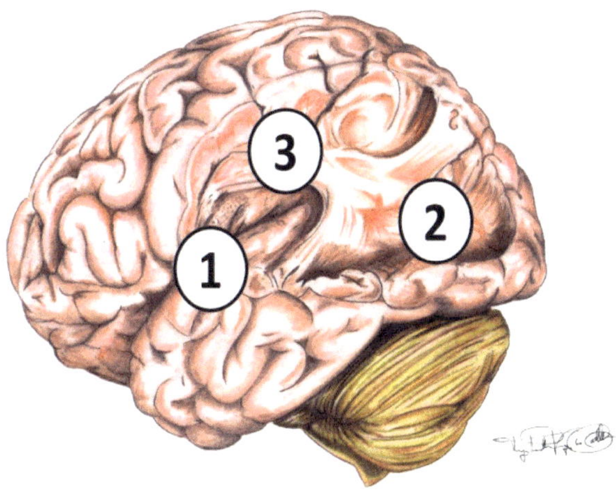

Figure 1. Language areas of the brain, 1: Broca's Area; 2: Wernicke's Area; 3: Arcuate Fasciculus. Illustration: Tiago Carvalho

The language process has been described through the neural mechanisms by which its main functions are performed. The sounds produced by speech require the incorporation of various pieces of information so as to generate the pattern of muscular activation necessary for verbal fluency.[31] The parts of the cerebral cortex used in the emission of speech deal with auditory information (temporal lobe), somatosensory information (parietal lobe), and motor information (frontal lobe). Together with the cerebellum, basal ganglia (primarily the thalamus), and the brain stem, interconnections between the brain's regions are responsible for the production of speech.[22,31-32]

One of the most studied functions in the field of neurolinguistics is a language and have the cognitive deficits associated with the mastery of this executive function. These deficits are responsible for the communication difficulties of patients who are injured in this cerebral area. Observations to this effect have motivated investigations into the importance of the relationship between cognitive mastery, executive functioning, and semantic knowledge. The use of computer programs in patient rehabilitation has been noteworthy, especially for patients with aphasia.[32]

In the field of healthcare, communication is incredibly relevant. Trauma victims with TBI may have cerebral injuries or may acquire neurological damage, resulting in language disorders that make the patients' social lives more difficult. It is in this context that communication between patients and healthcare professionals, primarily nonverbal communication, is fundamental and needs to be and understood by health professionals.[33]

Keeping in mind that speech production processes involve different activities in distinct areas of the cortical regions of the brain, it is understandable that disturbances in speech and language mechanisms appear as a consequence to different types of changes in the central nervous system, as in the examples of TBI and cerebrovascular accidents resulting in aphasia, apraxia, and dysarthria.[33]

These disorders can compromise the quality of life for both the patients and their families and therefore should continue to be studied throughout the patient's life. This will enable health-care professionals to provide improvements in treatment and better quality of life post-trauma.

The existence of man is only possible because of communication, and we are communicative beings par excellence. Communication permeates every aspect of our lives; from birth we exert influence on and are influenced by the environment in which we live. Over the course of our lives, the development of communication becomes more complex due to the necessity to master language, reading, the reasoning process, and an analysis of the world and of ourselves..[34]

3. Language

Language is an example of a higher brain function whose development is primarily based on a genetically-determined anatomic structure, but also on verbal stimulation from the outside environment. Language development depends not only on a perceptual motor reaction between perception and praxis, but also on a complex act involving cognition.[35-37]

A finite system of principles and rules that permits both the speaker to codify signs into sounds and the listener to decodify sounds into signs is based on the principles of language. This finite system, however, has the potential to be infinitely creative in the sense that it permits the speaker and the listener to create and understand an infinite set of new grammatical sentences.[35,38]

Neurolinguistics is the science of cerebral mechanisms underlying the comprehension, production, and abstract knowledge of language, be it spoken, signed, or written. The majority of studies on the physiology of language have focused on two chief entryways of linguistic information into the cognitive system: auditory input and visual input.[39]

Neurologically speaking, the term "language" seems to be grounded in thought itself, and sometimes it even seems to be synonymous with thought. The moment a new word is acquired has an impact on a child's development as this activity becomes a tool for analysis and synthesis and enables both an understanding of the child's surroundings and the self-regulation of the child's conduct.[40]

An interdisciplinary nature is attributed to language because it is the object of study in various branches of science. Moreover, language is also an instrument used in social interactions between individuals seeking to communicate in different contexts, and it permeates the thoughts of those who use it, mediates relationships between humans, and is responsible for the transmission of sociocultural customs and values.[41-42]

The use of language depends on multiple types of knowledge and includes linguistic, conceptual, and perceptual non-linguistic systems of information. Knowledge is expressed through phonological, syntactic, semantic, and discursive structures. Because it is an elaborate and highly complex process, alterations in language can happen on any of the aforementioned levels and can compromise both oral and graphic communication.[40]

It is through the visual and auditory systems that language reaches the brain, while the motor system produces spoken and written discourse. When speaking, we produce and articulate sounds that have meaning and are vehicles of ideational expression.[35]

Even though the left hemisphere normally dominates language, the right hemisphere has abilities that are significant for language comprehension. Moreover, the corpus callosum allows synergistic interactions between the hemispheres in order to produce language. Besides the interaction between Broca's and Wernecke's areas, this processing results in many capabilities, such as designation, articulation, comprehension, and the use of grammar.[43]

The left hemisphere analyses sequentially, and so, in the majority of people, it acts as the anatomical and physiological substrate of the language function. The right hemisphere, on the other hand, analyses spatially.[44]

Language is organized around a fundamental dual capacity: a lexical capacity (establishing; retaining in memory; receptively and productively using a significant amount of meaning-signifier-referent associations) and a grammatical capacity having to do with the organization on a sequencing level and structural dependencies between words (enunciated phrases) and sequences (paragraphs and discourse).[1]

This dual capacity is joined by an instrumental social dimension referred to as "the pragmatics of language".[45] Linguistic information can be transmitted in the form of discourse and written text, but it is the content of the message that is essential to the higher levels of language comprehension processes, such as the achievement of inferences and associations between pieces of textual information.[44]

With cranial traumas, what should be emphasized is oral language comprehension, which is defined as being related to that which is remembered from completed communication or from the application of information garnered from what has been heard, or which is connected to casual relationships established between elements in communication so as to give them coherence.[46]

Understanding involves a series of steps that, starting from a statement, allows one to rediscover the original idea. It is the process of [re]constructing meanings from spoken discourse, and through this process, the listener generally acquires information or knowledge through language. However, there is also oral communication with the ultimate goal of establishing and/or maintaining social relationships, of initiating interaction for the purpose of entertainment, seduction, pleasure, or even to confuse or deceive, forging a given understanding.[44]

An important concept in the representation of words is the mental lexicon, i.e., a mental stock of word-related data that includes information about semantics, syntax, pronunciation, and sound patterns. The process of accessing lexical representations (words) in the mental lexicon is influenced by the "auditory neighborhood" of words, which is defined as the number of words that differ from the target word by one phoneme.[47]

Words with more neighbours are identified more slowly, and there can be competition when activating different words during the recognition of discourse. Additionally, it is believed that the mental lexicon is organized into a network of specific information in which the organization of representations involves relationships between words so that words representing similar meanings are connected and tend to be grouped closer together.[48]

In order to understand words, individuals pass through pre-lexical processing stages: decodifying signals that enter acoustically and are translated into a phonological code. Subsequently, the lexical representation that best fits the auditory signal can be chosen from the mental lexicon (lexical selection). The form of the word selected initially activates the theme (storage of grammatical information) and, following that, the meaning of the word.[49]

Comprehension includes the following: the recognition or judicious guesswork of vocabulary; remembering orality-related systems, facts, and details (purposes embedded in the propositions of the speaker); and the identification and interconnections of ideas and the principles therein. In completing these tasks, listeners make use of their ability to draw inferences about content or perceived proposition.[50]

To do this, listeners must grasp, interpret, and evaluate oral information within the communicative context using their knowledge of the world and of vocabulary. Logical relationships, the chain of communicative functions, the use of gestures, and rhetorical resources all significantly influence the comprehension of oral language.[48]

In 1978, DeRenzi and Faglioni,[51] developed the Token Test, a tool widely used to quantify difficulties in listening comprehension in order to assess the receptive function of language. It makes use of short statements, and applying the test is both quick and easy.

This test is composed of 36 verbal instructions that demonstrate an increasing level of linguistic complexity, which are divided into six parts. To administer the test, twenty pieces are needed. These pieces are two different shapes (square and circle), two sizes (small and large), and four colours (black, yellow, green, and white). They are arranged according to the instructions provided by the test's authors. The participants respond to instructions such as: "Touch the circle"; "Touch the yellow square"; and "Touch the large black circle and small green square". The scores can range from 0 to 36 points and are obtained by totalling the scores of each test item. A score of 29 to 36 indicates comprehension impairment; a score of 25 to 28, mild impairment; a score of 17 to 24, moderate impairment; a score of 9 to 16, severe impairment; and a score of 0 to 8, very severe impairment.

Few studies[15,52-53] have investigated linguistic alterations in TBI victims, which means that sequelae are underestimated and underdiagnosed. The consequences of these alterations are not only physical but also psychosocial, potentially compromising patients' social relationships. Furthermore, as we have argued, oral comprehension skills are undoubtedly essential to everyday life, as they determine our ability to understand oral messages, a key feature of communication's effectiveness.

There has been confirmation, achieved through the most modern neuroimaging techniques, that the left hemisphere is responsible for the language process in virtually all right-handed individuals, as well as in more than half of left-handed and ambidextrous individuals.[54]

A TBI patient's communication can change according to language disorders. Some individuals may have difficulty understanding or producing oral and written language, or they may have difficulty with the more subtle aspects of communication, like body language or emotional and nonverbal cues.

Our study assessed 122 TBI patients between 14 and 83 years old who were admitted into a public trauma referral hospital located in a city in north-eastern Brazil in 2012. It revealed that most TBI victims presented some level of oral comprehension impairment after traumatic brain injury, ranging from mild to moderate alterations.[55]

4. Post-TBI communication disorders

Aphasia, dysarthria, and apraxias are among the chief alterations in communication that are caused by TBI-related disorders or neurological damage.[33]

There are also cognitive-linguistic disorders. These cognitive-communicative impairments were defined and classified by ASHA in 1988 as any change in communication resulting in cognitive deficits (such as memory, attention, and logical reasoning) that produce symptoms and difficulties in communication that are traditionally considered unclassifiable and that present within normal language during formal tests, such as the Boston Test. However, this type of data and the impact of these changes after the cerebral injury remain imprecise.[56-57]

Aphasia is a multimodal disorder that affects reading, writing, auditory comprehension, and orally-expressed language. It should not, however, be regarded as a specific disorder, as other cognitive processes, such as attention and short-term auditory memory, can also be involved.[58]

Thus, aphasia is essentially a linguistic processing disorder in which the mechanisms that transform thought into language are blocked. Furthermore, the disorder compromises initiative, creativity, and the ability to perform calculations, i.e., skills that call on the use of internal speech.[59]

Aphasia's cause is neurological in origin and could be associated with several aetiologies (vascular, infection, tumour, cranial trauma, degenerative disease, demyelinating diseases, and toxic disorders).[59-61] Its classification is described in Table 1 below.

APHASIAS	Emissive Aphasias	Broca's Aphasia
		Conduction Aphasia
		Transcortical Motor Aphasia
	Receptive Aphasias	Wernicke's Aphasia
		Transcortical Sensory Aphasia
		Anomic Aphasia
	Mixed Aphasias	Mixed Transcortical Aphasia
		Global Aphasia

Table 1. Classification of Aphasias

More common expressive or motor aphasia is associated with injuries involving the frontal language centre in the dominant hemisphere (Broca's area) and is therefore mainly associated with an inability to translate spoken concepts into meaningful sounds, or in other words, to produce speech. The result is speech that is not fluent, with pauses between words or phrases.

Conduction aphasia is characterized by phonemic paraphasias, anomies and semantic paraphasias during the conversation. The speech may appear with hesitation and self-corrections. A striking feature of this type of aphasia corresponds to errors found in the repetition test.

Another type of non-fluent aphasia is transcortical motor aphasia, whose main feature is the reduction of speech. Spontaneous language is extremely reduced, and its expression is slow and short.

Receptive or sensorial aphasia is related to injuries to the posterior language area in the dominant hemisphere. Wernicke's aphasia is the most serious comprehension aphasia associated with problems in the comprehension and formulation of speech.

In transcortical sensory aphasia, oral expression is fluent; at the same time, severe and moderate comprehension deficits appear; and there are semantic paraphasias, anomies and circumlocutions.

Anomic or amnesic aphasia is primarily characterized by semantic changes, paraphrases and anomies.

The mixed forms of aphasia are conditions that exhibit characteristics of the several manifestations described. As an example, there are: transcortical motor aphasia, in which oral expression is characterized by stereotypes and echolalia; and global aphasia, in which the patient has a severe impairment of oral expression and listening comprehension.

As the expressive aspects of speech depend upon the normal functioning receptive aspects, language expression may also be impaired in individuals with sensory aphasia, who may present with unintelligible words, changing words (paraphasia), and other expressive disorders related to speech production. Thus, the main differences between motor and sensory aphasia are in language comprehension, which is only slightly affected in the former but severely affected in the latter; and in speech, with non-fluent aphasia in motor aphasia and fluent aphasia in the sensory variety.[62]

It is important to emphasize and describe the changes in oral language abilities when dealing with aphasic syndrome.[63]

Verbal fluency: this criterion is mainly used to differentiate between fluent and non-fluent aphasic syndromes. With aphasia in which oral language is fluent, an ease of articulation can be seen even in long sentences; generally, this type of aphasia is the result of a posterior lesion. Non-fluent aphasia is characterized by a difficulty in initiating oral production, which causes strain. This type of aphasia is associated with anterior lesions.[64-66] Muteness would be the most extreme degree of reduction in fluency, whereas, logorrhoea is a marked increase in the number of words produced in a certain amount of time.[66]

Anomia: a difficulty or inability to recall names of objects, leading to a restriction in vocabulary. It is present at different levels in every type of aphasia and has several causes. Attempts to compensate for this difficulty often feature synonyms or circumlocution. While anomia does not exclusively affect substantive words, nouns are the most compromised class of words.[60,61-64] For example, sufferers have difficulty in recalling or retrieving the words in speech. The processes for retrieving the words are the same for aphasic and not aphasic, but for people with the disease, the operation becomes slow, cumbersome and often ineffective.

Paraphrase: occurs when a subject, while trying to say a word, substitutes the word with a phrase,[67] for example, what serves as combing for comb.

Circumlocution: expression which takes place when a patient can neither grasp the main theme of the enunciation nor discuss it. In his statement he touches upon the theme but does not manage to specifically discuss it,[67] for example, a wooden object which has a backrest, four legs and is used to sit on is a chair.

Repetition: one of the most basic mechanisms of human language; however in patients with aphasia, it can be impacted in different ways. The phonemic pathway is used to repeat not-words, while words are repeated by accessing the meaning. Repetition is kept intact in extrasylvian (transcortical) aphasia but harmed in perisylvian aphasia (Broca, Wernicke, conduction, and global).[64] An example of this condition is: "the house, the house, the house".

Auditory comprehension: a complex function resulting from the processing of speech sounds in Wernecke's area; occurs when concepts related to a registered word are activated and selected. The process involves several areas with different modalities and hierarchies that are distributed throughout the entire brain. Generally, it is more affected in fluent aphasia.[59,66,68]

Agrammatism: the disorganization of syntactic rules present in language, which leads to a significant reduction in an individual's statements. It is characterized by an almost telegraphic style, where prepositions, articles, conjunctions, and pronouns are omitted, but nouns, adjectives, and verbs (almost always in the infinitive) are preserved. A loss of prosody and a lack of declension for gender, time, and number can also be observed. It is a hallmark of Broca's aphasia.[60,63-64] An example of agrammatism is: "Father to lead to the college", note the absence of "my" "will go" "me" "to" and "a".

Stereotyping: perseverative and involuntary repetitions of a certain type of behaviour. Patients will use restricted verbal production, with or without linguistic meaning, every time that they attempt oral or written communication. It is present in Broca's aphasia.[60,63-64,66] For example, a patient produces this sound: eeeeeeeeeee, eeeeeee, eeeee…

Perseveration: maintaining the same response for distinct stimuli. Patients will use a word incorrectly right after it has been used in a different, more appropriate context. This is also associated with Broca's aphasia.[60,63-64]

Jargon: discourse that has no message, in which syntax and semantics are absent; it is language that is incomprehensible, without meaning, and spoken at a rapid pace. It is present in the more serious fluent aphasias.[60,63-64] For example, "It's going to rain upon noodle stones plantation."

Echolalia: the repetition of an interlocutor's sounds, in an unsolicited context, with no communicative purpose.[64,66] An example of echolalia is: chair, chair, chair...

Paraphasia: the substitution of letters, syllables, or words during discourse.

Phonological paraphasia: a wrong choice during the act of articulation, characterized by distortion in the production of phonemes. Patients substitute one phoneme for another, [60,64,66-67] for example, plants for pants.

Phonemic paraphasia: a change in the phonological level of language; it consists of substitutions influenced by production context or by similarity of certain traits. It can be manifested as a change, an omission, or an addition of phonemes or syllables and is present in a large number of fluent aphasias.[60,63,66-67] An example is shark for sharp.

Morphemic paraphasia: a change characterized by the substitution of words' grammatical morphemes,[67] for example: talk for talking.

Formal paraphasia: occurs when a swap, substitution, addition, or omission results in a different word in the language, without being characterized as a semantic swap, for example, goat for coat.

Verbal paraphasia: when the patient makes a substitution in an oral statement and cannot identify its relation to the content or form of the statement, for example, tiger for lion.

Semantic paraphasia: occurs when one word is substituted with another that has the same semantic context,[60] for example, pen for pencil.

Neologism: phonemic or graphemic sequences that obey a language's rules and resemble words, but do not exist in that language. When trying to say a word, patients will substitute the word with a sequence of meaningless sounds.[64] An example of this is, "The cake was eatful" instead of eatable.

Reduction: Decrease in the number of enunciations in a certain amount of time,[67] for example: "the girl's hair is beautiful" for "girl... beautiful hair".

http://dx.doi.org/10.5772/57321

Suppression: the complete absence of oral or graphic emission. This term can be considered a synonym for muteness when used in the context of oral statements.[67]

The sensorimotor sequelae of TBI can impair an individual's communication and affect the ability to produce intelligible speech. This happens when the trauma affects the areas of the brain that are responsible for the execution of movements necessary to produce speech and triggers a neuromotor disorder called dysarthria.[69] Dysarthria is characterized by slowness, weakness, and/or lack of muscle coordination related to the speech function. Its main consequence is a reduction in speech's intelligibility, which limits the speaker's communicative ability and social participation.[15,70]

It often develops after damage to the central or peripheral nervous system, which mainly affects laryngeal function, causing weakness or lack or muscle coordination during speech, as well as changes in the oral statement.[71]

For production to render intelligible speech, the phonoarticulatory apparatus needs to be working in perfect symphony; along with the oral cavity, the pulmonary, laryngeal, and pharyngeal structures form the apparatus, and any change to any of these will consequently impact speech intelligibility.[71-72] The dysarthria occurs when there is an impairment in the motor apparatuses necessary for oral production (which are: breathing, phonation, resonance, articulation, and prosody) following a central or peripheral neurological change. This fact justifies the emergence of the terms "dysarthrophonia" and "neurological dysphonia" as synonyms for dysarthria in order to describe this condition, which is not just a change in articulation.[73]

There are several types of dysarthria that vary according to the degree and location of the injury: flaccid dysarthria, spastic dysarthria, unilateral upper motor neuron dysarthria, hypokinetic dysarthria, hyperkinetic dysarthria, ataxic dysarthria, and lastly, mixed dysarthria[74]. The most common symptoms range from a decreased rate of speech, vagueness, articulation, slow, irregular speech to a lack of change in pitch or intensity.[75-76]

With TBI, flaccid dysarthria is the most common type. In flaccid dysarthria, the injury is located in the lower motor neuron, which is peripheral, but it can also emerge due to some cranial nerve lesions. Nerve conduction is impaired at a point between the cell body and the muscle, and the resulting changes are flaccidity, weakness, atrophy, and fasciculations.[73] Thus, motor function is changed, potentially resulting in muscle paralysis, a breathier and more monotone voice, hypernasality, imprecise articulation of consonants, diminished volume and predominantly pharyngolaryngeal resonance.[71] Loss of muscle mass is also common for this type of dysarthria.

Spastic dysarthria can also emerge after closed TBI and is caused by a bilateral lesion to the upper motor neuron, which causes an increase in muscle tone, spasticity, and weakness. Among its main characteristics are: a rough, stressed voice; tight, choked sounds; monotone; imprecise articulation of consonants; and hypernasality.[73]

Another type of dysarthria that can occur due to trauma injury is mixed dysarthria, in which changes typical of several types of dysarthria all emerge at the same time and have no sort of pattern. This is because the injuries involve multiple areas of the central and peripheral nervous system. This type can also occur in cases of stroke, degenerative, metabolic, and toxic diseases, and infectious diseases of the central or peripheral nervous system.

Research[77] conducted at the São Paulo Hospital (HSP), Brazil, developed a profile of dysarthric patients which showed that traumatic cranial lesions were the second most frequent aetiology related to dysarthria, with male patients prevailing due to both the higher rate of males in automobile accidents and the higher frequency of flaccid dysarthria in TBI patients. In another Brazilian study,[33] the prevalence of traumatic cranial injury in individuals attended to at the Acquired Neurological Disorders Outpatient Service of the Speech Pathology Department at UNIFESP was 75.6%. The majority of these cases also involved male patients, and 33% of patients were diagnosed with dysarthria.

After a diagnosis of dysarthria has been made and its aetiology determined, it is up to a speech therapist to evaluate the patient and connect phonological signs to neurological changes, so as to better determine short-, mid-, and long-term therapy practices that align with both the patient's prognosis and a general clinical view of the case. In severe cases of TBI, the occurrence of dysphagia associated with dysarthria is common. Therefore, it is necessary to have a clear and precise diagnosis so that adequate rehabilitation can be carried out.[73]

Apraxia should also be brought to attention and seen as an articulation disorder that leads to the loss of the ability to perform previously learned motor acts. In apraxia, there is a difficulty in associating the voluntary programming of the position of the muscles that form speech organs with the sequential movement of these muscle groups, which prevents the formation of appropriate language. Apraxic patients present these symptoms, although they exhibit no abnormalities in the motor and sensory systems, or in comprehension, cooperation, and attention skills. It is damage to the primary motor cortex that causes apraxia.[78]

They are, therefore, speech articulation disorders that result in a loss of the ability to programme and organize the position of the speech apparatus to voluntarily produce phonemes, or of the sequence of muscle movements to produce words; a loss which is not, however, accompanied by the weakness, slowness, or lack of coordination that affects these same muscles in reflexive or involuntary movements caused by cerebral lesion.[1]

5. Interdisciplinary approach

Interdisciplinary care for TBI victims performed by a multidisciplinary team is recommended during post-trauma treatment. It is necessary to evaluate the specific needs of the individual, aiming not only for preservation of live, but also for quality of life.

Family participation is vitally important in each treatment phase. A team with a speech therapist, doctor, nurse, occupational therapist, psychologist, physical therapist, social worker, and nutritionist, among other types of professionals, is highly suggested.

A speech therapist has the goal of helping patients improve their communicative abilities, in addition to addressing other concerns such as chewing and swallowing. It is important to start rehabilitation as soon as possible, as this encourages and optimizes the initial spontaneous recovery process.[14]

The main goal of a speech therapist, then, is to maximize a patient's communication.[79-80] In the initial phases, the objective of communication rehabilitation is to offer sufficient support in order to facilitate the recovery of the communicative function. Subsequently, the focus of the intervention becomes the generalization of communication skills in activities with varying contexts.[79]

The rehabilitation process involves combining suitable intervention approaches for each case, such as: behavioural approaches; skills and specific process training; guidelines; metacognitive approaches (tasks that require analysis of semantic similarities, main ideas and topics, and narrative schemes, among others); interventions focused on people living with the individual (training in the use of communication strategies and assistive technology, for example); and use of augmentative and alternative communication (AAC) that can be defined as "any resource that can be used to encode and transmit a message without requiring writing skills or vocalization"..[1,73,79,81]

Regarding the use of supplementary systems and alternative communication in the therapeutic process with patients with brain injury, although studies are scarce, the literature indicates these systems as facilitators of communication, i.e., as resources for the rehabilitation of speech and communication facilitation in day-to-day situations. There have also been descriptions of adaptations and the use of high-tech systems, mostly for cases of cerebral palsy.[1, 4-5]

In order for a TBI patient's communication to be optimized or to be made possible, cognitive stimulation should be started as soon as possible. The goal of an intervention is to maximize patients' potential and promote faster, more organized evolution through the stimulation of various sensory modalities, using familiar materials and resources.[82] With evolution on a cognitive level, the individual usually begins to communicate better and can start to speak, read, or write again. Such resources should be used in the process of care, as well as in the process of rehabilitating the linguistic and cognitive aspects that changed as a result of TBI.[73]

It is essential that patients in a state of mental confusion are oriented about the circumstantial, spatial, and temporal aspects that can help to contextualize their situation, such as: what happened, where they are, their location in time, what will be done at that moment, and other relevant pieces of information.[83]

When patients can control their attention, even for a limited amount of time, certain strategies can be employed that make message comprehension easier for the patients. These can include: always facing the patient while speaking, using repetition and redundancy, favouring short and direct sentences, speaking more slowly, minimizing the presence of noise or other stimuli, and using warning signals that gear the patient's attention towards receiving information.[83] These strategies can also be used by other members of the team and by the patient's relatives.[73,83]

In dysarthria cases where communication is impacted, rehabilitation should also take into account an intervention concerning specific aspects of speech production, with the aim of optimizing the patient's intelligibility.[79,84]

Most neurological damage resulting from TBI does not occur at the time of the injury, but rather over the course of the hours and days after the accident, thus it can be prevented or treated.[85]

Early indicators of a bad prognosis should be identified in the clinical history and, most importantly, during a physical examination of the patient. The doctor should have some knowledge of the patient so that he can try to return to use the window of time between the trauma and the ensuing damage in which to prevent future consequences, and in doing so prevent secondary damage.[86]

The nurse also provides care to these victims and should be able to obtain data on the patient's history, perform a physical examination, and provide prompt treatment in order to preserve life and prevent secondary damage.[87] The systematization of nursing care is not only necessary, but almost of utmost importance.

There is little scientific evidence concerning physical therapy intervention and the improvement of communication disorders in individuals with traumatic brain injury, but this area of healthcare has a range of behaviours that complement the treatment of the voice, speech, and language functions within an interdisciplinary context. To achieve adequate vocal performance, it is necessary to teach breathing.

Patients with head trauma commonly suffer losses in respiratory efficiency and effectiveness; when using accessory muscles, more energy is expended per respiratory cycle. The ideal is that the gas exchange process take the least possible toll on their energy reserves, and for this, focus is put on the diaphragm, extensive musculature adapted to breathing. Respiratory therapy is highly recommended for lung restructuring, maintaining the force and strength of respiratory musculature, increasing gas exchange efficiency, and eliminating the use of accessory muscles.[88-89] Techniques vary from manual manoeuvres to release the diaphragm and diaphragmatic breathing to the use of devices known to motivate inspiratory flow and volume.[90]

Another problem for cranial injury victims besides respiratory disorders is aphasia. One method that has evolved over the past decade is Constraint-Induced Language Therapy. This therapy consists of restricting the aphasic patient's communication to verbal communication and prohibiting any other method of communication, either gestural or written. The therapy session is long and tiring, held for an average of two or three hours a day over a period of about two weeks. It has been well-received in the clinical setting, and results have been promising.[91] Another advantage is that regardless of whether the aphonic presents with an acute or chronic condition, therapy brings about clinical improvement, and studies have shown changes of 28% to 47% in patients with chronic aphasia.[92]

Communication goes far beyond spoken language. Interaction using gestures and signs is crucial for those who have lost voice function and/or hearing after a traumatic brain injury. An injury to the precentral gyrus of the frontal lobe of the cerebrum, for example, can affect the primary movements, compromising the execution of gestures and signs. Physical therapy works to develop gross and fine motor skills, vision skills, sensory processing skills, movement coordination, and appropriateness of tone – key points for communication.[93] In situations where various muscle groups are affected by spasticity, physical therapists may recommend systemic drug therapy to contribute to the reduction of tone, in addition to recommending physical measurements, exercise and orthotics.[94]

From a psychological perspective, there are several areas that can directly affect communication. For over four decades, psychology has developed, through studies on language acquisition, into different models of theoretical and methodological concepts that serve as linguistic techniques and behaviour. The field of psycholinguistics aims to address conduct or behaviour related to language in the context of a subject's psychological and social functioning.

Neuropsychology is an area of psychology that allows greater dialogue between different professionals in the neurosciences for the purpose of establishing knowledge about cognitive functions, behaviour, and the functioning of the Central Nervous System (CNS).

This area holds extensive knowledge about clinical and research practices working mainly with the assessment (neuropsychological tests) and treatment (neuropsychological rehabilitation) of disorders that can occur in the Central Nervous System, such as trauma.[95]

In Neuropsychology, each function of the Central Nervous System is detailed, and this enables more effectively-directed rehabilitation. With TBI, in which most accidents occur in the region of the Frontal Lobe (considered executive functions), there are several studies of the components involved in this area, such as working memory, planning, problem solving, decision making, fluency, and inhibition control, among others.[96]

The main evaluations performed that concern linguistic components are: fluency, repetition, comprehension, and naming. These components can lead to more exact diagnoses about the seriousness of the spoken, written, and comprehended language impairment of a patient who has Central Nervous System dysfunction,[97] and can also help the patient's rehabilitation and communication.

Patients with Central Nervous System Dysfunction require more intensive care from their families because, besides having difficulties with motor and cognitive functions, they experience changes in personality and behaviour. It is important that patients and families work in an interdisciplinary way, using humanized reception and the structure provided by help groups in order to supplement the patients' rehabilitation and the social stability of the family.

6. Conclusion

Communication disorders can occur regardless of the severity of an injury, and they have a major impact on the level of discourse and the social exchanges. They can have a negative effect on the patients' recovery process, reintegration into the community, independence, familial interactions, and professional and academic success.

In this context, we stress the importance of the interdisciplinary approach in the follow-up treatment of these victims, so that prognosis and rehabilitation, as well as post-TBI quality of life, may improve.

Author details

Edilene Curvelo Hora*, Liane Viana Santana, Lyvia de Jesus Santos, Gizelle de Oliveira Souza, Analys Vasconcelos Pimentel, Natalia Tenório Cavalcante Bezerra, Sylvia Rodrigues de Freitas Doria, Tiago Pinheiro Vaz de Carvalho, Afonso Abreu Mendes Júnior, Jessica Almeida Rodrigues, Renata Julie Porto Leite Lopes and Ricardo Fakhouri

*Address all correspondence to: edilene@ufs.br

Health Sciences Post-Graduate Programme, Academic Trauma League (LITRAUMA), Federal University of Sergipe, Brazil

References

[1] World Health Organization. Neurotrauma. 2013. http://www.who.int/violence_injury_prevention/road_traffic/activities/neurotrauma/en (accessed 05 July 2013).

[2] Gentile JKA, Himuro HS, Rojas SSO, Veiga VC, Amaya LEC, Carvalho JC. Condutas no paciente com trauma cranioencefálico. Revista da Sociedade Brasileira de Clínica Médica 2011;9(1). http://files.bvs.br/upload/S/1679-1010/2011/v9n1/a1730.pdf (accessed 05 July 2013).

[3] Aghakhani N, Azami M, Jasemi M, Khoshsima M, Eghtedar S, Rahbar N. Epidemiology of Traumatic Brain Injury in Urmia, Iran. Iranian Red Crescent Medical Journal 2013;15(2). http://www.ncbi.nlm.nih.gov/pmc/articles/PMC3652507/ (accessed 05 July 2013).

[4] Brasil, Ministério da Saúde. Estatística e mortalidade. Óbitos por ocorrência segundo causas externas de morbidade e mortalidade do Brasil, 2011. http://tabnet.datasus.gov.br/cgi/tabcgi.exe?sim/cnv/ext10uf.def (accessed 05 July 2013).

[5] Pinheiro AL, De Almeida FM, Barbosa IV, Mesquita ME, Borges SRM, De Figueiredo CZM. Principais causas associadas ao traumatismo cranioencefálico em idosos. Enfermería global 2011;10(22). http://scielo.isciii.es/pdf/eg/v10n22/pt_clinica4.pdf (accessed 05 July 2013).

[6] Cheng-Min H, Jeffrey CL, Miranda JJ, Hyder AA. Traumatismos causados por El tránsito en países em desarrollo: agenda de investigación y de acción. Revista Peruana de Medicina Experimental y Salud Pública 2010;27(2) 243-247.

[7] Mackenzie EJ. Epidemiology of injuries: current trends and future challenges. Epidemiol Rev 2000;22 112-119.

[8] Koizumi MS, Lebrão ML, Mello-Jorge MHP, Primerano V. Morbidity and Mortality from traumatic brain lesion in São Paulo, 1997. Arquivos de Neuropsiquiatria 2000;58 1-13.

[9] Kay A, Teasdale G. Head Lesion in the United Kingdom. World J Sur 2001;25 1210-1220.

[10] JE Boswell, McErlean M, Verdile VP. Prevalence of traumatic brain lesion in an ED population. Am J Emerg Med 2002;20 177-180.

[11] The Levati, Farina ML, Vecchi G, Rossanda M, Marrubini N. Prognosis of severe head injuries. J Neurosurg 1982;57 779-783.

[12] Junque C, Bruna O, Marató M. Head trauma: an approach to neuropsychology and speech therapy - practical guide for professionals and families. 1st ed. São Paulo (SP): Santos Publisher; 2001.

[13] Hora EC, Sousa RMC, Ribeiro COM, Nunes MS, Araújo RCV, Santos ACFS et al. Traumatic Brain Injury: consequences and family needs. In: Agrawal A (ed) Brain Injury: functional aspects, rehabilitation and prevention. Rijeka: InTech; 2012 161-174. http://www.intechopen.com/books/brain-injury-functional-aspects-rehabilitation-and-prevention (accessed 28 September 2013).

[14] Junqué C, Bruna O, Mataró M. Traumatismos Cranioencefálicos: uma abordagem da neuropsicologia e da fonoaudiologia. São Paulo: Editora Santos; 2001.

[15] American Speech Hearing Association. http://www.asha.org/public/speech/disorders/TBI/ (accessed 20 August 2012).

[16] Ferreira ABH. Dicionário Aurélio Básico da Língua Portuguesa. Rio de Janeiro: Nova Fronteira; 1999.

[17] Souza TNU, Payão LMC. Apraxia da fala adquirida e desenvolvimental: semelhanças e diferenças. Revista da Sociedade Brasileira de Fonoaudiologia 2008;13(2) São Paulo April/June.

[18] Mansur LL. Linguagem. In: Malloy-Diniz, Fuentes D, Mattos P, Abreu N e cols. Avaliação neuropsicológica. Porto Alegre: Artmed; 2010.

[19] Almeida EC, Duarte PM. Consciência fonológica. Rio de Janeiro: Revinter; 2003.

[20] Mayor A. La pragmática del lenguaje: consideraciones para la intervención. Leng. Comun Madri 1991;(7) 17- 21.

[21] Hage SRV, Resegue MM, Viveiros DCS, Pacheco EF. Análise do perfil das habilidades pragmáticas em crianças pequenas normais. Pró-Fono Revista de Atualização Científica 2007, Barueri (SP); 19(1) 49-58.

[22] Gil R. As afasias. São Paulo: Editora Santos; 2002.

[23] Piaget J-P. O nascimento da inteligência na criança. 4ª ed. Rio de Janeiro: LTC Editora; 1987. Edição original de 1936.

[24] Aguiar VT. O verbal e o não verbal. São Paulo, Editora UNESP; 2004.

[25] Corraze J. As comunicações não verbais. Rio de Janeiro, Editora Zahar; 1982.

[26] Vygotsky LS. Pensamento e linguagem. São Paulo: Martins Fontes; 2000.

[27] Ardila R. Verbal Behavior de B.F. Skinner: sua importância no estudo do comportamento. Revista Brasileira de Terapia Comportamental Cognitivista 2007;9(2) 195-197.

[28] Pinker S. Tábula rasa: a negação contemporânea da natureza humana. São Paulo: Companhia das Letras; 2004.

[29] Alvarez AMMA, Sanchez ML, Carvalho IAM. Neuroaudiologia e linguagem. In: Fuentes, D, Malloy-Diniz LF, Camargo CHP, Consenza, RM e cols. Neuropsicologia teoria e prática. Porto Alegre: Artmed; 2008.

[30] Consenza RM, Fuentes D, Malloy-Diniz LF. A evolução das ideias sobre a relação entre cérebro, comportamento e cognição. In: Fuentes D, Malloy-Diniz LF, Camargo CHP, Consenza RM e cols. Neuropsicologia teoria e prática. Porto Alegre: Artmed; 2008.

[31] Girodo CM, Silveira VNS, Girodo GAM. Afasias. In: Fuentes D, Malloy-Diniz LF, Camargo CHP, Consenza RM e cols. Neuropsicologia teoria e prática. Porto Alegre: Artmed; 2008.

[32] Nicholas M, Sinotte MP, Helm-Estabrooks N. C-Speak Aphasia Alternative Communication Program for People with Severe Aphasia: Importance of Executive Functioning and Semantic Knowledge. Neuropsychological Rehabilitation 2011; 21(3) 322–366.

[33] Talarico TR, Venegas MJ, Ortiz KZ. Perfil populacional de pacientes com distúrbios da comunicação humana decorrentes de lesão cerebral, assistidos em hospital terciário. Revista CEFAC 2011;13(2) 330-339.

[34] Silva MJP. Comunicação tem remédio: a comunicação nas relações interpessoais em saúde. São Paulo: Editora Gente; 1996.

[35] Castaño J. Bases neurobiológicas del lenguaje y sus alteraciones. Revista de Neurologia 2003; 36(8) 781-785

[36] Johansson B, Berglund P, Ronnback L. Mental fatigue and impaired information processing after mild and moderate traumatic brain injury. Brain Inj 9009; 23(13) 1027-1040.

[37] Norrie J, Heitger M, Leathem J, Anderson T, Jones R, Flett R. Mild traumatic brain injury and fatigue: a prospective longitudinal study. Brain Injury 2010; 24(13) 1528-1538.

[38] Podell K, Gifford K, Bougakov D, Goldberg E. Neuropsychological assessment in traumatic brain injury. Psychiatric Clinics of North America 2010; 33(4) 855-876.

[39] Jean-Francois D, Thierry G, Cardebat D. Renewal of the Neurophysiology of Language Functional Neuroimaging Physiol 2005; Rev 85 49 –95.

[40] Luria AR, Yudovich FI. Linguagem e desenvolvimento intelectual na criança (Trad.: José Cláudio de A. Abreu). Porto Alegre: Artes Médicas, 1985.

[41] Eysenck M, Keane M. Limitações da atenção e do desempenho. Manual de psicologia cognitiva. 5th ed. Tradução de: Magda França Lopes. Porto Alegre: Artmed; 2007

[42] Flanagan DP. The achievement test desk reference (ATDR). Boston: Allyn and Bacon; 2008.

[43] Bear MF, Connors BW, Paradiso MA. Neurociências: desvendando o sistema nervoso. 3ª edição. Porto Alegre: Artmed; 2008.

[44] Puyuelo M, Rondal J. Manual de desenvolvimento e alterações da linguagem na criança e no adulto. Porto Alegre, Artes Médicas; 2007.

[45] Fonseca RP, Ferreira GD, Liedtke FV, Muller JL, Sarmento TF, Parente MAMP. Alterações cognitivas, comunicativas e emocionais após lesão hemisférica direita: em busca de uma caracterização da síndrome do hemisfério direito. Psicologia USP 2006;17(4): 241-262.

[46] Joly MCRA, Dias AS. Evidências de validade de uma prova informatizada de linguagem oral – Bilo. Psicologia: Teoria e Prática 2009;11(2) 50-68.

[47] Rassovsky Y, Satz P, Alfano MS, Light RK. Functional outcome in TBI I: neuropsychological, emotional, and behavioral mediators. Journal of Clinical and Experimental Neuropsicologia 2006; 28 (4) 567-580.

[48] Gazzaniga MS, Ivry RB, Mangun GR. Neurociência cognitiva. Porto Alegre, RS: Artmed; 2006.

[49] Munjal SK, Panda NK, Pathak A. Audiological deficits after closed head injury. J Trauma 2010;68(1) 13-18.

[50] Cavalcanti A, Galvão C. Terapia Ocupacional: fundamentação e prática. Rio de Janeiro: Guanabara Koogan; 2007.

[51] DeRenzi E, Vignolo LA. The Token Test: A sensitive test to detect receptive disturbances in aphasics. Brain 1962;85 556-678.

[52] Lesser R. Verbal and non-verbal memory components in the Token Test. Neuropsychologia 1976;14(1) 79-85.

[53] Nelson T, Galvão OF. Discrepancies between the Token Test Procedure Original and Reduced Brazilian Adaptation. Neurobiology 2010;73(1) 155-159.

[54] Marin L, Queiroz MS. The actuality of traffic accidents in the age of speed: an overview. Cad Public Health 2000;16(1) 7-21.

[55] Santana LV, Hora EC. Impairment in oral language comprehension among victims of Traumatic Brain Injury. Dissertação de Mestrado, NPGME UFS; 2013.

[56] Ortiz KZ, Araújo AA. Traumatismo craniencefálico: Avaliação e reabilitação fonoaudiológica. In: Ortiz KZ (org). Distúrbios neurológicos adquiridos: linguagem e cognição. São Paulo: Manole; 2010. p284-300.

[57] Marini A, Galetto V, Zampieri E, Vorano L, Zettin M, Carlomagno S. Narrative Language in traumatic brain injury. Neuropsychologia 2011;49 2904-2910.

[58] Berthier ML. Poststroke aphasia: epidemiology, pathophysiology and treatment. Drugs Aging 2005;22 163-182.

[59] Amasio AR. Aphasia. N Engl Med 1992;326: 531-9.

[60] Cupello RCM, Miranda ABR. Rupturas em trajetos cerebrais subjacentes a alguns sinais neurolingüísticos encontrados em diversos tipos de afasia. Fono atual 2003;23 42-59.

[61] Mac-Kay APMG. Afasia. In: Mac-Kay APMG, Assencio-Ferreira VJ, Ferri-Ferreira TMS. Afasias e demências: avaliação e tratamento fonoaudiológico 2003. São Paulo: Santos;47-59.

[62] Murdoch BE. Desenvolvimento da fala e distúrbios da linguagem: uma abordagem neuroanatômica e neurofisiológica. Rio de Janeiro: Revinter; 1997. p113-123.

[63] Jakubovicz R, Cupello R. Introdução à afasia: elementos para o diagnóstico e terapia. 6th ed. Rio de Janeiro: Revinter; 1996, 276p.

[64] Engelhart E, Laks J, Rozenthal M. Neuropsicologia. VII – Distúrbios da linguagem. Afasia – aspectos neuroclínicos/neuropsicológicos. Revista Brasileira de Neurologia 1996;32 21-26.

[65] Goodglass H, Kaplan E. The assessment of aphasia and related disorders. Philadelphia: Lea&Febiger; 1972..

[66] Mansur LL, Senaha MLH. Distúrbios de linguagem oral e escrita e hemisfério esquerdo. In: Nitrini R, Caramelli P, Mansur LL. Neuropsicologia: das bases anatômicas à reabilitação. São Paulo: Clínica Neurológica do Hospital das Clínicas da Faculdade de Medicina da Universidade de São Paulo; 1996. p183-201.

[67] Ortiz KZ. Distúrbios neurológicos adquiridos: linguagem e cognição. Rio de Janeiro: Revinter; 2005.

[68] Benson DF. Aphasia. In: Heilman KM, Valenstein E. Clinical neuropsychology. 3rd ed. New York: Oxford; 1993. p17-36.

[69] Dykstra AD, Hakel ME, Adams SG. Application of the ICF in reduced speech intelligibility in dysarthria. Semin Speech Lang 2008;28(4) 301-311.

[70] Guo YE, Togher L. The impact of dysarthria on everyday communication after traumatic brain injury: a pilot study. Brain Inj 2008; 22(1) 83-97.

[71] Carrilo L, Karin ZO. Vocal analusis (auditory – perceptual and acoustic) in dysarthrias. Pró-Fono Revista de Atualização Científica 2007;19(4) 381-386.

[72] Barreto SS, Ortiz KZ. Medidas de inteligibilidade nos distúrbios da fala: revisão crítica da literatura: [revisão]. Pró-Fono 2008;20(3) 201-206.

[73] Ortiz KZ. Distúrbios neurológicos adquiridos: fala e deglutição. Barueri: Manole: 2010.

[74] Ortiz KZ. Distúrbios neurológicos adquiridos: fala e deglutição. Barueri: Manole; 2006.

[75] Tjaden K, Wilding GE. Rate and loudness manipulations in dysarthria: acoustic and perceptual findings. J Speech Lang Hear Res 2004;47(4) 766-783.

[76] Kent RD, Duffy JR, Slama A, Kent JF, Clift A. Clinicoanatomic studies in dysarthria: review, critique, and directions for research. J Speech Lang Hear Res 2001;44(3) 535-551.

[77] Ribeiro AF, Ortiz KZ. Populational profile of dysarthric patients assisted in a tertiary hospital. Rev Soc Bras Fonoaudiol 2009;14(3) 446-453.

[78] Souza TNU, Payão LMC. Apraxia da fala adquirida e desenvolvimental: semelhanças e diferenças. Rev Soc Bras Fonoaudiol 2008;13(2) 193-202

[79] Cicerone KD, Dahlberg C, Kalmar K, Langenbahn DM, Malec JF, Bergquist TF, et al. Evidence-based cognitive rehabilitation: recommendations for clinical practice. Arch Phys Med Rehabil 2000;81(12) 1596-1615.

[80] MacDonald S, Wiseman-Hakes C. Knowledge translation in ABI rehabilitation: a modelo for consolidating and applying the evidence for cognitive-communication interventions. Brain Inj 2010;24(3) 486-508.

[81] Wallace T, Bradshaw A. Technologies and strategies for people with communication problems following brain injury or stroke. NeuroRehabil 2011; 28(3) 199-209.

[82] Chabok SY, Kapourchali SR, Leili EK, Saberi A, Mohtasham-Amiri Z. Effective factors on linguistic disorder acute phase following traumatic brain injury in adults. Neuropsychologia 2012;50 1444-1450.

[83] Shelton C, Shryock M. Effectiveness of communication/interaction strategies with patients who have neurological injuries in a rehabilitation setting. Brain Inj 2007; 21(12) 1259-1266.

[84] Sellars C, Hughes T, Langhorne P. Speech and language for dysarthria due to non-progressive brain damage. Cochrane Database Syst Ren 2005;20(3) 2088.

[85] Brain Trauma Foundation. Guidelines for the management of severe traumatic brain injury, 3rd Edition. Journal of Neurotrauma 2007; 24.

[86] Brain Trauma Foundation. Early indicators of prognosis in severe traumatic brain injury. 2000.

[87] Pereira N et al. O cuidado do enfermeiro à vítima de traumatismo cranioencefálico: uma revisão da literatura. Revista Interdisciplinar NOVAFAPI 2011;4(3) 60-65.

[88] Gething AD, Williams M, Davies B. Inspiratory resistive loading improves cycling capacity: a placebo controlled trial. British Journal of Sports Medicine 2004;38 730–736.

[89] Silva AMO, Boin IFS, Pareja JC, Magna LA. Analysis of respiratory function in obese patients submitted to fobi-capella surgery. Revista do Colégio Brasileiro de Cirurgiões 2007;34 314-320.

[90] Irwin S, Teckilin J. Fisioterapia Cardiopulmonar. São Paulo: Manole; 2003

[91] Pulvermuller F, Neininger B, Elbert T, et al. Constraint-induced therapy of chronic aphasia after stroke. Stroke 2001;32(7) 1621–1626.

[92] Moss A, Nicholas M. Language rehabilitation in chronic aphasia and time postonset: a review of single-subject data. Stroke 2006;37(12) 3043–3051.

[93] Gunel MK. Rehabilitation of children with cerebral palsy from a physiotherapist's perspective. Acta Orthop Traumatol Turc 2009;43(2) 173-180.

[94] Gracies JM, Elovic E, McGuire JR, Nance P., Simpson DM. Tradicional phamacologic treatments for spasticity part II: systemic treatments. In: Mayer NH, Simpsnon DM (ed.). Spasticity: we move self – study activity; 2002. p65-93.

[95] Consenza RM, Fuentes D, Malloy-Diniz LF. A evolução das ideias sobre a relação entre cérebro, comportamento e cognição. In: Fuentes D, Malloy-Diniz LF, Camargo CHP, Consenza RM e cols. Neuropsicologia teoria e prática. Porto Alegre: Artmed; 2008.

[96] Malloy-Diniz L, Sedo M, Fuentes D, Leite WB. Neuropsicologia das funções executivas. In: Fuentes D, Malloy-Diniz LF, Camargo CHP, Consenza RM e cols. Neuropsicologia teoria e prática. Porto Alegre: Artmed; 2008.

[97] Girodo CM, Silveira VNS, Girodo GAM. Afasias. Fuentes D, Malloy-Diniz LF, Camargo CHP, Consenza RM e cols. Neuropsicologia teoria e prática. Porto Alegre: Artmed; 2008.

Chapter 21

Long-Lasting Mental Fatigue After Traumatic Brain Injury – A Major Problem Most Often Neglected Diagnostic Criteria, Assessment, Relation to Emotional and Cognitive Problems, Cellular Background, and Aspects on Treatment

Birgitta Johansson and Lars Rönnbäck

Additional information is available at the end of the chapter

http://dx.doi.org/10.5772/57311

1. Introduction

Fatigue after traumatic brain injury (TBI) is common, but often overlooked. But for people fighting their fatigue after brain injury day after day, fatigue is a major problem. This post-injury mental fatigue is characterized by limited energy reserves to accomplish ordinary daily activities. Persons who have not experienced this extreme exhaustion which may appear suddenly, and without previous warning during mental activity, do not understand the problem. This is especially difficult to understand as the fatigue may appear even after seemingly trivial mental activities which, for uninjured persons, are regarded as relaxing and pleasant, as reading a book or having a conversation with friends. A normal, well-functioning, brain performs mental activities simultaneously throughout the day, but after a brain injury, it takes greater energy levels to deal with cognitive and emotional situations.

In this chapter, we highlight mental fatigue after TBI. In the case of long-lasting mental fatigue, it could be the only factor that keeps people from returning to the full range of activities that they pursued prior to their injury with work, studies and social activities. We describe mental fatigue and suggest diagnostic criteria and we also give a theoretical explanation for this. At the end of the chapter, we discuss treatment strategies and give some examples of possible therapeutic alternatives which may alleviate the mental fatigue.

Normally, the brain works in an energy-efficient manner and prominent energy reserves are present. This is due to well-functioning ion channel and amino acid transport systems and other effective physiological processes. After brain injury, some of these systems are down-regulated, and when mental energy requirements are high the physiological processes do not function to their full capacity; these cease to function efficiently with a resultant energy loss. This may be an explanation as to why the mental fatigue appears.

1.1. When does mental fatigue occur?

Annually, about 100-300/100 000 individuals sustain a TBI, and most of the injuries are mild in severity [1]. A majority of patients recover within one to three months following mild TBI [2, 3].

Fatigue is one of the most important long-lasting symptoms following TBI, and is most severe immediately after head injury. However it is difficult to arrive at any clear figure as to how common fatigue or, in particular, mental fatigue is. The reason for this is that different results have been obtained, and these are attributable to differences in definitions and differences in the methodology in the various studies. In follow-up studies, the frequency of prolonged fatigue varies from 16 up to 73 % [4-6]. There is no correlation between persistent fatigue and severity of the primary injury, age of the person at injury or time since injury [7, 8]. For those suffering from fatigue 3 months after the accident the fatigue remained relatively stable during longer periods [9]. In particular, for those subjects who were suffering from the syndrome one year after the accident improvement in the fatigue was limited [10].

In the above reports, fatigue is discussed in terms of a single construct, i.e. not differentiated between the physical or mental aspects. In this chapter, we consider mental fatigue as a separate construct and we discuss its relationship to cognitive and emotional symptoms.

1.2. Mental fatigue is not a separate diagnostic entity

Mental fatigue is not an illness, rather it represents a mental sequel, probably due to a disturbance of higher brain functions, either physical or psychological in origin. It is included in, and defined within the diagnoses Mild cognitive impairment (F06.7), Neurasthenia (F48.0) and Posttraumatic brain syndrome (F07.2) [11].

1.3. Typical characteristics of mental fatigue

A typical characteristic of pathological mental fatigue after TBI is that the mental exhaustion becomes pronounced during sensory stimulation or when cognitive tasks are performed for extended periods without breaks. There is a drain of mental energy upon mental activity in situations in which there is an invasion of the senses with an overload of impressions, and in noisy and hectic environments. The person feels that their brain is overloaded after a tiny load. Another typical feature is a disproportionally long recovery time needed to restore the mental energy levels after being mentally exhausted. The mental fatigue is also dependent on the total activity level as well as the nature of the demands

http://dx.doi.org/10.5772/57311

of daily activities. Fatigue often fluctuates during the day depending on the activities carried out. Thus, this fatigue is a dynamic process with variations in the mental energy level. The fatigue can appear very rapidly and, when it does, it is not possible for the affected person to continue the ongoing activity. Common associated symptoms include: impaired memory and concentration capacity, slowness of thinking, irritability, tearfulness, sound and light sensitivity, sensitivity to stress, sleep problems, lack of initiative and headache [12].

For many persons, this mental fatigue is the dominating factor which limits the person's ability to lead a normal life with work and social activities. For most people, fatigue subsides after a period of time while, for others, this pathological fatigue persists for several months or years even after the brain injury has healed. Interestingly, however is that as many as 30% of family or friends interpreted fatigue as laziness [9].

Theories as to the mechanisms accounting for mental fatigue including our own theory, suggest that cognitive activities require more resources and are more energy-demanding after brain injury than usual [13, 14]. Thus, more extensive neural circuits are used in TBI victims compared to controls during a given mental activity [15]. This indicates an increased cerebral effort after brain injury.

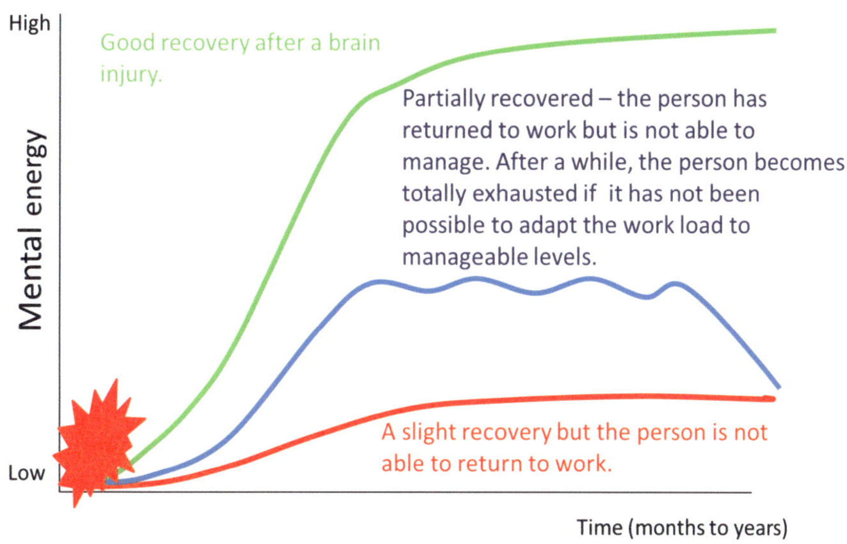

Figure 1. Schematic representation of recovery of mental energy after TBI. The green line represents a full recovery while the blue and red lines represent impaired recovery in terms of the mental energy levels. Persons whose recovery follows the blue line recover partially. On their return to work and daily activities, they are not able to manage and they become exhausted. Persons whose recovery follows the red line do not recover and are not able to return to work and daily activities.

Therapist Luann Jacobs describes mild TBI and the lack of energy and lack of endurance that many can experience. As they are able to do what is normal and what appears normal, they run the risk that their symptoms will be misunderstood [16].

"Mild brain injury is a real misnomer, as it conveys the idea that nothing much is a problem when quite the opposite is more often true. It is called "mild" because, in fact, the mildly brain injured can walk, talk, eat and dress independently, often times drive a car, shop, cook, go to school, or even work.

What the term fails to account for is the inherent limits of how often, for how long (endurance), and the all-important, how consistently (e.g., every day, once a week) these activities can be performed. Even more elusive is the concept of how many of these daily activities can be done sequentially in a given day as is normal in the lives of people who are not brain injured.

The fatigue they feel defies description, going far beyond and far deeper than anything a non-brain-injured person would consider profound exhaustion."

2. Theoretical explanation of mental fatigue

The cause of this extreme fatigue is not known. However, there are speculations that the symptom may be caused by dysfunction of the astrocytes, the most common supporting cells in the brain [17, 18]. As a consequence, nerve cell communications do not function properly.

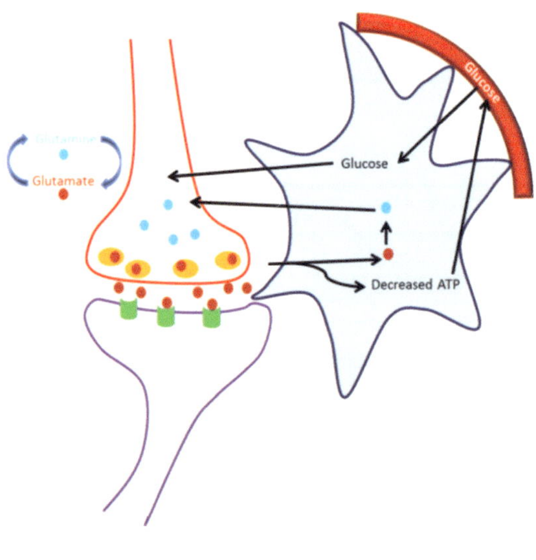

Figure 2. Schematic drawing of a synapse with glutamate as the transmitter and an astrocyte with processes surrounding the synaptic terminal. After being released from the presynaptic terminal (pre-syn; this is shown in red in the figure), glutamate interacts with glutamate recognizing receptors on the postsynaptic membrane (post-syn; shown in green in the figure). After stimulation of the postsynaptic neuron, glutamate is taken up by glutamate transporting systems on the astrocyte processes. Glutamate is converted to glutamine in the astrocyte and transported back to the presynaptic terminal where glutamine is converted back to glutamate. During this process, and with decreasing ATP levels as the signal, glucose is taken up from the blood to supply neurons and astrocytes with energy.

http://dx.doi.org/10.5772/57311

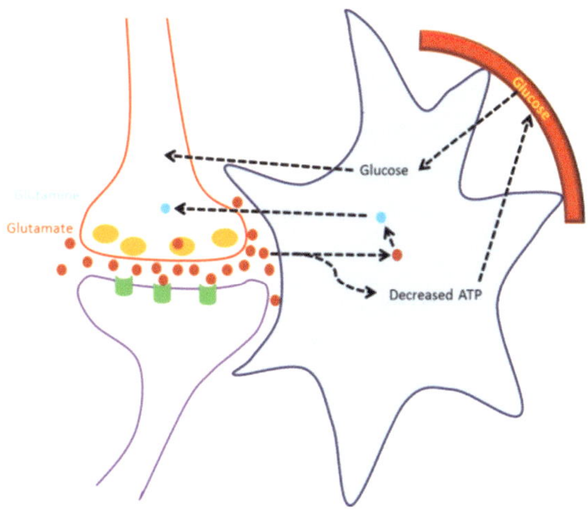

Figure 3. Following TBI there is a neuroinflammation with down-regulation of astroglial glutamate transport systems. If this state is not restored completely, there will be an impaired extracellular glutamate clearing with slightly increased extracellular glutamate levels, slight astrocyte swelling and impaired glucose uptake. Neuronal activity, if long-lasting, may result in energy crisis.

2.1. On cellular mechanisms probably underlying mental fatigue

Following TBI there is a low-grade neuroinflammation with down-regulation of astrocyte glutamate transporters and Na+/K+ ATPase activity [19, 20]. If these physiological systems are not restored completely there will be a dysfunctional support of the glutamate transmission. Glutamate signaling is essential for information processing, including learning and memory formation. Low levels and fine-tuning of extracellular glutamate are necessary to maintain high precision in information processing, and thereby high efficiency in the information handling within the CNS. Our hypothesis implies that such dysfunction could underlie the mental fatigue at the cellular level. From experimental data, the astroglial cells are considered the most important cells for clearing the extracellular space from glutamate during glutamate transmission. In addition, it is well-accepted from the experimental data that this clearing capacity is attenuated by substances or conditions associated with brain dysfunction or pathology (see [17]).

If the capacities of these processes are not fully restored, neuronal function is impaired in at least two ways: 1) extracellular glutamate levels increase upon neuronal activity leading to unspecific signaling and 2) lack of energy. In the event of a high mental load with high neuronal activity, these factors may lead to a metabolic collapse of neuronal circuits – we have previously called this a "dead-lock" situation, which may take a long time to restore.

We consider this metabolic failure as one probable explanation for the prominent and abrupt exhaustion that the TBI victims with mental fatigue can experience. The long restoring time at a cellular level corresponds to the long time it takes for the TBI victims to restore mental activity.

One way to restore this dysfunction is to stimulate Na^+/K^+-ATPase along the dopaminergic circuits which regulate attention and executive functions. Possible candidates are methylphenidate and the dopaminergic stabilizer OSU6162 (see below under the heading, 'Treatment').

3. Assessment of mental fatigue

There is an abundance of scales for assessing fatigue in general and several of these scales are designed for use in different diseases [21, 22]. The scales include questions relating to feelings of fatigue, perceived impact on activities, affective feelings and mental or cognitive effects. Many of the scales are self-reported on a Likert or an ordered scale, with the following response alternatives: Never, Sometimes, Regularly, Often or Always.

We have developed and used the Mental Fatigue Scale (MFS) during the last five years. We decided to construct this scale since we were not able to find an assessment scale adapted to mental fatigue. The MFS is a multidimensional questionnaire containing 15 questions. It incorporates affective, cognitive and sensory symptoms, duration of sleep and daytime variation in symptom severity. The questions concern the following: fatigue in general, lack of initiative, mental fatigue, mental recovery, concentration difficulties, memory problems, slowness of thinking, sensitivity to stress, increased tendency to become emotional, irritability, sensitivity to light and noise, decreased or increased sleep as well as 24-hour symptom variations. The questions in the scale are based on common activities and we have related the estimation to exemplified alternatives. It is also possible to provide estimations in-between two alternatives. The intention was to make the scale more consistent between individuals and also between ratings for the same individual. The exemplified alternatives can help the person to respond in a similar way despite the present state of fatigue or emotional state. The MFS is designed in a similar way as The Comprehensive Psychopathological Rating Scale (CPRS). The CPRS also includes exemplified alternatives and it is used to record changes in psychopathology over a comparatively short period [23]. The questions included in the MFS are based on symptoms described following longitudinal studies of patients with TBI, brain tumours, infections or inflammations in the nervous system, vascular brain diseases, and other brain disorders, which indicates that an acquired brain injury or disorder can result in similar symptoms [24-26]. The scale is free to use and can be downloaded at www.mf.gu.se (both in Swedish and English). We have transcribed one of the questions in the MFS, below:

Mental fatigue

Does your brain become fatigued quickly when you have to think hard? Do you become mentally fatigued from things such as reading, watching TV or taking part in a conversation with several people? Do you have to take breaks or change to another activity?

0	I can manage in the same way as usual. My ability for sustained mental effort is not reduced.
0.5	
1	I become fatigued quickly but am still able to make the same mental effort as before.
1.5	
2	I become fatigued quickly and have to take a break or do something else more often than before.
2.5	
3	I become fatigued so quickly that I can do nothing or have to abandon everything after a short period (approx. five minutes).

Figure 4 shows how healthy controls and subjects suffering from mild TBI, TBI and stroke have rated separate questions on the MFS. The brain injury victims were divided into different groups according to their total rating on MFS. When a person rates low on one question, the total rating on most of the separate questions will also be low, while persons rating high on one question on the MFS, will also rate most of the questions on a high level.

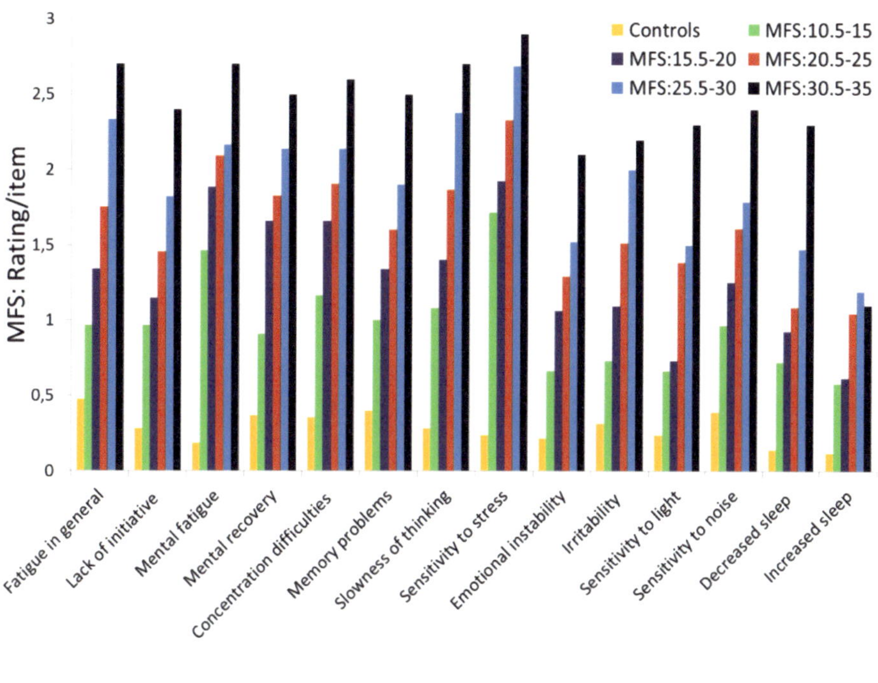

Figure 4. Rating on separate items on the Mental Fatigue Scale for controls and brain injured subjects. Brain injured subjects are divided into groups according to their total rating on MFS.

3.1. The use of MFS and results from the studies

The rating on MFS by healthy controls and people who suffered mild TBI or TBI did not reveal any significant differences between females and males, and there was no correlation between the results on MFS and age or education of the TBI victims (figure 5). Furthermore, we did not find any correlation for the TBI participants concerning time since injury and their rating on MFS. We have, in our studies worked with participants with mental fatigue lasting for six months or periods greater than six months. At this stage, we do not have any data relating to ratings early after TBI or mild TBI. This accounts for the fact that the rating may lack correlation to time since injury.

The control group rated MFS significantly lower than mild TBI and TBI victims. The participants included for the analysis were healthy controls and participants who had suffered mild TBI or TBI without major depression. The participants were between 20-67 years of age.

We recommend a cutoff score on the MFS at 10.5. A score of 10.5 on the MFS was found to deviate significantly from the control sample and is also above the 99th percentile for the control group. A score above 10.5 implies a problem for the person, although a serious problem is not always the case. However, such a score implies the need for the person to consider the current situation with their work and/or social life. The MFS had a high internal consistency and all separate items were rated significantly higher among brain injured subjects compared with healthy controls (see also figure 5).

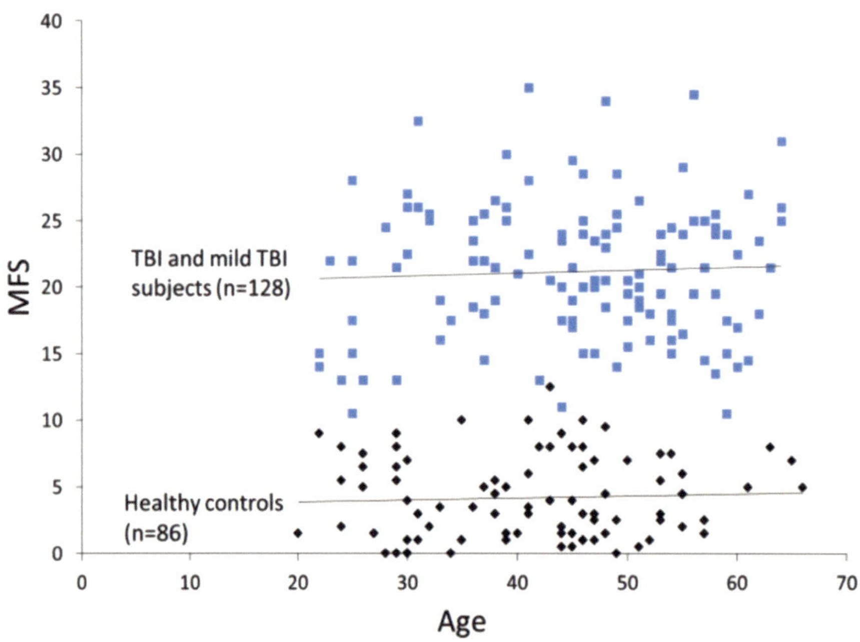

Figure 5. Correlation with age and rating on MFS for healthy controls and subjects with long-lasting mental fatigue after brain injury.

4. Mental fatigue and connection to cognitive functions

It has been proposed that subjective mental fatigue after TBI or mild TBI correlates to poor performance in attention tests and reduced processing speed [13, 27, 29-34]. We also found that information processing speed, attention and working memory were significantly reduced for the brain injury victims (both mild TBI and TBI) compared to controls. Furthermore, the tests correlated significantly to the results on the MFS (figure 6). Among the cognitive functions, processing speed was found to be a significant predictor for the rating on MFS [27].

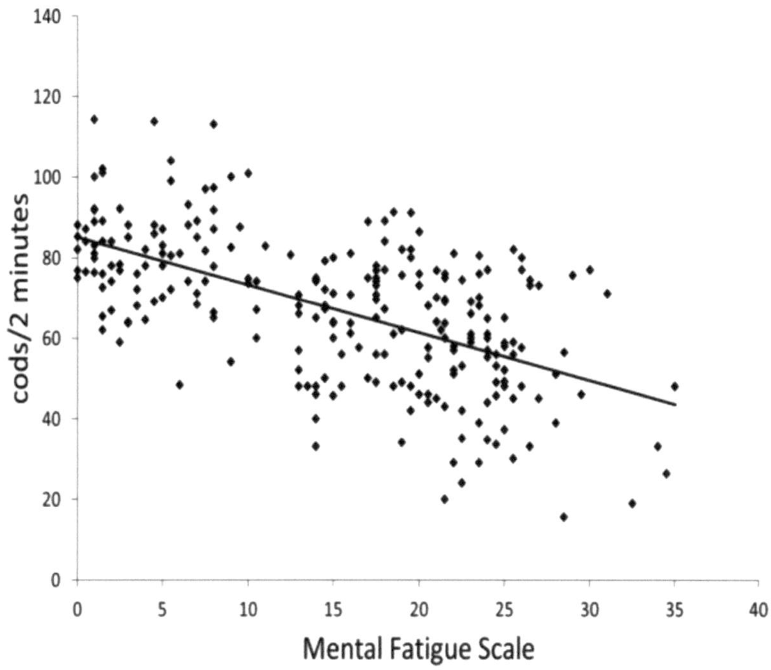

Figure 6. Correlation between Mental Fatigue Scale and information processing speed (Digit Symbol-Coding).

5. Mental fatigue and connection to emotional functions

In the population of TBI victims, depression is elevated although there is a wide variation in frequency, depending on methodological differences [35-37]. In our studies, we have included participants who complained of mental fatigue after TBI and we excluded subjects affected by major depression, as it was our intention to explore the mental fatigue component. Despite this, we found, with the use of the CPRS/MADRS, that there was an elevation in the rating of depression items for TBI subjects compared to controls. The CPRS scale includes both a depression and an anxiety scale [23, 38]. The CPRS depression scale is also called the Montgomery-Åsberg Depression Rating Scale (MADRS) [39].

However, there are overlapping items in the MFS and CPRS. The overlapping items include the following: lack of initiative, concentration difficulties, irritability and decreased sleep. With a factor analysis, the items were separated into a mental fatigue component and a depression and anxiety component. Irritability was placed in the depression-anxiety component and the other three items in the mental fatigue component. With an analysis using the new components, we found that by adjusting the mental fatigue component this removed the difference observed between the brain injured subjects and controls in the depression-anxiety component. However, by removing the depression-anxiety component this did not have an effect on the difference observed between the brain injured subjects and controls in the mental fatigue component.

In this subject sample, we were able to demonstrate that a significant effect on the difference observed between the brain injured subjects and controls in the scores for depression can result in an overestimation if the effect of the mental fatigue component is not taken into consideration. This indicates that mental fatigue and depression must be treated as separate constructs and it is also important to make this distinction for the purposes of therapeutic strategies.

6. Definition and diagnostic criteria for long-lasting mental fatigue

The diagnostic criteria for posttraumatic brain syndrome include most of the symptoms that are often present along with mental fatigue. However, we suggest mental fatigue to be a central symptom after a brain injury reflecting an inefficient support to the neuronal networks.

Mental fatigue is a lack of mental energy with impaired cognitive, emotional and sensory functioning. Mental fatigue is characterized by an unusual feeling of fatigue or malaise. There is a drain on the person's mental energy upon mental activity. The result is a diminished attention and concentration capacity. Situations which involve high levels of external cues and an overload of impressions are strenuous. Failing energy levels and excessively long recovery times are the result of over-exertion. The condition impairs the person's ability to function in their work, studies and gatherings with family and friends.

6.1. How to recognize long-lasting mental fatigue

- The mental fatigue has persisted for at least 1 month;
- The sum of scores from the MFS is 10.5 points or above.

Typical symptoms include:

- An unusually rapid drain of mental energy upon mental activity;

http://dx.doi.org/10.5772/57311

- Impaired attention and concentration capacity over time;
- Following over-exertion, a long recovery time disproportionate to the exertion level;
- Diurnal variation of the fatigue symptom with the fatigue often being better in the mornings and worse in the afternoons and evenings; variations from one day to the next;
- Usually one or several associated symptoms (see below):

The following additional or associated symptoms are common:

- Mood swings, irritability and stress intolerance;
- Trouble with memory;
- Sleep problems;
- Sensitivity to, or intolerance of light and loud noise;
- Headaches following over-exertion.

Sleep problems most often occur in the following way: either a shorter duration of sleep with interrupted wake-ups or sleeping more than usual. If the person becomes more mentally fatigued, the sleep will most often become worse, and if the person rests for some days the sleep can become improved again. The emotional load may increase the severity of the fatigue, but if mental fatigue exists, it will remain even once the emotional components, as depression or anxiety have been treated. However, it is important to treat the emotional problems. In this way, the mental fatigue may, to some extent be relieved.

6.2. If the fatigue is not acknowledged in time, this may result in the following

- A total and almost paralyzing fatigue;
- Longer periods of rest may be needed, often over several days;
- A worsening of symptoms over time;

6.3. Mental fatigue symptoms are often present in the following situations

- Situations in which there is an invasion of the senses with an overload of impressions, and noisy and hectic environments such as crowded events, also the hustle and bustle of shopping centers, and travelling by bus, etc.;
- Reading books and newspapers;
- Conversations with people – this becomes more of a struggle when more people are involved;
- Unexpected events.

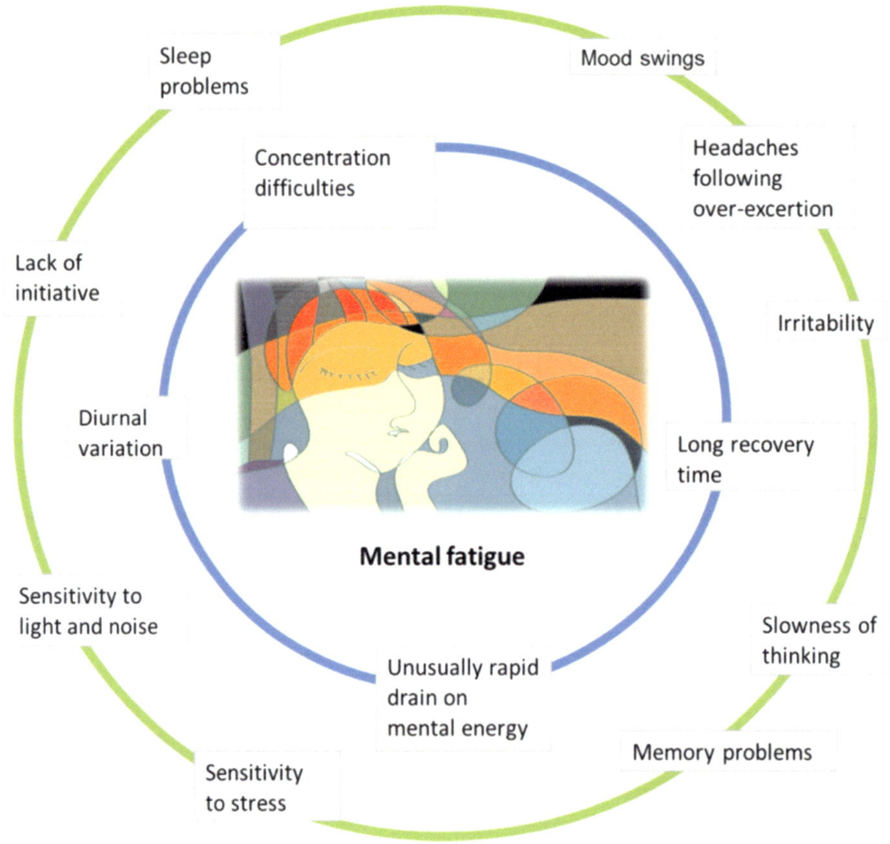

Figure 7. The figure illustrates mental fatigue. Characteristic symptoms are seen on the blue circle and associated symptoms on the green circle. (The figure in the middle is illustrated by Kristina Edgren Nyborg).

7. Treatment

There is currently no effective treatment for mental fatigue. For many people, there is an increased risk of doing too much and becoming even more fatigued. Today, the most important recommendations are to adapt to the energy available by doing one thing at a time, resting regularly and not overdoing things.

When mental fatigue is present, it is important to adapt work as well as daily activities to levels that the brain can manage. However, this is challenging for most people and it may take a long time, even years, to adapt to a sustainable level. It may also be difficult for the person to learn by himself/herself and it can take several years of considerable struggle, frustration, despair and depression, to find the right balance between rest and activity. Professional support is required but this can be hard or impossible to find especially when mental fatigue continues for many years.

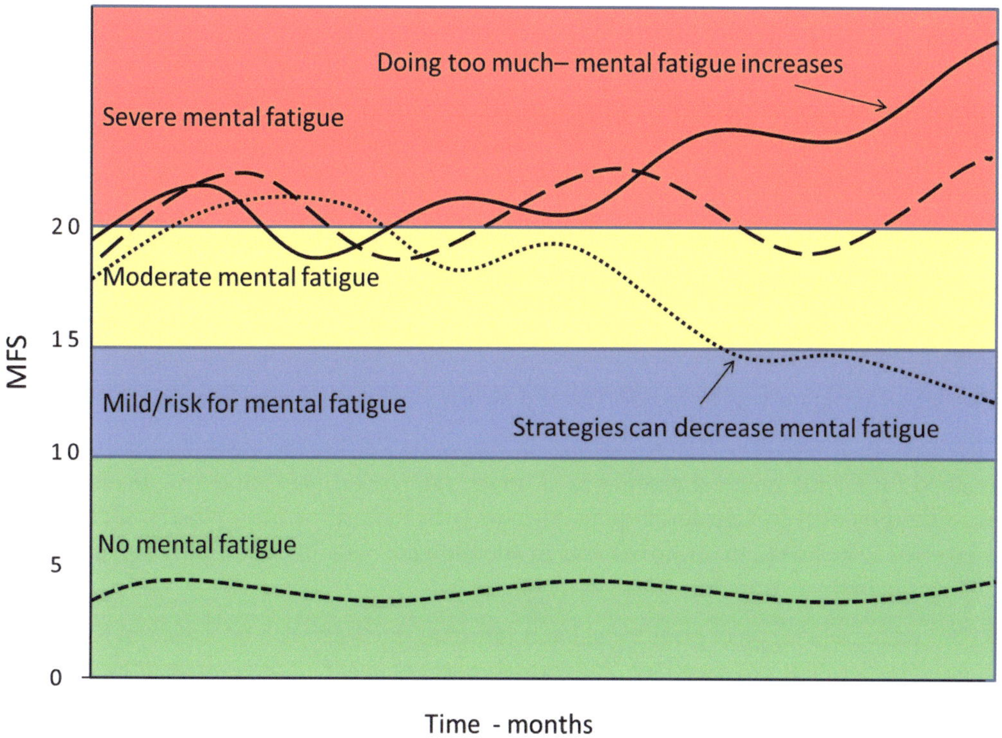

Figure 8. The figure illustrates levels and fluctuations in mental fatigue measured with the MFS after TBI and variations over time. Most mild TBI victims recover completely (green field) and do not exceed 10 points on the MFS. People within the blue, yellow or red fields suffer from metal fatigue to varying extents. It is also shown that treatment strategies decrease the mental fatigue, while over-exertion leads to increased rates on the MFS.

7.1. Treatment strategies

- Take regular breaks;
- Encourage rest before becoming over-tired;
- Try to work at a steady pace, taking one task at a time with short working periods, and prioritize the tasks;
- Plan the days' activities or the activities for the week in a diary or journal. Avoid over-exertion.

The use of strategies is important. By resting the brain as much as possible the mental energy will be alleviated. However, the brain and the individual also need positive experiences and stimulation to ensure wellbeing. It is difficult to achieve this balance between rest and stimulation.

7.2. Treatment studies for alleviating mental fatigue

When mental fatigue becomes a prolonged problem, it is essential to be able to alleviate the symptoms. We have reported on significantly reduced mental fatigue after treatment using the mindfulness-based stress reduction (MBSR) program [40, 41]. We have also reported on possible therapeutic strategies to reduce mental fatigue by means of pharmacological treatments, using neurostimulant substances as methylphenidate [42] which affects dopamine and norepinephrine signaling. We have also reported on a new substance not currently available on the market, (-)-OSU6162, which is a dopamine and serotonin stabilizer [43].

7.2.1. Mindfulness

The MBSR program was tested on TBI and stroke victims suffering from long-term mental fatigue [40]. MBSR is a clinically-effective

method for a wide range of conditions as stress, depression, pain, and fatigue after cancer, with the potential to help individuals to cope with their difficulties [44-47]. MBSR is also suggested to be linked to improvements in attention and cognitive flexibility [48] and also to changes in brain neuronal connectivity [49].

MBSR includes a range of both formal and informal practices. The intervention is based on Kabat Zinn's MBSR program [50]. The formal practices in MBSR are described by M. Cullen 2011 [51] and these include gentle Hatha yoga with an emphasis on mindful awareness of the body, a body scan designed to systematically, region by region, cultivate an awareness of the body without the tensing and relaxing of muscle groups associated with progressive relaxation, and sitting meditation with an awareness of the breath as well as a systematic widening of the field of awareness to include all four foundations of mindfulness: awareness of the body, feeling tone, mental states and mental contents. As such, the intention of MBSR is much greater than simple stress reduction. The program consists of eight weekly group sessions which are each approximately 2.5 hours long, one day-long silent-led retreat between sessions six and seven and home practice of about 45 minutes, six days a week. The participants receive guided instructions and CDs for home practice.

We found a significantly reduced mental fatigue after the MBSR program and participants improved their processing speed significantly compared to control on waitlist [40]. Improvement was independent of gender, time since injury and age. Another recent study with MBSR for mild TBI patients showed a similar result with significant improvement in quality of life, perceived self-efficacy, working memory and attention [52]. Furthermore, a small-scale study of 10 mild TBI subjects included in the MBSR program over a 12-week period also showed a significantly improved quality of life and decreased depression rating [53]. The effects were maintained one year later among the seven contactable participants. They also noted an improvement in reported energy levels at the follow-up [54]. However, after TBI, a short MBSR program over a 4-week period did not result in any cognitive or emotional changes [55].

The results demonstrate that mindfulness practice may be a therapeutic method well-suited to subjects suffering from mental fatigue after brain injury. One reason why MBSR was effective may be that this treatment offers strategies to better handle stressful situations

http://dx.doi.org/10.5772/57311

appropriately and economize with mental energy. Despite the problem of ensuring that participants stay awake, which is one of the fundamental aspects of meditation, it was possible to adjust mindfulness to suit the needs of mental fatigue subjects and to improve their wakefulness as well as reducing their mental fatigue levels.

7.2.2. Methylphenidate

Methylphenidate inhibits dopamine and noradrenalin reuptake resulting in increased extracellular concentration of dopamine and noradrenalin [56]. Methylphenidate has been used for many years in the treatment of ADHD in children, in the first instance to increase wakefulness, attention and concentration capacity. Methylphenidate has also been tested on TBI victims with positive effects on information processing speed and, to some extent on working memory and attention [57-63]. Guidelines for use of methylphenidate for deficits of attention and processing speed after TBI have been suggested [64], while no such guidelines exist for fatigue following TBI.

In an open randomized study, methylphenidate significantly improved mental fatigue dose-dependently as assessed with the MFS [42]. The item, pain was also studied and we found that this item was rated high by most of the subjects in our study as the participants were recruited on the basis of the items, TBI and pain. However, no significant alleviation of pain was reported as a result of methylphenidate treatment. However, it is important to note that pain can hide posttraumatic brain injury symptoms or mental fatigue which is not always connected to the actual pain. We also found that there was no interaction between the pain and the mental fatigue in those participants treated with methylphenidate. These findings indicate that, not only is it necessary to treat patients for the pain for which they are primarily referred to the clinic, but also for the mental fatigue, if present.

Methylphenidate was well-tolerated by TBI subjects. However, tolerance of methylphenidate differed between subjects and we therefore recommend starting treatment with an initial low dose.

7.2.3. (-)-OSU6162

The monoaminergic stabilizer OSU6162 interacts with both dopaminergic and serotonergic systems. It appears to act as an antagonist on a binding site of the D2 receptor. More recent research has demonstrated that OSU6162 also exerts a stabilizing effect on serotonergic neuronal circuits, acting as a partial 5-HT2A agonist [65, 66].

In two randomized, double-blind and placebo-controlled studies we found statistically significant alleviation of mental fatigue after a stroke or TBI by OSU6162 during 4 weeks' treatment with active drug [43]. However, the numbers of patients in these studies were small (21 TBI and 19 stroke victims). Further studies are needed, with a larger number of patients and, in particular longer treatment periods as mental fatigue may be long-lasting. Adverse reactions were mild and could be avoided by dose adjustment. Several patients experiencing such adverse reactions expressed the wish to receive continued treatment with the drug.

Similar results were detected for methylphenidate and OSU6162. These drugs were shown to have the effect of both alleviating mental fatigue and increasing information processing speed.

8. Conclusions

Mental fatigue can become a prolonged and distressing problem after TBI having considerable effect on life and wellbeing. It is important to acknowledge and assess mental fatigue when discussing the options regarding therapeutic methods as the mental fatigue has been the result of a TBI.

After TBI, mental energy levels are failing, and the brain needs to rest. It is not possible to improve the mental energy with training in order to perform more mental activities. In fact, training with a view to resting the brain is what is important. Suitably-adapted and energy-saving strategies are important and most patients need support in order to achieve an enduring balance between activities and rest as this is difficult, it takes a long time and may be frustrating.

The treatment studies we reported on are aimed at helping the person to manage their life better. However, it is important to stress that there is a risk that the medication can compel the person to do more than is appropriate. The reason for this is that, most often they want to carry out activities in a similar way as before the injury and have been longing for the chance to be able to do this. The problem is that, for most persons suffering from long-term mental fatigue after TBI, the activity levels are close to the threshold of what they are able to sustain. This makes them susceptible if they increase their activity levels too much. With mindfulness most participants reported on more energy, but they also became more pleased and happy with life. Mindfulness also gave them a tool to use and they could take command over their own lives; how it is here and now, not longing for a better life or ruminating over what has been. This also saves energy! A combination with neurostimulants and mindfulness may be a good therapeutic strategy.

In the future, research is warranted for early treatment with the intention to reduce the development of long-term mental fatigue. We also need to better elucidate and carry out an in-depth analysis of mental fatigue.

Author details

Birgitta Johansson and Lars Rönnbäck

*Address all correspondence to: birgitta.johansson2@vgregion.se

Department of Clinical Neuroscience and Rehabilitation, Institute of Neuroscience and Physiology, Gothenburg University, Sweden

References

[1] Holm, L., et al., Summary of the WHO Collaborating Centre for Neurotrauma Task Force on Mild Traumatic Brain Injury. J Rehabil Med, 2005. 37(3): p. 137-141.

[2] Carroll, L.J., J.D. Cassidy, and P.M. Peloso, Prognosis of mild traumatic brain injury: results of the WHO collaborating Centre Task Force on Mild Traumatic Brain Injury. J Rehabil Med, 2004. 43(suppl): p. 84-105.

[3] Lundin, A., et al., Symptoms and disability until 3 months after mild TBI. Brain Injury 2006. 20(8): p. 799-806.

[4] Stulemeijer, M., et al., Recovery from mild traumatic brain injury: a focus on fatigue. J Neurol Neurosur Ps, 2006. 253(8): p. 1041-1047.

[5] Naalt, J.v.d., et al., One year outcome in mild to moderate head injury: the predictive value of acute injury characteristics related to complaints and return to work. J Neurol Neurosur Ps, 1999. 66: p. 207-213.

[6] O'Connor, C., A. Colantonio, and H. Polatajko, Long term symptoms and limitations of activity of people with traumatic brain injury: a ten-year follow-up. Psychological reports, 2005. 97: p. 169-179.

[7] Belmont, A., et al., Fatigue and traumatic brain injury. Ann Readapt Med Phys, 2006. 49: p. 283-288.

[8] Reitan, R.M. and D. Wolfson, The two faces of mild head injury. Arch Clin Neuropsych, 1999. 14(2): p. 191-202.

[9] Norrie, J., et al., Mild traumatic brain injury and fatigue: a prospective longitudinal study. Brain Injury, 2010. 24(13-14): p. 1528-15-38.

[10] Bushnik, T., J. Englander, and J. Wright, Patterns of fatigue and its correlates over the first 2 years after traumatic brain injury. Head Trauma Rehabil, 2008. 23(1): p. 25-32.

[11] ICD 10: International Statistical Classification of Diseases and Related Health (ICD-10-SE). 2nd edition ed. 2010, Stockholm.

[12] Rönnbäck, L. and B. Johansson, Long-Lasting Mental Fatigue After Recovery from Meningitis or Encephalitis - A Disabling Disorder Hypothetically Related to Dysfunction in the Supporting Systems of the Brain in Essential Notes in Psychiatry, V. Olisah, Editor. 2012, InTech: Rijeka, Croatia. p. 561-564.

[13] Azouvi, P., et al., Divided attention and mental effort after severe traumatic brain injury. Neuropsychologia, 2004. 42: p. 1260-1268.

[14] Zomeren, A.H.v. and W. van den Burg, Residual complaints of patients two years after severe head injury. J Neurosurg Psychiatry, 1985. 48(1): p. 21-28.

[15] Kohl, A.D., et al., The neural correlates of cognitive fatigue in traumatic brain injury using functional MRI. Brain Injury, 2009. 23(5): p. 420-432.

[16] Jacobs, L. Mild Brain Injury: Implications for Independence [cited 2013.

[17] Rönnbäck, L. and E. Hansson, On the potential role of glutamate transport in mental fatigue. J Neuroinflam, 2004. 1: p. 22.

[18] Rönnbäck, L. and B. Johansson, Long-lasting mental fatigue after traumatic brain injury or stroke – e new perspective 2012, Saarbrucken: LAP Lambert Academic Publishing

[19] Frencham, K.A.R., A.M. Fox, and M.T. Maybery, Neuropsychological studies of mild traumatic brain injury: a meta-analytical review of research since 1995. J Clin Exp Neuropsychol, 2005. 27(3): p. 334-351.

[20] Block, L., et al., A new concept affecting restoration of inflammation-reactive astrocytes. Neuroscience and Biobehavioral Reviews, 2013. in press.

[21] Christodoulou, C., The assessment of fatigue, in Fatigue as a window to the brain, J. DeLuca, Editor. 2005, The MIT Press: Cambridge, Massachusetts. p. 19-35.

[22] Dittner, A.J., S.C. Wessely, and R.G. Brown, The assessment of fatigue. A practical guide for clinicians and researchers. J Psychosomatic Res, 2004. 56: p. 157-170

[23] Åsberg, M., et al., A comprehensive psychopathological rating scale. Acta Psychiatr Scand 1978. 271(suppl): p. 5-27.

[24] King, N.S., et al., The Rivermead post concussion symptoms questionnaire: a measure of symptoms commonly experienced after head injury and its reliability. J Neurol Neurosur Ps, 1995. 24: p. 587-592.

[25] Lindqvist, G. and H. Malmgren, Organic mental disorders as hypothetical pathogenetic processes. Acta Psychiatr Scand, 1993. 88(suppl 373): p. 5-17.

[26] Rödholm, M., et al., Asteno-emotional disorder after aneurysmal SAH: reliability, symptomatology and relation to outcome. Acta Neurol Scand, 2001. 103: p. 379-385.

[27] Johansson, B., P. Berglund, and L. Rönnbäck, Mental fatigue and impaired information processing after mild and moderate traumatic brain injury. Brain Injury, 2009. 23(13-14): p. 1027-1040.

[28] Johansson, B., et al., A self-assessment questionnaire for mental fatigue and related symptoms after neurological disorders and injuries. Brain Injury, 2010. 24(1): p. 2-12.

[29] Belmont, A., N. Agar, and P. Azouvi, Subjective fatigue, mental effort, and attention dificits after severe traumatic brain injury. Neurorehabil Neural Repair, 2009. 23(9): p. 939-944.

[30] Ziino, C. and J. Ponsford, Selective attention deficits and subjective fatigue following traumatic brain injury. Neuropsychology 2006. 20: p. 383-390.

[31] Ziino, C. and J. Ponsford, Vigilance and fatigue following traumatic brain injury. J Int Neuropshychol Soc, 2006. 12(1): p. 100-110.

[32] Ashman, T.A., et al., Objective measurement of fatigue following traumatic brain injury. J Head Trauma Rehabil 2008. 23(1): p. 33-40.

[33] Park, N.W., M. Moscovich, and I.H. Robertson, Divided attention impairments after traumatic brain injury. Neuropsychologia 1999. 37(10): p. 1119-1133.

[34] Ponsford, J., et al., Long-term outcomes after uncomplicated mild traumatic brain injury: a comparison with trauma controls. J. Neurotrauma, 2011. 28: p. 937-946.

[35] Silver, J.M., J.W. Mc Allister, and D.B. Arciniegas, Depression and cognitive complaints following mild traumatic brain injury. Am J Psychiatry, 2009. 166(6): p. 653-661.

[36] Bombardier, C.H., et al., Rates of Major Depressive Disorder and Clinical Outcomes Following Traumatic Brain Injury. JAMA, 2010. 303(19): p. 1938-1945.

[37] Ashman, T.A., et al., Psychiatric challenges in the first 6 years after traumatic brain injury: Cross-sequential analyses of axis I disorder. Arch Phys Med Rehabil, 2004. 85(Suppl 2): p. S36-S42.

[38] Svanborg, P. and M. Åsberg, A new self-rating scale for depression and anxiety states based on the Comprehensive Psychopathological Rating Scale. Acta Psychiatr Scand, 1994. 89(1): p. 21-28.

[39] Montgomery, S.A. and M. Asberg, A new depression scale designed to be sensitive to change. Br J Psychiatry, 1979. Apr (134): p. 382-389.

[40] Johansson, B., H. Bjuhr, and L. Rönnbäck, Mindfulness-based stress reduction (MBSR) improves long-term mental fatigue after stroke or traumatic brain injury. Brain injury, 2012. 26(13-14): p. 1621-8.

[41] Johansson, B., H. Bjuhr, and L. Rönnbäck, Evaluation of an advanced mindfulness program following a mindfulness-based stress reduction program for participants suffering from mental fatigue after acquired brain injury. Mindfulness, DOI: 10.1007/S12671-013-0249-z.

[42] Johansson, B., et al., Evaluation of dosage, safety and effects of methylphenidate on posttraumatic brain injury symptoms with focus on mental fatigue and pain, Brain Injury, accepted.

[43] Johansson, B., et al., Placebo-controlled cross-over study of the monoaminergic stabiliser (−)-OSU6162 in mental fatigue following stroke or traumatic brain injury. Acta Neuropsychiatrica, 2012. 24(5): p. 266-274.

[44] Smith, B.W., et al., A pilot study comparing the effects of mindfulness-based and cognitive-behavioral stress reduction. J Altern Complement Med, 2008. 14(3): p. 251-258.

[45] Grossmana, P., et al., Mindfulness-based stress reduction and health benefits: A meta-analysis. Journal of Psychosomatic Research 57 (2004) 35–43, 2004. 57: p. 35-43.

[46] Kabat-Zinn, J., L. Lipworth, and R. Burney, The clinical use of mindfulness meditation for the self-regulation of chronic pain Journal of Behavioral Medicine, 1985. 8(2): p. 163-190.

[47] Carlson, L.E. and S.N. Garland, Impact of mindfulness-based stress reduction (MBSR) on sleep, mood, stress and fatigue symptoms in cancer outpatients International Journal of Behavioral Medicine 2005. 12(4): p. 278-285.

[48] Moore, A. and P. Malinowski, Meditation, mindfulness and cognitive flexibility. Consciousness and Cognition, 2009. 18: p. 176-186.

[49] Kilpatrick, L.A., et al., Impact of mindfulness-based stress reduction training on intrinsic brain connectivity. Neuroimage, 2011. 1(56): p. 290-298.

[50] Kabat-Zinn, J., Full catastrophe living: How to cope with stress, pain and illness using mindfulness meditation. 2001, London, 15th ed.: Piatkus Books.

[51] Cullen, M., Mindfulness-Based Interventions: An Emerging Phenomenon. Mindfulness, 2011. 2: p. 186-193.

[52] Azulay, J., et al., A Pilot Study Examining the Effect of Mindfulness-Based Stress Reduction on Symptoms of Chronic Mild Traumatic Brain Injury/Postconcussive Syndrome. Journal of Head Trauma Rehabilitation, 2013.28(4):323-331.

[53] Bedard, M., et al., Pilot evaluation of a mindfulness-based intervention to improve quality of life among individuals who sustained traumatic brain injuries. Disabil Rehabil, 2003. 25(13): p. 722-731.

[54] Bedard, M., et al., A Mindfulness-Based Intervention to Improve Quality of Life Among Individuals Who Sustained Traumatic Brain Injuries: One-Year Follow-Up. The Journal of Cognitive Rehabilitation, 2005. spring: p. 8-13.

[55] McMillan, T., et al., Brief mindfulness training for attentional problems after traumatic brain injury: A randomised control treatment trial Neuropsychological Rehabilitation, 2002. 12(2): p. 117-125.

[56] Leonard, B.E., et al., Methylphenidate: a review of its neuropharmacological, neuropsychological and adverse clinical effects. Hum Psychopharmacol Clin Exp, 2004. 19: p. 151-180.

[57] Alban, J.P., M.M. Hopson, and W. J, Effect on methylphenidate on vital signs and adverse effects in adults with traumatic brain injury. Am J Phys Med Rehabil 2004. 83: p. 131-137.

[58] Plenger, P.M., et al., Subacute methylphenidate treatment for moderate to moderately severe traumatic brain injury: A preliminary double-blind placebo-controlled study. 1996,77:536-540. Arch Phys Med Rehabil 1996. 77: p. 536-540.

[59] Speech, T.J., et al., A double-blind controlled study of methylphenidate treatment in closed head injury. Brain Injury 1993. 7(3): p. 333-338.

[60] Whyte, J., et al., Effects of methylphenidate on attentional function after traumatic brain injury: a randomized, placebo-controlled trial. American journal of Physical Medicine & Rehabilitation, 1997. 76(6): p. 440-450.

[61] Whyte, J., et al., Effects of methylphenidate on attention deficits after traumatic brain injury: a multidimensional, randomized, controlled trial. Am J Phys Med Rehabil, 2004. 83(6): p. 401-420.

[62] Willmott, C. and J. Ponsford, Efficacy of methylphenidate in the rehabilitation of attention following traumatic brain injury: a randomized, corssover, double-blind, placebo controlled inpatient trial. J Neurol Neurosur Ps, 2009. 80: p. 552-557.

[63] Kaelin, D.L., D.X. Cifu, and s.B. Matthie, Methylphenidate effect on attention deficit in the acutely brain-injured adult. Arch Phys Med Rehabil, 1996. 77(1): p. 6-9.

[64] Warden, D.L., et al., Guidelines for the pharmacologic treatment of neurobehavioral sequelae of traumatic brain injury. J Neurotrauma, 2006. 23(10): p. 1468-1501.

[65] Burstein, E.S., et al., II. In vitro evidence that (-)-OSU6162 and (+)-OSU6162 produce their behavioral effects through 5-HT2A serotonin and D2 dopamine receptors. J Neural Transm, 2011. 118(11): p. 1523-1533.

[66] Carlsson, M.L., et al., I. In vivo evidence for partial agonist effects of (-)-OSU6162 and (+)-OSU6162 on 5-HT2A serotonin receptors. J Neural Transm, 2011. 118(11): p. 1511-1522.

Chapter 22

Attentional Processing in Traumatic Brain Injury and Posttraumatic Stress Disorder

Tobias Wensing, Jean Levasseur-Moreau,
Alexander T. Sack and Shirley Fecteau

Additional information is available at the end of the chapter

http://dx.doi.org/10.5772/57185

1. Introduction

The objective of this work is to review studies that examine deficits in sustained and selective attentional processing in adults diagnosed with Traumatic Brain Injury (TBI) or Posttraumatic Stress Disorder (PTSD). Deficits in these two attentional mechanisms are assumed to underlie the most prevalent behavioral consequences of TBI and PTSD and can result in deleterious consequences for an individual's everyday life.

TBI is considered a neurological condition that involves critical physical damage to brain tissue. PTSD is a psychiatric disorder that is accompanied by symptoms including intrusive re-experiencing of the traumatic event, avoidance and hyper-arousal. Individuals diagnosed with PTSD have been exposed to a traumatic event with highly threatening physical or psychological harm [1]. Consequently, the causing events underlying TBI, such as a motor vehicle accident or military assault, might also lead to a trauma that develops into a diagnosis of PTSD. Clinical reports described case studies, in which individuals showed symptoms of PTSD following a traumatic event. Moreover, neuropsychological assessments revealed that these symptoms were also accompanied by specific deficits in sustained and selective attentional processing [2] and related cognitive task performances [3]. A vast amount of literature emphasizes that impairments in cognitive processes may play a key role in several neurological and psychiatric disorders. For instance, cognitive deficits have been reported in both TBI and PTSD with considerable overlap including difficulties to concentrate, mental fatigue and hyper-arousal (see [4] for a review). Specifically, comorbid occurrence of TBI and PTSD in military personnel returning from conflicts in Afghanistan and Iraq approaches a prevalence of 50% [1]. Additionally, recent reviews on TBI and PTSD imply that the presence of TBI in an individual might accelerate subsequent development of PTSD diagnosis [4, 5]. Thus, the relationship between TBI and PTSD is pronounced in a one-way direction with PTSD occurring as a consequence of TBI, but not vice versa. This is mainly attributed to a shared causative event [4]. However, damage to critical brain tissue in TBI that is involved in emotional and cognitive processing after injury might also promote successive development of PTSD. Recent neuroimaging studies identified shared neurobiological deficits that include portions of the prefrontal cortex, as well as subcortical regions, such as the hypothalamus or pituitary gland (for a review, see [4]).

Importantly, despite potential symptomatic overlaps, TBI and PTSD require a detailed picture of cognitive deficits that underlie the most debilitating symptoms. With regards to the complex nature of attention and its prominent implication in symptoms observed in both TBI and PTSD, it appears crucial to identify and discuss similarities and differences in attentional processing, which may help to distinguish between both medical conditions. Thus, a comparison of attentional processing towards stimuli independent of emotional valence, either shared by or exclusive to TBI and PTSD, may promote specialized subsequent treatment and rehabilitation approaches.

Attentional processes have been studied for decades. However, we have yet to entirely grasp to which extent attention affects our behaviour. Attention is a central cognitive process that is considered as a precursor of a large majority of other higher cognitive functions. Although attention underlies a variety of behavioral impairments related to neurological and psychiatric disorders, its nature is rather complex and can be observed throughout numerous distinct subtypes, such as sustained attention, selective attention, and inhibition. Accordingly, these subtypes of attention are governed by the implication of different brain regions including the posterior parietal [6], inferior frontal [7], medial frontal and dorsolateral prefrontal cortex [8]. In this work, we will address processing of sustained and selective attention, since we believe that these subtypes of attention most likely resemble mechanisms, which underlie shared symptoms in TBI and PTSD, such as difficulties to concentrate or mental fatigue. Moreover, we will focus on reviewing studies, which solely implement neutral stimuli within their experimental designs. The ability to adequately process neutral stimuli is a vital cognitive requirement for coping with our environment. Deficits in this basic function may have a critical impact on an individuals' well being and even his survival.

2. Sustained attentional processing in TBI

Sustained attention (vigilance) is the ability to maintain attention for sporadic critical events during long periods of time [9]. Maintaining vigilant while performing everyday routines, such as driving a vehicle, is crucial to react to changes and unforeseen events in the environment. To date, behavioral findings on deficits in sustained attention for TBI are inconclusive. Varying findings have been reported for almost every behavioral aspect related to sustained attention, including response time (RT), accuracy, and intra-individual variability in responses (Table 1). In the studies reviewed here, RT is defined as the latency between the onset of a stimulus and the response given by a subject; accuracy mainly reflects errors in the correct execution of task instructions; intra-individual variability in responses refers to fluctuations in subjects' accuracy or RT over the course of a cognitive task.

Study	TBI (n)	Post-Trauma	Severity	Trauma Type	Cognitive Test	Main Findings
Bloomfield et al. [15]	44	71,5 Months	Mild - Severe	Combat; RTA; Falls	SART	More commission errors in "poor sleepers" group
Bonnelle et al. [18]	28	28 Months (Range: 3-73)	Mild - Severe	Combat; RTA; Falls; Sports	CRT	Decrement in RT over time; No differences in accuracy; Higher IIV
Dockree et al. [12]	29	34,81 Months	Mild - Severe	Combat; RTA; Falls; Work	SART;DART	No differences in RT; More commission errors; Higher IIV
Kim et al. [19]	18	51,1 Months	Moderate - Severe	-	VSAT	No differences in RT; No differences in accuracy
McAvinue et al. [10]	18	38,83 Months (Range: 2-132)	Mild - Severe	-	SART	No differences in RT; More commission errors
Sinclair et al. [16]	20	133 Days-13,4 Years	Mild - Severe	RTA; Falls	PVT	Slower RT; Lower accuracy; Higher IIV
Slovarp et al. [20]	9	3,6 Years	Severe	-	SAT	Slower RT; Lower hit rate; Higher IIV for TBI subgroup
Willmott et al. [14]	40	68,38 Years (Range: 12-462)	Moderate - Severe	RTA	SART	No differences in accuracy

Table 1. Summary of methodologies and main findings of studies investigating sustained attention in TBI (Note: CRT = Choice Reaction Time Task; DART = Dual-task Attention to Response Task; IIV = Intra-individual Variability; PVT = Psychomotor Vigilance Task; RT = Response Time; RTA = Road Traffic Accident; SAT = Sustained Attention Task; SART = Sustained Attention to Response Task; VSAT = Visual Sustained Attention Task; TBI = Traumatic Brain Injury; "-" indicates missing or not reported information).

McAvinue et al. [10] compared individuals with mild to severe TBI to gender-, age- and education-matched healthy individuals on a Sustained Attention to Response Task (SART; [11]). The SART is a cognitive test measuring sustained attention following the Go/NoGo paradigm. Subjects were visually presented with a series of single digits ranging from 1 to 9. Presentation of the digits is followed by a mask and subjects were subsequently asked to press a response button as accurately and fast as possible at the occurrence of every digit but "3" (for a detailed description on the SART, see [11]). To further assess the role of error awareness and feedback on task performance, McAvinue et al. [10] divided the SART into different conditions. Subjects were presented with randomized digits and asked to withhold their response as soon as the digit "3" appeared. Also, three altered versions of the original SART were implemented in this study: an "awareness" condition, requiring subjects to give a verbal report as soon as they made a commission error (i.e., pressing the response button although no target is present); a "feedback" condition in which subjects were given auditory feedback when making an error; and a "fixed" condition, in which digits were not presented randomly but in a fixed order. Overall, individuals with TBI made more commission errors than healthy controls. However, both groups showed similar RT. Moreover, results of the "awareness" condition suggest that individuals with TBI were less aware of the errors they made during the task. Also, error awareness, that is, the number of errors a subject was unaware of, was positively correlated with the number of commission errors made by individuals with TBI. Behavioral results of the "feedback" condition indicated that both groups benefited from the provided error feedback and that both groups were able to decrease their amount of commission errors to the same extent. Another study [12] combined the SART with a continuous performance task (CPT; [13]) resulting in the Dual-task Attention to Response Task (DART). In the original CPT, subjects were asked to respond via button press to random occurrences of any letter of the alphabet, except for "x" ("x"-condition). Additionally, a more demanding task required subjects to withhold response to the letter "x" only if preceded by the letter "a" ("ax"-condition). In the classical CPT, outcome measures are levels of accuracy, assessed in terms of commission and omission errors (i.e., not pressing the response button although a target is present). According to the authors, a combination of the SART and CPT would result in a cognitively more demanding task that challenges sustained attention. In the DART, subjects are required to indicate the random occurrence of grey colored digits additionally to the classical design of the SART described previously. Dockree et al. [12] compared individuals with TBI, who had mild to severe injuries, to gender-, age- and education-matched healthy subjects on both the classical SART and the newly developed DART. Compared to the SART, the cognitively more demanding DART led to an increase in commission errors for both study groups. Individuals with TBI and healthy individuals did not differ in terms of RT on both tasks. However, individuals with TBI made more commission errors and showed higher intra-individual variability in responses during both the SART and the DART. Within a series of attentional tasks, Willmott et al. [14] also used a classical SART paradigm to test sustained attention in moderate to severe TBI. This study revealed no difference between individuals TBI and healthy individuals regarding commission errors. However, individuals with TBI tended, though not statistically significant, to make more omission errors than healthy controls (i.e., missing responses to any digit but "3").

Bloomfield et al. [15] incorporated the SART in a study that compared individuals with mild to severe TBI with respect to absent or present sleeping difficulties. The authors emphasized that individuals with TBI and those with sleep disorders constantly report concentration and attentional impairments. Thus, a deficit in sustained attention might be even more pronounced in individuals with TBI suffering from sleep disorders [15]. Based on a battery of sleep measurement indices, subjects were assigned to either the "poor sleepers" or "good sleepers" group. In accordance with the authors' hypotheses, poor sleepers made more commission errors than good sleepers at the SART. Sinclair et al. [16] compared individuals with mild to severe TBI to age- and gender-matched healthy individuals on the Psychomotor Vigilance Task (PVT; [17]). The PVT is a cognitive test that requires subjects to respond as fast as possible to a randomly presented auditory target stimulus. This task was originally developed to assess sustained attention in individuals with sleep disturbances. Similar to Bloomfield et al. [15], the authors hypothesized that TBI might be accompanied by an increase in mental fatigue and sleeping difficulties. This study examined individuals with TBI that reported fatigue and sleep difficulties and compared them to a healthy control sample. Compared to controls, individuals with TBI showed increased RT, reduced accuracy and larger intra-individual response variability. However, effects were diminished or vanished completely when controlling for reported sleep disturbances and fatigue, respectively. Nonetheless, Sinclair et al. [16] accentuated the close relation between sustained attention and fatigue and sleep disturbances that could at least in part account for the observed results in this study.

A study by Bonnelle et al. [18] assessed sustained attention with a simple Choice Reaction Time task (CRT). In the CRT, subjects were asked to respond as accurately and as fast as possible to a target arrow that pointed either in the right or left direction with respective button presses. An increase in RT over the course of the task would be indicative of possible deficits in sustained attention. The examined TBI group suffered from mild to severe brain injuries. Although individuals with TBI displayed intact accuracy, they had increased RT and larger intra-individual variability compared to healthy controls. Additionally, the same group showed an increase in response speed over time with RT being significantly slower in the last third of the task assessment than in the first one. The authors proposed that this decrease in performance speed might reflect a deficit in maintaining vigilance over time and, thus, sustained attention [18].

Kim et al. [19] assessed individuals with moderate to severe TBI and healthy individuals using a Visual Sustained-Attention Task (VSAT). In the VSAT, subjects were asked to discriminate pairs of vertical lines as either of same or different size, and press a response button only if lines were equal. However, individuals with TBI and healthy controls did not statistically differ in terms of accuracy or RT, although there was a trend towards longer RT and poorer accuracy in individuals with TBI.

Another recent study by Slovarp et al. [20] focused on sustained attention processing in individuals with severe TBI. These individuals and an age-matched control group were tested on a Sustained Attention Task (SAT). Here, subjects were presented with a display of four letters with one letter appearing in the upper half and three letters arranged in a row in the lower half of the screen. The subjects' task was to indicate by button press, if the upper letter matched one of the lower letters on the same display. Outcome measures included levels of accuracy (false alarms/hit rates) and RT. Findings revealed that individuals with TBI made more errors in terms of lower hit rates and were comparably slower in giving responses. Although there was no evidence for an increase in RT variability over time on a group level, Slovarp et al. [20] observed substantial intra-individual response variability over the course of the task in a subgroup of TBI subjects.

3. Sustained attentional processing in PTSD

Deficits in sustained attention have also been reported in studies assessing individuals diagnosed with PTSD (Table 2). Vasterling et al. [21, 22] conducted different studies examining groups of Gulf and Vietnam War veterans, respectively. The researchers compared individuals, which had developed PTSD after combat with a group of PTSD-free veterans. Within a series of cognitive tasks, subjects also performed a computerized version of the CPT [13], which is assumed to measure sustained attention. The authors reported both more commission [21] and omission errors [22] on the "ax"-condition in veterans with PTSD when compared to combat-exposed individuals without PTSD. Sustained attention has also been assessed as part of a larger study on cognitive functioning in veterans of the Bosnian War [23]. Veterans diagnosed with PTSD were compared to a similarly combat-exposed but PTSD-free control group. Both groups were assessed on the SART, which has been widely used in the study of attentional deficits in individuals with TBI. Compared to controls, veterans with PTSD made more commission as well as omission errors in the SART. However, the authors did not find any differences in RT. Another study that measured sustained attention in combat-exposed individuals used the CPT in Vietnam War veterans with PTSD and trauma-free controls [24]. In this study, the CPT was presented visually as well as auditory. In addition to the visual "x"- and "ax"-conditions, subjects performed the same task while listening to letter strings presented via headphones. Task performance was measured in terms of accuracy and RT. While there were no differences between groups in the visual CPT condition, PTSD subjects made less correct responses in the auditory condition. Moreover, subjects did not differ in RT in the auditory and visual CPT. However, the authors did not report any differences in commission errors for both modalities.

Study	PTSD (n)	Control Group	Trauma Type	Cognitive Test	Main Findings
Jenkins et al. [25]	15	Trauma Survivors	Sexual Violence	CPT; PASAT	More omission errors
Koso et al. [23]	20	PTSD-free Veterans	Combat	SART	More commission and omission errors
Shucard et al. [24]	23	Healthy Controls	Combat	CPT	No differences in RT for both modalities; Auditory: lower accuracy
Stein et al. [27]	17	Healthy Controls; Trauma Survivors	Sexual Violence	PASAT	Lower scores in PTSD compared to healthy, but not trauma controls
Vasterling et al. [21]	19	PTSD-free Veterans	Combat	CPT	More commission errors
Vasterling et al. [22]	26	PTSD-free Veterans	Combat	CPT	More omission errors

Table 2. Summary of methodologies and main findings of studies investigating sustained attention in PTSD (Note: CPT = Continuous Performance Task; PASAT = Paced Auditory Serial Attention Task; PTSD = Posttraumatic Stress Disorder; SART = Sustained Attention to Response Task).

Jenkins et al. [25] assessed survivors of sexual violence with and without subsequent development of PTSD, as well as a healthy control group on several tests of cognitive functioning. In order to examine sustained attention in PTSD for both the visual and auditory modality, the authors used two different kinds of tasks. Besides the CPT for visual attention, they presented their subjects with a Paced Serial Auditory Addition Test (PASAT; [26]). Subjects were presented with auditory single digits and instructed to add each new digit to that of a directly preceding trial. Cognitive performance was measured with respect to time needed reporting a sum of digits, as well as the amount of correctly produced sums of digits. Results showed that individuals with PTSD made more omission errors at the CPT and the PASAT compared to trauma and healthy controls. Another study utilized the PASAT for measurement of auditory sustained attention in PTSD [27]. Survivors of sexual violence showed decreased PASAT overall scores compared to healthy individuals, who matched individuals with PTSD in terms of age, education and socioeconomic status. However, there was no apparent difference in accurate task performance between individuals with PTSD and a respective trauma control group.

In sum, we identified shared processing deficits in sustained attention in TBI and PTSD. Specifically, these deficits are pronounced in terms of reduced accurate task performance, while decrements in RT and variability in task performance over time seem to be associated with processing deficits in TBI only.

4. Selective attentional processing in TBI

Selective attention is the ability to focus attentional resources oriented towards a given stimulus despite the presence of distracting or competing stimuli. Difficulties in maintaining concentration and experiences of fatigue are among the most prevalent symptoms in individuals suffering from TBI. In order to cope with general tasks that involve more than one source of stimuli, it is important that individuals are able to separate task- or content-relevant from irrelevant input. Thus, selective attention is usually assessed by tasks requiring subjects to focus on relevant stimuli while ignoring irrelevant and distracting information (Table 3).

In a study by Goethals et al. [28], individuals with severe TBI completed a Stroop Colored Word Test (abbreviated as "Stroop" subsequently; [29]). In the Stroop task, subjects were presented with three different sorts of cards: the first card contained black-printed words of highly distinguishable colors ("blue", "green", "red", "yellow"); the second card depicted rectangles printed in blue, green, red or yellow color ("color-naming" condition); the last card of the Stroop task showed the same words as card one. However, words were not printed in black, but in a color that did not resemble its actual content ("incongruent" condition; i.e., the word "blue" printed in green). Performance on the first card served as a measure of reading abilities. For assessment of selective attention, subjects were compared on a Stroop interference index calculated by subtracting RT of the incongruent condition from those in the color-naming condition. Thus, deficits in selective attention would be reflected by Stroop interference indices, which largely differ from the value 0. In this study by Goethals et al. [28], impairment in Stroop performance was classified as 1 standard deviation (SD) above mean RT of a matched normative sample. Results showed that individuals with TBI showed deficits in the incongruent, but not the color-naming Stroop condition, which might indicate difficulties in selectively

Study	TBI (n)	Post-Trauma	Severity	Trauma Type	Cognitive Test	Main Findings
Belmont et al. [35]	27	3-26 Months	Severe	-	Go/NoGo-Task	Slower RT; More omission errors
Goethals et al. [28]	9	-	Severe	-	Classical Stroop-Task	Impairment on Stroop-Task
Mayer et al. [32]	22	3-20 Days	Mild	Combat; RTA; Falls	Modified Stroop-Task	No differences in RT
Ries et al. [36]	20	12-360 Months	Severe	-	Negative Priming Paradigm	Slower RT and lower accuracy; Negative priming effect on RT for degraded words
Smits et al. [31]	21	30,6 Days	Moderate - Severe	-	Modified Stroop-Task	Moderate: no differences to controls; Severe: lower accuracy on Stroop-Task
Soeda et al. [30]	5	12-84 Months	Severe	RTA	Modified Stroop-Task	No differences in accuracy
Willmott et al. [14]	40	68,38 Days (Range: 12-462)	Moderate - Severe	RTA	SAT; 2&7	Both tests: slower RT; 2&7: lower accuracy; Larger RT differences between SAT tests
Ziino et al. [33]	46	21-1153 Days	Mild - Severe	-	C-SAT	Slower RT; Lower accuracy; no differences in IIV

Table 3. Summary of methodologies and main findings of studies investigating selective attention in TBI (Note: 2&7 = Ruff 2&7 Selective Attention Task; C-SAT = Complex Selective Attention Task; IIV = intra-individual variability; RT = Response time; RTA = Road Traffic Accident; SAT = Sustained Attention Task; TBI = Traumatic Brain Injury; "-" indicates missing or not reported information).

attending towards task-relevant information and to ignore simultaneously occurring and distracting input.

Soeda et al. [30] used a modified version of the Stroop paradigm to assess selective attention in severe TBI. Unlike in the original version of the Stroop task, subjects were asked to validate congruent word and color combinations via button press. This alteration enabled the authors to directly measure levels of accuracy as well as RT. When compared to age- and education-matched healthy controls, individuals with TBI tended to make more errors, although this difference did not reach statistical significance. Another alteration of the original Stroop task has been used by Smits et al. [31], who examined selective attention in two distinct samples of individuals with moderate and severe TBI. In their "counting" Stroop task, subjects were presented with two task conditions: in the neutral condition, subjects had to respond via button press to the displayed amount of words showing animal names; in the incongruent condition, subjects had to indicate the amount of number words that have been presented on the screen, rather than their value, if the word was presented in normal letters (i.e., at the occurrence of three times the word "five", subjects should press a button for "3"). However, in this condition there were deviant trials in which the word representing a number was given in capital letters.

In that case, subjects were asked to validate the numeric value this word was representing (i.e., at the occurrence of three times the word "FIVE", subjects should press a button for "5"). Comparison to a group of healthy controls revealed that individuals with severe but not with moderate TBI made more errors in the incongruent counting Stroop task condition. Mayer et al., [32] varied the Stroop paradigm in a way as to assess selective attention across visual and auditory modalities. The task consisted of congruent and incongruent pairs of simultaneously presented visual and auditory stimuli of written out and pronounced numbers, respectively. Both visual and auditory stimuli occurred at either a high or a low frequency condition in order to induce an increase in cognitive task demands. A preceding multimodal cue instructed subjects to attend to numbers presented in one modality while ignoring the other. Congruent trials referred to identical pairs of simultaneously occurring stimuli, whereas incongruent trials comprised different numbers presented at the same time. When compared to an age- and education-matched healthy control sample, individuals with TBI showed a non-significant trend towards longer RT to incongruent auditory stimuli at high frequency. However, no other effects regarding experimental group, modality or frequency were observed.

Besides the classical or modified versions of the Stroop paradigm, several other cognitive tasks have been used to assess selective attention in TBI. Ziino et al. [33] developed a Complex Selective Attention Task (C-SAT) in order to examine individuals with mild to severe TBI and an age- and education-matched healthy control group on selective attentional processing. The C-SAT required subjects to respond to combinations of red- or green-colored letters and numbers with specific button presses (i.e., right button for green letter/red number; left button for right letter/green number). Subjects were compared on measures of accuracy by means of errors (pressing the wrong button) and misses (not pressing any button during a predefined period of time), as well as RT. The authors argued that this task is especially difficult for subjects suffering from TBI, since it requires a larger working memory load compared to rather simple selective attention tasks [33]. Individuals with TBI showed lower levels of accuracy with significantly more produced errors than healthy individuals, as well as slower RT. Both groups were also compared on variability in RT. However, there was no apparent greater response variability for individuals with TBI as a group.

Alongside a variety of other cognitive functions, Willmott et al. [14] tested selective attention in individuals with moderate to severe TBI using two different behavioral tasks: the Ruff 2 and 7 Selective Attention Test (2&7; [34]) and the Selective Attention Task (SAT; [33]). The 2&7 is a paper-and-pencil task that required subjects to manually cross out the digits "2" and "7", which were presented within a series of either letters or other digits. The procedure of the SAT was similar to that of the C-SAT reported previously [33] with the alteration that Willmott et al. [14] added a simple version to the SAT (SSAT). The difference between SSAT and C-SAT lay within its rather low cognitive demand; subjects were asked to respond with two distinct button presses if a letter or number was shown in a specific color (i.e., right button for yellow, left button for brown targets). For both tasks, subjects were measured on accuracy and RT. Results of the 2&7 revealed that individuals with TBI were less accurate and slower cancelling out digits embedded within letters, as well as within series of other digits. However, the difference between both task conditions was larger in the healthy control group. In the SAT, individuals with TBI showed slower RT than healthy individuals. Moreover, differences in RT between the SSAT and C-SAT were larger for individuals with TBI indicating that this group had problems with the more cognitively demanding version of the SAT. Belmont et al. [35] implemented a basic Go/NoGo paradigm in their study in order to assess selective attention in a sample of individuals with severe TBI. The task required subjects to respond to target letters via button press and withhold their response at the occurrence of non-targets. Performance of individuals with TBI was compared to that of gender-, age- and education-matched healthy individuals by means of accuracy, that is, the amount of omission errors, as well as RT. The authors found that individuals with TBI made more omission errors and were generally slower in responding than healthy individuals.

Ries et al. [36] used a negative priming paradigm that took into account response inhibition as part of a selective attentional process. Thus, selective attention would be reflected by both facilitation of the selected target as well as suppression of distracting information [37]. In their behavioral task, the authors presented individuals with severe TBI and healthy controls with two subsequent screens per trial, which consisted of a priming display followed by a probe display. Both displays showed two words in two different colors indicating targets and distractors. Additionally to targets or distractors, words were visually degraded in several ways and presented on 50% of the probe displays only. The other 50% of the probe displays consisted of intact word stimuli. Subjects were asked to respond to target words on the prime displays verbally by reading them out loudly. The authors hypothesized that negative priming of intact target words on priming displays would result in inhibitory difficulties on subsequent probe trials in individuals with TBI. Results showed that individuals with TBI generally had slower RT than healthy individuals. Moreover, word degradation slowed RT in both experimental groups, whereas this effect was even significantly more pronounced in individuals with TBI. For intact stimuli, there was a negative priming effect on RT for healthy individuals, but not individuals with TBI. Nonetheless, negative priming affected RT for both groups when stimuli were degraded. Also, individuals with TBI were less accurate than healthy controls. However, neither stimulus degradation nor negative priming had an effect on accuracy of both groups. Ries et al. [36] interpreted their findings as being indicative of inhibitory difficulties in individuals with severe TBI.

5. Selective attentional processing in PTSD

Interference of emotional or threat-related stimuli on attentional processing has been commonly reported in individuals diagnosed with PTSD [38-43]. A range of studies aimed to investigate if this interference could be translated to a context of stimuli with solely neutral valence as well. Problems in selecting task relevant cues while ignoring distracting information might reflect a general attentional deficit in PTSD, which could not be addressed to emotional saliency per se (Table 4).

Study	PTSD (n)	Control Group	Trauma Type	Cognitive Test	Main Findings
Bryant et al. [45]	14	Non-traumatized Controls	Interpersonal Violence; RTA	Oddball Paradigm	No differences in RT
Jenkins et al. [25]	15	Trauma Survivors	Sexual Violence	Posner Visual Selective Attention Task	No differences in RT
Vasterling et al. [21]	19	PTSD-free Veterans	Combat	Stroop-Task	No differences in Stroop Interference Scores
Vasterling et al. [22]	26	PTSD-free Veterans	Combat	Stroop-Task	No differences in Stroop Interference Scores

Table 4. Summary of methodologies and main findings of studies investigating selective attention in PTSD (Note: PTSD = Posttraumatic Stress Disorder; RT = Response time; RTA = Road Traffic Accident).

Studies on selective attention deficits in PTSD incorporating neutral stimuli are rather rare. In two consecutive studies, Vasterling et al. [21, 22] assessed selective attention in Vietnam and Gulf War veterans with PTSD and healthy controls on a Stroop task. For both studies, PTSD subjects did not show any differences to healthy individuals on Stroop interference scores. In a series of cognitive tests, Jenkins et al. [25] instructed subjects to perform the Posner Visual Selective Attention Task [44]. In this task, subjects were asked to respond to a target with a button press, which was preceded by two different sorts of cues indicating the position of the target on a computer screen. Valid cues predicted the actual later position of the target, whereas invalid cues appeared at the target's opposite location (i.e., cue on left side of the screen, subsequent target on the right side). Survivors of sexual violence with PTSD were compared to a group of survivors of similar trauma events without PTSD and a group of healthy individuals on RT measures. Results revealed no difference in RT between corresponding groups. Bryant et al. [45] used an oddball paradigm to assess selective attention in individuals with PTSD and trauma-free controls. Auditory stimuli consisted of standard and target tones differing in frequency with target tones occurring at a probability of 15% during the experiment. There was no statistical evidence for differences in RT between both groups.

Put together, TBI appears to be accompanied by deficits in selective attention. These deficits are mainly expressed in terms of reduced accuracy in cognitive tasks, as well as a general slowing in RT. However, no study on selective attention in PTSD revealed corresponding findings, which might be related to a shortage of empirical studies addressing selective attentional processing in PTSD.

6. Discussion

Results of the here reviewed studies suggest a shared processing deficit in sustained attention towards stimuli with neutral valence for TBI and PTSD (Figure 1). This attentional deficit is expressed in terms of reduced accuracy levels when cognitive task performance of individuals with TBI or PTSD is compared to that of healthy controls. In contrast to a shared deficit in accuracy for sustained attention tasks, findings on selective attention in TBI and PTSD point towards a processing deficit of neutral stimuli that is only observed in TBI. Studies assessing individuals with TBI reported reduced accuracy in task performance, as well as a general slowing in RT across different cognitive tasks, levels of severity and trauma types. However, no such corresponding findings were demonstrated in studies that assessed individuals diagnosed with PTSD on selective attention tasks. In order to understand why sustained attention tasks reveal shared attentional deficits in TBI and PTSD, but selective attention tasks do not, we have to consider several aspects that might help explaining our main findings such as the general mechanisms incorporated in execution of cognitive tasks, the subjects' states of arousal, as well as neural correlates underlying attentional processing in TBI and PTSD.

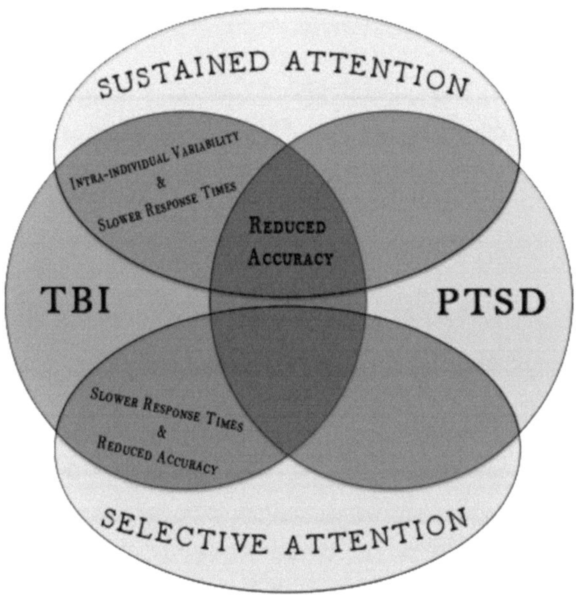

Figure 1. Schematic overview of shared and exclusive attentional processing deficits towards stimuli with neutral valence in Traumatic Brian Injury (TBI) and Posttraumatic Stress Disorder (PTSD).

6.1. The role of response inhibition

Sustained attention is mainly assessed by tasks following the Go/NoGo paradigm (i.e., SART and CPT). Here, subjects are instructed to maintain responsive to a set of targets via button press while withholding responses to a predefined and randomly occurring type of stimulus. Thus, accurate performance of a Go/NoGo-task requires subjects' ability of response inhibition. According to this rationale, we might question if a shared deficit resulting from sustained attention tasks is solely pronounced by reduced levels of accuracy, or would also reflect impairments in response inhibition. However, our findings do not support this idea. There is an obvious pattern between the here reviewed studies, which shows that studies assessing TBI and PTSD reported more commission errors only in tasks following the Go/NoGo paradigm. Commission errors are also considered false responses/alarms and occur when subjects fail to inhibit responses to a specific stimulus. However, it does not necessarily follow from these findings that a deficit in sustained attention reflects impairments in response inhibition. To the contrary, studies also reported increases in omission errors while assessing individuals with TBI and PTSD. Omission errors usually reflect missing responses to present targets and, therefore, do not require the ability to withhold responses. Additionally, other studies, which do not utilize a Go/NoGo paradigm within their cognitive tasks, report generally reduced accuracy in terms of lower hit rates towards target detection (which is similar to the concept of omission errors) in TBI and PTSD [16, 20]. Thus, it seems tempting, but unlikely that deficits in sustained attention can be simply addressed to impairments in response inhibition. Instead, there appears to be a deficit in sustained attention in terms of accuracy that is present in both TBI and PTSD.

Studies assessing selective attention point towards a processing deficit of neutral stimuli that is only observed in TBI (Figure 1). Tasks implemented in these studies follow rationales that require selection of content- and task-relevant information while inhibiting simultaneously occurring and competing sensory input (i.e., Stroop task; [28, 30-33, 36]. Thus, selective attention deficits expressed by reduced accuracy and slowing of RT might be related to an underlying impairment to inhibit distracting simultaneous information. Unlike commission errors in Go/NoGo-tasks, there is no equivalent specific accuracy measure, which might indicate an apparent deficit in response inhibition for individuals with TBI while performing selective attention tasks. It might be reasonable to assume that deficits on tasks such as the Stroop display impairments in inhibitory processes [29]. However, given that selective attention deficits have also been reported for TBI in tasks that do not require inhibition of simultaneously occurring input [14], results on these tasks do not necessarily explain selective attention deficits in individuals with TBI.

6.2. The impact of fatigue, arousal and emotion

Since neither sustained nor selective attentional processing deficits are fully explained by impairments of inhibitory mechanisms, we might argue that attentional deficits could possibly result from other phenomena that might be shared by or exclusive to TBI and PTSD. TBI has previously been associated with states of mental fatigue and sleep disturbances [46]. Two of the studies reviewed here extent this view by contrasting performance of sleep-deprived and normal sleeping individuals with TBI on a sustained attention task [15], as well as individuals with TBI and healthy individuals on a cognitive task that has been implemented in sleep studies previously [16]. Both studies propose a sustained attention deficit that is even more pronounced in the presence of sleep disturbances in individuals with TBI. Although Bloomfield et al. [15] reported more commission errors in "poor" sleepers with TBI when compared to "good" sleepers, very recent findings by Sinclair et al. [16] point towards possible floor effects of sleep difficulties in TBI. Since differences between groups in this study diminished as soon as the authors controlled for reported fatigue and sleep disturbances in their statistical analyses, it is hard to argue that these factors might exacerbate already present accuracy deficits in sustained attention for individuals with TBI. The inclusion of a healthy control group with or without sleeping difficulties in a study similar to that by Bloomfield et al. [15] might shed some light on the question in how far sleep disturbances or mental fatigue possibly influence the severity of sustained attention deficits in individuals with TBI. Sleep disturbances are also considered one of the hallmarks in PTSD [1]. However, sleep disturbances in PTSD are mainly pronounced in terms of reoccurring nightmares that resemble aspects of the experienced traumatic event. Therefore, sleep disturbances in PTSD might be rather related to mechanisms that enter the domain of memory and affective states than basic attentional processing (for a review on mechanisms of sleep disturbances in PTSD, see [47]). To our knowledge, there is only one recent study that examined the relation between attentional processing deficits and sleep disturbances in PTSD [48]. In this study, individuals with disaster-related PTSD-symptomatology were assessed on the PASAT. Behavioral outcomes, that is, the number of correct responses in the PASAT, were then related to self-reported symptom severity, including subscales of sleep disturbances. Although results revealed that deficits in sustained attention increased with symptom severity, there was no evidence for an impact of sleep disturbances on attentional processing. The authors argued that the lack of impact of sleep disturbance might be indicative of a solely attentional deficit in PTSD.

Although results from studies reviewed here are somewhat inconsistent with respect to how or if sleep disturbances and fatigue might influence performance on sustained attention tasks, we nonetheless see the urge to consider them in explaining a shared deficit for TBI and PTSD. Cognitive tasks assessing sustained attention are rather simple and modest in terms of duration. The original SART was administered over a period of 4.3 minutes [11]. Other sustained attention tasks, such as the PVT, show high validity when administered over a period of 10 minutes [17]. One might argue that tasks lasting shorter than 15 min might not be able to display the ability to maintain vigilance over time. However, we need to keep in mind that individuals with TBI, as well as those diagnosed with PTSD, might suffer from severe cognitive impairments, which could account for findings on sustained attention presented here. Symptoms of fatigue and difficulties to concentrate have been reported for both TBI and PTSD [4]. Thus, we propose that a shared deficit in sustained attentional processing might reflect a common underlying symptom in both medical conditions.

In selective attention tasks, processing deficits were observed for individuals with TBI, but not for those diagnosed with PTSD. Since deficits in selective attention for individuals with TBI are expressed in terms of decreased accuracy, as well as slower RT, we might again argue that impairments in TBI could be a possible consequence of states of fatigue. Studies reported that errors in selective attention task highly correlated with measures of self-reported fatigue [33, 35]. This effect even remained when results were controlled for mood by depression or general anxiety assessment scales. However, we did not find any apparent selective attentional processing deficits for studies on PTSD. Noteworthy, cognitive tasks assessing selective attention require larger mental effort than sustained attention tasks. Thus, selective attentional processing deficits only observed in TBI might be linked to increases in task difficulty and cognitive demands. Vice versa, an absent selective attention deficit in PTSD might reflect intact arousal, that is, a general reactivity and alertness to stimuli with neutral valence. This would be in line with symptoms of hyper-arousal, which are underlying the diagnosis of PTSD [1]. Nonetheless, the studies reviewed here did not report any facilitating effects towards stimuli for individuals with PTSD either. It seems more reasonable that states of hyper-arousal are closely related to stimuli with emotional valence. A vast amount of studies suggests that in studies on PTSD, individuals' performance on cognitive tasks is highly affected by interference of emotionally related stimuli [38-43]. Consequently, intact selective attentional processing might serve as a marker for subsequent development of hyper-arousal, and thus, would in part explain absent deficits on selective attention in individuals with PTSD reviewed here.

6.3. Neural correlates of sustained and selective attention in TBI and PTSD

It is rather difficult to draw an overall picture related to the impact of brain damage in TBI on behavioral results in this review. One major drawback in studies assessing attentional processing in TBI concerns missing reports of lesion sites in TBI. It seems quite obvious that damage to different brain regions may affect various cognitive networks and might subsequently result in a variety of behavioral and cognitive impairments. Within the studies reviewed here, only one systematically reported lesion sites for each participating individual [12]. However, given a large heterogeneity in lesion sites within the study's sample of individuals with TBI, the authors did not make any further relations between damaged brain regions and behavioral outcomes.

Regardless of challenges to make inferences of neurobiological deficits based on individual damage to the cortex, a handful of studies made an attempt to relate attentional deficits to alterations or specific states of cortical activity in individuals with TBI. Studies on attention in TBI reported deactivation of cortical areas relevant for accurate task performance [32], as well as changes in cerebral blood flow of the brain in resting state [19], that is, cortical activity and functional connectivity without stimulus-induced processing demands. The brain in resting state comprises a default mode network (DMN), which includes medial portions of the parietal, frontal and temporal cortex [49]. In individuals with TBI, studies indicate that insufficient deactivation of the DMN during cognitive tasks, as well as functional connectivity between areas incorporated in the DMN might be a neurological marker of behavioral deficits observed in sustained attention tasks [18, 32]. Similar to TBI, we identified only few studies that directly linked cortical abnormalities accompanying PTD to deficits in sustained attentional processing. Here, study results indicated that stress induced in combat-related situations might affect sustained attention task performance by a decrease in midbrain activity [50]; an effect, which seemed to diminish after multiple testing sessions, providing evidence for recovery mechanisms related to combat PTSD. However, studies examining the direct relations between attention to neutral stimuli and neural processing in PTSD are still rare. Hence, it seems quite assumptive to derive specific implementations for treatment or cognitive improvements from single study results.

Studies assessing selective attentional processing in TBI reported both decreases [30] or increases [31] in activity of the anterior cingulate cortex (ACC). Since this area is assumed to be involved in conflict monitoring [51], a mechanism essential for specific selective attention tasks such as the Stroop, decreases in task-related activity might reflect underlying impairments in cortical processing that could partially explain selective attentional processing deficits in TBI. However, studies did not distinguish between dorsal or rostral portions of the ACC. The former is assumed to be involved in general cognitive processing, whereas the latter shows responsiveness to stimuli with emotional valence, which might possibly account for these contradictory findings. Accordingly, increased activity in the ACC has been reported in studies on selective attention in PTSD [45, 52], including both dorsal and rostral portions. Thus, processing of stimuli in selective attention tasks appears to be unaffected at least in the ACC for individuals diagnosed with PTSD. Contrasting these results with decreases in ACC activity for TBI might serve as an indicator why selective attentional processing deficits were only observed in individuals with TBI.

6.4. Attentional deficits in comorbid TBI and PTSD

Historically speaking, it was long debated if TBI and PTSD could arise in a comorbid fashion, since TBI is accompanied by at least partial loss of consciousness and posttraumatic amnesia [12, 16, 19, 20, 32, 33]. According to this theory, amnesia or loss of consciousness would circumvent construction of a trauma-related memory, which is considered to facilitate PTSD-related symptoms, such as intrusive thoughts and re-experiencing. However, recent research proposed that TBI and PTSD share certain diagnostic features (for an extensive review, see [53]). As stated previously, it is assumed nowadays that TBI and PTSD occur at a prevalence rate of almost 50% [1]. Moreover, several case studies provided evidence that comorbid TBI and PTSD are not only theoretically related, but require accurate and reliable rehabilitation and treatment plans [2, 3]. Thus, we see a necessity to further explore attentional processing in populations with comorbid TBI and PTSD. To our knowledge, there is only one study that examined attentional processing in individuals with combat-related comorbid TBI and PTSD [54]. Individuals with comorbidity diagnosis were compared to two control groups, a group consisting of individuals with PTSD only and healthy controls on measures of the Attentional Network Task (ANT; [55]). The authors used the ANT to assess attentional processing on three distinct metrics, each reflecting a different aspect of attention: (1) *alerting*; (2) *orienting*; (3) *executive*. Although the ANT's theoretical framework slightly differs from other cognitive tasks reviewed in this work, we might link its alerting and executive networks to mechanisms of sustained and selective attention, respectively [54]. Study results indicated that individuals with comorbid TBI and PTSD were less accurate and slower in responding, and showed higher variability in responses compared to both the PTSD and healthy control group. Respective effects were also observable between individuals with PTSD and healthy controls, although without statistical significance. These findings were in line with several findings of attentional processing deficits described above. While there appeared to be a clear pattern for deficits in sustained and selective attention in TBI, this picture is less consistent in PTSD. Barlow-Odgen et al. [54] pointed out that none of the participating subjects had an exclusive diagnosis of TBI, but was always accompanied by PTSD. More systematic research on comorbid TBI and PTSD and even the inclusion of a group that only consists of individuals with TBI might further progress our understanding about shared and exclusive attentional processes in TBI and PTSD. Besides a considerable prevalence rate of comorbid TBI and PTSD, individuals with PTSD are very likely to meet criteria of at least one other psychiatric disorder [1]. Correspondingly, psychiatric disorders, which also occur at a high comorbidity rate with PTSD, including other anxiety disorders, depression [56] and substance abuse disorder [57], might exacerbate the impact of attentional processing in PTSD.

6.5. Implications for rehabilitation and treatment

Based on results discussed here, we believe that attentional processing deficits need to be particularly considered when developing and conducting treatment and rehabilitation programs for both TBI and PTSD. To date, cognitive training is highly recommended during TBI in a post-acute phase, including attention training and metacognitive training, but not at an acute stage [58]. Hence, results reviewed here might aid an even further detailed and specialized cognitive training, which takes into account specific deficits in sustained and selective attentional processing for individuals with TBI. Moreover, despite the concrete implementations for cognitive training, the general finding of sustained and selective attention deficits across trauma severity, the time since traumatic event and type of trauma might also affect other types of treatment approaches in TBI, such as physical and ergonomic rehabilitation programs, as well as speech or psychological therapy. Individuals with TBI consistently display difficulties to maintain attention during a cognitive task (e.g., [10, 12]). It seems reasonable that this deficit might also influence performance on other everyday situations that require the ability to stay vigilant over a specific period of time. Findings of studies reviewed here might support that adjustments of rehabilitation or training sessions to apparent impairments on sustained attention tasks could result in improvements of treatment outcome. We also think that findings of this review might especially promote treatment of comorbid TBI and PTSD or PTSD following TBI, respectively. Clinical case studies revealed that sustained and selective attentional difficulties are already tackled in intervention plans for PTSD following a traumatic accident [2]. Providing evidence that impairments in sustained attention are present in both TBI and PTSD populations and selective attention is preserved in the latter might serve as a blueprint for future treatment approaches, especially in terms of session duration and distribution over time. This might, for instance, be achieved by adjustment of intervention plans of cognitive training for shorter, but more frequent training sessions.

7. Limitations

Although we were able to shed some light on shared and rather exclusive attentional processing deficits in TBI and PTSD, there are some limitations to this review and selected studies that need to be taken into account. First of all, there is an extensive amount of literature on attentional processing in TBI using neutral stimuli. However, this is not the case for PTSD, where a vast amount of studies on attentional processing focuses on emotion-related stimuli and research and, thus, tends to neglect corresponding experimental designs that assess similar mechanisms related to stimuli with neutral valence. One might argue that this is not really surprising, since it appears more likely to find considerable differences between individuals with PTSD and control groups when emotion or even trauma-relevant content is involved. However, this tendency could lead to a publication bias that, in turn, might result in lost opportunities to identify basic cognitive mechanisms involved in the onset and development of PTSD. Another point of limitation refers to heterogeneity in study designs, which is mostly pronounced for those studies examining attentional processing in TBI. Given that most studies differ in terms of TBI severity, type of trauma, and time elapsed since traumatic events, it is unlikely to identify a pattern that might point towards possible recovery effects or impact of damage to specific brain regions. Regarding type of trauma, the same inconsistencies in etiology and causalities between studies could be observed in PTSD.

We would also like to point out that there is a rather low verge between continuous vigilant task performance, as reflected by sustained attention, and higher cognitive functions, such as working memory. Some of the studies reviewed here used cognitive tasks or alterations of task designs to recruit higher cognitive demands, which also incorporated working memory capacities [33]. We might question if accurate and fast performance on these kinds of tasks still exclusively reflects attentional processing. Regardless of mechanisms of attention, working memory deficits are a comprehensively investigated cognitive mechanism in TBI. Even within those studies presented here, tasks on working memory were part of larger setups of cognitive tasks [19, 20, 31]. Similar to performance on sustained attention tasks, results on working memory function revealed reduced accuracy levels for TBI, with task performance getting worse with an increase in working memory load [31].

We would like to emphasize that the purpose of this review was to provide improved insight on attentional processing, which might be shared by or be rather exclusive to TBI and PTSD. Thus, one might question the ecological validity of the cognitive tasks that were used throughout the studies reviewed here. Identification of overlap in potential attentional processing deficits should ultimately result in implications on how to provide suitable treatment and rehabilitation approaches. However, experimental settings assessing attention with cognitive tasks are usually highly artificial. Parsons et al. [59] made an interesting attempt towards an increased ecological setting for such tasks. The authors developed a virtual reality Stroop task that exposes individuals to an everyday-life or work-related environment. This approach might serve as one step towards more genuine experimental designs. In turn, these designs may elicit attentional processing mechanisms, which are more closely related to real-life situations for affected individuals in both TBI and PTSD. Consequently, improving the ecological validity of cognitive tasks will not only provide a more accurate picture of the underlying mechanisms in attentional processing in both medical conditions, but might also provide a more seamless translation between cognitive assessment (i.e., characterization of cognitive mechanisms or deficits) and rehabilitation approaches.

8. Conclusion

To this end, in this work we reviewed studies that assessed individuals with TBI, as well as those diagnosed with PTSD, on measures of sustained and selective attention to stimuli with neutral valence. Results of cognitive tasks measuring sustained attention suggest a shared deficit for TBI and PTSD, which is mainly characterized by a reduction in accurate task performance. Additionally, deficits in sustained attention for TBI are characterized by a general slowing in responses, as well as variability in RT and accuracy over time. In contrast to a shared accuracy deficit in sustained attention for TBI and PTSD, the studies reviewed here point towards impairments in selective attention only observed for TBI as reflected by reduced levels of accuracy and slowing in RT. However, no behavioral deficits for selective attentional processing were found in PTSD. In order to unravel such specific cognitive processing mechanisms in TBI and PTSD, we strongly encourage the conduction of future studies explicitly considering the impact of factors, such as (1) the type of traumatic event, (2) lesion severity and lesion sites in TBI, and (3) comorbid occurring TBI and PTSD on behavioral and rehabilitation outcomes. We also see a need for studies focusing on cognitive processes in PTSD to neutral rather than only emotionally loaded stimuli. Our review proposes that shared and exclusive attentional processing deficits in TBI and PTSD should be considered when developing or improving already existing post-trauma treatment approaches.

Acknowledgements

The authors would like to thank Centre Interdisciplinaire de Recherche en Réadaptation et Intégration Sociale (CIRRIS), Quebec City, Canada, the Canada Research Chair in Cognitive Neuroplasticity, Laval University, and the Faculty of Psychology and Neuroscience, Maastricht University, Maastricht, The Netherlands for supporting this work.

Author details

Tobias Wensing[1,2], Jean Levasseur-Moreau[1,3,4], Alexander T. Sack[2,5] and Shirley Fecteau[1,3,4*]

*Address all correspondence to: shirley.fecteau@fmed.ulaval.ca

1 Centre Interdisciplinaire de Recherche en Réadaptation et Intégration Sociale, Canada

2 Faculty of Psychology and Neuroscience, Maastricht University, Maastricht, The Netherlands

3 Centre de Recherche Universitaire en Santé Mentale de Québec, Quebec, Canada

4 Medical School, Laval University, Quebec, Canada

5 Maastricht Brain Imaging Center, Maastricht University, Maastricht, The Netherlands

References

[1] American Psychiatric Association. Diagnostic and statistical manual of mental disorders. 5th ed. Washington, D.C.: American Psychiatric Association; 2013.

[2] Williams WH, Evans JJ, Wilson BA. Neurorehabilitation for two cases of post-traumatic stress disorder following traumatic brain injury. Cognitive Neuropsychiatry 2003;8(1) 1–18.

[3] Ryan PB, Lee-Wilk T, Kok BC, Wilk JE. Interdisciplinary rehabilitation of mild TBI and PTSD: A case report. Brain Injury 2011;25(10) 1019–1025.

[4] Stein M, McAllister T. Exploring the convergence of posttraumatic stress disorder and mild traumatic brain injury. American Journal of Psychiatry 2009;166 768–776.

[5] McAllister TW. Neurobiological consequences of traumatic brain injury. Dialogues in Clinical Neuroscience 2011;13(3) 287-300.

[6] Corbetta M, Shulman GL. Control of goal-directed and stimulus-driven attention in the brain. Nature Reviews Neuroscience 2002;3(3) 201–215.

[7] Petersen SE, Posner MI. The Attention System of the Human Brain: 20 Years After. Annual Review of Neuroscience 2012;35(1) 73–89.

[8] Wager TD, Jonides J, Reading S. Neuroimaging studies of shifting attention: a meta-analysis. NeuroImage 2004;22(4) 1679–1693.

[9] Warm JS, Parasuraman R, Matthews G. Vigilance Requires Hard Mental Work and Is Stressful. Human Factors: The Journal of the Human Factors and Ergonomics Society 2008;50(3) 433–441.

[10] McAvinue L, O'Keeffe FM, McMackin D, Robertson IH. Impaired sustained attention and error awareness in traumatic brain injury: Implications for insight. Neuropsychological Rehabilitation 2005;15(5) 569–587.

[11] Robertson IH, Manly T, Andrade J, Baddeley BT, Yiend J. " Oops!": Performance correlates of everyday attentional failures in traumatic brain injured and normal subjects. Neuropsychologia 1997;35(6) 747-758.

[12] Dockree PM, Bellgrove MA, O'Keeffe FM, Moloney P, Aimola L, Carton S, et al. Sustained attention in traumatic brain injury (TBI) and healthy controls: enhanced sensitivity with dual-task load. Experimental Brain Research 2006;168(1-2) 218–229.

[13] Rosvold HE, Mirsky AF, Sarason I, Bransome ED Jr, Beck LH. A continuous performance test of brain damage. Journal of Consulting Psychology 1956;20(5) 343-350.

[14] Willmott C, Ponsford JL, Hocking C, Schönberger M. Factors contributing to attentional impairments after traumatic brain injury. Neuropsychology 2009;23(4) 424–432.

[15] Bloomfield IL, Espie CA, Evans JJ. Do sleep difficulties exacerbate deficits in sustained attention following traumatic brain injury? Journal of the International Neuropsychological Society 2010;16(1) 17-25.

[16] Sinclair KL, Ponsford JL, Rajaratnam SMW, Anderson C. Sustained attention following traumatic brain injury: Use of the Psychomotor Vigilance Task. Journal of Clinical and Experimental Neuropsychology 2013;35(2) 210–224.

[17] Dinges DF, Powell JW. Microcomputer analyses of performance on a portable, simple visual RT task during sustained operations. Behavior Research Methods, Instruments, & Computers 1985;17(6) 652–655.

[18] Bonnelle V, Leech R, Kinnunen KM, Ham TE, Beckmann CF, De Boissezon X, et al. Default Mode Network Connectivity Predicts Sustained Attention Deficits after Traumatic Brain Injury. Journal of Neuroscience 2011;31(38) 13442–13451.

[19] Kim J, Whyte J, Patel S, Europa E, Slattery J, Coslett HB, et al. A Perfusion fMRI Study of the Neural Correlates of Sustained-Attention and Working-Memory Deficits in Chronic Traumatic Brain Injury. Neurorehabilitation and Neural Repair 2012;26(7) 870–880.

[20] Slovarp L, Azuma T, LaPointe L. The effect of traumatic brain injury on sustained attention and working memory. Brain Injury 2012;26(1) 48–57.

[21] Vasterling JJ, Brailey K, Constans JI, Sutker PB. Attention and Memory Dysfunction in Posttraumatic Stress Disorder. Neuropsychology 1998;12(1) 125–133.

[22] Vasterling JJ, Duke LM, Brailey K, Constans JI, Allain AN, Sutker PB. Attention, learning, and memory performances and intellectual resources in Vietnam veterans: PTSD and no disorder comparisons. Neuropsychology 2002;16(1) 5–14.

[23] Koso M, Hansen S. Executive function and memory in posttraumatic stress disorder: a study of Bosnian war veterans. European Psychiatry 2006;21(3) 167–173.

[24] Shucard JL, McCabe DC, Szymanski H. An event-related potential study of attention deficits in posttraumatic stress disorder during auditory and visual Go/NoGo continuous performance tasks. Biological Psychology 2008;79(2) 223–233.

[25] Jenkins MA, Langlais PJ, Delis D, Cohen RA. Attentional Dysfunction Associated with Posttraumatic Stress Disorder Among Rape Survivors. Clinical Neuropsychology 2000;14(1) 7–12.

[26] Gronwall DMA. Paced Auditory Serial-Addition Task: A Measure of Recovery from Concussion. Perceptual and Motor Skills 1977;44 367–373.

[27] Stein MB, Kennedy CM, Twamley EW. Neuropsychological function in female victims of intimate partner violence with and without posttraumatic stress disorder. Biological Psychiatry 2002;52(11) 1079–1088.

[28] Goethals I, Audenaert K, Jacobs F, Lannoo E, Van de Wiele C, Ham H, et al. Cognitive neuroactivation using SPECT and the Stroop Colored Word Test in patients with diffuse brain injury. Journal of Neurotrauma 2004;21(8) 1059–1069.

[29] Stroop JR. Studies of interference in serial verbal reactions. Journal of Experimental Psychology 1935;18(6) 643-662.

[30] Soeda A, Nakashima T, Okumura A, Kuwata K, Shinoda J, Iwama T. Cognitive impairment after traumatic brain injury: a functional magnetic resonance imaging study using the Stroop task. Neuroradiology 2005;47(7) 501–506.

[31] Smits M, Dippel DWJ, Houston GC, Wielopolski PA, Koudstaal PJ, Hunink MGM, et al. Postconcussion syndrome after minor head injury: Brain activation of working memory and attention. Human Brain Mapping 2009;30(9) 2789–2803.

[32] Mayer AR, Yang Z, Yeo RA, Pena A, Ling JM, Mannell MV, et al. A functional MRI study of multimodal selective attention following mild traumatic brain injury. Brain Imaging and Behavior 2012;6(2) 343–354.

[33] Ziino C, Ponsford JL. Selective attention deficits and subjective fatigue following traumatic brain injury. Neuropsychology 2006;20(3) 383–390.

[34] Ruff RM, Niemann H, Allen CC, Farrow CE, Wylie T. The Ruff 2 and 7 Selective Attention test: A Neuropsychological Application. Perceptual and Motor Skills 1992;75 1311–1319.

[35] Belmont A, Agar N, Azouvi P. Subjective Fatigue, Mental Effort, and Attention Deficits After Severe Traumatic Brain Injury. Neurorehabilitation and Neural Repair 2009;23(9) 939–944.

[36] Ries M, Marks W. Selective Attention Deficits Following Severe Closed Head Injury: The Role of Inhibitory Processes. Neuropsychology 2005;19(4) 476–483.

[37] Neill WT. Inhibitory and facilitatory processes in selective attention. Journal of Experimental Psychology: Human Perception and Performance 1977;3(3) 444-450.

[38] McNally RJ, Kaspi SP, Riemann BC, Zeitlin SB. Selective processing of threat cues in posttraumatic stress disorder. Journal of Abnormal Psychology 1990;99(4) 398–402.

[39] Buckley TC, Blanchard EB, Neill WT. Information processing and PTSD: a review of the empirical literature. Clinical Psychology Review 2000;20(8) 1041–1065.

[40] Constans JI, McCloskey MS, Vasterling JJ, Brailey K, Mathews A. Suppression of Attentional Bias in PTSD. Journal of Abnormal Psychology 2004;113(2) 315–323.

[41] Bardeen JR, Orcutt HK. Attentional control as a moderator of the relationship between posttraumatic stress symptoms and attentional threat bias. Journal of Anxiety Disorders 2011;25(8) 1008–1018.

[42] El Khoury-Malhame M, Reynaud E, Soriano A, Michael K, Salgado-Pineda P, Zendjidjian X, et al. Amygdala activity correlates with attentional bias in PTSD. Neuropsychologia 2011;49(7) 1969–1973.

[43] Fani N, Tone EB, Phifer J, Norrholm SD, Bradley B, Ressler KJ, et al. Attention bias toward threat is associated with exaggerated fear expression and impaired extinction in PTSD. Psychological Medicine 2011;42(03) 533–543.

[44] Posner MI., Cohen Y. Components of visual orienting. In: Bouma H., Bonwhuis (eds.) Attention and performance X: Control of language processes. Hillsdale, N. J.: Erlbaum; 1984. p551-556.

[45] Bryant RA, Felmingham KL, Kemp AH, Barton M, Peduto AS, Rennie C, et al. Neural Networks of Information Processing in Posttraumatic Stress Disorder: A Functional Magnetic Resonance Imaging Study. Biological Psychiatry 2005;58(2) 111–118.

[46] Fichtenberg NL, Zafonte RD, Putnam S, Mann NR, Millard AE. Insomnia in a postacute brain injury sample. Brain Injury 2002;16(3) 197–206.

[47] Germain A. Sleep disturbances as the hallmark of PTSD: where are we now? American Journal of Psychiatry 2013;170(4) 372–382.

[48] Meewisse M-L, Nijdam MJ, de Vries G-J, Gersons BPR, Kleber RJ, van der Velden PG, et al. Disaster-related posttraumatic stress symptoms and sustained attention: Evaluation of depressive symptomatology and sleep disturbances as mediators. Journal of Traumatic Stress 2005;18(4) 299–302.

[49] Raichle ME, MacLeod AM, Snyder AZ, Powers WJ, Gusnard DA, Shulman GL. A default mode of brain function. Proceedings of the National Academy of Sciences of the United States of America 2001;98(2) 676–682.

[50] van Wingen GA, Geuze E, Caan MWA, Kozicz T, Olabarriaga SD, Denys D, et al. Persistent and reversible consequences of combat stress on the mesofrontal circuit and cognition. Proceedings of the National Academy of Science 2012;109(38) 15508–15513.

[51] Botvinick M, Nystrom LE, Fissell K, Carter CS, Cohen JD. Conflict monitoring versus selection-for-action in anterior cingulate cortex. Nature 1999;402(6758) 179–181.

[52] Shin LM, Bush G, Milad MR, Lasko NB, Brohawn KH, Hughes KC, et al. Exaggerated activation of dorsal anterior cingulate cortex during cognitive interference: a monozygotic twin study of posttraumatic stress disorder. American Journal of Psychiatry 2011;168(9) 979–985.

[53] Bryant RA. Posttraumatic stress disorder and traumatic brain injury: can they co-exist? Clinical Psychology Review 2001;21(6) 931–948.

[54] Barlow-Ogden K, Poynter W. Mild traumatic brain injury and posttraumatic stress disorder: Investigation of visual attention in Operation Iraqi Freedom/Operation Enduring Freedom veterans. Journal of Rehabilitation Research and Development 2012;49(7) 1101-1114.

[55] Fan J, McCandliss BD, Sommer T, Raz A, Posner ML. Testing the Efficiency and Independence of Attentional Networks. Journal of Cognitive Neuroscience 2002;14(3) 340–347.

[56] Ginzburg K, Ein-Dor T, Solomon Z. Comorbidity of posttraumatic stress disorder, anxiety and depression: A 20-year longitudinal study of war veterans. Journal of Affective Disorders. Elsevier B.V 2010;123(1-3):249–57.

[57] Berenz EC, Coffey SF. Treatment of Co-occurring Posttraumatic Stress Disorder and Substance Use Disorders. Current Psychiatry Reports 2012;14(5):469–77.

[58] Cicerone KD, Langenbahn DM, Braden C, Malec JF, Kalmar K, Fraas M, et al. Evidence-Based Cognitive Rehabilitation: Updated Review of the Literature From 2003 Through 2008. Archives of Physical Medicine and Rehabilitation 2011;92(4) 519–530.

[59] Parsons TD, Courtney CG, Arizmendi B, Dawson ME. Virtual Reality Stroop Task for neurocognitive assessment. Studies in Health Technology and Informatics 2011;163 433–439.

Chapter 23

Traumatic Brain Injuries and Sleep/Wake Disorders

Narayan P. Verma and Arunima Verma Jayakar

Additional information is available at the end of the chapter

1. Introduction

Traumatic brain injury (TBI) and sleep/wake disorder/s have a complex relationship [1]. A sleep disorder may make a person more prone to TBI by making him or her drowsy or inattentive and therefore more prone to fall or have an accident [2]. A sleep disorder may also make a person with concussion more prone to develop prolonged concussion or post-concussion syndrome in which symptoms last more than 3 weeks or even more than 3 months [3-7]. Likewise a sleep disorder may make them more prone to future concussions and cumulative injury. [7]

Less known and more common and recently recognized is the sleep/wake disorder caused by TBI itself, most simply termed the post traumatic sleep disorder [8-10] We discourage the use of acronym **PTSD**, however, lest it be confused with post-traumatic stress disorder. We propose the acronym **PTSLD**. This chapter is dedicated to delineating this disorder.

2. TBI

TBI is a problem of significant and increasing proportions-recently described as a silent epidemic [1]. The number of individuals with TBI is expected to climb with the return of Iraq and Afghanistan veterans back to USA. Just like atom bomb induced cancer was the signature injury of World War II and Agent Orange the signature injury of Vietnam War, TBI is the signature injury of Iraq and Afghanistan wars. [1]

Current estimates indicate that the TBI occurs in 100–400 per 100,000 people per year in North America and Europe. Men are more often affected than women. The most common age group which suffers from traumatic brain injury is 15–35 years. It is the most frequent cause of death between the ages 1-15. It accounts for one third of all injury related deaths in the USA. [8]

The TBI may result from fall, domestic violence, street violence, during birth, motor vehicle accidents, war related injuries, a work related injury or due to sports. Falls and motor vehicle accidents are the most common causes in civilian practice. In developing countries such as India with smaller land size, more population density, lax driving law enforcement and booming motor vehicle growth per capita, motor vehicle accidents are as much as 100 times more common than developed countries such as UK or USA.

TBI could be due to a blunt or a penetrating trauma. The trauma may be direct or indirect such from a nearby explosion-as many as 59 percent soldiers exposed to improvised explosive devices (IEDs) develop TBI. [1]

Contrary to popular belief a significant loss of consciousness (LOC) is not always necessary to make a diagnosis especially in so called mild TBI. Per Center of disease control (CDC), a history of clear cut LOC is seen in only less than 10 percent of patients with concussion. [8]

TBI is often described as acute, subacute and chronic-arbitrarily according to the time elapsed. It is also rated as mild, moderate or severe. Although there is no consensus, many prevailing criteria exist. The departments of defense and veterans affairs [3] have attempted to do this as follows:

	GCS	**PTA**	**LOC**
Mild	13–15	<1 day	0–30 minutes
Moderate	9–12	>1 to <7 days	>30 min to <24 hours
Severe	3–8	>7 days	>24 hours

GCS-Glasgow coma scale, **LOC**-loss of consciousness, **PTA**-post-traumatic amnesia

Table 1. Severity of traumatic brain injury

Only thing mild about **mild TBI** is the name. It accounts for 75 percent of all TBI. It may potentially be associated with significant, enduring and sometimes devastating consequences, greater likelihood of injury from a repeat concussion and long-term risk of Parkinson's disease, dementia, depression, suicide or homicide. [8] The incidence of mild TBI, also called concussion, is believed to be 6 per 1000 but this may be an underestimate. Per the center of disease control (CDC), a total of 1.4 million visits to hospitals/ER per year in USA are related to concussion/mild TBI. Additional 1.6-3.8 million never visit hospital or the ER. Mild TBI is often considered to occur when Glasgow coma score (GCS) at 24 hours after trauma is 13-15, LOC is 0-30 min and post-traumatic amnesia (PTA) is less than a day. It is often used interchangeably with the term concussion. Sports teams often use the acute concussion evaluation (ACE) questionnaire to evaluate the symptoms of concussion in the sport field.

The ACE questionnaire scores the individual on characteristics of Injury (such as severity and type of trauma, LOC, amnesia and presence or absence of seizures), presence and severity of physical symptoms (such as headache, photophobia, dizziness, nausea, blurred vision etc), cognitive symptoms (fogginess, confusion, forgetfulness, perseveration, slow cerebration, lack of concentration), emotional symptoms (irritability, sadness, emotional lability, nervousness) and sleep-related symptoms (such as insomnia, hypersomnia, drowsiness, daytime sleepiness, hyperarousal, flashbacks, nightmares), whether they are increased by exertion and how the person feels as compared to before injury on a seven point sale 0-6) and risk factors such as previous history of concussion, headaches, depression anxiety, sleep disorder or developmental disorder such as ADHD or learning disability. This may also be used for serial follow up.

The military equivalent of this scale is called MACE. A score of 25 or above is considered indicative of concussion in MACE. Symptoms persist from 3 weeks to 3 months and are called post-concussive syndrome if they persist beyond 3 months. Although it is believed that concussion results from biomechanical alterations in the brain and there is no structural damage, neuropathological and MRI data with tractography (diffuse tensor imaging or more sophisticated constrained spherical deconvolution) refute this thesis. If *conventional* neuroimaging such as head CT or MRI of the brain is abnormal, the TBI is no longer mild. However moderate to severe TBI may occur with or without abnormal conventional neuroimaging.

Moderate TBI is defined as LOC greater than 30 min but less than 24 hours, GCS 9-12 and/or PTA 1-7 days. It is variously graded by Global assessment of functioning or GAF scale [4] scored from 0-100) or some regional scales such as Rancho Los Amigos Scale [5] which assesses head injured patients on 8 levels of cognitive functioning (LOCF). The GAF score would be expected to be 51-60 with moderate TBI.

Severe TBI is defined as LOC greater than 1 day, GCS 3-8 and/or PTA greater than 7 days. Seizure occurring acutely during the head trauma does not necessarily make the TBI severe but chronic seizure disorder starting 3 months to several years after the head trauma certainly qualifies the TBI as a severe injury. Likewise macro injury, infarction, encephalomalacia, hematoma, persistent focal or lateralized neurological signs, dementia, severe personality change or new onset severe psychiatric disorder stamp an injury as severe. GAF score of 50 or below will indicate a severe head injury even when the neuroimaging is negative. There were 5.3 million people in USA living with severe TBI by 1999 [6]. The number must certainly be higher now. Severe TBI is of 2 types: closed and penetrated.

TBI may occur alone or may be associated with involvement of not just the brain but also skull, scalp, meninges, eyes, ears, sinuses and other neighborhood structures as well as injuries to neck and body.

3. TBI and sleep/wake disorders

Sleep/wake disorders may, in fact, potentially make folks more prone to TBI by making them sleepy and/or inattentive and therefore more likely to be subject of an injury or accident. A preexisting sleep disorder also makes the likelihood of concussion being prolonged and persistent. However, this chapter will mainly deal with the issue of sleep wake disorder/s caused by TBI, a far more common and as yet not well defined problem.

Sleep related problems secondary to chronic TBI have been described anecdotally or in case-report format since 1941. [11-19]. Some commonly reported disorders include hypersomnia, narcolepsy, delayed sleep phase, insomnia, fatigue, alteration of sleep-wake schedule, and movement disorders. It has been found clinically that, insomnia [20], hypersomnia [21-28] and excessive daytime sleepiness (EDS) are common [29, 30] in TBI and may at times occur in the same patient at different intervals from traumatic insult (see below). Only more recently in last 30 years, attempts have been made to explore this relationship in detail. Guilleminault et al in 1982 [13] described impaired daytime functioning and somnolence in 98 percent of all patients with TBI and further expanded their findings in year 2000, [21] extending their observations to even those with cervical whiplash and commenting on the medico-legal dilemma.

Post-traumatic sleep/wake disorders may significantly impair the rehabilitation potential of an injured individual and need to be accurately diagnosed and treated. Organized literature in this important area is sparse and fragmented. An organized account of these disorders is essential not only to improve the rehabilitation potential of these unfortunate individuals but to protect their medical coverage from auto insurers, as they encounter significant skepticism from adjusters regarding their sleep/wake issues to be causally related to their accidents and injury.

To be perfectly accurate, the sleep/wake disorder may not only result from head injury but neck and bodily injuries may cause or contribute to sleep/wake issues equally or even predominantly. [21]

The post-traumatic sleep/wake disorders may evolve, recede or be persistent after TBI. In a prospective study [31], there was found to be a high prevalence of sleep disorders (46%) and of excessive daytime sleepiness (25%) in 87 subjects at least 3 months after TBI [23]. 47% of the subjects in the aforementioned study was found to have a sleep disorder: OSA (23%), PTH (11%), narcolepsy (6%), or PLMS (7%) and 26% of the subjects had EDS [23]. In immediate post-traumatic period, hypersomnia may be common in hospitalized patients due to medications and interrupted nocturnal sleep due to pain and frequent nursing evaluations. Later on, it may be replaced by insomnia. Parasomnias may occur as well. *TBI is now known to cause nearly the entire spectrum of any or all sleep disorders and further may aggravate a pre-existing sleep/wake disorder by potential mechanisms enumerated elsewhere in this chapter.*

4. Acute TBI and sleep-related symptoms

Watson et al in 2007 [32] found in a prospective study of 514 patients that sleep related symptoms are common during acute phase of TBI. As much as 54 percent patients have daytime somnolence, more in those with more severe injury. They result in daytime somnolence which in turn may lead to poor daytime performance, altered sleep-wake schedule, heightened anxiety, and poor individual sense of well-being, insomnia, and depression. Half of these individuals are still sleepy at the end of one year. Relationship with severity or localization of head injury was disputed by Baumann et al [28] who evaluated patients prospectively as well, but their study ended at 6 months instead of one year in Watson's study.,. However, similar to Watson study, they also found that quality of life was impaired by these symptoms.. CSF hypocretin-1 was found to be significantly reduced levels in those patients with excessive daytime sleepiness (EDS) symptoms.

5. Chronic TBI and sleep related symptoms

Verma et al in 2007 [10] in a retrospective study found that sleep changes and deranged sleep architecture are common in *chronic* TBI patients, arbitrarily defined as 3 months to 2 years after head trauma.. The sleep disorders seen in this population are similar to those seen in the general population but individual percentages are higher. Hypersomnia accounted for 50 percent of all patients and insomnia and parasomnia for quarter each. Global assessment of functioning (GAF) scores correlated with some (stage N1 percentage, impaired sleep efficiency and wake during sleep), but not all (stage shifts and wake before sleep) measures of sleep disruption, indicating a complex and multifactorial pathogenesis.

6. Pathogenesis

The possible pathogenetic mechanisms of TBI causing sleep disorders include: direct brain injury, indirect brain injury, collateral damage to neck and back and resulting pain interfering with sleep, [13] weight gain (secondary to head trauma or medications used to treat head trauma or its sequelae such as posttraumatic mood, anxiety or stress disorder), pre-existing genetic propensity for narcolepsy, which may be clinically aggravated or precipitated by head-trauma [11], a pre-existing anatomical abnormality of sleep-related brain mechanisms, oropharyngeal abnormality aggravated by head trauma or resulting weight gain, anatomical abnormalities caused by head trauma such as jaw dislocation, TMJ problems, and brainstem and forebrain lesions induced by TBI.

Direct brain injury was first described by Strich in 1961 [22] as diffuse degeneration of white matter subsequently termed the diffuse axonal injury (DAI). This was later determined in animal experiments to be the consequence of inertial loading of the head by prolonged coronal angular acceleration [23] with brunt of abnormality in septum pellucidum, corpus callosum, deep gray matter and dorso-lateral pons and midbrain, areas closely associated with sleep-wake mechanisms. The biochemical basis of this injury is excitotoxicity, [24] inflammation, [25] free radicals/eicasanoids, [9] hyperglycolysis, [26] hyperglycemia, [26] and apolipoprotein E e4 synthesis. [27] These mechanisms most likely operate in sleep disorders associated with mild head injury. MRI with tractography may provide a direct evidence of such injury. It has also been hypothesized that the hypocretin system may be partly responsible for the patho-physiology of sleep wake disturbances present post TBI [28].

7. Classification: We propose the following classification of post-traumatic sleep/wake disorders

1. Post-traumatic sleep wake disorder/s-resulting from TBI
2. Post-traumatic sleep/wake disorder/s resulting from neck and/or bodily injuries
3. Post-traumatic sleep/wake disorder/s resulting from both-TBI and neck and/or bodily injuries

Each group has 2 subtypes:

Primary: This group consists of patients who never had any sleep/wake related issues what so ever prior to the accident.

Secondary: These patients have a preexisting sleep/wake disorder which is either aggravated or altered by the accident and in fact may have contributed to the occurrence of accident by making patient inattentive and therefore prone to have an accident. A pre-existing sleep disorder also makes the likelihood of an enduring concussion more as well as increases the proneness to further and cumulative deterioration after repeat concussion.

8. Clinical features

Insomnia: This is the most common consequence of the TBI. It is pretty much universal in all patients with mild TBI at least in initial stages. It is associated with headache, dizziness, mood changes, imbalance and blurred vision and flashbacks in various combinations. It usually resolves in 3 weeks to 3 months in most mild cases but sometimes may be nagging and persistent. It may be sleep onset or sleep maintenance or associated with premature awakening in the morning-the so called the " terminal" insomnia. It may be contributed to by associated anxiety and depression as shown by Verma et al [10] based on Hamilton Anxiety Scale (HAS-appendix 3) and Beck's Depression Inventory (BDI-appendix 4). Nightmares and flashbacks may also contribute. At times it reflects more serious pathology such as sleep apnea caused by TBI and/or neck or spinal injury or periodic limb movements induced by medications used to manage the patient. The medications such as topiramate, methylphenidate etc themselves may aggravate or cause insomnia. Circadian rhythm disorder may complicate insomnia or cause it either due to direct injury to the biological clock or patient's sleep hygiene suffering from frequent examinations by nurses, therapists and other workers, not going to work or office and irregular bedtime and wake up time.

Many patients with more severe head injury initially have hypersomnia due to medications, TBI itself, complicating sleep apnea or narcolepsy but later on after several months or even years develop insomnia. Same factors as listed above operate. Reverse is also true. Patients with mild TBI may develop hypersomnia/ parasomnia or narcolepsy later on even though they had insomnia to begin with. Thus the natural history of sleep/wake disorders is more complicated than the sleep/wake disorders in general as the type of disorder may switch over time.

Although, sleep studies are not generally indicated in patients with most cases of insomnia from other etiologies, the post-traumatic insomnia requires a sleep study for many reasons. It requires documentation as the adjusters frequently look for objective confirmation of subjective symptoms. Also post-traumatic insomnia may not be just pure insomnia but contributed to by sleep apnea, narcolepsy, periodic limb movements, parasomnias such as the REM behavior disorder or a circadian rhythm disorder. In addition, the insomnia may be replaced by hypersomnia or parasomnia and/or even nocturnal seizures later on. The polysomnogram (PSG) should be done with expanded EEG montage with simultaneous video-taping. Traditional investigating methods such as sleep diary, actigraphy and scales such as Hamilton anxiety scale (HAS -appendix 3) and Beck's depression inventory (BDI -appendix 4) also help in dissecting, intellectualizing and treating the issue at hand.

Hypersomnia: This is the second most common consequence. It is quite universal in acute stages of moderate to severe injury as patient is often kept intubated and sedated, on pain medications and primary brain and brainstem pathology from TBI may also contribute. Weight gain from medications to use the consequences of TBI such as antidepressants, anxiolytics or anticonvulsant medications, lack of activity, not working or going to office, being sedentary due to severe TBI and/or neck/bodily injury, overeating due to injury to satiety center of the brain may result in development of obstructive sleep apnea even when there are no oropharyngeal anatomical risk factors. Direct or indirect injury to sleep centers and breathing centers may also contribute. Lowered hypocretin levels may be operative as stated elsewhere. Neck injury may impair diaphragmatic function and add insult to injury. Associated high spinal cord injury may be devastating but cervical whiplash itself is known to cause obstructive sleep apnea [21]. Narcolepsy may be precipitated in a person who is genetically predisposed for it or even be caused by TBI, sometimes even mild TBI. After several years it may be replaced by insomnia in some patients. Hypersomnia with prolonged sleep and even Klein-Levin like syndrome might occur. The periodic limb movements (PLMs) are common either due to medications used to treat TBI or due to unknown reasons such as inactivity or complex chemical changes/alterations, not yet known. Circadian rhythm disorders may contribute. Video-polysomnography with expanded EEG montage and frequently a multiple sleep latency test (MSLT) is addition if the Epworh Sleepiness Scale (ESS-appendix 1) is 11 or more is essential for the diagnosis and should be in-lab and not portable. Actigraphy and sleep diary might help as well.

Parasomnias: These are third most common complications and often co-exist with hypersomnia or insomnia. Each patient may have more than one parasomnia. They may also develop as a remote complication of TBI. As repeated concussions are known to predispose to Parkinson's disease and Parkinson's disease is associated with or even preceded by REM behavior disorder (RBD) by as much as 3 years, this is not entirely unexpected. Sleepwalking, nocturnal eating disorder, nocturnal seizures, nocturnal enuresis either as a part of post-traumatic OSA or due to TBI itself and confusional arousals all are seen and common. In-lab video-polysomnography (video-PSG) with expanded EEG montage is essential for the documentation and diagnosis. The family may also be encouraged to use their smart phones to record these events to help in diagnosis.

9. How to approach a patient with PTSLD

Detailed history is important. One should carefully ascertain if sleep related symptoms started after TBI or preceded that. If latter, document any changes in severity of symptoms or change in symptoms. The routine scales administered in our practice are: Mini mental state examination, ACE questionnaire, Hamilton anxiety Scale (appendix 3), Beck's depression inventory (appendix 4), Epworth sleepiness Scale (appendix 1) and Berlin questionnaire (appendix 2). Careful determination of LOC, PTA and GCS is done based on hospital, ER and other previous records. Computerized psychological and neuropsychological testing (easily administered by even a medical assistant using the 'neurotrax' system) to determine the global assessment of cognitive functioning and levels of anxiety and depression is important to establish a baseline and future follow up. Physical examination should pay careful attention to the HEENT examination, TMJ, neck size, chin (prognathia, retrognathia, micrognathia), oropharyngeal examination for tonsillar size from 1-4 and Mallampati score 1-4, focal and lateralized neurological signs and cardiopulmonary examination. Sleep/wake related history should include details about snoring, witnessed pauses in breathing, bedtime, wake up time, circadian rhythm, gasping and choking in sleep, hypnagogic hallucinations, hypnapompic hallucinations, nightmares, nocturnal incontinence, seizures, sleep walking and acting out of dreams, any falls from bed, restless legs and periodic limb movements (by asking questions such as do you have creepy crawling sensation in your limbs and feet which improve by movement). Wakefulness should be evaluated for alertness, drowsiness, dozing, napping, daydreaming and automatic behaviors. The ESS (appendix 1) is helpful in quantitating sleepiness and Berlin questionnaire (appendix 2) about the probability of sleep related breathing disorder. Sometimes the sleepiness scales are not reliable in patients with severe head injury. Caregiver's input is needed in those situations. Current medication list is critical.

Ancillary tests include an MRI of the head with tractography (diffuse tensor imaging or preferably constrained spherical deconvolution), EEG, an overnight in lab video-PSG with an expanded EEG montage and a 5 nap daytime multiple sleep latency test (MSLT) if ESS is greater than 10. A seven day sleep diary and actigraphy is obtained in those with insomnia or circadian rhythm issues. CSF hypocretin levels may be useful.

Initial follow up visits are monthly for 3 months and then 3 monthly times two. Six monthly visits are obtained after that. Annual ancillary evaluation is more limited and defined by patient's clinical symptomatology. However, sleep disorders may change their characteristics during the course and re-evaluation may need to be tailored accordingly. Therefore a cook book approach is not useful. Maintenance of wakefulness test may be useful in quantitating residual daytime sleepiness.

10. PTSLD

We propose this term as an acronym for post-traumatic sleep/wake disorders to distinguish it from PTSD or post-traumatic stress disorder.

1. Post-traumatic sleep-related breathing disorder:

This is fairly common in general population affecting 2-4 percent of all adults. In patients with TBI, sleep apnea defined as apnea-hypopnea index (AHI) of 10 or greater may be present in up to 30 percent of all patients. Seventy five percent of apneas and hypopneas are obstructive in nature. This condition may present as hypersomnia, insomnia or may only be seen on laboratory evaluation as an unexpected finding. It may cause a secondary REM behavior disorder (RBD -a parasomnia) which may potentially be injurious to the patient if not recognized and treated and further compound the TBI. Mechanisms of post-traumatic sleep related breathing disorder are several. Patient may have pre-existing anatomical abnormalities which were insufficient to cause the sleep related breathing disorder prior to TBI but the occurrence of TBI provides a sufficient milieu for it to clinically manifest. Sedative medications such as clonazapam may potentiate apnea, antidepressants such as sertraline and mitrazepine and anticonvulsants such as valproic acid cause weight gain which is a known risk factor for this condition. Weight gain may also result from physical inactivity and direct damage to hypothalamic centers related to feeding and satiety. Tracheostomy, if done, during the acute management of TBI may further increase the risk especially in children by causing tracheomalacia, as seen in one child by the senior author of this chapter. In addition to known risks of this condition such as premature death, hypertension, heart attack, stroke, dementia and diabetes, this condition may impair the control of patient's seizure disorder if present. Newborns with perinatal head trauma and abused children may develop central apnea due to direct injury to the breathing centers in the brain. In general, more severe the head injury, higher the apnea hypopnea index and hypoxia are. ESS (appendix 1) may be unreliable in those with moderate to severe head injury and should not be used as a sole criterion to order or not order the sleep studies. [10] Berlin questionnaire (appendix 2) also helps in predicting the probability of sleep related breathing disorder.

2. Post-traumatic narcolepsy:

The incidence of narcolepsy in general population is about 1:2000 in the USA. Hormonal change and minor head trauma at puberty are long known to be initiating factors for narcolepsy in neurology text books as the genetic propensity of narcolepsy usually manifests clinically at or after puberty 90 percent of the time. In addition, most patients with narcolepsy remain the same throughout their lifetime. Post-traumatic narcolepsy is different in that it is far more common than general population (it is seen in up to 6-9 percent of all patients with TBI) [10, 11] and down the road, after several years in our experience, symptoms may sometimes abate and even be replaced by insomnia. Hypocretin levels are known to be reduced by TBI and may well play a pathogenetic role. New onset cataplexy might occur after TBI, increasing the risk of falls and therefore repeated TBI. Nightmares are common in TBI and careful history is needed to distinguish them from hypnagogic and hypnapompic hallucinations seen as auxiliary symptoms of narcolepsy. Occurrence of narcolepsy is not correlated with the degree of TBI.

Polysomnogram (PSG) will nearly always show some degree of sleep disruption such as increased percentage of N1, frequent awakenings, reduced sleep efficiency (less than 85 percent), reduced N3 (delta) percentage (less than 15 percent) and sometimes SOREMP (sleep onset REM period or REM sleep occurring with 15 min of sleep onset). The MSLT will show a sleep latency of 8 min or less and 2 or more SOREMPs in 5 naps. [33]

3. Post-traumatic hypersomnia.

Idiopathic hypersomnolence syndrome is long known to often follow TBI and conditions such as Guillaine Barre Syndrome (GBS) or infectious mononucleosis (IM). The PSG shows relatively normal or even improved sleep efficiency sometimes greater than 95 percent, increased or normal delta percentage, relatively few awakenings, minimal sleep disruption if any but severe daytime somnolence on MSLT with no SOREMPs. Medication effect needs to be excluded and one has to be careful not to misdiagnose 15 percent cases of narcolepsy in whom MSLT is initially negative as post-traumatic hypersomnia. The PSG features described above are helpful in distinction and there is no history of hypnagogic or hypnapompic hallucinations or cataplexy. Autonomic symptoms are sometimes present. [33]

4. Post-traumatic periodic limb movement disorder (PLMD):

These may a solitary abnormality on PSG but more often are associated with other conditions such as OSA and narcolepsy. They may have been present premorbidly but are often worsened by medications used for the treatment of TBI. They may be asymptomatic and may not require any treatment or cause significant patient insomnia or spousal discomfort and need treatment. Lower extremities are affected and sometimes only one side but upper extremities may be involved in addition or alone. Neurologic deficit such as hemiparesis or paraparesis may worsen or cause this condition. Anemia of chronic disease may compromise the serum ferritin level and may compound the issue. They are considered significant if more than 15/hour in an adult or 5/hr in a child. [33] Up to 30 percent of all patients may have this condition. [10] They may or may not report RLS in addition when awake.

5. Post-traumatic REM behavior disorder (RBD):

This condition was first described to occur in cats when lesions were created in peri-locus coeruleous area to interrupt impulses going down the ventral reticulopsinal tract to spinal motor neurons in REM sleep [10]. Similar mechanisms are operative in humans with TBI. Associated Parkinson's, alcoholism and medications may also contribute. Up to 13 percent patients show symptoms of RBD and/or show increased tone in chin EMG on PSG [10]. Patients typically act out their dreams during last third of sleep at night when REM percentage is the highest and potentially fall from bed, climb out of windows or walk out in freezing weather. It may be secondary to post-traumatic OSA and then it responds to CPAP. If not, RBD precautions and medications are necessary. It may be a precursor of Parkinsonism in patients with punch-drunk syndrome and may precede that condition by as much as 3 years. It should be distinguished from sleep walking which usually occurs during the first half of sleep and there is no dream recall. It should also be distinguished from NREM sleep related confusional arousals which are similar to night terrors.

6. Other post-traumatic parasomnias:

They include sleep paralysis, cataplexy, sleep walking, nightmares, sleep enuresis and nocturnal eating disorder. [33] All parasomnias (these plus RBD) occur in 25 percent of all patients with TBI, either by themselves on in addition to other disorders. Birth injuries to the head may be associated with head banging disorder. [16]

7. Post-traumatic insomnia:

At least one quarter of all patients with TBI have insomnia either sleep onset or sleep maintenance or a combination thereof. [10] Hamilton anxiety scores (appendix 3) are typically elevated in those with sleep onset insomnia and Beck's depression inventory scores (appendix 4) in those with sleep maintenance insomnia. Physical factors such as frequent examination by nurses and respiratory therapists during ICU stay may be the cause at least in acute TBI. Medications such as bronchodilators, anticonvulsants such as topiramate or stimulants such as methylphenidate may cause insomnia. Circadian rhythm abnormalities and not going to regular work and physical inactivity leading to frequent daytime naps may cause insomnia at night. Post traumatic sleep related breathing disorder both of obstructive type or central type may cause insomnia as well. Severe restless leg syndrome (RLS) may cause sleep onset insomnia and PLMs, sleep maintenance insomnia. Patients with post-traumatic narcolepsy may sometimes present initially as insomnia at night and only a careful history uncovers the diagnosis. For example, a patient treated by the senior author was treated for 2 years as insomnia by various physicians, until she disclosed to the author additional history of severe daytime sleepiness and napping since TBI and disturbing "nightmares" (actually hypnagogic hallucinations) with automatic behavior which prevented her from holding onto any job and not succeed in her new marriage. She responded beautifully to sodium oxybutate and was immensely grateful.

8. Post-traumatic circadian rhythm disorder:

This is a fairly common complication. The disorders include a delayed sleep phase syndrome (DSPS), irregular sleep wake cycle, advanced sleep phase syndrome (ASPS) and non-24 hour sleep wake cycle. [33] Irregular sleep wake cycle is common during the acute phase of TBI in moderate to severe cases as nurses and respiratory therapists check on the patient frequently and patient is on medications including sedatives or anesthetic agents such a propofol. It is also common in chronic TBI patients with a psychiatric disorder or blindness. Otherwise DSPS is the most common complication related to circadian rhythm abnormality in patients with TBI and a direct injury to suprachiasmatic nucleus may well be the cause. ASPS is quite frequent in elderly patients with TBI. Non-24 hour cycle is a rare complication in some patients showing a stepladder pattern on actigraphy. [33]

11. Differential diagnosis

Lack of pre-existing history of sleep related symptoms is critical for the diagnosis of the primary post-traumatic sleep disorder, although the TBI may aggravate a pre-existing sleep/

wake disorder or make it more difficult to treat. Secondary gain may need to be excluded but objective confirmation from tests as outlined above, obviates that possibility. Interviewing the family members, friends, golf buddies etc and previous medical records are helpful in determining whether the disorder is primary or secondary. Previous anatomical abnormities do not automatically exclude a primary post traumatic sleep/wake disorder as patient may have been compensated before and the TBI may have been the "last straw which broke the camel's back". Likewise, a positive HLA testing does not automatically make narcolepsy pre-existing or genetic, as head trauma, even mild or minimal is known to be the initiating factor for narcolepsy. Please refer to the international classification of sleep disorders [33] edition 3 for further help in differential diagnosis of post-traumatic sleep disorder/s from non-traumatic etiologies as detailed discussion of that would be tangential to the intent of this chapter.

Treatment : Once it is realized that TBI may cause or aggravate a pre-existing sleep/wake disorder, management is simple. It is treated like any other sleep disorder of another etiology by adding medications, reduction of medications, meditation, machines, devices or behavioral techniques. Treatment is important as it will interfere with rehabilitation potential of the patient unless addressed head on. The treatment modalities outlined below are well described in standard text books of sleep medicine and in the practice guidelines of the American Academy of Sleep Medicine [34] and would only be briefly outlined below without individually referring each modality to prevent unnecessary expansion of the reference list and dilute the intent of this chapter.

Medications and reduction of medications: Mild sleep apnea may be managed with proptriptyline. Periodic limb movements may be helped by Clonazepam, dopamine agonists, gabaergic agents, tonic water and vitamin D. Nocturnal eating disorder often responds to topiramate. REM behavior disorder responds well to Clonazepam and/or melatonin. Nocturnal seizures require anticonvulsants. Carbamazepine is most effective. Hypnagogic hallucinations and nightmares may require clonazepam, imipramine or more complex pharmacological remedies as NMDA agonists.

Reduction of medications may also help by reducing weight, decreasing excessive daytime sleepiness, lessening the aggravation of OSA or reducing PLMs. Some medications such as amitriptyline may induce or aggravate RBD in TBI patients and that might improve by this strategy. Sometimes, reducing topiramate, certain antidepressants, stimulants and wakefulness promoting agents etc may improve insomnia.

Meditation: simple meditation techniques such as Hong-Sau which do not require any special equipment or posture or more complex such techniques such as yoga exercises requiring exercise mats and lotus position may be useful at times in reducing anxiety and improving sleep onset insomnia.

Behavioral techniques: 11 principles of sleep hygiene (appendix 5), 6 Bootzin's principles (appendix 6) and cognitive behavioral therapy for insomnia (appendix 7) are often useful for the management of insomnia. In fact most enduring relief of post-traumatic insomnia comes with non-pharmacological behavioral techniques.

Patients with RBD may require padding around their beds, alarms and double locks at doors and boarding up of windows (RBD precautions). Those with nocturnal eating disorder, may need to have a lock on the refrigerator.

Phototherapy: It is useful in the treatment of post-traumatic delayed sleep phase syndrome and advanced sleep phase syndrome, non 24 hr sleep/wake cycle and insomnia caused by post-traumatic depression. Morning exposure to a standard 2500-10,000 lux lamp from a distance of about 18-24 inches for 30-60 minutes is used in all of these conditions except ASPS in which evening exposure is required. Nausea and queasiness may occur and the duration of exposure may need to be optimized upwards gradually.

Chronotherapy. It will be useful in managing circadian rhythm disorders resulting from head trauma. Advancing the bedtime by 3 hrs a day may help the treatment of DSPS over 8-10 days. It may be used alone or in conjunction with phototherapy and/or melatonin.

CPAP: Continuous positive airway pressure it is the mainstay of treatment for moderate to severe OSA. Patients may experience difficulty in using this device due to facial or TMJ injuries or chest trauma or CHF or if they are claustrophobic.

BIPAP and BIPAP-ST (NIPV)- Bi-level treatment is necessary in some of the patients who have co-morbid muscle disease or CHF. It provides a lesser pressure to facilitate exhalation in an individual with weak muscles or CHF. The inhalation pressures IPAP) are at least 2 cm of H_2O or more than the pressure for exhalation (EPAP). BIPAP-ST or NIPV provides additional protection by backing up ventilation if the respiratory rate falls below a predetermined rate such as 10/min.

Adaptive servo ventilation (ASV): is often useful when nothing else works in complex or central sleep apnea caused by TBI by overwhelming the apneas not only simply by pressure but also volume of the inhaled air and mostly obviates the need for tracheostomy except in acute stages. This device is quite expensive but well worth it as the senior author has never prescribed tracheostomy for sleep-related breathing disorder ever since this device has been commercially available. Prior to that, it was needed at least in one patient every year in our clinic.

Jaw advancement devices: may be useful in those with mild post-traumatic OSA with TMJ issues.

Ongoing follow up is essential by at least 3-6 monthly office visits and yearly sleep studies since post-traumatic sleep disorders are notorious to change during their natural history and may require altogether different treatment as the time passes by.

12. Discussion

A spectrum of sleep disorders are a common finding after the acute phase of TBI [9]. They result in daytime somnolence which in turn may lead to poor daytime performance, altered sleep-wake schedule, heightened anxiety, and poor individual sense of well-being, insomnia and depression [10] Sleep changes and deranged sleep architecture are more common in *chronic* TBI patients as compared to the general population [10]. Sleep disturbances can compromise the rehabilitation process and the ability to return to work. [20] A high index of suspicion may lead to a diagnosis and subsequent treatment of these disorders and contribute to physical and cognitive rehabilitation of these patients. [10] A proper diagnosis and greater awareness of this complication protects patient's rights for medical care under auto-insurance laws in states such as Michigan [2]. This will also be critical in the management of TBI related symptoms of returning veteran of Iraq and Afghanistan war since TBI is the signature injury of those wars and has become a silent epidemic. [1]

13. Directions for future research and efforts:

The future research and efforts should concentrate on primary prevention of TBI, better delineation of premorbid sleep/wake status by some scales (similar to those which predict premorbid IQ), early identification and accurate diagnosis of post-traumatic sleep/wake disorder, its exact impact on physical, cognitive and occupational rehabilitation, convincing the auto-insurances not to be stingy in the care of these unfortunate individuals and look at sleep/wake related complaints as a medical issue and not a malingering issue, and the US government to provide greater research and medical funds for this important medical condition.

Appendix 1. The epworth sleepiness scale

0 = no chance of dozing	
1 = slight chance of dozing	
2 = moderate chance of dozing	
3 = high chance of dozing	
SITUATION	**CHANCE OF DOZING**
Sitting and reading	___________
Watching TV	___________
Sitting inactive in a public place (e.g. a theater or a meeting)	___________
As a passenger in a car for an hour without a break	___________
Lying down to rest in the afternoon when circumstances permit	___________
Sitting and talking to someone	___________
Sitting quietly after a lunch without alcohol	___________
In a car, while stopped for a few minutes in traffic	___________

Appendix 2. Berlin questionnaire

Top of Form

1. Body Mass Index Information:

Height (in inches):

Weight (in pounds):

CATEGORY 1 QUESTIONS

2. Do you snore?

Yes **

No

I don't know

3. How loud is your snoring?

My snoring is as loud as breathing

My snoring is as loud as talking

My snoring is louder than talking **

My snoring is very loud **

4. How frequently do you snore?

Almost every day **

3-4 times per week **

1-2 times per week

1-2 times per month

Never or almost never

5. Does your snoring bother other people?

Yes **

No

6. How often have your breathing pauses been noticed?

Almost every day **

3-4 times per week **

1-2 times per week

1-2 times per month

Never or almost never

CATEGORY 2 QUESTIONS

7.Are you tired after sleeping?

Almost every day **

3-4 times per week **

1-2 times per week

1-2 times per month

Never or almost never

8. Are you tired during waketime?

Almost every day **

3-4 times per week **

1-2 times per week

1-2 times per month

Never or almost never

9. How often do you nod off or fall asleep while driving?

Almost every day **

3-4 times per week **

1-2 times per week

1-2 times per month

Never or almost never

CATEGORY 3 QUESTIONS

10. Do you have high blood pressure?

Yes **

No

I don't know

BMI (body mass index)

BMI > 30 **

Weight

BMI = -------------------- X 703

Height X Height

Weight in pounds, height in inches OR Weight in kilograms, height in meters

Berlin Scoring Results

Any answer followed by double asterisks (**) is a positive response.
Category 1 is positive with 2 or more positive responses to questions 2 through 6
Category 2 is positive with 2 or more positive responses to questions 7 through 9
Category 3 is positive with 1 or more positive responses and/or a BMI>30
2 or more positive categories indicates a high likelihood of sleep apnea

---

Bottom of Form

Appendix 3. Anxiety rating scales

1. Background
    1. Authored by Max Hamilton in 1959
    2. Public domain anxiety rating scale
2. Symptom Rating Scale (0=Not Present, 4=Disabling)
    1. Anxious Mood
        - 1.1. Worries
        - 1.2. Anticipates worst
    2. Tension
        - 2.1.Startles
        - 2.2. Cries easily
        - 2.3. Restless
        - 2.4. Trembling
    3. Fears
        - 3.1. Fear of the dark
        - 3.2. Fear of strangers
        - 3.3. Fear of being alone
        - 3.4. Fear of animal
    4. Insomnia
        - 4.1. Difficulty falling asleep or staying asleep
        - 4.2. Difficulty with Nightmares
    5. Intellectual

    - 5.1. Poor concentration
    - 5.2. Memory Impairment
6. Depressed Mood
    - 6.1. Decreased interest in activities
    - 6.2. Anhedonia
    - 6.3. Insomnia
7. Somatic Complaints: Muscular
    - 7.1. Muscle aches or pains
    - 7.2. Bruxism
8. Somatic Complaints: Sensory
    - 8.1. Tinnitus
    - 8.2. Blurred vision
9. Cardiovascular Symptoms
    - 9.1. Tachycardia
    - 9.2. Palpitations
    - 9.3. Chest Pain
    - 9.4. Sensation of feeling faint
10. Respiratory Symptoms
    - 10.1. Chest pressure
    - 10.2. Choking sensation
    - 10.3. Shortness of Breath
11. Gastrointestinal symptoms
    - 11.1. Dysphagia
    - 11.2. Nausea or Vomiting
    - 11.3. Constipation
    - 11.4. Weight loss
    - 11.5. Abdominal fullness
12. Genitourinary symptoms
    - 12.1. Urinary frequency or urgency
    - 12.2. Dysmenorrhea
    - 12.3. Impotence
13. Autonomic Symptoms
    - 13.1. Dry Mouth
    - 13.2. Flushing

        - 13.3. Pallor
        - 13.4. Sweating
    - 14. Behavior at Interview
        - 14.1. Fidgets
        - 14.2. Tremor
        - 14.3. Paces

3. Interpretation
    1. Above 14 symptoms are graded on scale
        - 1.1. Not present: 0
        - 1.2. Very severe symptoms: 4
    2. Criteria
        - 2.1. Mild Anxiety (minimum for Anxiolytic): 18
        - 2.2. Moderate Anxiety: 25
        - 2.3. Severe Anxiety: 30
4. Other Anxiety Scales
    1. Zung Self Rating Scale for Anxiety
    2. Beck Anxiety Scale
    3. GAD-7

Appendix 4. Beck depression inventory

1. Background
    1. Twenty-one question survey completed by patient
    2. Answers scored on 0 to 3 scale
        - 1.1. Minimal: 0
        - 1.2. Severe: 3
2. Questions
    1. Sadness
    2. Hopelessness
    3. Past failure
    4. Anhedonia
    5. Guilt
    6. Punishment
    7. Self-dislike

   8. Self-blame
   9. Suicidal thoughts
   10. Crying
   11. Agitation
   12. Loss of interest in activities
   13. Indecisiveness
   14. Worthlessness
   15. Loss of energy
   16. Insomnia
   17. Irritability
   18. Decreased appetite
   19. Diminished concentration
   20. Fatigue
   21. Lack of interest in sex

3. Interpretation
   1. Score <15: Mild Depression
   2. Score 15-30: Moderate Depression
   3. Score >30: Severe Depression

Resources: Beck Depression Inventory
   1. General
      1.1. Intended for use by licensed professionals only
      1.2. Copyrighted by the Psychological Corporation
   2. Available for purchase from Psychological Corporation
      2.1. http://www.psychcorp.com/

5. Reference
   1. Beck (1996) Beck Depression Inventory, Harcour

Appendix 5. Eleven principles of sleep hygiene

1. Wake up and go to bed at about the same time every night. Bedtime and wake-up time should not differ from working days to weekend nights by more than approximately an hour.

2. Avoid sleeping in on weekends to "catch up" on sleep. This makes it more likely that you will have problems falling asleep.

3. If you take naps, they should be short (no more than an hour) and scheduled in the early to midafternoon. However, if you have a problem with falling asleep at night, napping during the day may make it worse and should be avoided.

4. Spend time outside every day. Exposure to sunlight helps to keep your body's internal clock on track.

5. Exercise regularly. Exercise may help you fall asleep and sleep more deeply.

6. Use your bed for sleeping only. Don't study, read, listen to music, watch television, etc., on your bed.

7. Make the 30–60 minutes before a quiet or wind-down time. Relaxing, calm, enjoyable activities, such as reading a book or listening to calm music, help your body and mind slow down enough to let you get to sleep. Don't study, watch exciting/scary movies, exercise, or get involved in "energizing" activities just before bed.

8. Eat regular meals and don't go to bed hungry. A light snack before bed is a good idea; eating a full meal in the hour before bed is not.

9. Avoid eating or drinking products containing caffeine from dinner time on. These include caffeinated sodas, coffee, tea, and chocolate.

10. Do not use alcohol. Alcohol disrupts sleep and may cause you to awaken throughout the night.

11. Smoking disturbs sleep. Don't smoke at least one hour before bed (and preferably, not at all!).

Appendix 6. Six Bootzin's principles for stimulus control in the treatment of insomnia

1. Go to bed when sleepy.

2. Use the bed for sleeping; do not read, watch television or eat in bed.

3. If you are unable to fall asleep, get up and move to another room; stay up until you are really sleepy, then return to bed; if sleep still does not come easily, get out of bed again. The goal is to associate bed with falling asleep quickly.

4. Repeat step 3 as necessary throughout the night.

5. Set the alarm and get up at the same time every morning regardless of how much you slept through the night. This helps the body acquire a constant sleep-wake rhythm.

6. Do NOT nap during the day.

Appendix 7. Cognitve behavioral therapy for insomnia: weekly for 8-10 weeks:

Sleep hygiene-see above-appendix 5

Stimulus control-see above-appendix 6

Sleep restriction or curtailment

Relaxation, meditation, hypnosis

Reducing muscle tension and hyperarousal by biofeedback

Relapse prevention:

1. Don't compensate for sleep loss
2. Start stimulus control procedures immediately
3. Re-engage sleep restriction should the insomnia persist beyond a few days.

Author details

Narayan P. Verma[1*] and Arunima Verma Jayakar[2]

*Address all correspondence to: narayangod@aol.com

1 Oakland University William Beaumont School of Medicine, Oakland, USA

2 BG Tricounty Neurology and Sleep Clinic PC, USA

References

[1] Theodorou AA, Rice SA. Is the silent epidemic keeping patients awake? JCSM 2007;3:347-348

[2] Verma NP. Importance of traumatic brain injury. J Clin Sleep Med 2008; 4: 177.

[3] Department of Defense and Department of Veterans Affairs (2008). "Traumatic Brain Injury Task Force".http://www.cdc.gov/nchs/data/icd9/Sep08TBI.pdf

[4] 4.. Modified Global Assessment of Functioning – Revised (mGAF-R)-adapted in 2004 by the Florida DCF Functional Assessment Workgroup from the original M-GAF reported by S. Caldecott-Hazard & R.C.W. Hall, 1995

[5] Gouvier, WD, Blanton, PD.LaPorte, KK, Nepomuceno, C.. Reliability and validity of the Disability Rating Scale and the Levels of Cognitive Functioning Scale in monitor-

ing recovery from severe head injury. *Archives of Physical Medicine and Rehabilitation* 1987; 68(2), 94-97.

[6] Thurman D, Alverson C, Dunn K, Guerrero J, Sniezek J. Traumatic brain injury in the United States: a public health perspective. J Head Trauma Rehabil 1999;14(6):602-615.

[7] Faul M, Xu L, Wald MM, Coronado VG. Traumatic brain injury in the United States: emergency department visits, hospitalizations, and deaths. Atlanta (GA): Centers for Disease Control and Prevention, National Center for Injury Prevention and Control; 2010.

[8] Injury prevention and control: traumatic brain injury: http://www.cdc.gov/traumatic-braininjury/

[9] Parker F, Andrews PJD, Azouvi P, Aghakhani N, Perrouin-Verbe B. Acute Traumatic brain injury. Continuum2001;7:7–31.

[10] Verma A; Anand V; Verma NP. Sleep disorders in chronic traumatic brain injury. *J Clin Sleep Med 2007*; 3(4):357-362.

[11] Gill AW. Idiopathic and traumatic narcolepsy. Lancet 1941;1:474–6.

[12] Amico G, Pasquali F, Pittaluga E. Pickwickian-narcoleptic disorders after brain concussion. Riv Sper Freniatr Med Leg Alien Ment 1972;29:74–85.

[13] Guilleminault C, Faull K, Miles L, Van den Hoed J. Post-traumatic excessive daytime sleepiness: a review of 20 patients. Neurology. 1982;33:1584–1589.

[14] Prigatano GP, Stahl ML, Orr WC, Zeiner HK. Sleep and dreaming disturbances in closed head injury patients. J Neurol Neurosurg Psychiatry 1982;45:78–80.

[15] Parsons LC, Ver Beek D. Sleep-awake patterns following cerebral concussion. Nurs Res 1982;31:260–264.

[16] Drake ME. Jactitio nocturna after head injury. Neurology 1986;36:867–868.

[17] Patten SB, Lauderdale WM. Delayed sleep disorder after traumatic brain injury. J Am Acad Child Adolesc Psychiatry 1992;31:100–102.

[18] Biary N, Singh B, Bahou Y, al Deeb SM, Sharif H. Post-traumatic paroxysmal nocturnal hemidystonia. Mov Disord 1994; 9:98–99.

[19] Lankford DA, Wellman JJ, O'Hara C. Post-traumatic narcolepsy in mild to moderate closed head injury. Sleep. 1994;17(suppl 8):S25–S28.

[20] Ouellet MC, Savard J, Morin CM. Insomnia following traumatic brain injury: a review. Neurorehabil Neural Repair.2004;18:187–198.

[21] Guilleminault C, Yuen KM, Gulevich BA, Karadeniz D, Leger D, Philip P. Hypersomnia after head-neck trauma-a medicolegal dilemma. Neurology. 1999;54:653–659

[22] Strich SJ. Shearing of nerve fibers as a cause for brain damage due to head injury. Lancet. 1961;2:443–448.

[23] Copes WS, Champion HR, Sacco WJ. Progress in characterizing anatomic injury. J Trauma. 1990;30:1200–1207.

[24] Mody I, MacDonald JF. NMDA receptor-dependent excitotoxicity: the role of intracellular calcium release. Trends Pharmacol Sci. 1995;16:356–359.

[25] McKeating EG, Andrews PJ. Cytokines and adhesion molecules in acute brain injury. Br J Anaeth. 1998; 80:77–84.

[26] Hovda DA. Effect of hyperglycemia after cortical injury. Crit Care Med. 1997;25:1267–1268.

[27] Teasdale GM, Nicoll JR, Murray G, Fiddes M. Association of apolipoprotein E polymorphism with outcome of head injury. Lancet. 1997;350:1069–1071.

[28] BaumannCR, Werth, E, Stocker R, Ludwig S, Bassetti C L Sleep–wake disturbances 6 months after traumatic brain injury: a prospective study. *Brain 2007;130*(7), 1873-1883.

[29] Rao V, Rollings P. *Sleep disturbances following traumatic brain injury. Curr Treat Options Neurol 2002;4:77-87.*

[30] Cohen M, Oksenberg A, Snir D, Stern MJ, Groswasser Z. *Temporally related changes of sleep complaints in traumatic brain injured patients. J Neurol Neurosurg Psychiatry 1992;55:313-315.*

[31] Castriotta R. J., Wilde, MC, Lai J M, Atanasov S, Masel, BE, Kuna, S T. Prevalence and consequences of sleep disorders in traumatic brain injury. *Journ Clin Sleep Med* 2007;3(4), 349.

[32] Watson NF, Dikmen S, Machamer J, Doherty M, Temkin N. Hypersomnia following traumatic brain injury. JCSM 2007; 3:363-368.

[33] International classification of sleep disorders. Third edition (ICSD-III). http://www.aasmnet.org/library/default.aspx?id=9

[34] American Academy of Sleep Medicine-Practice guidelines. http://www.aasmnet.org/practiceguidelines.aspx

Chapter 24

Memory Deficits and Transcription Factor Activity Following Traumatic Brain Injury

Chris Cadonic and Benedict C. Albensi

Additional information is available at the end of the chapter

http://dx.doi.org/10.5772/57447

1. Introduction

Traumatic brain injury (TBI) is a serious condition and a leading cause of death and disability [1]. No two head injuries are alike and multiple complications are common in TBI. The most serious aspect of TBI is that of cognitive impairment as evidenced by animal and clinical studies focusing on synaptic plasticity and memory [2-5]. However, post trauma effects also include communication problems, sensory deficits, emotional and behavioral problems, physical complications and pain, increased suicide risk, and an increased risk for chronic CNS diseases, such as Alzheimer's disease [6, 7].

In this chapter we provide an introduction to the study of TBI and how it affects memory functioning. In addition, we survey some of the existing evidence that describes how TBI leads to memory impairment as measured in animal models and also the evidence for how TBI results in memory impairment as seen in human studies. Given that specialized proteins called transcription factors are required for the formation of long term memories, we also explore the major transcription factors that are involved in long term synaptic plasticity and long term memory. Finally, we discuss the experimental studies that investigate the effect of TBI on transcription factor regulation and the associated consequences on memory.

2. Traumatic brain injury (TBI)

TBI is a type of acquired brain injury. Acquired brain injuries can be subcategorized into either non TBI or TBI types. Examples of non TBI include anoxia, strokes, and brain infections, to name a few. TBI can be further divided into open brain injury or closed brain injury. In general, open and closed types of TBI can occur as a result of assaults, falls, motor vehicle accidents, blast injuries, and sports injuries etc. However, open brain injury is specifically caused by penetrating injuries, whereas closed brain injury is a result of internal pressure and shearing associated with blunt trauma to the head. Young men in their twenties and the elderly are most at risk for TBI.

The process of TBI is further characterized by the physical and neurochemical changes that are subjected upon the brain, which occur in a time dependent manner. In other words, we call the primary injury the causative event that occurs at the moment of injury, such as a baseball hitting the skull or a bullet penetrating the brain. These sorts of events are on the order of seconds; whereas so-called secondary injury is characterized by biochemical and neurological changes that drive pathophysiological processes in the weeks to months following the primary injury. These changes include vascular alterations, astrocyte swelling and astrogliosis, glutamate excitotoxicity, calcium overload, mitochondrial dysfunction, protease activation, cytoskeletal breakdown, cytokine release and inflammatory responses, the initiation of cell death programs, and cognitive impairment, to name a few.

3. Classifying TBI and memory functioning

Numerous studies report that TBI frequently results in impaired functioning in a wide range of cognitive tests [8, 9]; most commonly affecting processes such as attention [10-12], memory [10, 13] and information processing [14-16]. Despite the widespread cognitive sequelae following TBI [17], there exists a prominent body of work devoted to detailing its consequences for memory functioning [18-20]. However, given the heterogeneous nature of characterizing TBI, such as separating diagnoses by injury severity or duration of loss of consciousness [21-23], variability in memory impairments following TBI are quite common in the literature [9, 19, 24]. Thus, it is important to consider the manner in which TBI is classified and the paradigms used to gauge cognitive impairments [9, 19]. Furthermore, the extent to which memory functioning becomes impaired independent of attention processes or executive functioning becomes problematic to determine, and thus the integrative nature of memory must be taken into account during interpretation [19]. Given these considerations in interpreting the impairments that accompany TBI, it is nonetheless necessary to conceptualize the deficits within a simplistic framework for non-injured memory functioning.

4. Characterization of memory function

Memory has long been known to represent much more than a single, localized functional system [25, 26], but rather a diffuse system of cognitive processes that collectively amount to the internal processing, storage and retrieval of information for ongoing and/or future use [27]. Categorizing the constituents of this diffuse system involves identifying two dimensions: the time frame at which storage of information moves to retrieval, i.e., short-term or long-term memory; and the nature of this information, e.g., explicit or implicit memory [27]. Short-term memory systems regard information that is maintained transiently, and thus only retain information in an accessible state for a short period of time [28]. A related system known as working memory, that may be distinguished from short-term memory only functionally, represents the active system of information that is operated upon and processed during behavior [28, 29]. The working memory system involves three primary components: an active verbal (or speech-based) information subsystem, referred to as the phonological loop; an active visuospatial information subsystem, referred to as the visuospatial sketch pad; and a processing center that coordinates, controls and schedules mental operations as well as cognitive resource allocation for these operations, referred to as the central executive [30]. Beyond the timeframe of seconds on which short-term memory systems operate, memory becomes selectively transferred into a system known as long-term memory, which not only operates on the order of hours to days to months, but also with a capacity far beyond that of short-term memory [27, 28].

Long-term memory may be categorized with regard to the nature of the information stored; explicit memory (or declarative memory), which regards information that is consciously learned and accessible; and implicit (or nondeclarative memory), which regards information that does not require conscious awareness to learn or access [27]. Explicit memory may be further divided into episodic and semantic memory; episodic memory represents knowledge regarding personal events or information in one's life, such as what one had to eat for dinner last Friday; while semantic memory represents learned facts, meanings, understandings or general knowledge about the world, such as what the definition of an island is [31]. In addition to specialized categorization, the processing of explicit memory has been identified to involve four different operations: encoding, consolidation, storage, and retrieval [27]. Encoding involves incorporation of incoming information with existing information in memory, and thus more efficient encoding processes result in more effective memory functioning, referred to as deep encoding [32]. Consolidation refers to the stabilization of transiently encoded information to facilitate transition to permanent storage; this process is incurred by the activity of transcription factors and protein synthesis that mediate long-term potentiation (LTP), a molecular correlate of memory, and synaptic connectivity [33]. Storage of explicit memory refers to the mechanisms by which information has been processed and stabilized, and thus has been allocated to long-term memory [27]. Finally, the process of retrieval involves accessing stored memories from long-term memory and making the information usable in working memory at the time of retrieval [27]. Implicit memory may also be further subcategorized into priming and procedural skills [27, 34, 35]; priming represents the facilitation of memory processing for an item following previous exposure to either that item or another item similar on some dimension; while procedural skills involve the learning of a skill or sequence of action. It's important to note that although long-term memory may be divided into explicit and implicit systems, these systems don't always necessarily operate independent of each other [20]. This is apparent when considering a system that may be evaluated through either explicit or implicit memory tests, such as source memory or context [19], which refers to knowledge regarding any background information that accompanied the presentation of an item or event [19].

Though memory involves a much more intricate and integrative cognitive network than discussed above, memory impairments following TBI often fall within the scope of the proposed framework, and any supplementary knowledge may follow from the results and/or experimental designs to be discussed. Major avenues of investigation for these impairments typically separate into experimental designs involving induced or simulated TBI in animal models [36] and experimental or clinical evaluation of human brain injury at various time points post-injury [19, 20].

5. Evidence for memory impairment from animal models

5.1. Animal models for assessing TBI

To accommodate the variability often seen in human brain injury, numerous animal models have been developed to elucidate the typical patterns of cognitive and neurobehavioural dysfunction, as well as both the biochemical aspects of primary and secondary injury and more recently, the neurobiological consequences of head trauma [36-38]. Commonly used models for simulating human brain injury in animals include the controlled cortical impact model (CCI) [39, 40], the central or lateral fluid percussion injury model (CFP or FPI) [41], the weight-drop model [42] and the blast injury model [43, 44]. Commonly used to supplement these models as a paradigm for approximating memory impairments are memory tests that measure either: spatial memory, as measured by the Morris water maze (MWM) [45], the Barnes maze [46] or the Olton radial arm maze [47]; or associative learning, as measured by passive-avoidance [48] or operant conditioning paradigms [49].

The CCI model of TBI involves the use of an impact device to deliver a controlled strike to an exposed area of the dural surface [38]. The physical parameters of the strike, such as velocity and depth of impact, are easily controlled in the CCI model [38] and is thus a useful model for detecting the biomechanical consequences of TBI [36, 37]. In the FPI model of TBI, injury is incurred by a pendulum striking a fluid reservoir resulting in a calculated increase of intra-cranial pressure, which varies as the height and force of strike are altered, leading to deformation of neural tissue [50]. The weight-drop model involves dropping a weight onto an immobilized animal [42], with injury severity adjusting proportionately with alterations in the mass of the weight [38]. In blast models of TBI, the effects of blast waves from an explosion are emanated at varying locations and carried through shock tubes (or open exposure, as in [44]) to an immobilized animal [51, 52]. The blast model provides an accurate representation of TBI incurred by explosives devices such as improvised explosive devices (IEDs) [44].

The Morris water maze (MWM) task is a paradigm that is commonly used to assess spatial memory functioning [53-55]. The MWM task involves placing an animal into a large water tank that contains a platform submerged in an opaque liquid, so as to conceal the location of the platform. The animal is free to swim in the liquid until it either discovers the platform, or reaches a pre-determined maximum time allotment for a single trial, at which point the animal is placed on the platform for a short amount of time. During the acquisition phase, the animal progresses to learn the location of the platform relative to environmental cues (e.g., visual cues)

http://dx.doi.org/10.5772/57447

and will show a decrease in path length and time spent locating the platform using these cues in retained spatial memory systems [45, 56]. The Barnes maze similarly examines spatial memory [46], but rather than implementing a water tank filled with opaque liquid, the Barnes maze utilizes a large *table based* open field platform with numerous holes dispersed across it. One of the holes on the platform allows the animal to escape, and thus the animal will learn to find which hole allows escape from the open area only after utilizing environmental cues to find its relative position [38]. The Olton radial arm maze assesses spatial memory as well, but instead involves the use of a maze with eight arms extend outward from a center platform. Each of the arms are experimentally determined to either contain food or not to contain food, and after placing an animal on the center platform, measuring the number of visits to arms without food provides an indicator for errors in reference memory [38]. Additionally, animals with intact working memory will visit the arms with food and avoid those without food. Finally, measuring associative memory in animals following TBI may be carried out by implementing operant conditioning procedures by having reinforcement be contingent on pressing a bar only in a specific location [13]. Associative memory may also be measured through avoidance conditioning. This is done by placing an animal in one of two connected chambers, where one is entirely black and the other is entirely white; animals placed in the white side have the propensity to cross over to the black side, which is accompanied by a mild foot shock. A measure of acquired avoidance thus becomes a correlate of the latency the animal shows before crossing to the black side, and poor memory performance will show little to no increase in latency [38].

5.2. Memory impairment in animal models

Cognitive impairment in the CCI model shows a high degree of variability and inconsistency, since not only do the methodological and analytical protocols for many studies disagree, but the number of studies that report simulated injury severity amongst CCI studies is variable [38]. Upon taking this variability into consideration, however, many studies were shown to have demonstrated that TBI in rodents show a deficit in spatial memory following TBI induction using the CCI model [53-55, 57]. Interestingly, mild injury produced by the CCI model show no physical damage to the cortex or hippocampus, but still show deficits in both acquisition and retention in the MWM task [55]. Using a variation of the Morris water maze designed to measure working memory, Kobori and Dash [58] showed that significant and long-lasting working memory impairment followed CCI-induced TBI. Soblosky et al. [59] showed no significant working memory impairments following CCI-induced TBI, although deficits in reference memory were significant.

The FPI model of TBI provides a consistent measure of memory impairment following variations of injury severity and experimental paradigm alteration [38]. MRI studies of TBI in rats have shown a temporal evolution of brain injury incorporating both cytotoxic and vasogenic forms of edema where injury extends to the hippocampal formation, a region associated with new memory formation [60]. Bramlett et al. [61] demonstrated not only that impairment in retention occurs on a standard MWM paradigm occurring before TBI induction, but also impairment in acquisition when being retrained on the MWM task following TBI induction. Furthermore, by altering the MWM paradigm to include a cue on the platform throughout the acquisition stage, deficits remained, indicating effects outside of hippocampal functioning occurred [61]. Whiting and Hamm [62] utilized the FPI model to induce TBI and measure memory impairment using the MWM task. Whiting and Hamm found that there was no significant change in spatial memory impairment for 4, 8 and 24 hour post-training conditions, but when conducting the MWM prior to introducing FPI-induced TBI either 1 or 14 days post-training, cognitive impairment was significantly increased in the injured animals, only recovering when being trained on the MWM once again. The work of Whiting and Hamm indicates that the primary deficit following FPI-induced TBI may be centralized in task acquisition, but not long-term memory retention. In using the Barnes maze as a measure of spatial memory impairment, Lima et al. [63] showed that cognitive testing 1 month and 3 months prior to Barnes maze training resulted in escape latencies that were significantly increased for FPI-induced TBI animals in contrast to healthy controls. In adopting a similar testing paradigm as Whiting and Hamm, Lyeth et al. [64] trained animals in the Olton radial arm maze and subsequently introduced mild and moderate FPI-induced TBI. Contrary to evidence from CCI model data [59], FPI-induced TBI resulted in no significant impairment for reference memory, but resulted in working memory deficits in both mild and moderate TBI groups, though the severity correlated with recovery time with regard to working memory. To reconcile this discrepancy, Chown et al. [38] discuss the cortical damage present in only the CCI-induced TBI, which may account for the reported errors in reference memory that were found in only CCI-induced TBI in addition to several studies using FPI-induced TBI where cortical damage was found, reconsolidating the results of Soblosky et al. [65, 66]. Gorman et al. [13] measured associative learning functionality in FPI-induced TBI rats by training the subjects to depress an operant lever by location, and to neglect another bar that was not previously rewarding. Gorman et al. found that FPI-induced TBI performed significantly worse than controls, showing more prominent dysfunction when inter-trial times increased. Interestingly, Gorman et al. also reported that shortly after the FPI procedure, long-term memory deficit was significant in a visual discrimination task, only to return to baseline after repeated test sessions. In avoidance conditioning, Hamm et al. (1993) were unable to show acquisition deficits in animals given avoidance condition 9 days after FPI-induced TBI, even though Yamaguchi et al. (1996) were able to produce a significant deficit. What may reconcile this difference is that the timing at which animals were trained relative to post-injury time-frames, aren't as necessarily representative of the relative condition of TBI between groups they are designed to characterize [67]. Similar to the cognitive consequences found in FPI-induced TBI studies, blast models show a range of general impairments and frequently demonstrate a deficit in spatial memory functioning [44]. Interestingly, however, Rubovitch et al. [44] additionally reported recovery following TBI-induction was common for low pressure blast waves, but memory impairment was persistent at higher pressure blast waves.

Weight-drop models of TBI have been shown to induce severe retrograde amnesia impairments, showing a reduced deficit for increasing time delays between avoidance conditioning and subsequent TBI induction [68]. Zhou and Riccio also demonstrated that this induced amnesia was alleviated when rats were presented with a pre-test reminder cue, which was argued to signify that memory impairment following TBI induction reflected a deficit for memory retrieval, rather than a deficit in encoding or consolidation.

Although animal models have provided evidence for a general framework for the cognitive sequelae following TBI, particularly regarding spatial learning, and acquisition and retention deficits in memory, rarely have animal models investigated memory functioning beyond spatial memory assessments [38]. Thus a supplemental discussion beyond that of spatial memory deficits evident from animal models of TBI may be readily apparent in clinical and experimental models for TBI in the human populace.

6. Evidence for memory impairment from clinical and experimental models

Memory impairments are not only one of the most consistently reported cognitive deficits following TBI [67], but also one of the most persistent deficits, showing slower recovery than other cognitive functions [69] and in some cases continuing several years later [70]. In considering the pervasiveness of these memory impairments, however, it is necessary to consider the injury severity with which these deficits may be correlated, since variations of severity often correlate well with degree of recovery [20] and variations of clarity with regard to defining the nature of memory impairments [19]. Thus, discussing memory impairments that occur in mild TBI separately from those found in moderate or severe TBI may help to not only identify the overarching memory impairments found in general TBI, but to also detail the persistence of memory impairments.

6.1. Mild TBI memory impairments

TBI severity is characterized by one of, or more commonly, a combination of three measures [19]: the Glasgow Coma Scale (GCS), which measures a collection of motor, verbal and attentive responses to assess conscious activity [71]; the period over which consciousness has been lost (LOC); and Post Traumatic Amnesia (PTA), which represents the timeframe over which current events are not properly processed and stored [72]. Mild TBI (mTBI) typically falls within the range of a GCS score of 12 – 15, length of coma shorter than 20 minutes, and PTA length shorter than 1 day; while moderate TBI corresponds to a GCS score of 9 – 12, length of coma between 20 minutes and 36 hours, and PTA length between 1 – 7 days; and finally severe TBI corresponds to a GCS of 3 – 8, length of coma longer than 36 hours, and a PTA length of longer than 7 days [19].

Though memory deficits appear to be one of the most prevalent concerns for patients recovering from mTBI, up to 90% of these patients show recovery within 3 months post-injury and typically only show chronic memory dysfunction alongside cognitive function impairment [20]. Upon recovery to ostensibly normal cognitive functioning, re-emergence of general impairment may become apparent under appropriate variations to test conditions, such as stress induction [73] or modality of presentation [14]. Imaging studies have shown that this recovery may only be a symptomatic alleviation, however, as specific areas of frontal cortex activity rise in mTBI despite exhibiting recovered cognitive functioning [74], thus showing that although cognitive performance appears normal, the compensatory mechanisms responsible for this recovery commonly results in fatigue and similar neurobehavioural concerns [20]. For mTBI patients that have shown no such recovery, many studies have agreed in addressing the resulting memory deficits are rarely direct disruptions of explicit memory storage processing (i.e., consolidation and storage), but are frequently a result of dysfunction in executive processing; such as strategies for effective memory retrieval; information grouping strategies, known as semantic clustering; selective attention; and processing efficiency [20].

Nolin [75] showed that patients with mTBI were impaired on not only a standard free recall memory test, but also showed a high incidence of incorrect word reports and false-positives on a subsequent recognition test, and were much more susceptible to interference by a distractor word list. This test performance coincides with that of patients with amnesia since both groups show difficulty with memory retrieval, but while patients with amnesia show no improvement when given appropriate cues during the recognition test, patients with mTBI do, indicating that retrieval may be recovered in mTBI patients when central executive process reliance is reduced by avoiding retrieval strategy selection [20]. To further test the consequences for working memory following mTBI, McDowell et al. [76] assessed performance on a visual reaction time test when singly presented, and while concurrently being presented with a digit span test, which examines short-term memory span ability for a sequence of digits recalled both forward (digits forward) and backward (digits backward). Patients with mTBI showed much slower reaction times on the visual reaction time test, and a larger decrement for reaction time performance when presented with both tasks simultaneously, further illustrating the role for mTBI in central executive processing. Following evidence indicating a greater importance for working memory functioning for the visuospatial sketch pad than the phonological loop, patients with mild executive dysfunction exhibited more pronounced and persistent impairment for visual memory than that for verbal memory [20, 77].

It appears that changes in the modulation of working memory or executive cognitive processes due to mTBI may be largely responsible for the memory deficits seen [78], as argued above. And with this, there becomes a wide-reaching set of memory processes that may be affected by the observed impairment of the central executive. This may be readily apparent in memory functions that: are reliant on attention processes [76, 79], involve planning and selection of cognitive strategies for memory functioning, involve creating and maintaining scheduled plans, regard goal-directed behavior such as multitasking or involve primarily visuospatial memory [20, 77, 78].

6.2. Moderate to severe TBI memory impairments

As in the case of mTBI, much research has been devoted to addressing a basis for memory dysfunction following moderate to severe TBI [19]. Similar consequences for working memory as a result of mTBI are prevalent in moderate to severe TBI as well. Using digit span tests, Brooks [80] identified a deficient central executive but intact phonological loop following TBI, a result corroborated by Levin et al. [81] specifically in visual short-term memory. Haut et al. [82] found a deficiency in the processing speed for short-term memory by employing Sternberg's paradigm, which involves presenting the participant with a set of digits of length two, four or six. Participants are subsequently presented with a digit that is either consistent or inconsistent with the set shown, and reaction times are recorded and corrected to signify a measure of short-term memory scanning time. TBI patients exhibited longer reaction times than for healthy controls, indicating a requirement for longer short-term memory scanning to complete the task. Research regarding verbal and visual modalities in working memory following moderate to severe TBI show a similar result as in mTBI, but have yet to be directly contrasted. Zec et al. [70] employed a battery of verbal memory tests, and found consistent impairment across all tests in severe TBI patients in contrast to performance of a spinal cord injury group. Logical memory and association processing were found to be significantly impaired in TBI patients [80, 83]. Haut et al. [84] found no difference in the sensitivity of meaning of information units in TBI patients and controls in a logical memory test derived from the WMS-R (see [85]). Kersel et al. [86] employed an auditory verbal learning test in severe TBI patients six months and one-year post injury, which showed a significant impairment on all test trials for both post-injury time points. Similar to the result from the mTBI section, noticeable improvement was observed for verbal memory in both time points. Turning to visual memory, Brooks [80] utilized a variety of visual memory tests that all indicated impairment following TBI. Under only alterations of the testing paradigm used, many studies reliably corroborated this deficit in visual memory [70, 87, 88]. Shum et al. [89] used a visual learning test composed of Chinese characters, a standard Rey AVLT verbal memory test (see [90] for a review) and a spatial memory test; the results showed a significant impairment on learning rate and visual pattern recognition score, but did not show a difference in susceptibility to interference (dissimilar to the results found in mTBI; see [75]) or a substantial difference in spatial memory performance (dissimilar to the results from animal studies; see [55]. Skelton et al. [91], however, found a significant spatial memory deficit on a computer generated arena maze. This discrepancy may reside in methodology employed in each maze setup, which would inherently be different and thus may contribute to any alterations in performance. And with this is mind, it appears that both visual and verbal memory systems are impaired in TBI, although there remains more work in directly contrasting the modalities under similar conditions.

The rate at which learning occurs (i.e., the learning rate) following TBI has been posited to be affected in a similar manner as memory, and thus a slower learning rate post-injury is expected. Many studies have found slower rates in TBI patients contrasted with controls [70, 92, 93], TBI patients contrasted with controls on verbal, visual or both verbal and visual presentation modalities [94]. To explain the source of this slowed learning effect often observed in TBI patients, Paniak et al. [95] and Blackstein et al. [96] interpreted the rate deficiency as inefficient organization and learning strategies (consistent with mTBI; see [75]). Interestingly, Vakil and Oded [97] compared learning rates in TBI patients on both free recall and cued recall, which indicated that the deficit in learning was apparent only in free recall. This can be accounted for by recalling that memory retrieval organization appears to be impaired in both mTBI and TBI, and thus facilitating memory retrieval through cuing relieves the executive processes of carrying out any retrieval plan operations. On the opposite end of the spectrum, studies show that the rate at which information is lost (i.e., forgotten) is faster in TBI patients [82, 84, 98], and this effect is more pronounced in free recall paradigms [99]. To explain this effect, DeLuca

[100] argued that this rapid forgetting rate found in TBI patients may be an issue of encoding, but not consolidation nor retention. Organization of meaning in memory, otherwise known as semantic organization, has so far produced largely inconsistent results in TBI patients. Attempts to reconcile these discrepancies have resulted in detailing these varied results by noting that TBI patients will have difficulty in tasks that require applying and/or learning a strategy, but will not have such difficulty when no active strategy or an automatic/passive strategy is necessary [98, 101, 102].

Implicit memory consists of priming and procedural skill learning, which have sub-categorization. Priming studies have shown that impairment typically follows only deep encoding, while explicit memory tasks showed impairment regardless of level of encoding [103]. In many subsequent studies, priming effects have shown to occur similarly in TBI patients and controls [104-106], but a deficit would then become apparent under a variation of divided attention [106, 107]. Priming alterations between conceptual priming (i.e., priming on conceptual relations) and perceptual (i.e., priming on superficial characteristic relations) showed that conceptual priming consistently produced deficits in TBI patients [105]. Procedural skill learning has been previously accepted to have no alteration for well-practiced skills acquired prior to the injury [104, 108], but evidence demonstrating deficits post-injury remain inconsistent. Vakil [19] argues that inconsistencies in the literature regarding whether skill learning remains intact [109] or is impaired [110] following TBI is dependent on whether the testing methodology involves frontal lobe activity (as the frontal lobes are particularly vulnerable to TBI) or tasks that do not involve the frontal lobes.

To further elucidate effects on explicit and implicit memory following TBI, source memory can be surveyed explicitly, as through direct inquiry to recall background information for a specific event or situation, or indirectly, as through priming effects due to context [19]. Measuring different aspects of context and source memory, however, yields some varying results. When measured directly, source memory for spatial location was significantly impaired [111], as was frequency of occurrence for words from a study list [112]. Thus explicit measures of source memory were consistently impaired relative to controls [19]. Temporal order judgments for study lists did not yield consistent results, either showing no effect [113] or a significant impairment [111]. Further investigation into the effects of TBI on integrative memory concepts such as source memory and context may help provide a more definitive connection between explicit and implicit memory systems consequences.

7. Transcription factors and memory

Literally, hundreds of molecules have been shown to play a role in various forms of memory. In addition, specialized proteins known as transcription factors have also been implicated in memory. However, there are only a few families of transcription factors that are actively studied and that appear to be critically involved in long term synaptic plasticity and long term memory [114]. These include activating protein 1 (AP-1), CCAAT enhancer binding (C/EBP) protein, early growth response (Egr) factor protein, nuclear factor kappa B (NF-κB) protein, and cAMP response element-binding (CREB) protein.

http://dx.doi.org/10.5772/57447

The factor AP-1 is composed of proteins coded by several genes such as *c-fos, c-jun* and *ATF*. The C/EBP family of transcription factors is coded by six distinct genes: *C/EBPα, C/EBPβ, C/EBPγ, C/EBPδ, C/EBPε,* and *C/EBPζ*. Among the Egr family of transcription factor genes, the *zif 268* gene (a.k.a. *Egr-1, Krox24, NGF-I-A, TZs8* or *Zenk*) is probably the most well studied. There are several genes that contribute to the NF-κB transcription factor complex that include *NF-κB1, NF-κB2, c-Rel, RelA, RelB, IkBα, IκBβ, IκBε,* and *IκBζ*. The CREB (or CREB/ATF) family of transcription factor proteins is produced with three homologous genes, which are *creb*, *crem*, and *atf-1*.

Transcription factors are important for biological processes where they regulate the basal process of transcription, the selective activation of genes, and/or the repression of genes. More specifically, transcription factors control transcriptional regulation where information encoded in the DNA of each cell is copied into a molecule of RNA. Ultimately, transcription factors regulate multiple functions on different time scales and in different spatial regions. In some cases, transcription factors even initiate the expression of additional transcription factors, which hints at the multiphasic layering and the overall complexity of transcriptional regulation.

8. Transcription factor activity following TBI

The use of DNA gene microarrays has greatly increased our understanding of how genes are differentially regulated following TBI. In particular, these techniques enable the simultaneous evaluation of thousands of genes, which assist in protein expression profiling and the identification of molecular mechanisms that are involved in the pathophysiology of secondary injury in TBI. For example, alterations in the transcription of genes following TBI lend insight into a neuron's response to trauma. These responses involve both the initiation of programmed cell death and the restoration of compromised cell function. Understanding these complex responses no doubt is central to the discovery and development of therapeutic strategies for treating TBI.

Early studies using these methods [115] demonstrated alterations in several classes of genes following TBI, including neurotrophic factor genes, heat shock protein genes, cytokine genes, and immediate early genes (IEGs). IEGs (e.g.s., *c-fos* and *Egr-1*) received considerable attention in initial studies given these genes are responsible for the encoding of proteins that regulate growth factors, growth factor receptors, cytoskeletal proteins and transcription factors, etc. For example, in one gene array study *c-fos* and *Egr-1* mRNA expression levels were significantly increased in both ipsilateral and contralateral regions at 120 mins following TBI [116]. These findings are interesting in light of the fact that both genes code for transcription factors, thus controlling the expression of other genes. In another gene array study by Kobori et al. [117], which was broader in scope, CCI was induced in C57BL/6 mice and approximately 10,000 genes were evaluated. In this study, 7 functional classes of genes were found to be increased following CCI; this included transcription factors, signal transduction genes and genes coding for inflammatory proteins. Of these, the transcription factor, *c-jun* and the neurotrophic factor, *bdnf* mRNA levels increased as a result of TBI. Gene arrays have also been used following TBI in human subjects. For example, in a study by Michael et al. [118], global changes in gene expression were evaluated in 4 patients during surgery following TBI. These results showed that 4 genes previously shown to be associated with TBI (i.e., *c-Fos, Egr-1, Jun B,* and *HSP70,*) were all up-regulated in at least one TBI subject. Collectively, these studies show that IEGs are upregulated in both animal models and in human subjects following TBI suggesting IEGs play essential roles in secondary injury associated with TBI.

Other transcription factors have also been investigated following TBI. In a mouse study by Beni et al. [119], a closed head injury (CHI) model was utilized and the transcription factors NF-κB and AP-1 were evaluated in the presence of the pineal hormone melatonin. Besides being involved in pineal function, melatonin also acts as an antioxidant, which was being evaluated for its potential to attenuate the effects of TBI. Here it was found that CHI-induced TBI activated NF-κB and AP-1 at 24 hours following CHI. In particular, the study showed a transient activation of AP-1 and a longer activation of NF-κB after CHI. Interestingly, melatonin inhibited the late-phase activation of NF-κB and decreased AP-1 to below basal levels when measured at 8 days following CHI. These results suggested inhibition of NF-κB by melatonin was associated with improved outcome, whereas the prolonged activation of NF-κB after CHI was harmful. NF-κB activity has also been studied in other TBI models. In a study by Chen et al. [120], NF-κB and also TLR4, IL-β, TNF-α, IL-6 and ICAM-1 were upregulated in a weight drop model of TBI. However, when the cholesterol-lowering agent simvastatin was used, the induction of TLR4/NF-κB pathway was suppressed after TBI.

CREB activity has also been looked at following TBI. In an earlier study by Dash et al. [121] using rats, the phosphorylation of CREB was found increased just 5 minutes after lateral cortical impact, but decreased to control levels after 30 minutes. In addition, c-Fos and the AP-1 complex expression was found increased following CREB phosphorylation. Hu et al. [122] also looked at changes in CREB pathway signaling following TBI in a context of hippocampal mossy fiber reorganization, which occurs after various CNS pathological events or insults. In this study the FPI model was used in rats and it was found that signaling pathways of TrkB–ERK1/2–CREB/Elk-1 were robustly activated in association with mossy fiber organization. These results suggest that activation of the CREB signaling pathway may contribute to mossy fiber reorganization after the onset of TBI. However, some studies also demonstrate that CREB is downregulated following TBI. For example, Atkins et al. [123] used a parasagittal FPI model in rats to study CREB signaling. Here it was found that the activation of ERK and CREB after 30 seconds of glutamate stimulation or KCl depolarization was decreased in hippocampal slices from animals at 2, 8, or 12 weeks after TBI as compared to control rats. One reason for the apparent inconsistency among these CREB studies may have to do with the time course of the measurements. For example, deficits in CREB activation in the study by Atkins et al. may be due to synaptic loss *in the weeks* following TBI, as opposed to CREB measurements shortly after TBI. In addition, CREB family members can function as either transcriptional activators or repressors, and family members may have distinct functions under different conditions, which could also explain some of the differences seen in CREB activity following TBI. Another observed function of CREB following TBI may have to do with the regulation of apoptotic

http://dx.doi.org/10.5772/57447

activity. Wu et al. conducted a study [124] to see if CREM-1 was involved in CNS injury or repair, and performed TBI in rats. Here they looked at the association of CREM-1 with p-CREB on PC12 cells. Their results suggested that the association of CREM-1 with p-CREB was enhanced in apoptotic cells and therefore, CREM-1 might regulate neuronal death after TBI by interacting with CREB.

Some studies evaluating the C/EBP family of transcription factors in TBI have also been conducted. For example, in a study by Sandhir and Berman [125], C/EBP isoforms were evaluated since they are known to regulate the expression of proinflammatory genes. In this study, CCI was subjected on either younger adult (5-6 mos old) or older (21-24 mos old) C57BL/6 control mice and C/EBP mRNA and protein expression levels were evaluated during the first week following CCI. In this study it was found that protein and mRNA expression levels of C/EBP isoforms overall were similar in younger brains and in older brains before CCI. Following CCI, C/EBPα mRNA expression appeared to go down on day 1 in young adult and in older brains, but these results were not statistically significant. However, a significant increase in C/EBPα mRNA expression was seen on days 3 and 7 in the young adult brains and on day 7 in the older brain as compared to levels before CCI. Also, C/EBPα protein levels were significantly elevated on days 3 and 7 in young and older brains as compared to pre CCI levels. It was also found that a significant upregulation of C/EBPβ mRNA expression occurred on days 1 and 3 in both young and older adults, which was associated with significant increases in C/EBPβ protein levels on the same days in the same groups as compared to pre CCI levels. With regard to C/EBPδ mRNA levels, only on day 1 in older brains was there a significant increase in expression, whereas protein levels of C/EBPδ were significantly increased in both young and old brains on days 1, 3, and 7. Collectively, these results show clear differences in the temporal expression among the C/EBP isoforms. These results overall suggest that C/EBP transcription factors contribute to inflammatory responses following TBI in aged brains, where, the expression of C/EBPβ and δ appear to play roles in the early phase of the inflammatory response.

9. Conclusions

TBI is a serious condition resulting in disability or death. Currently, there is no standardized treatment. However, research has been attempted in animal models and human trials have been conducted showing the effects of TBI on various outcomes. In addition, a large amount of evidence has been collected that demonstrates that TBI is associated with cognitive impairment and memory dysfunction. A considerable amount of data also show that long term memory is associated with the activation of transcription factors, which regulate and initiate new gene expression. The protein products from this expression contribute to biological functions associated with the formation, retention, and reconsolidation of long term memories. However, following TBI numerous mechanisms associated with transcriptional regulation become affected. In fact, we now know that transcription factor regulation following TBI is complex where some transcription factors contribute not only to processes of memory formation, but also contribute to neurodegenerative processes. In other words, multiple signaling pathways exist and play various roles in inflammatory signaling, programmed cell death, mossy fiber reorganization, endogenous neuroprotection, and the initiation of neurodegenerative processes. It is hoped that by understanding the complexity of transcriptional regulation after TBI, that new targets can be identified which could be exploited for pharmacological intervention. In this regard, our understanding is still quite infantile and further research is necessary.

Acknowledgements

Dr. Benedict C. Albensi, Principal Investigator, is also the Everett Endowment Fund Chair and is supported in part by the Everett Endowment, which funds his Alzheimer's research at the St. Boniface Hospital Research Centre - Division of Neurodegenerative Disorders. Other funding supporting basic memory research in his lab is from NSERC. Dr. Albensi is also an Associate Professor in the Dept. of Pharmacology and Therapeutics, Faculty of Medicine and an Adjunct Professor in the Dept. of Electrical and Computer Engineering in the Faculty of Engineering both at the University of Manitoba. He is also a Research Affiliate at the Centre on Aging and a Scientist at the Manitoba Inst. of Child Health (MICH) at the Univ. of Manitoba.

Author details

Chris Cadonic[1,3] and Benedict C. Albensi[1,2,3*]

*Address all correspondence to: balbensi@sbrc.ca

1 Biomedical Engineering Program, Faculty of Engineering, University of Manitoba, Canada

2 Dept. of Pharmacology & Therapeutics, Faculty of Medicine, University of Manitoba, Canada

3 Division of Neurodegenerative Disorders, St. Boniface Hospital Research, Winnipeg, Manitoba, Canada

References

[1] Zubko, N., *TBI: the disability in disguise.* Behav Healthc, 2011. 31(5): p. 26, 28.

[2] Albensi, B.C. and D. Janigro, *Traumatic brain injury and its effects on synaptic plasticity.* Brain Inj, 2003. 17(8): p. 653-63.

[3] Brody, D.L. and D.M. Holtzman, *Morris water maze search strategy analysis in PDAPP mice before and after experimental traumatic brain injury.* Exp Neurol, 2006. 197(2): p. 330-40.

[4] De los Reyes Aragon, C.J., et al., *The effect of cognitive impairment on self-generation in Hispanics with TBI.* NeuroRehabilitation, 2012. 30(1): p. 55-64.

[5] Gordon, S.N., P.J. Fitzpatrick, and R.C. Hilsabeck, *No effect of PTSD and other psychiatric disorders on cognitive functioning in veterans with mild TBI.* Clin Neuropsychol, 2011. 25(3): p. 337-47.

[6] Graves, A.B., et al., *The association between head trauma and Alzheimer's disease.* Am J Epidemiol, 1990. 131(3): p. 491-501.

[7] Mortimer, J.A., et al., *Head trauma as a risk factor for Alzheimer's disease: a collaborative re-analysis of case-control studies. EURODEM Risk Factors Research Group.* Int J Epidemiol, 1991. 20 Suppl 2: p. S28-35.

[8] Mateer, C.A. and C.S. Sira, *Cognitive and emotional consequences of TBI: intervention strategies for vocational rehabilitation.* NeuroRehabilitation, 2006. 21(4): p. 315-26.

[9] Dean, P.J. and A. Sterr, *Long-term effects of mild traumatic brain injury on cognitive performance.* Front Hum Neurosci, 2013. 7: p. 30.

[10] Arciniegas, D., et al., *Attention and memory dysfunction after traumatic brain injury: cholinergic mechanisms, sensory gating, and a hypothesis for further investigation.* Brain Inj, 1999. 13(1): p. 1-13.

[11] Mangels, J.A., et al., *Effects of divided attention on episodic memory in chronic traumatic brain injury: a function of severity and strategy.* Neuropsychologia, 2002. 40(13): p. 2369-85.

[12] Chan, R.C., *Sustained attention in patients with mild traumatic brain injury.* Clin Rehabil, 2005. 19(2): p. 188-93.

[13] Gorman, L.K., B.L. Shook, and D.P. Becker, *Traumatic brain injury produces impairments in long-term and recent memory.* Brain Res, 1993. 614(1-2): p. 29-36.

[14] Duncan, C.C., M.H. Kosmidis, and A.F. Mirsky, *Closed head injury-related information processing deficits: an event-related potential analysis.* Int J Psychophysiol, 2005. 58(2-3): p. 133-57.

[15] Ferraro, F.R., *Cognitive slowing in closed-head injury.* Brain Cogn, 1996. 32(3): p. 429-40.

[16] O'Jile, J.R., et al., *Information processing following mild head injury.* Arch Clin Neuropsychol, 2006. 21(4): p. 293-6.

[17] Vanderploeg, R.D., G. Curtiss, and H.G. Belanger, *Long-term neuropsychological outcomes following mild traumatic brain injury.* J Int Neuropsychol Soc, 2005. 11(3): p. 228-36.

[18] Gronwall, D. and P. Wrightson, *Memory and information processing capacity after closed head injury.* J Neurol Neurosurg Psychiatry, 1981. 44(10): p. 889-95.

[19] Vakil, E., *The effect of moderate to severe traumatic brain injury (TBI) on different aspects of memory: a selective review.* J Clin Exp Neuropsychol, 2005. 27(8): p. 977-1021.

[20] Flynn, F.G., *Memory impairment after mild traumatic brain injury.* Continuum (Minneap Minn), 2010. 16(6 Traumatic Brain Injury): p. 79-109.

[21] Davis, A.E., *Cognitive impairments following traumatic brain injury. Etiologies and interventions.* Crit Care Nurs Clin North Am, 2000. 12(4): p. 447-56.

[22] Arciniegas, D.B., *The cholinergic hypothesis of cognitive impairment caused by traumatic brain injury.* Curr Psychiatry Rep, 2003. 5(5): p. 391-9.

[23] Heegaard, W. and M. Biros, *Traumatic brain injury.* Emerg Med Clin North Am, 2007. 25(3): p. 655-78, viii.

[24] Tellier, A., et al., *The heterogeneity of mild traumatic brain injury: Where do we stand?* Brain Inj, 2009. 23(11): p. 879-87.

[25] Cohen, R.L., C. Netley, and M.A. Clarke, *On the generality of the short-term memory/reading ability relationship.* J Learn Disabil, 1984. 17(4): p. 218-21.

[26] Squire, L.R. and S. Zola-Morgan, *Memory: brain systems and behavior.* Trends Neurosci, 1988. 11(4): p. 170-5.

[27] Kandel, E.S., J.; Jessell, T.; Siegelbaum, S. & Hudspeth, A., ed. *Principles of Neural Science.* 5 ed. 2013, McGraw-Hill Companies: New York. 1441-1459.

[28] Cowan, N., *What are the differences between long-term, short-term, and working memory?* Prog Brain Res, 2008. 169: p. 323-38.

[29] Becker, J.T. and R.G. Morris, *Working memory(s).* Brain Cogn, 1999. 41(1): p. 1-8.

[30] Baddeley, A., Hitch, G, *Working memory.* Recent advances in learning and motivation Vol. 8, ed. G. Bower. Vol. 8. 1974, New York, NY: Academic Press.

[31] Saumier, D. and H. Chertkow, *Semantic memory.* Curr Neurol Neurosci Rep, 2002. 2(6): p. 516-22.

[32] Craik, F.I., *Levels of processing: past, present. and future?* Memory, 2002. 10(5-6): p. 305-18.

[33] Davis, H.P. and L.R. Squire, *Protein synthesis and memory: a review.* Psychol Bull, 1984. 96(3): p. 518-59.

[34] Schacter, D.L., *Implicit expressions of memory in organic amnesia: learning of new facts and associations.* Hum Neurobiol, 1987. 6(2): p. 107-18.

[35] Goshen-Gottstein, Y. and M. Moscovitch, *Repetition priming for newly formed and preexisting associations: perceptual and conceptual influences.* J Exp Psychol Learn Mem Cogn, 1995. 21(5): p. 1229-48.

[36] Xiong, Y., A. Mahmood, and M. Chopp, *Animal models of traumatic brain injury.* Nat Rev Neurosci, 2013. 14(2): p. 128-42.

[37] Cernak, I., *Animal models of head trauma.* Neurorx, 2005. 2(3): p. 410-22.

[38] Chown, A., Noble, L. & Flint, R., *Traumatic Brain Injury and Memory Loss: A Review of Nonhuman Animal Models.* Journal of Behavioral and Neuroscience Research, 2010. 8(1): p. 30-48.

[39] Lighthall, J.W., *Controlled cortical impact: a new experimental brain injury model.* J Neurotrauma, 1988. 5(1): p. 1-15.

[40] Dixon, C.E., et al., *A controlled cortical impact model of traumatic brain injury in the rat.* J Neurosci Methods, 1991. 39(3): p. 253-62.

[41] Dixon, C.E., et al., *A fluid percussion model of experimental brain injury in the rat.* J Neurosurg, 1987. 67(1): p. 110-9.

[42] Marmarou, A., et al., *A new model of diffuse brain injury in rats. Part I: Pathophysiology and biomechanics.* J Neurosurg, 1994. 80(2): p. 291-300.

[43] Cernak, I., et al., *Involvement of the central nervous system in the general response to pulmonary blast injury.* J Trauma, 1996. 40(3 Suppl): p. S100-4.

[44] Rubovitch, V., et al., *A mouse model of blast-induced mild traumatic brain injury.* Exp Neurol, 2011. 232(2): p. 280-9.

[45] Morris, R., *Developments of a water-maze procedure for studying spatial learning in the rat.* J Neurosci Methods, 1984. 11(1): p. 47-60.

[46] Barnes, C.A., *Memory deficits associated with senescence: a neurophysiological and behavioral study in the rat.* J Comp Physiol Psychol, 1979. 93(1): p. 74-104.

[47] Brown, M.F., R.F. Farley, and E.J. Lorek, *Remembrance of places you passed: social spatial working memory in rats.* J Exp Psychol Anim Behav Process, 2007. 33(3): p. 213-24.

[48] Flint, R.W., Jr. and D.C. Riccio, *Pretest administration of glucose attenuates infantile amnesia for passive avoidance conditioning in rats.* Dev Psychobiol, 1997. 31(3): p. 207-16.

[49] Staddon, J.E. and D.T. Cerutti, *Operant conditioning.* Annu Rev Psychol, 2003. 54: p. 115-44.

[50] McIntosh, T.K., et al., *Traumatic brain injury in the rat: characterization of a lateral fluid-percussion model.* Neuroscience, 1989. 28(1): p. 233-44.

[51] Benzinger, T.L., et al., *Blast-related brain injury: imaging for clinical and research applications: report of the 2008 st. Louis workshop.* J Neurotrauma, 2009. 26(12): p. 2127-44.

[52] Thompson, J.M., K.C. Scott, and L. Dubinsky, *Battlefield brain: unexplained symptoms and blast-related mild traumatic brain injury.* Can Fam Physician, 2008. 54(11): p. 1549-51.

[53] Dixon, C.E., et al., *One-year study of spatial memory performance, brain morphology, and cholinergic markers after moderate controlled cortical impact in rats.* J Neurotrauma, 1999. 16(2): p. 109-22.

[54] Griesbach, G.S., et al., *Controlled contusion injury alters molecular systems associated with cognitive performance.* J Neurosci Res, 2009. 87(3): p. 795-805.

[55] Scheff, S.W., et al., *Morris water maze deficits in rats following traumatic brain injury: lateral controlled cortical impact.* J Neurotrauma, 1997. 14(9): p. 615-27.

[56] D'Hooge, R. and P.P. De Deyn, *Applications of the Morris water maze in the study of learning and memory.* Brain Res Brain Res Rev, 2001. 36(1): p. 60-90.

[57] Albensi, B.C., et al., *Cyclosporin ameliorates traumatic brain-injury-induced alterations of hippocampal synaptic plasticity.* Exp Neurol, 2000. 162(2): p. 385-9.

[58] Kobori, N. and P.K. Dash, *Reversal of brain injury-induced prefrontal glutamic acid decarboxylase expression and working memory deficits by D1 receptor antagonism.* J Neurosci, 2006. 26(16): p. 4236-46.

[59] Soblosky, J.S., et al., *Reference memory and allocentric spatial localization deficits after unilateral cortical brain injury in the rat.* Behav Brain Res, 1996. 80(1-2): p. 185-94.

[60] Albensi, B.C., et al., *Diffusion and high resolution MRI of traumatic brain injury in rats: time course and correlation with histology.* Exp Neurol, 2000. 162(1): p. 61-72.

[61] Bramlett, H.M., E.J. Green, and W.D. Dietrich, *Hippocampally dependent and independent chronic spatial navigational deficits following parasagittal fluid percussion brain injury in the rat.* Brain Res, 1997. 762(1-2): p. 195-202.

[62] Whiting, M.D. and R.J. Hamm, *Mechanisms of anterograde and retrograde memory impairment following experimental traumatic brain injury.* Brain Res, 2008. 1213: p. 69-77.

[63] Lima, F.D., et al., *Na+,K+-ATPase activity impairment after experimental traumatic brain injury: relationship to spatial learning deficits and oxidative stress.* Behav Brain Res, 2008. 193(2): p. 306-10.

[64] Lyeth, B.G., et al., *Prolonged memory impairment in the absence of hippocampal cell death following traumatic brain injury in the rat.* Brain Res, 1990. 526(2): p. 249-58.

[65] Lowenstein, D.H., et al., *Selective vulnerability of dentate hilar neurons following traumatic brain injury: a potential mechanistic link between head trauma and disorders of the hippocampus.* J Neurosci, 1992. 12(12): p. 4846-53.

[66] Smith, D.H., et al., *Persistent memory dysfunction is associated with bilateral hippocampal damage following experimental brain injury.* Neurosci Lett, 1994. 168(1-2): p. 151-4.

[67] Arcia, E. and C.T. Gualtieri, *Association between patient report of symptoms after mild head injury and neurobehavioural performance.* Brain Inj, 1993. 7(6): p. 481-9.

[68] Zhou, Y. and D.C. Riccio, *Concussion-induced retrograde amnesia in rats.* Physiol Behav, 1995. 57(6): p. 1107-15.

[69] Lezak, M.D., *Recovery of memory and learning functions following traumatic brain injury.* Cortex, 1979. 15(1): p. 63-72.

[70] Zec, R.F., et al., *Long-term consequences of severe closed head injury on episodic memory.* J Clin Exp Neuropsychol, 2001. 23(5): p. 671-91.

[71] Teasdale, G. and B. Jennett, *Assessment of coma and impaired consciousness. A practical scale.* Lancet, 1974. 2(7872): p. 81-4.

[72] Russell, W.R. and A. Smith, *Post-traumatic amnesia in closed head injury.* Arch Neurol, 1961. 5: p. 4-17.

[73] Alexander, M.P., *Mild traumatic brain injury: pathophysiology, natural history, and clinical management.* Neurology, 1995. 45(7): p. 1253-60.

[74] Kirkwood, M.W., et al., *Management of pediatric mild traumatic brain injury: a neuropsychological review from injury through recovery.* Clin Neuropsychol, 2008. 22(5): p. 769-800.

[75] Nolin, P., *Executive memory dysfunctions following mild traumatic brain injury.* J Head Trauma Rehabil, 2006. 21(1): p. 68-75.

[76] McDowell, S., J. Whyte, and M. D'Esposito, *Working memory impairments in traumatic brain injury: evidence from a dual-task paradigm.* Neuropsychologia, 1997. 35(10): p. 1341-53.

[77] Busch, R.M., et al., *Role of executive functioning in verbal and visual memory.* Neuropsychology, 2005. 19(2): p. 171-80.

[78] McAllister, T.W., et al., *Working memory deficits after traumatic brain injury: catecholaminergic mechanisms and prospects for treatment -- a review.* Brain Inj, 2004. 18(4): p. 331-50.

[79] Park, N.W., M. Moscovitch, and I.H. Robertson, *Divided attention impairments after traumatic brain injury.* Neuropsychologia, 1999. 37(10): p. 1119-33.

[80] Brooks, D.N., *Wechsler Memory Scale performance and its relationship to brain damage after severe closed head injury.* J Neurol Neurosurg Psychiatry, 1976. 39(6): p. 593-601.

[81] Levin, H.S., R.G. Grossman, and P.J. Kelly, *Short-term recognition memory in relation to severity of head injury.* Cortex, 1976. 12(2): p. 175-82.

[82] Haut, M.W., T.V. Petros, and R.G. Frank, *The recall of prose as a function of importance following closed head injury.* Brain Inj, 1990. 4(3): p. 281-8.

[83] Bennett-Levy, J.M., *Long-term effects of severe closed head injury on memory: evidence from a consecutive series of young adults.* Acta Neurol Scand, 1984. 70(4): p. 285-98.

[84] Haut, M.W., et al., *Short-term memory processes following closed head injury.* Arch Clin Neuropsychol, 1990. 5(3): p. 299-309.

[85] Wechsler, D., ed. *Wechsler Adult Intelligence Scale - Revised*. 1987, Psychological Corporation: New York.

[86] Kersel, D.A., et al., *Neuropsychological functioning during the year following severe traumatic brain injury.* Brain Inj, 2001. 15(4): p. 283-96.

[87] Brooker, A.E. and J.C. George, *Visual recognition memory of severely head-injured patients.* Percept Mot Skills, 1984. 59(1): p. 249-50.

[88] Hannay, H.J., H.S. Levin, and R.G. Grossman, *Impaired recognition memory after head injury.* Cortex, 1979. 15(2): p. 269-83.

[89] Shum, D.H., D. Harris, and J.G. O'Gorman, *Effects of severe traumatic brain injury on visual memory.* J Clin Exp Neuropsychol, 2000. 22(1): p. 25-39.

[90] Hawkins, K.A., D. Dean, and G.D. Pearlson, *Alternative forms of the Rey Auditory Verbal Learning Test: a review.* Behav Neurol, 2004. 15(3-4): p. 99-107.

[91] Skelton, R.W., et al., *Humans with traumatic brain injuries show place-learning deficits in computer-generated virtual space.* J Clin Exp Neuropsychol, 2000. 22(2): p. 157-75.

[92] Vakil, E. and H. Blachstein, *Rey Auditory-Verbal Learning Test: structure analysis.* J Clin Psychol, 1993. 49(6): p. 883-90.

[93] Vanderploeg, R.D., T.A. Crowell, and G. Curtiss, *Verbal learning and memory deficits in traumatic brain injury: encoding, consolidation, and retrieval.* J Clin Exp Neuropsychol, 2001. 23(2): p. 185-95.

[94] Constantinidou, F., et al., *Pictorial superiority during verbal learning tasks in moderate to severe closed head injury: additional evidence.* J Gen Psychol, 1996. 123(3): p. 173-84.

[95] Paniak, C.E., D.L. Shore, and B.P. Rourke, *Recovery of memory after severe closed head injury: dissociations in recovery of memory parameters and predictors of outcome.* J Clin Exp Neuropsychol, 1989. 11(5): p. 631-44.

[96] Blachstein, H., Vakil, E., & Hoofien, D., *Impaired learning in patients with closed-head injuries: An analysis of compoenents of the acquisition process.* Neuropsychology, 1993. 7: p. 530-535.

[97] Vakil, E. and Y. Oded, *Comparison between three memory tests: cued recall, priming and saving closed-head injured patients and controls.* J Clin Exp Neuropsychol, 2003. 25(2): p. 274-82.

[98] Vakil, E., et al., *Relative importance of informational units and their role in long-term recall by closed-head-injured patients and control groups.* J Consult Clin Psychol, 1992. 60(5): p. 802-3.

[99] Carlesimo, G.A., et al., *Forgetting from long-term memory in severe closed-head injury patients: effect of retrieval conditions and semantic organization.* Cortex, 1997. 33(1): p. 131-42.

[100] DeLuca, J., et al., *Acquisition versus retrieval deficits in traumatic brain injury: implications for memory rehabilitation.* Arch Phys Med Rehabil, 2000. 81(10): p. 1327-33.

[101] Levin, H.S. and F.C. Goldstein, *Organization of verbal memory after severe closed-head injury.* J Clin Exp Neuropsychol, 1986. 8(6): p. 643-56.

[102] Perri, R., et al., *Deficient intentional access to semantic knowledge in patients with severe closed-head injury.* Cortex, 2000. 36(2): p. 213-25.

[103] Shum, D.H., Harris, D., & O'Gorman, J. G., *Performance on verbal implicit and explicit memory tasks following traumatic brain injury.* Jouranl of Head Trauma Rehabilitation, 1996. 11: p. 43-53.

[104] Vakil, E., et al., *Head-injured patients and control group: implicit versus explicit measures of frequency of occurrence.* J Clin Exp Neuropsychol, 1994. 16(4): p. 539-46.

[105] Vakil, E. and J. Sigal, *The effect of level of processing on perceptual and conceptual priming: control versus closed-head-injured patients.* J Int Neuropsychol Soc, 1997. 3(4): p. 327-36.

[106] Watt, S., E.A. Shores, and S. Kinoshita, *Effects of reducing attentional resources on implicit and explicit memory after severe traumatic brain injury.* Neuropsychology, 1999. 13(3): p. 338-49.

[107] Schmitter-Edgecombe, M., *The effects of divided attention on implicit and explicit memory performance.* J Int Neuropsychol Soc, 1996. 2(2): p. 111-25.

[108] Schmitter-Edgecombe, M. and H.M. Nissley, *Effects of divided attention on automatic and controlled components of memory after severe closed-head injury.* Neuropsychology, 2000. 14(4): p. 559-69.

[109] Nissley, H.M. and M. Schmitter-Edgecombe, *Perceptually based implicit learning in severe closed-head injury patients.* Neuropsychology, 2002. 16(1): p. 111-22.

[110] Vakil, E., et al., *Impaired skill learning in patients with severe closed-head injury as demonstrated by the serial reaction time (SRT) task.* Brain Cogn, 2002. 50(2): p. 304-15.

[111] Vakil, e., & Tweedy, J. R., *Memory for temporal order and spatial position information: Test of the automatic-effortful distinction.* Neuropsychiatry, Neuropsychology, and Behavioral Neurology., 1994. 7: p. 281-288.

[112] Levin, H.S., et al., *Automatic and effortful processing after severe closed head injury.* Brain Cogn, 1988. 7(3): p. 283-97.

[113] Vakil, E., et al., *Direct and indirect memory measures of temporal order and spatial location: control versus closed-head injury participants.* Neuropsychiatry Neuropsychol Behav Neurol, 1998. 11(4): p. 212-7.

[114] Alberini, C.M., *Transcription factors in long-term memory and synaptic plasticity.* Physiol Rev, 2009. 89(1): p. 121-45.

[115] Marciano, P.G., et al., *Expression profiling following traumatic brain injury: a review.* Neurochem Res, 2002. 27(10): p. 1147-55.

[116] Awasthi, D., et al., *Early gene expression in the rat cortex after experimental traumatic brain injury and hypotension.* Neurosci Lett, 2003. 345(1): p. 29-32.

[117] Kobori, N., G.L. Clifton, and P. Dash, *Altered expression of novel genes in the cerebral cortex following experimental brain injury.* Brain Res Mol Brain Res, 2002. 104(2): p. 148-58.

[118] Michael, D.B., D.M. Byers, and L.N. Irwin, *Gene expression following traumatic brain injury in humans: analysis by microarray.* J Clin Neurosci, 2005. 12(3): p. 284-90.

[119] Beni, S.M., et al., *Melatonin-induced neuroprotection after closed head injury is associated with increased brain antioxidants and attenuated late-phase activation of NF-kappaB and AP-1.* FASEB J, 2004. 18(1): p. 149-51.

[120] Chen, G., et al., *Simvastatin reduces secondary brain injury caused by cortical contusion in rats: possible involvement of TLR4/NF-kappaB pathway.* Exp Neurol, 2009. 216(2): p. 398-406.

[121] Dash, P.K., A.N. Moore, and C.E. Dixon, *Spatial memory deficits, increased phosphorylation of the transcription factor CREB, and induction of the AP-1 complex following experimental brain injury.* J Neurosci, 1995. 15(3 Pt 1): p. 2030-9.

[122] Hu, B., et al., *Changes in trkB-ERK1/2-CREB/Elk-1 pathways in hippocampal mossy fiber organization after traumatic brain injury.* J Cereb Blood Flow Metab, 2004. 24(8): p. 934-43.

[123] Atkins, C.M., et al., *Deficits in ERK and CREB activation in the hippocampus after traumatic brain injury.* Neurosci Lett, 2009. 459(2): p. 52-6.

[124] Wu, X., et al., *Cyclic AMP response element modulator-1 (CREM-1) involves in neuronal apoptosis after traumatic brain injury.* J Mol Neurosci, 2012. 47(2): p. 357-67.

[125] Sandhir, R. and N.E. Berman, *Age-dependent response of CCAAT/enhancer binding proteins following traumatic brain injury in mice.* Neurochem Int, 2010. 56(1): p. 188-93.